Carbohydrate Metabolism in Cultured Cells

Carbohydrate Metabolism in Cultured Cells

Edited by

Michael J. Morgan
University of Leicester
Leicester, England

Plenum Press • New York and London

Library of Congress Cataloging in Publication Data

Carbohydrate metabolism in cultured cells.

Bibliography: p.
Includes index.
1. Cell metabolism. 2. Carbohydrates—Metabolism. 3. Cell culture. I. Morgan, Michael J.
QH634.5.C37   1986   574.87′61   86-12170
ISBN 0-306-42240-9

© 1986 Plenum Press, New York
A Division of Plenum Publishing Corporation
233 Spring Street, New York, N.Y. 10013

Printed in the United States of America

Contributors

RONALD A. COOPER   Department of Biochemistry, University of Leicester, Leicester LE1 7RH, United Kingdom

PELIN FAIK   Department of Biochemistry, University of Leicester, Leicester LE1 7RH, United Kingdom

ALAN H. FAIRLAMB   Laboratory of Medical Biochemistry, The Rockefeller University, New York, New York 10021

MICHAEL W. FOWLER   Wolfson Institute of Biotechnology, University of Sheffield, Sheffield S10 2TN, United Kingdom

JUANA M. GANCEDO   Instituto de Investigaciones Biomédicas, Consejo Superior de Investigaciones Científicas, Facultad de Medicina de la Universidad Autónoma, 28029 Madrid, Spain

PETER J. F. HENDERSON   Department of Biochemistry, University of Cambridge, Cambridge CB2 1QW, United Kingdom

HERMAN M. KALCKAR   Chemistry Department, Boston University, Boston, Massachusetts 02215

WILLIAM McCULLOUGH   Department of Biology, University of Ulster, Newtownabbey, Co. Antrim BT37 0QB, Northern Ireland

WALLACE L. McKEEHAN   W. Alton Jones Cell Science Center, Lake Placid, New York 12946

MICHAEL J. MORGAN   Department of Biochemistry, University of Leicester, Leicester LE1 7RH, United Kingdom

FRED R. OPPERDOES   Research Institute for Tropical Diseases, International Institute for Cellular and Molecular Pathology, B-1200 Brussels, Belgium

STEPHEN A. OSMANI   Department of Biochemistry, King's College London, London WC2R 2LS, United Kingdom

PIETER W. POSTMA   Laboratory of Biochemistry, B.C.P. Jansen Institute, University of Amsterdam, 1018TV Amsterdam, The Netherlands

CLIVE F. ROBERTS   Department of Genetics, University of Leicester, Leicester LE1 7RH, United Kingdom

ANTONIO H. ROMANO   Microbiology Section, The University of Connecticut, Storrs, Connecticut 06268

IMMO E. SCHEFFLER   Department of Biology, University of California at San Diego, La Jolla, California 92093

MICHAEL C. SCRUTTON   Department of Biochemistry, King's College London, London WC2R 2LS, United Kingdom

GAGIK STEPAN-SARKISSIAN   Wolfson Institute of Biotechnology, University of Sheffield, Sheffield S10 2TN, United Kingdom

DONNA B. ULLREY   Chemistry Department, Boston University, Boston, Massachusetts 02215

Foreword

It is perhaps obvious to any student of Biology that the discovery of chemical processes in whole organisms has usually preceded the elucidation of the component steps. However, it is perhaps less obvious that the unravelling of the sequences in which those chemical steps occur in living matter, of the precise mechanisms involved, and of the manner in which they are regulated, would have been achieved neither by the study of intact plants and animals nor even of extracts derived from them. Our ability to understand the nature and regulation of metabolism rests on two main premises: the postulate that life processes can indeed be validly investigated with individual cells and cell-free extracts, and the thesis that there is an essential "unity in biochemistry" (as Kluyver put it, 60 years ago) that enables events in one organism to be legitimately studied in another.

Of particular utility in this latter respect has been the use of cultures of single-celled organisms, growing in defined media—especially prokaryotes, such as *Escherichia coli,* and eukaryotes, such as *Neurospora* and *Saccharomyces* sp., to which both biochemical and genetical techniques could be applied. It was, of course, Pasteur's observations of bacterial fermentations that first overthrew the belief that oxygen was essential for all energy-yielding processes: his recognition that *"la fermentation. . . . . c'est la vie sans air"* laid the foundations of our knowledge of glycolysis. Investigations of mutants of *Neurospora* led Beadle and Tatum to develop the concept that any one enzyme was specified by its appropriate and designated gene; the code in which this specification was written was subsequently deciphered largely with the use of *E. coli.* And our knowledge of the manner in which carbohydrates are taken up by cells, and of the manner in which the operation of multitudinous metabolic processes are coordinated one with another so as to maintain the constancy of the interior milieu, rests also on work with *E. coli,* done largely in the laboratories of Monod in France and of Umbarger in the U.S.A. some 30 years ago.

Yet it is now also clear that Monod's epigram, that what applied to *E.coli* applies also to *E.lephants,* cannot be wholly true. We now know that, although the genetic code is indeed universal, the manner in which it is expressed in eukaryotic cells differs significantly and is more complex than the manner in which this expression occurs in prokaryotes. Eukaryotic cells furthermore contain membrane-bounded compartments that not only separate different cellular components but also segregate different cellular events; these are not found in prokaryotes. The component cells of plants and animals communicate with their neighbors in the tissues and organs of which they form a part, and with distant parts of the organism via chemical messengers: again, prokaryotes do not manifest to any great extent these biological capabilities. The study of intermediary metabolism in eukaryotic cells and of its regulation will thus inevitably reveal complexities that cannot be deduced from our knowledge of these processes, elucidated with single-celled free-living microorganisms.

The present book attempts to focus on this gap, and to bridge it, by bringing together up-to-date reviews of information obtained from "classical" studies with microorganisms with information obtained from plant and animal cells treated as if they were microorganisms. Expertise in these different areas is usually not common to cell physiologists: it is rare for animal or plant biochemists to be familiar with the potential inherent in microbial methodology, and (vice-versa) few microbiologists have any deep understanding of the problems as well as the promise of growing plant or animal cells in culture. I greatly welcome this move to bring together practitioners in these apparently disparate fields: the history of Biochemistry surely teaches us that major progress is made only when the methodology of one discipline is brought to bear on another.

Hans Kornberg

Preface

A book devoted to carbohydrate metabolism demands an explanation since any casual, but observant, biochemist knows that all that has to be said on carbohydrate metabolism is available in biochemistry textbooks. The less casual observer will know that there is more to carbohydrate metabolism than even the most comprehensive biochemistry text can convey or the most complete metabolic map describe. In this book I have elicited the help of various expert authors to bring to fruition a seed that was planted some twenty years ago when I began my work on carbohydrate metabolism (*Escherichia coli*), in the laboratory of Hans Kornberg in Leicester. For a few years I deserted the charms of carbohydrate metabolism for the more ephemeral delights of interferon research, but soon realized the error of my ways and returned, albeit in animal cells, to the study of carbohydrate metabolism.

During these travels I myself delved in a number of the directions represented in this book and began to learn the lessons of cross-fertilization and hybrid vigor. When I was approached with a view to producing a book on carbohydrate metabolism, I felt that something new had to be said if it were worth saying at all. I hope that this book, in which two main themes are explored, the regulation of carbohydrate metabolism and the use of cultured cells as an experimental system, does say something new, besides underlining established fact, and identifies areas worthy of further exploration and enlightenment.

In the first chapter Herman Kalckar and Donna Ullrey describe their studies on the regulation of hexose transport in fibroblasts and discuss in detail the experimental procedures used. Their chapter also illuminates one of the dominant subthemes running through the book, namely the use of well-characterized mutants to dissect metabolic pathways, in this case the regulation of transport. In the next chapter, Pelin Faik and I paint a broad canvas of the utilization of carbohydrates by animal cells, and again develop the idea of the use of bio-chemical mutants. We also emphasize the use of well-defined culture systems for exploring the regulation of differentiation and gene expression in animal cells. The use of well-defined mutants is further emphasized by Immo Scheffler's chapter on respiration-deficient mammalian cells, which illustrates how bio-chemical genetics are beginning to unravel the complexities of mitochondrial biogenesis. Wallace McKeehan presents convincing evidence for his hypothesis that glutamine metabolism is of major importance in the energy metabolism of mammals and that the series of reactions leading to glutamine breakdown, glu-taminolysis, is a metabolic pathway deserving further attention and investiga-tion.

These opening chapters on carbohydrate metabolism in animal cells are followed by chapters on plant cells and the African trypanosomes. Gagik Stepan-Sarkissian and Michael Fowler describe uptake and metabolism of the wide range of carbon sources that plant cells in culture will grow on. As they point out, this is a relatively unexplored facet of plant cell culture that no doubt will become of more importance as more cells find industrial uses. Alan Fairlamb and Fred Opperdoes cover the methods of cultivation of trypanosomes, the substrates used, and end products of their metabolism. They draw particular attention to that fascinating organelle, the glycosome, which contains many of the glycolytic enzymes, and to the possibilities of glycolysis as a target for chemotherapy against the trypanosomes, which cause such devastating illness.

Tony Romano continues with a description of the transport systems for sugars in baker's yeast and filamentous fungi. His chapter leads into the elegant contribution by Juana Gancedo on carbohydrate metabolism in yeast. She em-phasizes the methodological approaches, presents an overview of carbohydrate metabolism, and ends with an interesting and informative discussion of regulato-ry mechanisms. This is followed by a monumental treatment of carbon metabo-lism in filamentous fungi by William McCullough, Clive Roberts, Stephen Os-mani, and Mike Scrutton. This is an exhaustive treatment of special note since there have been no comparable reviews of this topic published recently. They manage to cover all aspects in great detail, including the breakdown of com-pounds, the enzymes involved, and the synthesis of carbohydrates.

Finally, there are three chapters on bacteria. Active transport of sugars into the Enterobacteriaceae falls naturally into two parts: that involving phosphoryla-tion of the sugar, and this is covered by Pieter Postma; and that involving the transport of unmodified sugar (active transport), which is covered by Peter Henderson. Both chapters present an exhaustive account of the state of the art and show the elegant methods of molecular biology and recombinant DNA now available for dissecting metabolic pathways. Finally, Ron Cooper develops the theme of convergent pathways of sugar catabolism in bacteria and shows how one can divide metabolism into branch lines and main trunk pathways.

It is hoped that the reader will first dive into those sections with which he or she is most familiar before swimming into less certain waters. However, the reader is encouraged to take the plunge and is certain of finding many familiar rocks on which to cling. Any good qualities that the book may have are entirely due to the authors; the faults must lie with me.

Finally, it gives me great pleasure to be able to thank those who have helped bring this book to completion. Special thanks must go to the staff of the Wellcome Trust, and especially to Wendy Jacobs and Sue Parkes for their cheerful forbearance on the word processor, to Siân Spry for keeping my filing trays clear, and to Derek Metcalfe for oiling the wheels. Without them this book would never have been finished. Finally, grateful thanks to my wife Pelin, without whom this book would never even have been started.

Michael J. Morgan

London

Contents

Chapter 2

*The Utilization of Carbohydrates by Animal Cells: An Approach to
Their Biochemical Genetics*

MICHAEL J. MORGAN AND PELIN FAIK

Chapter 3

Biochemical Genetics of Respiration-Deficient Mutants of Animal Cells

IMMO E. SCHEFFLER

Chapter 4

Glutaminolysis in Animal Cells

WALLACE L. McKEEHAN

Chapter 5

The Metabolism and Utilization of Carbohydrates by Suspension Cultures of Plant Cells

GAGIK STEPAN-SARKISSIAN AND MICHAEL W. FOWLER

Chapter 6

Carbohydrate Metabolism in African Trypanosomes, with Special Reference to the Glycosome

ALAN H. FAIRLAMB AND FRED R. OPPERDOES

Chapter 7

Sugar Transport Systems of Baker's Yeast and Filamentous Fungi

ANTONIO H. ROMANO

Chapter 8

Carbohydrate Metabolism in Yeast

JUANA M. GANCEDO

Chapter 9

Regulation of Carbon Metabolism in Filamentous Fungi

WILLIAM MCCULLOUGH, CLIVE F. ROBERTS, STEPHEN A.
OSMANI, AND MICHAEL C. SCRUTTON

Chapter 10

The Bacterial Phosphoenolpyruvate: Sugar Phosphotransferase System of Escherichia coli and Salmonella typhimurium

PIETER W. POSTMA

Chapter 11

Active Transport of Sugars into Escherichia coli

PETER J. F. HENDERSON

Chapter 12

Convergent Pathways of Sugar Catabolism in Bacteria

RONALD A. COOPER

1

Studies of Regulation of Hexose Transport into Cultured Fibroblasts

HERMAN M. KALCKAR and DONNA B. ULLREY

1. INTRODUCTION

The intent of this review is to discuss some physiological features of cellular regulation of glucose transport into cultured fibroblasts. It will be confined in the following ways:

1. The animal cells under discussion are cultured avian and mammalian fibroblasts in the postlogarithmic growth phase (or in tumorigenic cultures, medium to high density). The cultures are monolayer cultures, not suspension cultures.
2. The main theme will be the effects of carbohydrates and their metabolism in relation to the regulation of uptake or transport of hexoses into cultured fibroblasts, including special metabolic mutants. The downregulation of the hexose transport of the cultures will be called "the glucose-mediated transport curb."
3. The transport systems discussed are largely the facilitated diffusion of hexoses and its regulation by carbohydrate metabolism. Where amino acid transport systems (such as the A and L systems) are discussed, this is strictly related to carbohydrate transport and metabolism.
4. Since hexose transporters are membrane proteins, aspects of protein recycling will have to be considered, and hence also the effects of inhibitors of protein synthesis.

HERMAN M. KALCKAR and DONNA B. ULLREY • Chemistry Department, Boston University, Boston, Massachusetts 02215.

5. Some stress-related factors, such as glucose starvation, uncouplers of respiration, as well as oncological transformations and their effects on hexose transport regulation, will be discussed.

It is well known that cultured fibroblasts from chick embryos (CEF) as well as fibroblasts from other mammalian species (mouse 3T3, hamster NIL lines) show an enhancement of uptake of hexoses, if maintained for several hours in the absence of glucose (or if fructose replaces glucose) in a nutrient medium (Martineau *et al.*, 1972; Kalckar and Ullrey, 1973; Christopher *et al.*, 1976b,c).

The enhancement is arrested by cycloheximide, but once the enhancement has reached a certain level and glucose starvation is continued, addition of cycloheximide does not change this level (Martineau *et al.*, 1972; Christopher *et al.*, 1976a). Moreover, when down-regulation ensues upon glucose refeeding of fibroblast cultures (glucose-mediated "curb"), addition of cycloheximide prevents the mediated curb (Christopher *et al.*, 1976a, c).

Some of these features have been disclosed by other investigators who also examined some of the kinetic parameters of the cellular uptake system (Kletzien and Perdue, 1975). The use of 3-*O*-methylglucose for quantitative determination of hexose transport systems in fibroblast cultures has been extensive in CEF cultures, including cultures transformed with oncogenic viruses (Weber, 1973; Kletzien and Perdue, 1975; Christopher *et al.*, 1976a).

In this review we shall essentially confine ourselves to metabolically mediated transport regulations and only occasionally refer to studies of other types of regulation.

It was shown by two independent groups that glucose starvation of CEF cultures brings about a four- to fivefold increase of V_{max} with only minor differences in K_m (Kletzien and Perdue, 1975; Christopher *et al.*, 1976a). The same feature was recently demonstrated in membrane preparations from glucose-fed and glucose-starved CEF cultures (Yamada *et al.*, 1983). In mammalian cell cultures (NIL), V_{max} was found to be more than tenfold higher in glucose-starved than in glucose-fed cultures (Christopher, 1977).

2. HEXOSE UPTAKE OR TRANSPORT TESTS

The use of 3-*O*-methylglucose for determining transport rates into cells presents some drawbacks as a routine method, mainly due to the well-known fact that this ligand equilibrates across the membrane in less than a minute. Moreover, in order to obtain correct kinetic data, it is necessary to preequilibrate the cultures over 30 min with nonlabeled 3-*O*-methylglucose succeeded by repeated washings prior to the transport test itself. In spite of these time-consuming and isotope-wasting features it is advisable to use this ligand when faced with newly disclosed regulatory responses and also when working with new cell lines, especially mutants (see Kalckar *et al.*, 1979; Ullrey *et al.*, 1982).*

Is it possible to replace 3-*O*-methylglucose with other hexoses or hexose analogues? Labeled glucose has been used. However, it has been found that some glucose-fed cultures being ''induced'' also excrete lactic acid, which can give a highly distorted picture of the uptake of labeled glucose due to the excretion of labeled lactic acid (Lust *et al.*, 1975; Salter and Cook, 1976).

The use of 2-deoxyglucose as a probe for the hexose uptake and transport system has been analyzed in CEF cultures, fed or starved for glucose (Kletzien and Perdue, 1975). If the uptake tests were drawn within 1 min and at a concentration of labeled 2-deoxyglucose near the K_m for transport, i.e., 2 mM, the transport step was found to be rate-limiting, even in the glucose-starved cultures (Kletzien and Perdue, 1975). In this case, the kinetic constants for the transport step (K_m and V_{max}) were found to be of the same order of magnitude whether determined by labeled 3-*O*-methylglucose or by 2-deoxyglucose, i.e., V_{max} 90–95 for starved and 18–20 nmoles/min per mg cell protein for glucose-fed cultures (Kletzien and Perdue, 1975).

However, Christopher *et al.* (1976a) found clear evidence for the existance of two hexose transport systems, the low affinity system, usually tested with 3-OMG with a K_m of approx. 1 mM and a high-affinity system disclosed by the use of labeled 2-DOG. The latter showed a K_m of appr. 0.05–0.1 mM and it was sensitive to N-ethylmaleimide (Christopher *et al.*, 1976a).

In certain cases, 2-deoxyglucose should be avoided as a probe; e.g., when dealing with mutants that accumulate glucose-6-phosphate when fed glucose (such as a useful mutant to be described later that is unable to convert glucose-6-phosphate to fructose-6-phosphate and hence accumulates the former, Pouysségur *et al.*, 1980; Ullrey *et al.*, 1982). In such cases, the use of 3-*O*-methylglucose or of galactose is advisable (Ullrey *et al.*, 1982; Ullrey and Kalckar, 1982).

A useful study of the kinetics of galactose transport in NIL hamster cultures (in which the cellular galactokinase has been rendered inactive by means of carefully titrated sulfhydryl reagents) is relevant to this discussion. It was found (Christopher, 1977) that 0.2 and 1.0 mM *N*-ethylmaleimide (NEM) rendered NIL cells galactokinase-less (without making the cells leaky for labeled L-glucose), thus making it possible to test the hexose transport system with galactose. The design was as follows. Confluent monolayers of NIL fibroblasts—after being maintained in standard DME medium with glucose, or alternatively with

---

*The common assumption that 3-*O*-methylglucose is not metabolized after transport into cells has been challenged recently by the results of studies on clearance of labeled 3-*O*-methylglucose from heart muscle (Gatley *et al.*, 1984). These studies indicate that 3-*O*-methylglucose undergoes a distinct phosphorylation–dephosphorylation ''futile'' cycle in heart muscle from rat (Gatley *et al.*, 1984). Such a futile cycle would especially affect estimates of cellular space available to glucose.

fructose, over 18–24 hr—were washed with PBS. Subsequently, they were incubated for 15 min (at 37°C) with PBS containing 0.5 mM NEM (Christopher, 1977).

In the same study, two types of monitoring were performed. The first type was designed to show, by means of paper chromatography, that untreated fibro-blasts exposed 10 min to [^{14}C]galactose, accumulated galactose solely as phos-phorylated products (UDP-hexose and galactose-1-phosphate) without any free galactose left in the cells; this was indeed the case (Christopher, 1977; Christopher et al., 1977). Conversely, the NEM-treated cultures exposed 10 min to [^{14}C]galactose accumulated only free galactose without any trace of galac-tose-1-phosphate or its metabolic derivatives (Christopher, 1977). This strongly indicated that the kinase step was practically eliminated, whereas the transport step was left unimpaired.

The second type of monitoring, that of the galactose transport system, could now be performed by exposing the NEM-treated cultures to 10 sec of [^{14}C]galac-tose (incubation with [^{3}H]L-glucose at 25°C served as a control for leakiness as well as for nonspecific adsorption of hexose). Galactose transport in fructose-maintained cultures was about 20-fold higher than that of cultures maintained in glucose medium, and (with little change in K_m) the V_{max} was less than 10% of that of the fructose cultures. Moreover, the 20-fold enhancement of the transport observed in the fructose cultures, was also demonstrable in untreated cultures exposed to either labeled galactose or 3-O-methylglucose (Christopher, un-published observations). In the former case, 1-min exposures also showed the marked enhancement (Christopher, 1977).

Evidently, this serves to ensure that with proper reservations, galactose can serve as a very useful monitor for designing experiments on hexose transport regulation.

It is surprising that labeled galactose could be used over a span of 5 to 10 min in the uptake tests on near-confluent NIL cultures without significantly distorting the tests for regulatory features (Christopher et al., 1976b,c; Christo-pher, 1977). Sparse NIL cultures showing high uptake of galactose, show only small differences in uptake rates in glucose-fed versus starved cultures (Kalckar and Ullrey, 1973).

2.1. Effects of Cycloheximide on Hexose Transport Regulation

The down-regulation or curb of the hexose uptake system is affected in a very involved way by the well-known inhibitor of protein synthesis, cyclohexi-mide, as described by Christopher et al. (1976a,b,c) and Martineau et al. (1972).

In NIL hamster fibroblast cultures fed glucose, addition of cycloheximide in low concentrations (approx. 2 μg/ml) gradually brought about a loss of transport or uptake activity, which by 8 to 12 hr had dwindled to about 5% of the basic level (Christopher *et al.*, 1976b,c). This feature was also observed in CEF cultures (Christopher *et al.*, 1976a; Yamada *et al.*, 1983). As described above, glucose starvation of CEF or NIL cultures (or replacement of glucose with fructose) gives rise to an enhancement of the hexose uptake system, and this enhancement was retarded by cycloheximide (Martineau *et al.*, 1972; Christopher *et al.*, 1976a,b,c).

The first cycloheximide effect (i.e., the loss of hexose transport in glucose-fed cells, far below the basic value) was strikingly expressed at low cycloheximide concentrations (1–3 μM) whereas at higher concentrations (10–50 μM) the levels of hexose transporters were scarcely diminished (Christopher *et al.*, 1976b,c, 1979; Yamada *et al.*, 1983). Although protein synthesis at large was severely diminished by both concentrations of cycloheximide, Christopher invoked as a basis for the seemingly so aberrant cycloheximide regulation of the hexose transport system, the existence of a well-expressed turnover mechanism of the carriers. It was suggested that the synthetic phase of the turnover mechanism was inhibited by minute amounts of cycloheximide whereas the degradative phase (presumably catalyzed by intracellular cathepsins) became strongly inhibited when the cycloheximide concentration was increased (Christopher *et al.*, 1976a, 1979; Amos *et al.*, 1976; Christopher, 1977).

Pursuing the cycloheximide effects on the regulation of the hexose transport system in glucose-starved and subsequently refed NIL cultures, Christopher *et al.* (1976b,c) disclosed the following additional noteworthy features (also fully confirmed in CEF cultures by Yamada *et al.*, 1983). Upon refeeding the glucose-starved NIL cultures glucose (or replacing fructose with glucose), the enhanced uptake or transport levels of the glucose-starved cultures gradually went down, i.e., the hexose transport system was again subject to the glucose-mediated curb. Addition of cycloheximide (10 μg/ml) at the onset of glucose refeeding, arrested or greatly delayed the development of the glucose-mediated uptake curb (Christopher *et al.*, 1976b,c; Table I). This type of response was also observed after

Table I. Glucose-Mediated Transport Curb Is Prevented by Cycloheximide[a]

Culture conditions for NIL cells (MEM, with fructose or glucose, 10 mM)	Transport or uptake (37°C)	
	[^{14}C]-3-O-Methylglucose (10 sec) (rel.)	[^{14}C]-Galactose (10 min) (rel.)
Onset of glucose refeeding	1.0	1.0
Glucose refeeding, 20 hr	0.27	0.13
Glucose + cycloheximide (10 μg/ml), 20 hr	0.90	0.93

[a]Modified from Christopher *et al.* (1976c).

refeeding cultures, if the cycloheximide concentration used was of the order of 50 μg/ml (Yamada *et al.*, 1983).

Most of these intriguing responses could also be demonstrated by another inhibitor of protein synthesis, emetine (Tillotson *et al.*, 1984). Emetine and cycloheximide have been classified as inhibitors of the elongation step in protein synthesis (see Tillotson *et al.*, 1984). Glucose-fed CEF treated with very low concentrations of emetine (0.1 μM) suffered a heavy loss of 3-*O*-methylglucose transport activity whereas emetine in higher concentrations (10 μM) scarcely affected the transport activity (Tillotson *et al.*, 1984). These results bear a striking similarity to those of the cycloheximide experiments (Christopher *et al.*, 1976b,c; Yamada *et al.*, 1983; Tillotson *et al.*, 1984).

2.2. Is the Release of the Mediated Curb of the Hexose Transport System Dependent on Transcription?

In this review the expressions "down-regulation" of transport, or transport "curb" are used instead of "repression–derepression," since there seem but few indications that transcription is involved.

However, we shall not fail to report that camptothecin, a specific inhibitor of transcription, was found, in amounts as low as 2μg/ml, to markedly inhibit the release of the mediated transport curb (Kalckar *et al.*, 1980a,b). This slow response, which can be reversed, might well indicate that the nature of the release is in fact a derepression and that the glucose-mediated curb is a genuine repression. Additional research is needed to clarify this point.

3. METABOLIC PATHWAYS

Studies attempting to clarify the nature of metabolic vectors essential for bringing about the glucose-mediated curb of the hexose transport system have been conducted on CEF cultures as well as on cultures of various rodents (especially hamsters). In 1980, Pouysségur *et al.* isolated from the hamster lung fibroblast line 023 (Franchi *et al.*, 1978) an interesting metabolic mutant (DS-7) highly defective in the enzyme phosphoglucose isomerase [PGI; D-glucose-6-phosphate ketol isomerase (EC 5.3.1.9)].

Among the mutants available to us, we consider the PGI mutant the most useful, especially because the role of glucose is strikingly confined as far as metabolically mediated transport regulation is concerned.

The designs used in the analyses of the regulatory patterns of the hexose transport system in the PGI mutant have disclosed the complexity of the so-called "glucose-mediated" transport curb (Ullrey *et al.*, 1982).

The parental line 023 is PGI$^+$ (Pouysségur *et al.*, 1980), and much like the response of the NIL line (Kalckar et al., 1979), glucose as well as D-glucosamine brought about a marked transport curb (Ullrey et al., 1982). In contrast, the PGI mutant did not respond to D-glucosamine, whereas glucose still elicited a marked transport curb (Ullrey et al., 1982).

Before these studies had been completed, we believed that the transport curb could be instituted if two requirements were fulfilled: (1) the presence of a hexose (or a hexose analogue) that is a ligand of the hexose carrier; glucose and glucosamine are ligands, fructose is not (Ullrey and Kalckar, 1981; see also Lipmann and Lee, 1978); (2) generation of oxidative energy metabolism in which glucose or D-glucosamine was considered the main contributor (Kalckar et al., 1979).

The inability of the PGI mutant to respond to D-glucosamine, while still responding to glucose, forced us to revise the requirements for the development of the transport curb: glucose is specifically required (Ullrey et al., 1982; Ullrey and Kalckar, 1982).

The role of L-glutamine in energy metabolism of fibroblast cultures had been stressed by several groups of investigators (Donnelly and Scheffler, 1976; Zielke et al., 1978; Reitzer et al., 1979); Morgan and Faik, 1981).*

We had shown that it was possible to maintain NIL cultures for at least 24 hr in a medium without L-glutamine as well as glucose (Kalckar et al., 1980a). Addition of glucose brought about a strikingly marked transport curb (Kalckar et al., 1980a) This device was also tried on the PGI mutant, but in this case glucose was unable to elicit the transport curb unless it was supplemented with mannose or with pyruvate (Ullrey and Kalckar, 1982). The case with mannose deserves special comments.

It has long been known that mannose-6-phosphate can be converted to fructose-6-phosphate by a specific mannose-6-phosphate isomerase (Slein, 1950). In the PGI mutant, mannose, but not glucose, was able to serve in energy metabolism (anaerobic as well as aerobic); in fact, mannose was the only hexose that was able to generate large amounts of lactic acid in this cell line (Ullrey and Kalckar, 1982), and presumably also pyruvate. Yet, by itself mannose was unable to elicit a transport curb; the latter required as mentioned the simultaneous

---

*DME (Dulbecco modified Eagle's) medium is distributed in two forms—pyruvate present (usually 1.0 mM) or absent; the latter usually contains glucose at 25 mM whereas the former contains only 5.6 mM glucose. Some companies have different catalog numbers for the two DME media, and some are not explicit but may be willing to reveal the contents if contacted. The confusion is occasionally carried over to a guessing game regarding the presence of L-glutamine and its concentration, usually 4 mM. It would undoubtedly clarify matters if name and catalog numbers were furnished in publications, especially in studies on metabolic mutants. In any case, the presence or absence of pyruvate in DME media should be stated. For the incubation period (10–24 hr) it is usually stated whether or not pyruvate was present and at what concentration (see, e.g., Ullrey et al., 1982; Ullrey and Kalckar, 1982; and Table 1, Kalckar and Ullrey, 1984a).

presence of glucose as well (Ullrey and Kalckar, 1982; Kalckar and Ullrey, 1984a). Evidently, the PGI mutant is unable to use glucose-6-phosphate in energy metabolism (Pouysségur *et al.*, 1980; Ullrey and Kalckar, 1982), being confined either to the catabolic pentose shunt or to the anabolic pathway to glucosylation or galactosylation. At present, we are unable to state which of the many enzymes or intermediary products are crucial for the glucose effect. However, it may be important to bear in mind that galactose was able to replace glucose in initiating a transport curb together with mannose (Kalckar and Ullrey, 1984a). Galactose is phosphorylated to galactose-1-phosphate by a specific kinase, different from the hexokinase that catalyzes the phosphorylation of mannose as well as glucose (see Leloir, 1951). In the PGI mutant, the confined metabolism of glucose brings about an accumulation of glucose-6-phosphate (Pouysségur *et al.*, 1980), which in turn interferes with metabolism of mannose (Ullrey and Kalckar, 1982; Kalckar and Ullrey, 1984a). This metabolic interference with mannose metabolism did not ensue if galactose was used instead of glucose (Kalckar and Ullrey, 1984a).

Some of the revealing features of the mediated transport curb of the PGI mutant are illustrated in Tables II and III from our recent work (Ullrey *et al.*, 1982; Ullrey and Kalckar, 1982; Kalckar and Ullrey, 1984a,b).

Since it is obviously of great importance to distinguish between inhibitions of the hexose transport system and down-regulations of the nature of a metabolically mediated transport curb, we have frequently used inhibitors of oxidative energy metabolism or oxidative phosphorylation (see Kalckar *et al.*, 1979; Ullrey and Kalckar, 1981). If inhibitors such as DNP, oligomycin, or malonate circumvent the glucose-imposed down-regulation significantly without altering the transport rates in the absence of glucose, we regard the glucose effects, including the combined glucose–mannose effect, as a mediated curb (Kalckar *et al.*, 1979; Ullrey and Kalckar, 1981). This is illustrated in Tables II and III in which malonate, being the more gentle of the metabolic inhibitors in terms of preserving cell adhesion, was used.

The role of the glucose-6-phosphate dehydrogenase pathway for the metabolically mediated curb of the hexose transport system in CEF has been studied by Gay and Amos (1983). They first found an inverse relationship between hexose transport rates and the cellular levels of phosphoribosyl pyrophosphate (PRPP), i.e., the higher the PRPP levels in CEF, the more intense the glucose-imposed curb seemed to be. Conversely, absence of glucose or replacement of this with xylose, led to a lift of the transport curb and the levels of PRPP decreased (Gay and Amos, 1983). The same investigators observed an additional correlation between the glucose-mediated hexose transport curb and guanine or guanine nucleotide formation. Addition of guanine or hypoxanthine to CEF cultures incubated with glucose, brought about an additional down-regulation of the hexose transport system (Gay and Amos, 1983). In the latter case, i.e.,

Table II. Catabolic Down-Regulation of 3-O-
Methylglucose Transport by Glucose and
Glucosamine in the PGI mutant and Its Parental
Strain[a]

	[14C]-3O-Methylglucose transported (pmoles/mg cell protein per 20 sec)	
Conditioning medium[b]	DS-7 (PGI⁻)	023 (parental)
No hexose	13.0	17.8
No hexose + malonate	23.9	20.3
Glucose	3.6	4.6
Glucose + malonate	15.3	12.0
Glucosamine	10.7	5.7
Glucosamine + malonate	10.7	17.5
Fructose	15.2	18.2
Fructose + malonate	21.7	20.8

[a]From Ullrey et al. (1982).
[b]Concentrations in medium: glucose, 22 mM; glucosamine, 5 mM; fructose, 22 mM; malonate, 24mM; DME medium with 4 mM L-glutamine. Dialyzed fetal calf serum, 10%. Transport was measured at 23°C.

incubation with glucose plus one of the purines, the cellular levels of PRPP were much lower than those observed with glucose alone, indicating that the combined incubation may have led to the synthesis of purine nucleotides (Gay and Amos, 1983).

In some very recent studies (Amos et al., 1984) it was observed that CEF

Table III. Uptake Curb and Lactate Formation in the PGI Mutant, Maintained in Medium Devoid of L-Glutamine and Pyruvate[a]

Substrate[b]	[14C]-Galactose uptake (nmoles/mg protein per 10 min)	Lactate formed (μmoles/mg cell protein per 20 hr)
Mannose	3.45	33.70
Glucose	3.39	1.60
Mannose + glucose	1.38	6.83
Mannose, glucose, + malonate	2.68	11.84

[a]From Kalckar and Ullrey (1984a).
[b]The concentrations of substrates were: 20 mM mannose, 2 mM glucose, and 25 mM malonate.

cultures grown for two and three passages in media deprived of nicotinamide or niacin lost the glucose-mediated transport curb. Later addition of niacin, nicotinamide or its mononucleotide (NMN) to the deprived cultures had no effect; however, addition of NAD did restore the response of the cultures to impose the glucose-mediated transport curb (Amos *et al.*, 1984).

The striking effects of nicotinamide deficiency on hexose transport regulation was also observed in NIL hamster cultures (Amos *et al.*, 1984). Removal of nicotinamide or niacin permitted slow growth with increasing generation time until cessation of growth ensued, yet preservation of viability persisted. Addition of nicotinamide brought about a resumption of growth. Yet the normal glucose-imposed transport curb was only restored by incubation with $NAD(10^{-5}M)$: NMN and nicotinamide remained inactive in this regard.

The authors left various interpretations concerning the possible role of NAD in transport regulation open for discussion. One option considered was a possible action by a split product, adenosine diphosphoribose (ADPR), since many fibroblast cultures contain a hydrolase that converts NAD to ADPR. Another option emphasized was the action of NAD in oxidative energy metabolism, since the latter had been shown to be involved in the establishment of the transport curb (Kalckar *et al.*, 1979; Gay and Amos, 1983).

In 1979 Howard *et al.* were able to describe a marked insulin stimulation of glucose entry into human skin fibroblast cultures (Howard *et al.*, 1979). The Km of transport of 3-0MG was not affected by the presence of insulin, but the V_{max} was increased from 14 to 23 by the presence of insulin in concentrations of 2×10^{-7} M (Howard *et al.*, 1979). One of the factors which may have been important for the consistant effect of insulin at these concentrations, is perhaps the composition of the so-called IM medium, a mineral-amino acid-vitamin medium, with albumin replacing serum. During the 18 hr preincubation of the cultures in IM medium, glucose was absent (Howard *et al.* 1979).

There is an abundant literature about the effect of insulin on the hexose uptake in mammalian fibroblast cultures. The recent work by Germinario *et al.* (1984a) seems of special interest to our own, because the insulin was used in concentrations of the same order of magnitude on human fibroblasts as Howard *et al.*, yet with a different outcome. Germinario *et al.* (1984a) found that glucose in the medium brought about a curb of the hexose transport system, but also increased insulin binding as well as insulin stimulation of the uptake of 2-DOG. The latter effect seems to overcome the curb.

The response to insulin in nearly corresponding concentrations (0.1 U/ml or approximately 5×10^{-7} M) was also studied. In bicarbonate-buffered CEF cultures maintained over 24 hr in 0.5% calf serum and 5.5 mM glucose, insulin stimulated glucose transport by a factor of 3.5 to 4-fold (Amos *et al.*, 1976). Glucose-starved CEF cultures in standard bicarbonate medium showed a starvation enhancement in the absence of insulin. These enhanced transport rates of the starved cultures were not affected by insulin. However, in the absence of CO_2– bicarbonate, the greatly lowered uptake rates were somewhat stimulated by insulin, at least in the starved state. In Hepes buffers the rates of the glucose-fed cultures were often less than 10% of the glucose-starved ones (Amos *et al.*, 1976, Howard *et al.*,1979, Germinario *et al.*,1984).

The hexose uptake system is affected not only by the buffer used, but also by the type of cations. Bader (1976) found that CEF cultures exposed to lowered sodium ion concentrations in the range from 80 to 150 mM showed lowered uptake rates of 2-deoxyglucose; cultures transformed by RSV, Bryan strain, showed the usual increased uptake capacity and this was not affected by the lowered level of Na^+ (Bader, 1976).

The stimulation of the hexose uptake system by serum addition to serum-deprived CEF cultures has been observed by many investigators, starting with Sefton and Rubin (1971; cited by Perdue, 1976, 1978). Removal of serum from untransformed cultures is usually not extended beyond 6 hr which brings about a fall in hexose transport to about half that of the original capacity (Perdue, 1976); the addition of fetal calf serum (5%) produced an increase in the uptake capacity of 50% within 10 min and this rapid increase was not arrested or delayed by addition of cycloheximide (Perdue, 1976).

Withdrawal of serum or L-glutamine, or both, from the maintenance medium for NIL cultures, over a period of 24 hr, did not affect the uptake rates of glucose-starved cells very much but the transport capacity of glucose-fed cells was drastically curtailed (Kalckar *et al.*, 1980a). Since addition of DNP or malonate counteracted the intense glucose-mediated transport curb very effectively, energy generation seems needed for this intense type of curb as well (Kalckar *et al.*, 1980a).

4. CHARACTERIZATION OF CERTAIN CELLULAR MACROMOLECULES AND STRUCTURES

4.1. Enzyme Assays of the Hexose Uptake System in Lysed Cells

Since hexokinase as well as galactokinase are part of the cellular uptake systems in cultured fibroblasts, studies of the levels of these enzymes in lysed fibroblasts have been conducted. Hexokinase levels in homogenates of CEF have been determined under near-optimal conditions, using glucose as well as 2-deoxyglucose as substrates, and spectrophotometric as well as chromatographic methods, respectively (Kletzien and Perdue, 1975). Analyses on homogenates from CEF cultures starved of glucose for 24 hr did not show hexokinase levels any higher than those of the homogenates from glucose-fed cultures (Kletzien and Perdue, 1975). BHK cultures did not show any difference in hexokinase activity either, whether starved or fed (Kletzien and Perdue, 1975).

As part of the studies of regulation of galactose uptake in NIL cultures, lysed fibroblasts were analyzed as to the galactokinase levels under near-optimal conditions (Christopher et al., 1976c). The method used was that of Blume and Beutler (1975). The galactokinase levels found in starved NIL cultures were not higher than those found in glucose-fed cultures (Christopher et al., 1976b). In fact, both hexokinase and galactokinase levels in the lysed cells were about 15% higher in the glucose-fed cells than those deprived of glucose (Kletzien and Perdue, 1975; Christopher et al., 1976c).

4.2. Membrane-Associated Glucose-Binding Proteins Released without Cell Lysis

Just as sugar-binding proteins in prokaryotic cells play a role in glucose or galactose transport (reviewed by Boos, 1974), so it seems that CEF also make use of a related device (Lee and Lipmann, 1977, 1978). It was first observed that membrane preparations from confluent CEF cultures contain a glucose-binding protein, which was relatively loosely bound to the membranes. In CEF transformed by RSV, the binding protein was found to be more tightly bound, but could be extracted from the membrane fraction by Triton X-100 and fractionated through a Sephadex G-200 column (Lee and Lipmann, 1977). The binding protein was also found to promote the uptake of 2-deoxyglucose in quiescent cultures over and above that of the serum effect (Lee and Lipmann, 1977).

In later work a fractionation procedure for the separation of the glucose-binding protein and the transport-promoting factor was developed (Lee and Lipmann, 1978). The mild extraction used a buffer mixture containing 10 mM Na octanoate and 5 mM triethanol-amine-HCl (pH 7.4), 100 mM NaCl/2 mM EDTA. The cells detached but did not lyse and could be spun down, the wash could be concentrated and applied to a Sephadex G-50 column.

Pooled fractions of proteins were incubated for 1 hr with [^{14}C]glucose (1.5 × 10^{-5} M) and, after ultrafiltration, applied to a Sephadex G-200 column equilibrated with a Tris buffer (pH 7.4) containing 2 mM EDTA, 50 mM NaCl, and 0.02% Triton X-100. Elution with this solution released three protein bands containing [^{14}C]glucose. The last band showed a pratically noteworthy stimulation of [^{3}H]-2-deoxyglucose uptake, as tested on CEF cultures. A sharp separation of the two functions was achieved by mounting some of the peaks from the Sephadex G-200 column (after desalting) on a DEAE–cellulose column (Lee and Lipmann, 1978). Glucose uptake rates assessed by the proper column fractions on CEF, grown in fructose, showed about fourfold higher rates than found on glucose-grown CEF. The molecular weight of the glucose-binding protein was also assessed. Gel electrophoresis in SDS did not deprive the protein of its ability to bind glucose. Hence, [^{14}C]glucose could serve as a direct marker of the binding protein and its subunits. In this way, it was found that the main peak corresponded to a molecular weight of 73,000. Since the binding activity persisted even after 2 hr of boiling, gel electrophoresis indicated the existence of a monomer of molecular weight 18,000 as well as a dimer (Lee and Lipmann, 1978).

This study indicates that the fraction that stimulated uptake of [^{14}C]-2-deoxyglucose left about 90% of the labeled [^{14}C]-2-deoxyglucose as the 6-phosphate ester. This is presumably a result of a stimulation of the transport step. Competition between various hexoses has also been examined and it was found that glucose and glucosamine, as well as mannose, competed both in the binding and in the transport test; galactose, fructose, and L-glucose did not. There were indications that binding tends to shift with time from loose to tight binding (Lipmann and Lee, 1978).

The interesting glucose-binding protein shows much higher affinity for glucose than that of the cytochalasin-sensitive hexose carrier protein and their molecular weights are also different (see Lee and Lipmann, 1978; Pessin *et al.*, 1982). An earlier report on the existence of a small population of a high-affinity glucose-binding protein in CEF (Christopher *et al.*, 1976a) should be kept in mind.

4.3. Studies on Isolated Plasma Membrane Preparations in Regard to Their Hexose Transport Population and Identification of Specific Transport Proteins

Identification of the hexose transporter population has been possible through the use of a potent, reversible inhibitor of glucose transport, cytochalasin B (for early references on the effect of cytochalasin B on glucose transport in cultured fibroblasts, see Cushman and Wardzala, 1980). In contrast, cytochalasin E possesses affinity for cytoskeleton but not for the hexose transporters, and these differences have constituted the basis for a useful new technique (see Cushman and Wardzala, 1980).

The following recent technique for characterization of the hexose transporters in the plasma membranes of large-scale cultures of CEF deserves at least a brief description here (see Pessin *et al.*, 1982). Crude plasma membranes were obtained by the nitrogen cavitation method (Inui *et al.*, 1979; Pessin *et al.*, 1982; Yamada *et al.*, 1983), resuspended in 1 mM Tris, 1 mM Hepes, pH 8.2 (buffer A), and centrifuged at 100,000g for 60 min. The pellets were resuspended in buffer A + 0.5 mM MgCl$_2$ and layered onto a discontinuous gradient of 10% and 20% dextran (both w/w) in buffer A + 0.5 mM MgCl$_2$. The 10% top interface was collected and diluted with buffer B, which has a pH of 7.5 and contains

EDTA (1 mM), and centrifuged again at 100,000g for 1 hr. This second pellet was resuspended in buffer B to give a final concentration of 2–3 mg protein/ml. This membrane fraction was enriched seven- to tenfold for 5′-nucleotidase compared with the original cell homogenate (see Pessin *et al.*, 1982).

The membrane suspension was first mixed with cytochalasin E (2 μM), and either glucose or sorbitol (500 mM) was present in buffer B. To this mixture was then added various concentrations of [^{3}H]cytochalasin B and after a brief incubation the samples were spun (150,00g, 30 min at 4°C) and the pellet dissolved and counted (see Cushman and Wardzala, 1980; Pessin *et al.*, 1982).

A new effective way of labeling the hexose transport carriers covalently by photoaffinity labeling, using [^{3}H]cytochalasin B, was developed by Pessin *et al.* (1982). In short, an excess of [^{3}H]cytochalasin B, ie. concentrations of 0.5μM (with or without D-glucose or sorbitol) was incubated with plasma membrane preparations (approx. 1 mg protein/ml) at 40°C in quartz cuvettes. The samples were then irradiated for 10 min with UV light from a strong mercury arc lamp. The samples were subjected to ultracentrifugation and the pellet redissolved and electrophoresed on a 9% polyacrylamide gel containing SDS. The results of these procedures showed the following. Preparations from glucose-fed cells in the presence of 0.5 μM [^{3}H]cytochalasin B revealed a covalent labeling of one polypeptide of molecular weight 46,000. The same procedures applied to glucose-starved cells resulted in a more intense covalent labeling of two polypeptides of molecular weight 46,000 and 52,000 (Pessin *et al.*, 1982). The photoaffinity labeling of either of the polypeptides was inhibited by D-glucose, 2-deoxyglucose, or 3-O-methylglucose, but not by L-glucose or D-sorbitol (Pessin *et al.*, 1982). Addition of 500 mM D-glucose combined with 0.1 μM [^{3}H]cytochalasin B inhibited by close to 80% the specific labeling of membrane preparations from starved cells (Pessin *et al.*, 1982, see also Shanahan *et al.*, 1982). Of the two, the polypeptide of molecular weight 52,000 was more sensitive to the competition by glucose (Pessin *et al.*, 1982).

Isolated plasma membrane vesicles retain the facilitated diffusion system of hexose transport (Inui *et al.*, 1980; Zala and Perdue, 1980). Membrane vesicles, stemming from CEF subjected to glucose starvation, or to refeeding with glucose, were the subject of kinetic analyses of labeled glucose influx, the initial rate of which was secured by sampling after 5 sec (Yamada *et al.*, 1983). The K_m values for all the membrane preparations were of the same order of magnitude, i.e., approximately 10 mM. However, the V_{max} of the hexose transport system of the membrane vesicles was found to be much higher in vesicles from glucose-starved CEF than in those from glucose-fed CEF, the values being 6 and 1.5 nmoles/mg per 5 sec (Yamada *et al.*, 1983). Moreover, the finding that cycloheximide in low concentration (2 μg/ml) was most effective in decreasing the hexose carrier population in glucose-fed NIL cultures (Christopher *et al.*, 1976b,c) was confirmed in a most satisfactory way by these kinetic studies.

Preparations from fed cells subjected to 2 μg/ml cycloheximide showed the lowest V_{max} about 1.0 nmole/mg per 5 sec (Yamada et al., 1983).

At present, data concerning regulation of transcriptional or translational events remain too scattered to lend themselves to a discussion here.

5. EFFECTS OF GLUCOSE STARVATION ON A VARIETY OF PLASMA MEMBRANE PROTEINS

Glucose starvation of cultured mammalian fibroblasts does not only affect the hexose transport system but also seems to affect amino acid transport. Two amino acid transport systems, called A and L (Christensen, 1975), were found to be present in NIL hamster fibroblasts (Isselbacher, 1972; Ullrey et al., 1975; Christopher et al., 1976b; Parnes and Isselbacher, 1978). The A system catalyzes active transport of amino acids like L-alanine and the analogue γ-aminoisobutyric acid (AIB) and requires the presence of sodium ions (see Christensen, 1975). Both systems are active transport systems. The only aspects that will be discussed here are the somewhat unexpected observations showing that the regulations of the A and L transport systems in NIL hamster fibroblasts are markedly affected by glucose feeding or glucose starvation (Ullrey et al., 1975; Christopher et al., 1976b; Nishino et al., 1978) and are also affected by cycloheximide (Nishino et al., 1978; Christopher et al., 1979).

The rate of transport of L-leucine was found to be 4- to 4.5-fold higher in glucose-starved NIL fibroblasts than in glucose-fed cells (Christopher et al., 1976b). The AIB transport system, most of which depends on sodium ions, was found to be 3.5-fold higher in NIL fibroblasts maintained in fructose medium than in those maintained in glucose medium (Kalckar et al., 1980b). Although pool effects were ruled out (Kalckar et al., 1980b), it was deemed important to try to demonstrate the effects of glucose feeding of NIL fibroblasts on the A transport system in plasma membrane preparations. It was found that membrane preparations from NIL cultures maintained in fructose medium showed a Na^+-dependent AIB transport corresponding to a V_{max} of 5.32, whereas membrane preparations from glucose-fed cultures showed AIB transport rates of only 1.25 (Nishino et al., 1978). Moreover, the effects of cycloheximide on glucose-starved (or fructose-fed) NIL cultures as well as the effects on glucose refeeding (Christopher et al., 1976b,c) were mirrored in a most noteworthy way by the membrane preparations in regard to Na^+-dependent AIB transport (Christopher et al., 1979).

Interpretations of the glucose effects on amino acid transport systems are not at hand. However, it should be kept in mind that hexoses interfere with the activity of certain amino acid transport systems of the intestinal mucosa (Saunders and Isselbacher, 1965).

Isolation of plasma membrane preparations from glucose-starved avian or mammalian fibroblasts, pulse-labeled with radioactive amino acids, and analysis of the polypeptides in SDS–polyacrylamide gels has revealed a vigorous biosynthesis of certain proteins or glycoproteins associated with glucose depletion (Banjo and Perdue, 1976; Pouysségur *et al.*, 1977; Shiu *et al.* 1977; Franchi *et al.*, 1978). The two proteins synthesized had molecular weights of 95,000 and 72,000–75,000 and seem to be derivatives of some fully glycosylated proteins, synthesized in the glucose-fed fibroblasts (Shiu *et al.*, 1977). However, they are not components of the hexose transport system (Shiu *et al.*, 1977; Zala *et al.*, 1980). The heavier glycoprotein was not synthesized in a mutant of 3T3 Balb cells, AD6, unless the strain was fed *N*-acetylglucosamine (Pouysségur *et al.*, 1977). Incorporation of L-fucose into the protein fraction was also found to be abnormally low in AD6 when the medium was not fortified with *N*-acetylglucosamine. General transport of D-glucosamine or of L-fucose was not affected. The mutant AD6 was considered defective in the acetylation step of glucosamine since *N*-acetylglucosamine rendered the biosynthesis of glycoproteins and glycolipids normal (Pouysségur *et al.*, 1977).

6. THE GLUCOSE-MEDIATED CURB OF HEXOSE TRANSPORT REQUIRES OXIDATIVE ENERGY

The glucose-mediated transport curb seems to depend on an oxidative metabolism or oxidative phosphorylation (Kalckar *et al.*, 1979; Ullrey and Kalckar, 1981). It was first found that uncouplers or inhibitors of oxidative phosphorylation such as DNP or oligomycin release the glucose-mediated curb or hexose transport. In addition, curtailment of oxygen supply or interference with the tricarboxylic acid cycle by malonate also brought about a release from the curb (Ullrey and Kalckar, 1981; Ullrey *et al.*, 1982). The transport enhancement did not ensue in the absence of glucose or if fructose replaced glucose (Kalckar *et al.*, 1980a; Ullrey *et al.*, 1982). Some of these relationships are illustrated in Table IV.

The release of the transport curb by inhibitors or anaerobic metabolism has been discussed in terms of possible alterations in membrane conformations including the role of endocytosis, for the energy-requiring transport curb (Kalckar, 1983; Kalckar and Ullrey, 1984b). The transport enhancement brought about by the inhibitors of respiratory energy metabolism, has also served as a useful tool in analyzing and interpreting different types of alterations in transport regulations (Kalckar *et al.*, 1980a; Ullrey *et al.*, 1982). For example, many types of downregulation, such as metabolically mediated transport curbs, may look like plain inhibitions. Yet, one would not expect a transport inhibition to be converted into a transport enhancement by adding an additional inhibitor, like DNP or malonate (see Kalckar *et al.*, 1979; Ullrey *et al.*, 1982).

Table IV. Effect of Glutamine Deprivation on Derepression of [^{14}C]-Galactose Uptake by NIL Cells in Serum-Free Medium[a]

	[U-^{14}C]-Galactose uptake (nmoles/mg protein)	
Additions	Without glutamine	With glutamine (4 mM)
Glucose (22 mM)	0.26	0.16
Glucose (22 mM) + DNP[b]	1.90	2.18
Fructose (22 mM)	1.37	0.92

[a]NIL cells were grown to near-confluence in standard DME medium with 10% calf serum. The cultures were then rinsed twice, and maintained for 20 hr in modified DME *without serum* and with or without glutamine and additions as indicated. All samples contained 0.1% DMSO. They were then rinsed twice with PBS and assayed for galactose uptake: [U-^{14}C]galactose, 1×10^{-4} M, 10 min at 37°C (Kalckar *et al.* (1980a).
[b]DNP 2×10^{-4} M in 0.1% DMSO.

In spite of the useful data obtained from the studies of thePGI mutant, the nature of the induced curb of the aldohexose transport system by glucose still remains a puzzle. Any new attempts to interpret the mechanism of this type of regulation will, however, have to take into account some recent unanticipated observations. Two supposedly non-metabolizable sugars, 3-OMG and D-allose have shown a pronounced ability to induce a transport curb of the uptake of 2-DOG in human fibroblast cultures (Germinario *et al.*, 1985). Moreover, the transport of 3-OMG into the cultured hamster fibroblast lines, 023 and DS-7 (the PGI mutant), if exposed to D-allose for 6–18 hr respond with a remarkably strong curb (Ullrey and Kalckar, 1986). This curb is typical, inasmuch as it can be abolished by metabolic inhibitors like malonate or cycloheximide; in the PGI mutant D-glucose and D-allose are the only D-aldoses which are able to bring about a typical transport curb (Ullrey and Kalckar, 1986).

D-Allose is an all-cis hexose. Tritiated D-allose has been shown to behave like a typical transport ligand of the hexose transport system of adipose fat cells (Loten *et al.*, 1976).

7. NUCLEOSIDE TRIPHOSPHATE LEVELS IN CULTURED FIBROBLASTS AS A FUNCTION OF GENERAL METABOLISM AND NUTRITION

The steady-state levels of ATP, GTP, CTP, and UTP as a function of the nutrition and the energy metabolism of cultured NIL hamster fibroblasts have been determined (Rapaport *et al.*, 1979, 1983). Replacement of a glucose maintenance medium by a fructose medium did not lower the levels of the four NTPs in NIL or in transformed NIL cultures (Rapaport *et al.*, 1979). Absence of any hexose did lower the NTP levels, especially those of UTP and GTP in NIL as well as in PyNIL (Rapaport *et al.*, 1979). This loss of triphosphates was not only prevented but also overcompensated when protein synthesis was arrested during the hexose starvation period (Rapaport *et al.*, 1979, 1983). This paradoxical effect was also very pronounced in the transformed cultures (Rapaport *et al.*, 1979).

Curtailment of protein synthesis seems to stabilize the triphosphates at high levels. Thus, maintenance of NIL cultures in media devoid of L-glutamine [showing curtailed levels of protein synthesis (Ullrey and Christopher, unpublished)] seemed to stabilize ATP at high levels (Rapaport *et al.*, 1983). Addition of DNP to NIL fibroblasts, maintained in the curtailed medium (deprived of L-glutamine), did not lower the high ATP levels (and a high ATP/ADP ratio of 6 persisted) and yet showed the usual dramatic release of the hexose transport curb (Table V). The lack of correlation between triphosphate levels and the imposition of the transport curb was also evident when D-glucosamine replaced glucose (Rapaport *et al.*, 1979). In this case, the galactose uptake into NIL cells was down-regulated to less than 25% by a mere 5 mM D-glucosamine over 18 hr, yet the only triphosphate level markedly affected was that of UTP (Rapaport *et al.*, 1979). Prolonged exposure to D-glucosamine brought about an impressive accumulation of UDP-*N*-acetylglucosamine and UDP-glucosamine (Rapaport *et al.*, 1979).

The maintenance of the ATP levels in the PGI mutant has been studied by Plesner *et al.* (1985). In the presence of L-glutamine ATP levels were found to be high with glucose as well as mannose. In the absence of L-glutamine mannose was able to stabilize high ATP levels over 24 hr in contrast, glucose was only able to maintain ATP cellular concentration of 18–20% of the levels found in mannose-fed cultures (Plesner *et al.*, 1985). As mentioned earlier, mannose-6-phosphate is converted to fructose-6-phosphate by a specific isomerase programmed by a gene different from the one affected in the PGI mutant. The preservation of high levels of ATP by mannose feeding of the PGI mutant was

Table V. ATP Levels in Near-Confluent NIL Hamster Fibroblast Monolayer Cultures, Maintained in Medium Devoid of L-Glutamine and Serum[a]

Incubation medium	ATP (nmoles/mg protein)	ATP/ADP	[14C]-Galactose uptake (nmoles/mg protein per 10 min)
I. Glucose	33.8	5.29	0.27
II. Glucose + DNP[b]	34.0	5.88	1.54
III. No sugar	32.4	6.36	2.30

[a]From Rapaport *et al.* (1983).
[b]DNP, 0.2 mM final concentration.

also reflected in the ATP/ADP ratio which was found to be 8–8.5 in mannose-fed PGI⁻ cells but only 1–1.5 in glucose-fed PGI⁻ cells. If the mutant cells were supplied with L-glutamine the ATP/ADP ratio remained above 8 with glucose as well as mannose. In the parental 023 cell line, the ATP levels as well as the ATP/ADP ratio remained high even with glucose.

Donnelly and Scheffler (1976) studied energy metabolism in Chinese hamster fibroblast cultures (CCL16) and in respiration defective mutants. They calculated that in CCL16, about 60% of the energy maintaining ATP levels was derived from glucose through glycolysis, whereas 40% stemmed from respiration, using L-glutamine as a source, and this satisfied the energy requirements (Donnelly and Scheffler, 1976). In a respiratory mutant, energy stems only from glycolysis (DeFrancesco et al., 1976; Donnelly and Scheffler, 1976).

Regarding the role of energy metabolism, it would be interesting to study the regulation of glucose transport in other metabolic mutants isolated from fibroblast cultures. Among the most interesting mutants, one might list the respiration-deficient mutants isolated by Scheffler and co-workers (Scheffler, 1974; DeFrancesco et al., 1976; Donnelly and Scheffler, 1976), as well as a glycolytic mutant of Chinese hamster ovary (CHO) cells.* Morgan and Faik (1980) have isolated a PGI defective mutant from CHO cells that has an additional defect in phosphoglycerate kinase (see Chapter 2).

8. HEXOSE TRANSPORT REGULATION AND ONCOGENIC TRANSFORMATION OF CULTURED FIBROBLASTS

During the years 1969 to 1975 when culture techniques of avian and mammalian fibroblasts became more established and techniques for mass transformation with oncogenic viruses had been developed, it was discovered that oncogenic transformation greatly enhanced the glucose transport system. This extensive field has fortunately been reviewed up to the late 1970s and we will therefore refer the reader to reviews and overviews up to 1978–1979: Hatanaka (1976), Kalckar (1976) (only a brief overview), Perdue (1978), Parnes and Isselbacher (1978) (largely for amino acid transport); this brief list does not contain all of the earlier reviews. For instance, "The Fogarty International Conference on Cellular Regulation of Transport and Uptake of Nutrients" (H. M. Kalckar, ed.) was published in a special volume of the *Journal of Cellular Physiology* (Vol. 89, No. 4) and at least 130 pages are devoted to discussions on oncogenic transformations and uptake regulations; moreover, other reviews may have been overlooked.

Later work on isolated plasma membrane vesicles from transformed and

*We have so far been unable to detect a pronounced effect of glucose on the hexose transport system in CHO cells (Kalckar and Ullrey, unpublished observations).

untransformed cells has focused on the enhancement of glucose transport induced by the transforming virus. It has been asserted, for instance, that phosphorylation, not transport, is enhanced in 3T3 fibroblasts transformed by SV40 [Colby and Romano (1975); Singh *et al.* (1978) showed that isozyme II or hexokinase was increased after transformation]. This assertion has remained unchallenged for some time, since some of the first vesicle studies on SV40-transformed 3T3 cells did not show any rise in V_{max} of the hexose transport system (Lever, 1976).

Recent work on membrane vesicles from SV40-transformed 3T3 cells, using faster kinetics and including influx and efflux studies and countertransport as well, has demonstrated a 2.5 to 3-fold increase in V_{max} of glucose transport into the membrane vesicles of transformed fibroblasts as compared with nontransformed cultures (Inui *et al.*, 1979). Moreover, studies of glucose transport in plasma membrane vesicles from CEF transformed by sarcoma viruses (Decker and Lipmann, 1981; Salter *et al.*, 1982) have fully confirmed the early results of Weber (1973) and Kletzien and Perdue (1975) concerning up-regulation of the glucose transport step after transformation.

In both cases the CEF population was transformed by temperature-sensitive sarcoma viruses, i.e., avian sarcoma virus (ASV) (Decker and Lipmann, 1981) or RSV (Salter *et al.*, 1982). Semipurified membrane preparations from the transformed cells showed an increase of V_{max} of the order of 3 as compared with those from untransformed preparations.

With the membrane preparations from ASV-transformed CEF, Decker and Lipmann (1981) extended the kinetic studies into glucose concentrations much higher than 5 mM. In this way the existence of nonsaturable, yet stereospecific transport systems was disclosed (Decker and Lipmann, 1981). These authors recalled earlier studies by Hatanaka (1976) in which mouse embryo fibroblasts, transformed by mouse sarcoma virus (H-MSV), showed biphasic kinetics when the concentrations of glucose, mannose, or galactose went beyond 5 to 10 mM. In the case of the membrane preparations, the nonsaturable system, although more conspicuous in the preparations from transformed cells, seemed also present in the untransformed cell preparations (Decker and Lipmann, 1981; see also Singh *et al.*, 1978).

The approach of Salter *et al.* (1982) was a different one: encouraged by the observations that human erythrocyte glucose transporter cross-reacts immunologically with certain proteins from mammalian cells, they used antibody against the erythrocyte glucose transporter (Sogin and Hinkle, 1980, cited in Weber *et al.*, 1984b) on membrane preparations from transformed and untransformed CEF.

CEF exposed to temperature-sensitive RSV (SR-RSV, tsNY68; Kawai and Hanafuse (1971), cited in Salter *et al.*, 1982) at the permissible as well as nonpermissible temperature, were incubated with [35S]methionine, lysed in detergent-containing buffer, and reacted with antiserum prepared against human erythrocyte glucose transporter. After several other steps (binding of the immune complex to protein A–Sepharose, washing, boiling in SDS with mercaptoethanol), the antibody complexes were electrophoresed on polyacrylamide gels. The autoradiograms from the gels showed two immunospecific bands of molecular weights 41,000 and 82,000, the latter suspected of being a dimer of the former. These bands were not precipitated by nonimmune serum; moreover, their precipitation was blocked by purified erythrocyte glucose transporter. The specific 41,000 and 82,000 bands were much stronger (of the order of 3.5 to 4-fold) in gels from plasma membranes of transformed than from nontransformed CEF (Salter et al., 1982; Weber et al., 1984b).

The glucose-mediated transport curb seems to be well preserved in most tumor cell lines and even in lines from transformed cultures (Kletzien and Perdue, 1975, 1976; Ullrey et al., 1975; Christopher et al., 1977).

The transport studies on transformed cell cultures by Kletzien and Perdue (1975, 1976), who determined the V_{max} of 2-deoxyglucose uptake of glucose-fed and starved tsNY68-infected CEF at permissible and nonpermissible temperatures, represent a convincing example. The glucose-mediated curb was marked at 41°C, the nonpermissible temperature for transformation; yet at 37°C, the transport curb, although more moderate, still persisted (Table VI).

Other types of tumorigenic cultures, such as the hamster lung fibroblast line 023, showed a very marked glucose-mediated curb, which could also be mediated by D-glucosamine (Ullrey et al., 1982).

However, it should be recalled that a number of nontumorigenic mammalian cell lines do not show significant glucose effects on their hexose transport system (see Kalckar, 1976). Moreover, human fibroblast cultures have shown

Table VI. The Effect of Hexose Starvation on the V_{max} for Sugar Transport in tsNY68-Infected Chick Embryo Fibroblasts[a]

Culture temperature	V_{max} for sugar transport[b] (nmoles/mg protein per min)	
	Standard medium	Hexose-free medium
37°C	39	90
41°C	15	85

[a]From Kletzien and Perdue (1976).
[b]V_{max} values were determined from Lineweaver–Burk plots of initial rates of sugar uptake btween 0.3 and 3.0 mM 2-deoxyglucose. Cultures were maintained at the appropriate temperature for 30 hr in hexose-free medium prior to the transport measurement at 37°C.

only moderate to feeble glucose transport effects (Salter and Cook, 1976; Ullrey and Kalckar, unpublished).

The role of L-glutamine in hexose uptake regulation has been studied in tumorigenic fibroblasts. An interesting example worth discussing is a polyoma-transformed BHK variant (the GIV strain) that was found to grow in media devoid of L-glutamine. In spite of its high tumorigenic potential, the metabolism of hexoses was found to be only moderate (Gammon and Isselbacher, 1976). Hexose uptake was also found to be relatively low and it varied according to the presence or absence of L-glutamine (Kalckar et al., 1976). Addition of L-glu-tamine to GIV cultures brought about a down-regulation of amino acid transport (L system) whereas galactose uptake was up-regulated (Kalckar et al., 1976). Conversely, removal of L-glutamine from the medium over 12–20 hr, reversed the pattern, i.e., the L system was up-regulated whereas the galactose uptake was down-regulated (Kalckar et al., 1976). Pool effects were ruled out. These inverse regulations were slow, of the order of 10 to 20 hr (Kalckar et al., 1976). It seems worth pointing out that GIV fibroblasts, although malignant and meta-static (Gammon and Isselbacher, 1976), showed only a relatively slow uptake of galactose (Kalckar et al., 1976). In contrast, the rate of cycloleucine transport was high, especially in the absence of L-glutamine (Kalckar et al., 1976).

Increase of amino acid transport in oncogenically transformed mammalian fibroblast cultures has been described (Foster and Pardee, 1969; Isselbacher, 1972).

The situation is different in the case of RSV-transformed CEF. Nakamura and Weber (1979) and Inui et al. (1980) found that only hexose transport was enhanced after transformation, whereas amino acid transport was not significantly changed. The increase of hexose transport in the transformed state was well preserved even at low serum concentrations (Perdue, 1976).

Having listed a number of cases in which oncogenic transformation brought about a clear-cut enhancement of the hexose transport system, it should also be mentioned that some exceptions have been encountered, especially in cases of SV40 infections of human or monkey cells (see Kalckar, 1976).

9. EVOLUTIONARY ASPECTS

The glucose-mediated curb of hexose transport is not confined to avian and mammalian fibroblasts. As early as 1970–1971, two groups (Scarborough, 1970; Neville et al., 1971) observed some closely related regulatory responses in conidia from Neurospora crassa. Withdrawal of glucose or replacement of glucose by fructose in the medium, brought about a striking enhancement of hexose transport activity (Scarborough, 1970; Neville et al., 1971). The type of hexose transport in the conidia may well be active, since the analogue 3-O-methylglucose added to conidia maintained in fructose medium was transported against a concentration gradient (Scarborough, 1970; Neville *et al.*, 1971).

ADDENDUM

Most recently, transport mutants from L6 rat myoblasts have been isolated and described (D'Amore *et al.*, 1986). After chemical mutagenization, a novel method was introduced for selection of transport mutants, the most effective being resistance to 2-deoxy-2-fluoro-glucose, called 2FG (D'Amore *et al.*, 1986). The growth medium with 2FG contained excess fructose, replacing the competitive glucose. The transport mutants isolated turned out to have a greatly decreased number of hexose transporters involved in high-affinity transport as tested with 2FG (D'Amore *et al.*, 1986; D'Amore and Lo, 1986). Apparently the low-affinity transport system (as assayed with labeled 3-O-methyl-glucose) was not affected (D'Amore and Lo, 1986).

REFERENCES

Amos, H., Christopher, C. W., and Musliner, T. A., 1976, Regulation of glucose transport in chick fibroblasts: Bicarbonate, lactate and ascorbic acid, *J. Cell. Physiol.* **89**:662–676.

Amos, H., Mandel, K. G., and Gay, R. 1984, Deprival of nicotinamide leads to enhanced glucose transport in chick embryo fibroblasts, *Fed. Proc.* **43**:2265–2268.

Bader, J. P., 1976, Sodium: A regulator of glucose uptake in virus-transformed and non-transformed cells, *J. Cell. Physiol.* **89**:677–682.

Banjo, B., and Perdue, J. F., 1976, Increased synthesis of selected membrane polypeptides correlated with increased sugar transport sites in glucose starved chick embryo fibroblasts, *J. Cell Biol.* **70**:270a.

Blume, K. G., and Beutler, E., 1975, Galactokinase in human erythrocytes, *Methods Enzymol.* **42**:47–53.

Boos, W., 1974, Bacterial transport, *Annu. Rev. Biochem.* **43**:123–146.

Christensen, H. N., 1975, *Biological Transport*, 2nd ed., Benjamin, New York.

Christopher, C. W., 1977, Hexose transport regulation in cultured hamster cells, *J. Supramol. Struct.* **6**:485–494.

Christopher, C. W., Kohlbacher, M. S., and Amos, H., 1976a, Transport of sugars in chick-embryo fibroblasts: Evidence for a low affinity system and a high affinity system for glucose transport, *Biochem. J.* **158**:439–450.

Christopher, C. W., Ullrey, D., Colby, W., and Kalckar, H. M., 1976b, Paradoxical effects of cycloheximide and cytochalasin B on hamster cell cultures, *Proc. Natl. Acad. Sci. USA* **73**:2429–2433.

Christopher, C. W., Colby, W., and Ullrey, D., 1976c, Depression and carrier turnover, evidence for two distinct mechanisms of hexose transport regulation in animal cells, *J. Cell. Physiol.* **89**:683–692.

Christopher, C. W., Colby, W., Ullrey, D., and Kalckar, H. M., 1977, Comparative studies of glucose-fed and glucose-starved hamster cell cultures: Responses in galactose metabolism, *J. Cell. Physiol.* **90**:387–406.

Christopher, C. W., Ullrey, D., and Kalckar, H. M., 1979, Regulation of amino acid and hexose transport in cultured animal cells, In: *Structure and Function of Biomembranes* (K. Yagi, ed.), pp. 39–50, Japan Scientific Societies Press, Tokyo.

Colby, C., and Romano, A. H., 1975, Phosphorylation but not transport of sugars is enhanced in virus transformed mouse 3T3 cells, *J. Cell. Physiol.* **85**:15–24.

Cushman, S. W., and Wardzala, L. J., 1980, Potential mechanism of insulin action on glucose transport in the isolated rat adipose cell: Apparent translocation of intracellular transport systems to the plasma membrane, *J. Biol. Chem.* **255**:4758–4762.

D'Amore, T., and Lo, T. C. Y., 1986, *J. Cell. Physiol.*, in press.

D'Amore, T., Duronio, V., Cheung, M. O., and Lo, T. C. Y., 1986, Isolation and characterization of hexose transport mutants in L6 rat myoblasts, *J. Cell. Physiol.* **126**:29–36.

Decker, S., and Lipmann, F., 1981, Transport of D-glucose by membrane vesicles from normal and virus-transformed chicken embryo fibroblasts, *Proc. Natl. Acad. Sci. USA* **78**:5358–5361.

DeFrancesco, L., Scheffler, I. E., and Bissell, M. J., 1976, A respiration deficient Chinese hamster cell line with a defect in NADH-coenzyme Q reductase, *J. Biol. Chem.* **251**:4588–4595.

Donnelly, M., and Scheffler, I. E., 1976, Energy metabolism in respiration-deficient and wild type Chinese hamster fibroblasts in culture, *J. Cell. Physiol.* **89**:39–51.

Foster, D. O., and Pardee, A. B., 1969, Transport of amino acids by confluent and non-confluent 3T3 and polyoma virus-transformed 3T3 cells growing on glass cover slips, *J. Biol. Chem.* **144**:2675–2681.

Franchi, A., Silvestre, P., and Pouysségur, J., 1978, Carrier activation and glucose transport in Chinese hamster fibroblasts, *Biochem. Biophys. Res. Commun.* **85**:1526–1534.

Gammon, M. T., and Isselbacher, K. J., 1976, Neoplastic potentials and regulation of uptake of nutrients. 1. A glutamine independent variant of polyoma BHK with a very high neoplastic potential, *J. Cell. Physiol.* **89**:759–764.

Gatley, S. J., Holden, J. E., Halama, J. R., DeGrado, T. R., Bernstein, D. R., and Ng, C. K., 1984, Phosphorylation of glucose analog 3-O-methyl-D-glucose by rat heart, *Biochem. Biophys. Res. Commun.* **119**:1008–1014.

Gay, R. J., and Amos, H., 1983, Purines as 'hyperrepressors' of glucose transport: A role for phosphoribosyl diphosphate, *Biochem. J.* **214**:133–144.

Germinario, R. J., Ozaki, S., and Kalant, N., 1984, Regulation of insulin binding and stimulation of sugar transport in cultured human fibroblasts by sugar levels in the culture medium, *Arch. Biochem. Biophys.* **234**:559–566.

Germinario, R. J., Chang, Z., Manuel, S., and Oliveira, M., 1985, Control of sugar transport in human fibroblasts independent of glucose metabolism or carrier interaction, *Biochem. Biophys. Res. Comm.* **128**:1418–1424.

Hatanaka, M., 1976, Saturable and nonsaturable process of sugar uptake: Effect of oncogenic transformation in transport and uptake of nutrients, *J. Cell. Physiol.* **89**:745–749.

Howard, B. V., Mott, D. M., Fields, R. M., and Bennett, P. H., 1979, Insulin stimulation of glucose entry in cultured human fibroblasts, *J. Cell. Physiol.* **101**:129–138.

Inui, K.-I., Moller, D. E., Tillotson, L. G., and Isselbacher, K. J., 1979, Stereospecific hexose transport by membrane vesicles from mouse fibroblasts: Membrane vesicles retain increased hexose transport associated with viral transformation, *Proc. Natl. Acad. Sci. USA* **76**:3972–3976.

Inui, K.-I., Tillotson, L. G., and Isselbacher, K. J., 1980, Hexose and amino acid transport by chicken embryo fibroblasts infected with temperature-sensitive mutant to Rous sarcoma virus, *Biochim. Biophys. Acta* **598**:616–627.

Isselbacher, K. J., 1972, Increased uptake of amino acids and 2-deoxy-D-glucose by virus-transformed cells in culture, *Proc. Natl. Acad. Sci. USA* **69**:585–589.

Kalckar, H. M., 1976, Cellular regulation of transport and uptake of nutrients: An overview, *J. Cell. Physiol.* **89**:503–516.

Kalckar, H. M., 1983, Regulation of hexose transport-carrier activity in cultured animal fibroblasts: Another confrontation with cellular recycling requiring oxidative energy generation?, *Trans. N.Y. Acad. Sci.* **41**:83–86.

Kalckar, H. M., and Ullrey, D., 1973, Two distinct types of enhancement of galactose uptake into hamster cells: Tumour virus transformation and hexose starvation, *Proc. Natl. Acad. Sci. USA* **70**:2502–2504.

Kalckar, H. M., and Ullrey, D. B., 1984a, Further clues concerning the vectors essential to regulation of hexose transport as studied in fibroblast cultures from a metabolic mutant, *Proc. Natl. Acad. Sci. USA* **81**:1126–1129.

Kalckar, H. M., and Ullrey, D. B., 1984b, Energy-requiring regulation of hexose transport, as studied in fibroblast cultures of a metabolic mutant, in: *Symposium "The Cell Membrane,"* (E. Haber, ed.).

Kalckar, H. M., Christopher, C. W., and Ullrey, D., 1976, Neoplastic potentials and regulation of uptake of nutrients. II. Inverse regulation of uptake of hexose and amino acid analogues in the neoplastic GIV line, *J. Cell. Physiol.* **89**:765–768.

Kalckar, H. M., Christopher, C. W., and Ullrey, D., 1979, Uncouplers of oxidative phosphorylation promote derepression of the hexose transport system in cultures of hamster cells, *Proc. Natl. Acad. Sci. USA* **76**:6453–6455.

Kalckar, H. M., Ullrey, D. B., and Laursen, R. A., 1980a, Effects of combined glutamine and serum deprivation on glucose control of hexose transport in mammalian fibroblast cultures, *Proc. Natl. Acad. Sci. USA* **77**:5958–5961.

Kalckar, H. M., Christopher, C. W., and Ullrey, D. B., 1980b, Long-term regulation of amino acid and hexose transport in cultured animal cells, in: *Advances in Pathobiology* (C. M. Fenoglio and D. W. King, eds.), pp. 350–364, Thieme, Stuttgart.

Kletzien, R. F., and Perdue, J. F., 1975, Induction of sugar transport in chick embryo fibroblasts by hexose starvation, *J. Biol. Chem.* **250**:593–600.

Kletzien, R. F., and Perdue, J. F., 1976, Regulation of sugar transport in chick embryo fibroblasts and in fibroblasts transformed by a temperature sensitive mutant of the Rous sarcoma virus, *J. Cell. Physiol.* **89**:723–728.

Lee, S. G., and Lipmann, F., 1977, Isolation from normal and Rous sarcoma virus transformed chicken fibroblasts of a factor that binds glucose and stimulates its transport, *Proc. Natl. Acad. Sci. USA* **74**:163–167.

Lee, S. G., and Lipmann, F., 1978, Glucose binding and transport proteins extracted from fast-grown chicken fibroblasts, *Proc. Natl. Acad. Sci. USA* **75**:5427–5431.

Leloir, L. F., 1951, The metabolism of hexosephosphate, in: *Phosphorus Metabolism* (W. D. McElroy and B. Glass, eds.), Vol. I, pp. 69–93, Johns Hopkins Press, Baltimore.

Lever, J. E., 1976, Regulation of amino acid and glucose transport activity expressed in isolated membranes from untransformed and SV40-transformed mouse fibroblasts, *J. Cell. Physiol.* **89**:779–788.

Lipmann, F., and Lee, S. G., 1978, A glucose binding transport factor isolated from normal and malignantly transformed chicken fibroblasts, in: *Microenvironments and Metabolic Compartmentation* (P. A. Srera and R. W. Estabrook, eds.), pp. 263–281, Academic Press, New York.

Loten, E. G., Regen, D. M., and Park, C. R., 1976, Transport of D-allose by isolated fat cells: An effect of adenosine triphosphate on insulin stimulated transport, *J. Cell. Physiol.* **89**:651–660.

Lust, W. D., Schwartz, J. P., and Passonneau, J. V., 1975, Glycolytic metabolism in cultured cells of the nervous system. I. Glucose transport and metabolism in CG glioma cell line, *Mol. Cell. Biochem.* **8**:169–176.

Martineau, R., Kohlbacher, M., Shaw, S., and Amos, H., 1972, Enhancement of hexose entry into chick fibroblasts by starvation: Differential effects on galactose and glucose, *Proc. Natl. Acad. Sci. USA* **69**:3407–3411.

Morgan, M. J., and Faik, P., 1980, The regulation of carbohydrate metabolism in animal cells: Isolation of a glycolytic variant of Chinese hamster ovary cells, *Cell Biol. Int. Rep.* **4**:121–127.

Morgan, M. J., and Faik, P., 1981, Carbohydrate metabolism in cultured animal cells, *Biosci. Rep.* **1**:669–686.

Nakamura, K. D., and Weber, M. T., 1979, Amino acid transport in normal and Rous sarcoma virus-transformed chicken embryo fibroblasts, *J. Cell. Physiol.* **99**:15–22.

Neville, M. M., Suskind, S. B., and Roseman, S., 1971, A depressible active transport system for glucose in *Neurospora crassa*, *J. Biol. Chem.* **246**:1294–1301.

Nishino, H., Christopher, C. W., Schiller, R. M., Gammon, M. T., Ullrey, D., and Isselbacher, K. J., 1978, Sodium dependent amino acid transport by cultured hamster cells: Membrane vesicles retain transport changes due to glucose starvation and cycloheximide, *Proc. Natl. Acad. Sci. USA* **75**:5048–5051.

Parnes, J. R., and Isselbacher, K., 1978, Transport alterations in virus-transformed cells, *Prog. Exp. Tumor Res.* **22**:79–122.

Perdue, J. F., 1976, Loss of post-translational control of nutrient transport in *in vitro* and *in vivo* virus-transformed chicken cells, *J. Cell. Physiol.* **89**:729–736.

Perdue, J. F., 1978, Transport across serum-stimulated and virus-transformed cell membranes, in: *Virus-Transformed Cell Membranes* (C. Nicolau, ed.), pp. 182–280, Academic Press, New York.

Pessin, J. F., Tillotson, L. G., Yamada, K., Gitomer, W., Carter, S. C., Mora, R., Isselbacher, K. J., and Czech, M. P., 1982, Identification of the stereospecific hexose-transporter from starved and fed chicken embryo fibroblasts, *Proc. Natl. Acad. Sci. USA* **79**:2286–2290.

Plesner, P., Ullrey, D. B., and Kalckar, H. M., 1985, Mutations in the phosphoglucose isomerase gene can lead to marked alterations in cellular ATP levels in cultured fibroblasts exposed to simple nutrient shifts, *Proc. Natl. Acad. Sci. USA* **82**:2761–2763.

Pouysségur, J., Shiu, R. P. C., and Pastan, I., 1977, Induction of two transformation-sensitive membrane polypeptides in normal fibroblasts by a block in glycoprotein synthesis or glucose deprivation, *Cell* **11**:941–947.

Pouysségur, J., Franchi, A., Salomon, J.-C., and Silvestre, P., 1980, Isolation of a Chinese hamster fibroblast mutant defective in hexose transport and aerobic glycolysis: Its use to dissect the malignant phenotype, *Proc. Natl. Acad. Sci. USA* **77**:2698–2701.

Rapaport, E., Christopher, C. W., Svihovec, S., Ullrey, D., and Kalckar, H. M., 1979, Selective high metabolic liability of uridine, guanosine and cytosine triphosphates in response to glucose deprivation and refeeding of untransformed and polyoma virus transformed hamster fibroblasts, *J. Cell. Physiol.* **104**:229–236.

Rapaport, E., Plesner, P., Ullrey, D. B., and Kalckar, H. M., 1983, 2,4-Dinitrophenol does not reduce ATP levels in starving hamster fibroblasts although hexose transport is markedly affected, *Carlsberg Res. Commun.* **48**:317–320.

Reitzer, L. J., Wice, B. M., and Kennell, D., 1979, Evidence that glutamine, not sugar, is the major energy source for cultured Hela cells, *J. Biol. Chem.* **254**:2669–2676.

Salter, D. W., and Cook, J. S., 1976, Reversible independent alterations in glucose transport and metabolism in cultured human cells deprived of glucose, *J. Cell. Physiol.* **89**:143–156.

Salter, D. W., Baldwin, S. A., Lienhard, G., and Weber, M. J., 1982, Proteins antigenically related to the human erythrocyte glucose transporter in Rous sarcoma virus-transformed cells, *Proc. Natl. Acad. Sci. USA* **79**:1540–1544.

Saunders, S. J., and Isselbacher, K. J., 1965, Inhibition of intestinal amino acid transport by hexoses, *Biochim. Biophys. Acta* 102:397–409.

Scarborough, G. A., 1970, Sugar transport in *Neurospora crassa*, *J. Biol. Chem.* 245:1694–1698.

Scheffler, I. E., 1974, Conditional lethal mutants of Chinese hamster cells: Mutants requiring exogenous carbon dioxide for growth, *J. Cell. Physiol.* 83:219–230.

Shanahan, M. F., Olson, S. A., Weber, M. J., Lienhard, G. E., and Gorga, J. C., 1982, Photolabeling of glucose-sensitive cytochalasin B binding proteins in erythrocytes, fibroblasts and adipocyte membranes, *Biochem. Biophys. Res. Commun.* 107:38–43.

Shiu, R. P. C., Pouysségur, J., and Pastan, I., 1977, Glucose depletion accounts for the induction of two transformation-sensitive membrane proteins in Rous sarcoma virus-transformed chick embryo fibroblasts, *Proc. Natl. Acad. Sci.* (USA) 74:3840–3844.

Singh, M., Singh, V., August, J. T., and Horecker, B. L., 1978, Transport and phosphorylation of hexoses in normal and Rous-sarcoma virus-transformed chicken embryo fibroblasts, *J. Cell. Physiol.* 97:285–292.

Slein, M. W., 1950, Phosphomannose isomerase, *J. Biol. Chem.* 186:753–761.

Tillotson, L. G., Yamada, K., and Isselbacher, K. J., 1984, Regulation of hexose transport of chicken embryo fibroblasts during glucose starvation, *Fed. Proc.* 43:2262–2264.

Ullrey, D. B., and Kalckar, H. M., 1981, The nature of regulation of hexose transport in cultured mammalian fibroblasts: Aerobic 'repressive' control by D-glucosamine, *Arch. Biochem. Biophys.* 209:168–174.

Ullrey, D. B., and Kalckar, H. M., 1982, Schism and complementation of hexose mediated transport regulation as illustrated in a fibroblast mutant lacking phosphoglucose isomerase, *Biochem. Biophys. Res. Commun.* 107:1532–1538.

Ullrey, D., Gammon, M. T., and Kalckar, H. M., 1975, Uptake patterns and transport enhancements in cultures of hamster cells deprived of carbohydrates, *Arch. Biochem. Biophys.* 167:410–418.

Ullrey, D. B., Franchi, A., Pouysségur, J., and Kalckar, H. M., 1982, Down regulation of the hexose transport system: Metabolic basis studied with a fibroblast mutant lacking phosphoglucose isomerase, *Proc. Natl. Acad. Sci. USA* 79:3777–3779.

Ullrey, D. B., and Kalckar, H. M., 1986, D-Allose promotes an energy-requiring transport curb in a fibroblast metabolic mutant, Abstract, Amer. Soc. Microbiology Ann. Meeting, March 1986.

Weber, M., 1973, Hexose transport in normal and in Rous sarcoma virus-transformed cells, *J. Biol. Chem.* 248:2978–2983.

Weber, M. J., Evans, P. K., Johnson, M. A., McNair, T. F., Nakamura, K. D., and Salter, D. W., 1984b, Transport of potassium, amino acids, and glucose in cells transformed by Rous sarcoma virus, *Fed. Proc.* 43:107–112.

Yamada, K., Tillotson, L. G., Isselbacher, K. J., 1983, Regulation of hexose carriers in chicken embryo fibroblasts: Effect of glucose starvation and role of protein synthesis, *J. Biol. Chem.* 258:9786–9792.

Zala, C. A., and Perdue, J. F., 1980, Stereospecific D-glucose transport in mixed membrane and plasma membrane vesicles derived from cultured chick embryo fibroblasts, *Biochim. Biophys. Acta* 600:157–172.

Zala, C. A., Salas-Prato, M., Yan, W.-T., Banjo, B., and Perdue, J. F., 1980, In cultured chick embryo fibroblasts the hexose transport components are not the 75000 and 95000 dalton polypeptides synthesized following glucose deprivation, *Can. J. Biochem.* 58:1179–1188.

Zielke, H. R., Ozand, P. T., Tildon, J. T., Sevdalian, D. A., and Cornblath, M., 1978, Reciprocal regulation of glucose and glutamine utilization by cultured human diploid fibroblasts, *J. Cell. Physiol.* 95:41–48.

2

The Utilization of Carbohydrates by Animal Cells
An Approach to Their Biochemical Genetics

MICHAEL J. MORGAN and PELIN FAIK

1. THE UTILIZATION OF CARBOHYDRATES

Animal cells in culture, in marked contrast to microorganisms (Chapters 10–12), are able to grow on only a limited number of carbohydrates other than glucose. The early literature contains a number of reports on the ability of carbohydrates to support growth of various cell types (Eagle *et al.*, 1958; Morgan and Morton, 1960) but some investigations are open to the criticism that they did not take fully into account the facts that many commercially available carbohydrates are contaminated with glucose, that serum itself contains glucose, and that the cells may thus really have been utilizing glucose.

In order to determine which carbohydrates can actually be utilized by cells in culture, it is essential first to ensure that media contain only the carbohydrate(s) under investigation (Faik and Morgan, 1977a) and, second, to define "growth": cells continue to divide for some time after transfer to medium lacking carbohydrate, and presumably cell division only ceases when the energy stores, including utilizable protein and biosynthetic precursors, become exhausted. Growth should, therefore, be measured through a number of cell divisions and be compared with appropriate carbohydrate-deficient control cultures. A number of investigators have met these criteria or variations upon them and have investigated which carbohydrates can be substituted for glucose and still support growth of animal cells through many cell divisions. A summary of these

MICHAEL J. MORGAN and PELIN FAIK • Department of Biochemistry, University of Leicester, Leicester LE1 7RH, United Kingdom.

results is presented in Table I, where it will be apparent that only a very limited number of carbohydrates support the growth of most cells in culture.

Glucose, mannose, fructose, and galactose are utilized by most cell types: glucose and mannose, in particular, appear to be metabolized at about the same rates and permit similar, rapid, growth rates with the production of abundant amounts of lactic acid; growth on fructose and galactose, in general, is much slower and does not result in the production of large amounts of lactic acid.

In 1953, Harris and Kutsky reported on the utilization of sugars by chick heart fibroblasts in dialyzed media. Extensive precautions were taken to use dialyzed plasma and serum and the amount of sugar remaining in the medium (as reducing substances) was determined. Cell proliferation was found to occur in sugar-free media as well as in medium supplemented with D-glucose. There were, however, a number of significant differences seen in the two conditions both in the rate of proliferation and in the actual morphology of the cells (their sugar-free media in fact contained of the order of 0.1 mM glucose equivalence). Fructose, mannose, maltose, and glycogen allowed essentially the same amount of growth as glucose, and lactic acid formation was approximately the same on glucose, mannose, maltose, or glycogen; there was little lactic acid production from fructose. A significant decrease in sugar concentration occurred in the medium of cells incubated with glucose, mannose, or fructose, but not in the case of L-glucose, galactose, ribose, arabinose, xylose, sucrose, lactose, or raffinose. Growth on maltose and glycogen was due to the presence in the serum and plasma of hydrolases: in the absence of cells, both maltose and glycogen were hydrolyzed to glucose (see Section 5.1.6). Harris and Kutsky raised the interesting possibility of adaptive changes in the carbohydrate requirements of cell populations. They were unable to find any adaptive shifts in the presence of the inert sugars they tested, although they did detect alterations in cell morphology. They suggested that it would be desirable to extend this exploration under culture conditions that would favor adaptive changes more strongly but apparently did not follow up these intriguing suggestions (Harris and Kutsky, 1953).

Bailey et al. (1959) investigated carbohydrate utilization in mouse lymphoblasts. Dialyzed serum was used and the sugars to be tested were analyzed by paper chromatography. Glucose and mannose were the best substrates for cell proliferation, followed by fructose and glucosamine, and lactic acid production was greatest on glucose and mannose. These studies apparently extended over about 35 days of continuous culture, and under the conditions used, sorbitol, glycerol, L-sorbose, lactic acid, and mannitol were able to support growth. As will become apparent later, a number of these substrates do not appear to be utilized by other cells and it is not clear whether the basis of the difference in the results was due to the cell type used or to some other component present in the basal medium.

Table I. *Growth Substrates of Cultured Cells*

Compounds supporting growth	References[a]
Hexoses	
D-Fructose	1–3, 5–7, 9
D-Galactose	1–3, 5–7, 9
D-Glucose	1–10
D-Mannose	1–3, 5–7, 9
Pentoses	
D-Ribose	2
D-Talose	2
D-Xylose	2, 8
Disaccharides	
Cellobiose	2
Melibiose	2
Turanose	2
Maltose	5
Trehalose	2, 5
Trisaccharide	
Maltotriose	5
Polysaccharides	
Amylose	5
Dextrin	5
Glycogen	5
Alcohol	
Sorbitol	2
Sugar phosphates	
G1P	2, 6
G6P	2, 5, 7, 9
Compounds not supporting growth	
Heptoses	
Glucoheptose	5
Mannoheptulose	5
Sedoheptulose	5
Hexoses	
D-Galactose	1, 2, 5, 6
D- or L-fucose	2, 5
L-Sorbose	2, 5
L-Glucose	1, 5
L-Rhamnose	2, 5, 6
D-Gulose	2
D-Allose	2
D-Altrose	2
2-Deoxyglucose	3
3-O-Methylglucose	3
D-Fructose	6
D-Mannose	6
D-Rhamnose	5

(*continued*)

Table I. (Cont.)

Compounds supporting growth	References[a]
Pentoses	
D- or L-arabinose	2, 5, 6
D-Lyxose	2, 6
D-Ribose	1, 2, 5, 7–9
D-Xylose	1, 5, 6, 9
L-Xylose	5, 8
2-Deoxy-D-ribose	5
D-Talose	5
Disaccharides	
Cellobiose	5, 6
Melibiose	5, 6
Turanose	5, 6
Lactose	1, 5–7, 9
Maltose	1, 4, 6, 7, 9, 10
Palatinose	5
Sucrose	1, 2, 5–7, 9
Trehalose	6, 10
Trisaccharides	
Melizitose	5, 6
Raffinose	1, 5, 6
Polysaccharides	
Arabic acid	5
Arabinogalactan	5
Lactobionic acid	5
Mucic acid	5
Stachyose	5
Glycogen	1
Inulin	6
Starch	4,10
Alcohols	
D- or L-arabitol	5
m-Inositol	5
Mannoheptitol	5
Ribitol	5
Sorbitol	5
i-Erythritol	5, 6
Galactitol	5, 6
Mannitol	2, 5, 6
Xylitol	2, 5
Glycosides	
α- or β-methylglucoside	2, 5
α- or β-methylxyloside	5
β-Methylgalactoside	5
α-Methylmannoside	5, 6
Esculin	6
Salicin	6

Table I. (Cont.)

Compounds supporting growth	References[a]
Sugar acids	
L-Ascorbic acid	5
D-Glucaric acid	5
D-Glucoheptonic acid	5
D-Galacturonic acid	2, 5, 6
D-Glucuronic acid	2, 5, 6
D-Gluconic acid	2, 5
Lactones	
D-Galactolactone	5, 6
Glucoheptuno-1,4-lactone	5
D-Glucuronolactone	5
Ribonolactone	5
Nucleotide sugar	
CDP-Mannose	6
Hexosamines	
D-Glucosamine	3, 5
Mannosamine	6
Acetylmannosamine	6
Phosphates	
Galactose-6-phosphate	6
Mannitol-6-phosphate	6
Glucose-6-phosphate	5, 6
DL-Glycerophosphate	5, 6
α-Glycerophosphate	5
Phosphoglyceric acid	2
Six-carbon compound	
Citric acid	2, 5, 7, 8
Five-carbon compound	
α-Ketoglutaric acid	5
Four-carbon compounds	
Dihydroxymaleic acid	5
Dihydroxytartaric acid	5
Oxaloacetic acid	5
Tartaric acid	5
Fumaric acid	2, 5, 7, 9
Malic acid	2, 5, 7, 9
Succinic acid	2, 5, 7, 9
D-Erythrose	1, 2, 5
Three-carbon compounds	
Dihydroxyacetone	
Glyceraldehyde	5
Glycerol	2, 5, 9
Lactic acid	2, 5, 7, 9
Pyruvic acid	2, 5–7, 9
Two-carbon compound	
Acetic acid	2, 5

(*continued*)

Table I. (Cont.)

Compounds supporting growth	References[a]
Amino acids	
Alanine	7, 9
Arginine	7, 9
Aspartic acid	7, 9
Glutamic acid	7, 9
Glutamine	7, 9
Histidine	7, 9
Isoleucine	7, 9
Serine	7, 9
Threonine	7, 9
Valine	7, 9
Miscellaneous	
Glucuronamide	5
Triacetylglucal	5

[a]1, Harris and Kutsky (1953); 2, Eagle et al. (1978); 3, Melnykovych and Bishop (1972); 4, Rheinwald and Green (1974); 5, Burns et al. (1976); 6, Dahl et al. (1976); 7, Faik and Morgan (1976); 8, Demetrakopoulos et al. (1977); 9, Faik and Morgan (1977a); 10, Scannell and Morgan (1980).

In an extensive study, the utilization of carbohydrates by a variety of human cell cultures was examined (Eagle et al., 1958): glucose and mannose were shown to be essentially equally active in terms of the concentration required for optimal growth yield, rate of growth, and rate of lactic acid production. For a number of cell types, trehalose, turanose, fructose, galactose, glucose-1-phosphate, and glucose-6-phosphate as well as mannose permitted growth at rates comparable to that of glucose. Lactic acid was not produced by cells growing on fructose or galactose to any great extent, but was not measured on the other substrates. Xylose, ribose, and talose permitted slight growth, but less than with glucose, and a large number of other compounds that were tested did not support growth (Table I). It was not possible to detect any initial lag indicative of an adaptation prior to the initiation of growth with any of the substrates tested.

The amount of sugar utilized by growing cultures varied markedly with the specific substrate and its concentration, but was not related to either the type of cell or the amount of growth obtained (Eagle et al., 1958). This lack of correspondence between carbohydrate utilization and growth was noted by a number of workers and remained an enigma until more recent work revealed the role of glutamine (Chapter 4).

Eagle et al. (1958) proposed that glucose and mannose were readily transported into the cell and phosphorylated. Consequently, their rate of utilization would increase with their concentration in the medium to reach levels that would not only exceed those required for optimal growth, but would also exceed the oxidative capacity of the cell. They concluded that the accumulation of lactic acid observed with these two substrates would reflect the inability of the cell to use the available carbohydrate oxidatively, either because one or more enzyme systems were saturated or because essential cofactors were depleted. Marked differences were observed in the affinity of hexokinase in the variety of cells tested with the substrates glucose, mannose, fructose, and galactose: the overall order of hexokinase activity in HeLa, KB, and liver cell extracts decreased in the order glucose (= mannose) > fructose > galactose. (See Section 2.2 for further discussion of the role of hexokinase in the regulation of cell proliferation.)

Melnykovych and Bishop (1972) investigated the utilization of hexoses and synthesis of glycogen in two strains of HeLa cells using a basal carbohydrate-free MEM prepared in their laboratory and supplemented with the carbohydrate under investigation. In both cell strains, fructose and galactose were not equivalent to glucose in their capacity to support growth and, in addition, mannose was toxic to one cell strain. The basis of the mannose toxicity appeared to be due to the unusually poor utilization by the particular cell line (but see Section 5.1.3).

Burns et al. (1976) tackled the problem of glucose contamination of re-agents and conversion of carbohydrates to glucose by serum enzymes in a study in which culture media were supplemented with glucose oxidase and catalase. Thus, any glucose arising from contamination or breakdown of polysaccharide would be converted to the inert, nonutilizable, gluconic acid. The basal medium was glucose- and pyruvate-free Dulbecco–Vogt medium (see Section 3.1 for discussion of the importance of pyruvate). Ninety-three carbohydrates were surveyed for toxic effects on Chinese hamster ovary and mouse LM (TK $^-$) cells; 15 were toxic, 27 produced growth inhibitory effects in the presence of glucose, and several naturally occurring carbohydrates were either toxic or growth inhibitory (L-fucose, glucosamine, 2-deoxy-D-ribose, D-glucuronic acid, glucuronamide, and glycerophosphate). The 51 carbohydrates that were neither toxic nor growth inhibitory were assayed for their ability to replace glucose as a sole carbon source. Fifteen were able to support the proliferation of one or more of the cell lines, some requiring the additional presence of 1 mM pyruvate in order to support cell growth (see below). Their results are incorporated in Table I.

Faik and Morgan (1976) observed that glucose, mannose, and maltose supported the growth of the Chinese hamster cell line CHO-K1 at the same rate; glucose-6-phosphate, galactose, and fructose supported growth at slower rates whereas lactose, sucrose, ribose (in the presence or absence of pyruvate), pyruvate, citrate, succinate, fumarate, and malate did not support growth. These studies were extended to include lactate and a variety of amino acids in the presence and absence of a gluconeogenesis-inducing steroid (dexamethasone; Table II). A specially prepared, glucose-free Ham's F12 medium (Flow Laboratories) was used, supplemented with the appropriate compound (usually 10 mM) and the glucose-free macromolecular components of fetal bovine serum, pre-

Table II. Growth of CHO-K1 Cells on Various Energy Sources[a]

Compounds supporting growth		Doubling time (hr)	Compounds not supporting growth[b]	
Energy source	Concentration (mM)			
Glucose	10	13	Lactose	Valine
	1.0	13	Sucrose	Isoleucine
	0.1	18	Ribose	Serine
Fructose	10	52	Xylose	Threonine
Mannose	10	14	Pyruvate	Aspartate
Galactose	10	36	Citrate	Glutamate
Maltose	5	20	Succinate	Histidine
Glucose-6-phosphate	10	20	Fumarate	Arginine
			Malate	Proline
			Alanine	Glutamine

[a]The energy sources were treated in glucose-free Ham's F12 medium supplemented with the macromolecular components of fetal bovine serum.
[b]All tested at 10 mM in the presence and absence of dexamethasone (1×10^{-7} M).

pared by passage of the serum through a Sephadex G-50 column and elution with a saline solution (Faik and Morgan, 1977a). Each batch of macroserum was tested for its ability to support the growth of CHO-K1 cells in the presence and absence of added glucose (Fig. 1). In addition, each compound utilized was extensively tested for the presence of glucose by assay with glucose oxidase. The method detects glucose at a concentration greater than 10^{-6} M.

Dahl *et al.* (1976) reported that CHO-K1 cells grew at the same rate on both mannose and glucose, while two other strains of Chinese hamster cell lines, V79 and CHS, did not grow on mannose. Mannose metabolism was further investigated in CHO-K1 and V79 cells: hexokinase (with both mannose and glucose as substrates), phosphomannose isomerase, phosphoglucose isomerase, and glucose-6-phosphate dehydrogenase were measured in the two cell lines and no significant differences were found (Dahl and Morse, 1979). No differences were detected in the rates of uptake of mannose or release of carbon dioxide from mannose, nor could differences be found in the rate of synthesis of nuclear material. Differences were found in the relative plating efficiencies of the two cell lines in the presence of differing mannose/glucose ratios in the medium: mannose had essentially no effect on the growth of CHO-K1 cells but seriously inhibited growth of the V79 strain. It was also shown that CHO-K1 cell extracts would hydrolyze α-mannosides whereas V79 cells would not. It was proposed that this defect in polysaccharide degradation, due to a lack of α-mannosidase activity, might be the basis for the difference in mannose metabolism of the two cell lines.

Differences in the ability of cells to grow on sugars have been used to select

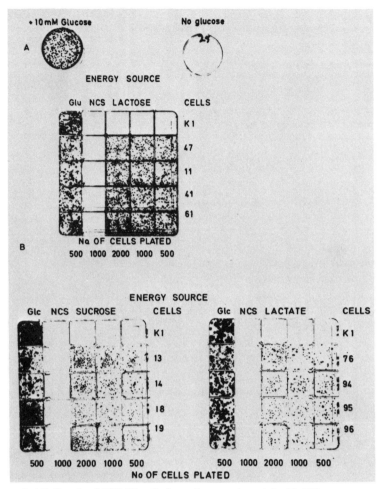

Figure 1. (A) Growth of CHO-K1 cells in medium supplemented with 10% macroserum in the presence and absence of glucose. (B) Growth of CHO-K1 cells and various "mutants" able to grow in media in which lactose, sucrose, or lactate substitutes for glucose; absence of growth is shown in media lacking an added carbohydrate. (After Faik and Morgan, 1977a.)

for differentiated function(s), changes in glycoprotein synthesis, or selection of cell type. These and other studies will be discussed in Section 5. It will also become evident that the studies so far described did not always clearly take into account the two roles that glucose, and other utilizable carbohydrates, play: (1) as an energy source, i.e., the ultimate provision of ATP, and (2) for the provision of essential precursors unobtainable from any other component in the medium. It would therefore be appropriate at this stage to describe some aspects of the intermediary metabolism of carbohydrates in rapidly dividing cells in cell culture.

2. GLYCOLYSIS

Changes in the rate of glucose utilization following application of various stimuli have frequently been observed in cultured cells. Thus, an increase in aerobic glycolysis has been noted following lymphocyte proliferation (Wang *et al.*, 1976), and when fibroblasts are exposed to agents known to stimulate proliferation such as serum, growth factors, and virus transformation (Fodge and Rubin, 1973; Diamond *et al.*, 1978; Kalckar and Ullrey, this volume). These changes have been ascribed to alterations in the activity of either the glucose transport system and/or enzymes of glycolysis, and are considered by some to play an important role in the regulation of cell growth. The first step in the metabolism of any carbohydrate in animal cells in culture is its transport into the cell. Information about the transport of carbohydrates is largely limited to hexose transport (mainly glucose and its analogues) and in particular to changes in the rates of hexose transport following treatment with various stimuli. A number of reviews on the transport process and its regulation have recently appeared and in this volume its regulation is discussed by Kalckar and Ullrey. We do not intend to discuss transport *per se*, but may have occasion to refer to specific aspects of it. Instead, we will discuss those glycolytic enzymes that may have an important place in the regulation of glycolysis in actively dividing cultured animal cells (Table III).

2.1. Glucokinase

Glucokinase (ATP:D-glucose-6-phosphotransferase, EC 2.7.1.2) is a glucose-6-phosphotransferase characterized by a low affinity for glucose, narrow substrate specificity, lack of product inhibition, and a tissue distribution apparently restricted to the hepatic parenchymal cell and the pancreatic beta cell. In both cells it is purported to occupy a specific role that enables the cell to respond rapidly to changes in circulating plasma concentrations of glucose. Glucokinase activity may be detected in freshly isolated hepatocytes and beta cells but does not appear to be retained in cultured cells. A primary role of glucokinase deficiency in diabetes is still an unproven hypothesis (Meglasson and Matschinsky, 1984). Halban *et al.* (1983) have shown the presence of glucokinase-like activity in intact rat islet cells, but an absence of activity in the insulin-producing rat cell line RINm5F. A high-affinity (hexokinase-like) glucose-phosphorylating activity was found in RINm5F cells and the authors propose that the absence of a glucokinase-like activity contributes toward the failure of glucose to stimulate insulin release in these cells.

Table III. Variation in the Activities (mUnits/mg Protein) of Some Glycolytic Enzymes

Enzyme stimulated	Wild type	Mutant	Normal	Transformed	Control	
Hexokiase	25.7[a]	31.1[a]	170[b]	317[b]	61.8[d]	58.3[d]
			48[c]	86[c]	70.0[e]	120[e]
			29[k]	52[k]		
Phosphoglucose isomerase	193[a]	9.3[a]			700[e]	920[e]
Phosphofructokinase	37.4[a]	71.3[a]	50[b]	171[b]	160[e]	280[e]
	32.0[f]	290[f]	188[c]	319[c]	31.6[g]	108.4[g]
					13.0[h]	75.0[h]
					9.0[i]	20.0[i]
Glyceraldehyde-3-phosphate dehydrogenase	2275[a]	1758[a]			1400[j]	2300[j]
Phosphoglycerate kinase	3464[a]	6.9[a]	4470[b]	12,750[b]	—	—
Pyruvate kinase	841[a]	555[a]	1340[c]	2550[c]	—	—
			280[k]	320[k]		
Lactate dehydrogenase	2951[a]	3505[a]	2280[c]	4480[c]	1580[e]	2340[e]
			1660[k]	2510[k]		

References: [a]Morgan and Faik (1980); [b]Gregory and Bose (1977); [c]Singh et al. (1974b); [d]Kletzien and Perdue (1974); [e]Pauwels et al. (1984); [f]Bishayee and Das (1981); [g]Diamond et al. (1978); [h]Wang et al. (1980); [i]Bruni et al. (1983); [j]Aithal et al. (1983); [k]Weber et al. (1984).

2.2. Hexokinase

Hexokinase (ATP:D-hexose-6-phosphotransferase, EC 2.7.1.1) is the first step in glycolysis proper and catalyzes the phosphorylation of a number of hexoses including glucose and mannose. It is found in all cells and is distinguished from glucokinase by its high affinity (low K_m) for glucose. Animal hexokinases vary in several respects between tissues and have been separated into at least three low-K_m isozymes. The distribution of the isozymes is tissue specific (Walker, 1966), but little is known of the variation in isozyme distribution in cultured cells. Most interest with respect to such distribution has centered on the possibility that a particulate form, associated primarily with mitochondria, is involved in the regulation of glycolysis (Bustamante and Pedersen, 1977) and may be less sensitive to inhibition by glucose-6-phosphate.

One characteristic of fast-growing cells is that they sustain high rates of glycolysis under aerobic conditions, thus converting glucose to lactic acid. It is still not clear why this apparent inefficient conversion should occur, but it has been suggested that hexokinase plays a pivotal role (Singh et al., 1974a; Bustamante and Pedersen, 1977; Wohlhueter and Plagemann, 1981; Table III). In particular, the form of hexokinase with a propensity for mitochondrial binding

may play a key role in the high rate of aerobic glycolysis of cancer cells (Bustamante *et al.*, 1981).

Such a peculiar role for hexokinase (or indeed any particular enzyme of the
glycolytic pathway) has been challenged by Racker who has emphasized that
there can be no glycolysis without ATPase (Racker, 1976). Extracts of Ehrlich
ascites cells did not glycolyze significantly unless an ATPase was added. The
addition of hexokinase had no effect unless phosphate was added. This was
particularly true when the endogenous ATPase was removed by precipitation
with calcium-ATP and under these conditions the addition of hexokinase had
little or no effect. Thus, in the absence of phosphate, or a phosphate-generating
ATPase activity, hexokinase activity would not be rate-limiting (in extracts
containing phosphate, of course, hexokinase and indeed other enzymes could be
rate-limiting). The point that Racker appears to be making follows from the
stoichiometry of glycolysis: for every mole of lactate formed from glucose, 1
mole of phosphate has to be regenerated (Racker, 1984). The regeneration of
phosphate and ADP for glycolysis may be due to the activity of the Na^+, K^+-
ATPase, which can be removed from the cytoplasm by precipitation with calcium-ATP (Racker *et al.*, 1983).

Colby and Romano (1974) concluded that changes in the uptake of 2-
deoxyglucose following transformation with tumor viruses were due to changes
in the activity of hexokinase rather than the activity of the putative transport
system. However, the suggestion that transport is rate-limiting for growth initiation and growth rate has come under attack and it seems likely that even if these
factors have a prime role in determining flux through glycolysis, then that flux is
not directly related to the rate of cell proliferation (Romano and Connell, 1982b).

2.3. Phosphoglucose Isomerase

Phosphoglucose isomerase (PGI; D-glucose-6-phosphate ketol isomerase,
EC 5.3.1.9) catalyzes the reversible conversion of glucose-6-phosphate to fruc-
tose-6-phosphate. It is present in a large excess compared with the rates of
metabolic flux, is involved in both glycolysis and gluconeogenesis, and is not
believed to be important in the control of glycolytic flux (Table III). There have,
therefore, been relatively few studies of the enzyme in cultured animal cells, but
mutants have been isolated (see below and Chapter 1) and its deficiency in
humans results in a form of nonspherocytic hemolytic anemia.

2.4. Phosphofructokinase

Phosphofructokinase   (PFK;   ATP:D-fructose-6-phosphate   1-phos-
photransferase, EC 2.7.1.11) catalyzes the phosphorylation of fructose-6-phos-
phate to fructose-1,6-bisphosphate. Regulation of the activity of PFK has long
been considered as one of the principal means of control of glycolysis. The

kinetic behavior *in vitro* of the enzyme is highly complex, there being mutual interactions between the two substrates, D-fructose-6-phosphate and ATP. Recently, a modulator of PFK activity, fructose-2,6-bisphosphate, has been described and probably plays a major role in the regulation of the enzyme activity *in vivo* (Hers and Van Schaftingen, 1982).

Various workers have described changes in the activity of PFK following infection of cells with tumor viruses or by stimulation with serum and growth factors (Table III). Fodge and Rubin (1973) attributed an increase in glycolysis following infection of chick embryo cells with Rous sarcoma virus (RSV), to an increase in the activity of PFK. Decreases in the concentrations of glucose-6-phosphate and fructose-6-phosphate and an increase in the concentration of fructose-1,6-bisphosphate were found following cell stimulation. These changes were best explained by proposing an increase in PFK activity. A similar increase in the fructose bisphosphate/fructose-6-phosphate ratio following infection with RSV was also demonstrated. This apparent change in PFK activity correlated with an increased utilization of glucose via the glycolytic route, but the *in vitro* activity of the enzyme in extracts of the cells apparently was not determined.

Schneider *et al.* (1978) were able to show that the addition of serum to quiescent cultures of mouse 3T3 cells markedly enhanced the specific activity of PFK assayed in cell homogenates. The activation appeared to be independent of protein synthesis, suggesting a modification of preexisting enzyme molecules— possibly by phosphorylation. Diamond *et al.* (1978) were able further to show that glycolysis in homogenates of 3T3 cells, stimulated by serum or other factors known to stimulate cell division, was persistently increased and was not blocked by the addition of inhibitors of either protein or RNA synthesis.

A similar shift in PFK activity was shown to occur in cultured mouse spleen lymphocytes following concanavalin A stimulation (Wang *et al.*, 1980). These workers were able to demonstrate a period following stimulation in which enzyme activation was dependent on protein synthesis and a second stage during which the increase in PFK activity became progressively less inhibited, suggesting the activation of preexisting enzyme molecules.

Bruni *et al.* (1983) assayed PFK activity in extracts of resting human fibroblasts following serum stimulation and simultaneously determined the cellular concentration of fructose-2,6-bisphosphate. A twofold increase in PFK activity was shown and this correlated with the increase in glucose uptake and lactate production. The concentration of fructose-2,6-bisphosphate increased following serum stimulation and fructose-2,6-bisphosphate was shown to increase the affinity of the enzyme (*in vitro*) for fructose-6-phosphate. However, the fructose-2,6-bisphosphate failed to affect the maximum velocity of fibroblast PFK and thus the observed increase in enzyme activity in the stimulated cells could not be ascribed to the higher content of fructose-2,6-bisphosphate (Bruni *et al.*, 1983).

A similar activation of PFK, through the agency of fructose-2,6-bisphosphate, has been proposed to explain the glucose-induced stimulation of glycolysis in pancreatic islet cells (Malaisse *et al.*, 1981).

Bishayee and Das (1981) investigated a variant of 3T3 cells (NR-6) that lacks epidermal growth factor receptors. The variant was found to be deficient in cytochrome *c* oxidase activity, resulting in a failure of the oxidative phosphorylation pathway. This deficiency was compensated for by an exceptionally high rate of aerobic glycolysis associated with an elevated level of PFK activity. The activity of other enzymes of glycolysis was not reported.

Gregory and Bose (1977) were unable to show an activation of PFK (or other glycolytic enzymes) following transformation of normal rat kidney cells with Kirsten sarcoma virus and attributed the changes in glycolytic activity to changes in hexose transport. PFK (and pyruvate kinase) activities were found to be elevated in crowded cultures of transformed cells (Gregory and Bose, 1979), whereas no density-dependent changes in enzyme activity were detected in tumor cells grown *in vivo*. Gregory and Bose (1979) thus cautioned that while *in vitro* cell cultures might be ideal for studying various metabolic effects, they might be less than ideal in studying the regulation of growth *in vivo*.

A twofold increase in PFK activity, and comparable increases in hexokinase, pyruvate kinase, lactate dehydrogenase, and PGI were observed in neuroblastoma cells subject to hypoxic culture conditions for 48 hr (Pauwels *et al.*, 1984).

2.5. Glyceraldehyde-3-phosphate Dehydrogenase

Glyceraldehyde-3-phosphate dehydrogenase [G3PD; D-glyceraldehyde-3-phosphate: NAD^+ oxidoreductase (phosphorylating), EC 1.2.1.12] catalyzes the reversible conversion of glyceraldehyde-3-phosphate to 1,3-diphosphoglycerate. The enzyme is present in large quantities, operates near equilibrium, and is involved in both glycolysis and gluconeogenesis.

The activity of the enzyme is not usually regarded as an important factor in determining the flux through glycolysis. Recent results (Aithal *et al.*, 1983), however, show that a stimulation of G3PD activity occurs during the initiation of renal epithelium cell growth following stimulation by exposure of cells to medium containing low concentrations of potassium. The effect is present *in vivo* since rats fed a potassium-deficient diet exhibit accelerated kidney growth accompanied by an enhanced activity of G3PD. High-density quiescent cultures of the BSC-1 renal cell line exposed to low-potassium medium responded with a 60% increase in G3PD activity within 2 hr (Table III). Treatment of the cells with calf serum also increased the enzyme's activity. PFK activity was not altered in cells exposed to low-potassium medium, but results were not reported on the effect of serum on PFK activity.

The increase in G3PD activity in renal epithelium cell cultures was apparently mediated by a new cytosolic protein with an apparent molecular weight of 62,000 (Aithal *et al.*, 1983). It will be interesting to learn if this increase in enzyme activity is peculiar to the kidney system or is the first example of a new general regulatory mechanism.

2.5.1. Molecular Genetics

Genomic clones of the glyceraldehyde phosphate dehydrogenase gene have been examined from a variety of species. Eleven introns are present in the chicken gene and appear to divide it into mononucleotide-binding, NAD-binding, and catalytic domains (Stone *et al.*, 1985). Although only one gene coding for the enzyme is known to be functional in human, mouse, rat, and chicken, there appears to be an abundance of pseudogenes. Chicken appears to be unique, containing only one glyceraldehyde phosphate dehydrogenase related sequence. Human, hare, guinea pig, and hamster appear to contain 10 to 30 pseudogenes and mouse and rat appear to contain greater than 200 (Piechaczyk *et al.*, 1984). The reason for this amplification is not known. (The gene has been localized to chromosome 12 in humans and chromosome 6 in mouse.)

2.6. Phosphoglycerate Kinase

Phosphoglycerate kinase (PGK; ATP: 3-phospho-D-glycerate 1-phosphotransferase, EC 2.7.2.3) catalyzes the reversible conversion of 1,3-diphosphoglycerate to 3-phosphoglycerate, one of the two ATP-generating steps in the glycolytic pathway. The enzyme is present in large amounts in all tissues examined and is believed not to have a major role in the regulation of glycolysis. PGK deficiency in humans causes a rare form of X-linked hereditary nonspherocytic hemolytic anemia associated with severe mental disorders (Miwa *et al.*, 1972) and fibroblasts can be isolated that contain the deficient enzyme species (Yoshida and Miwa, 1974). A variant of Chinese hamster ovary (CHO) cells has been described that lacks PGK activity (Morgan and Faik, 1980; Table III); its properties are described below.

2.6.1. Molecular Genetics

cDNA clones for human PGK, one of which is full length, have been isolated independently by two groups (Michelson *et al.*, 1983; Singer-Sam *et al.*, 1983) and are being used to investigate the organization of the gene in humans (Hutz *et al.*, 1984) and in Chinese hamster cells (Faik *et al.*, 1986). The human PGK cDNA has been introduced into the PGK-deficient line of CHO cells

(Morgan and Faik, 1980) by DNA-mediated gene transfer using either the cal-cium phosphate–DNA coprecipitation method or by fusion of bacterial pro-toplasts containing the PGK gene encoded in a plasmid (Faik et al., 1986). Transformants were isolated following cotransformation with the antibiotic re-sistance marker AGPT, which confers selectable resistance to the antibiotic G418, or by a direct selection for the expression of the PGK cDNA clone.

The PGK-deficient cell line is unable to metabolize mannose, and indeed, mannose is toxic (Morgan and Faik, 1981). Restoration of PGK activity should restore the ability of the cells to grow on mannose and/or limit toxicity. Both methods of DNA-mediated gene transfer resulted in the isolation of PGK-positive transformants and these were shown to behave as would be predicted from the properties of the wild-type cells: they catabolized mannose to lactic acid and mannose was no longer toxic. Expression of the human form of PGK in the transformants was demonstrated by starch gel electrophoresis of the enzyme and by Southern blot analysis (Faik et al., 1986). This would appear to be an interesting model system for studying the regulation of expression and structure of a glycolytic enzyme.

2.7. Other Glycolytic Enzymes

Singh et al. (1974a,b) analyzed normal and RSV-infected chick embryo cells for the activities of glycolytic enzymes and intermediates. Crossover points in the levels of glycolytic intermediates were determined and the authors con-cluded that the increase in glucose flux following virus transformation was due to increases in the activities of hexokinase, pyruvate kinase, and PFK (Singh et al., 1974a). Further analysis showed that these enzymes were indeed present at elevated levels in extracts of the transformed cells (Singh et al., 1974b). These results (Singh et al., 1974a,b) were recently confirmed by Webber et al. (1984) who demonstrated an increased activity of hexokinase and lactate dehydrogenase activity following transformation of chick embryo fibroblasts with RSV (Table III).

Ardawi and Newsholme (1982) measured the maximum activities of some glycolytic enzymes in lymphocytes isolated from fed and starved rats. They observed a significant increase in the activity of pyruvate kinase but only modest increases in the other glycolytic enzymes measured. Stimulation of the lympho-cytes in a graft-versus-host reaction increased the activities of hexokinase, PFK, lactate dehydrogenase, and two nonglycolytic enzymes (citrate synthase and glutaminase). The authors concluded that glycolysis has an important place in the response of the lymphocyte to immunological challenge.

Enzyme and coenzyme levels were measured in synchronized cultures of Chinese hamster cells and the activities of hexokinase, phosphoglucomutase, glucose-6-phosphate dehydrogenase, 6-phosphogluconate dehydrogenase and PGI were shown to increase continuously during the interphase period of the cell cycle (Blomquist *et al.*, 1971). This result is consistent with a stable, basal cellular metabolism during interphase.

Cooper *et al.* (1983) showed that three glycolytic enzymes were phosphorylated following transformation of chick cells with RSV. The enzymes, enolase, phosphoglycerate mutase, and lactate dehydrogenase, were phosphorylated at tyrosine residues. Aldolase, triose phosphate isomerase, PGK, and pyruvate kinase were found not to be phosphorylated. The authors speculated on the role of phosphorylation in the regulation of glycolysis. As the enzymes do not appear to be activated by phosphorylation, and are not considered to be important in determining the flux through glycolysis, it was suggested that phosphorylation could perhaps modify the organization of the glycolytic enzymes within the cytoplasm. Potentially, phosphorylation of only a small fraction of molecules could thus have large effects on the spatial organization and thereby on glycolysis. Several attempts have been made to show that enzymes of glycolysis may interact with one another and such organization has been demonstrated in the glycosome particle found in parasites (Chapter 6). Reference has been made (Section 2.2) to the possible association of hexokinase with mitochondria and there have been some studies suggesting an interaction between glyceraldehyde phosphate dehydrogenase and aldolase (Ovadi and Keleti, 1978). Such associations would be highly likely to play a role in the activity of the glycolytic pathway, but little is yet established.

3. THE PROVISION OF ENERGY

It is now generally accepted that glucose is not the only major energy source present in cell culture medium, and that a varying proportion of the total energy requirement is derived from the oxidation of glutamine (Chapters 3 and 4). Analysis of glycolysis-deficient mutants (Pouysségur *et al.*, 1980; Morgan and Faik, 1980) has shown that cells can obtain their energy needs entirely from respiration. One mutant (Pouysségur *et al.*, 1980) is deficient in PGI, while the other (Morgan and Faik, 1980) is deficient in both PGI and PGK. Both mutants grow on glucose almost as well as the wild-type cells, but neither produces lactic acid. The oxidation of [1-^{14}C]glucose is normal whereas that of [6-^{14}C]glucose is negligible. Both mutants are rapidly killed by the application of respiratory inhibitors such as cyanide and oligomycin at concentrations that do not kill the wild-type cells, and both oxidize glutamine extensively (Morgan *et al.*, 1981). Thus, both cell types rely on respiration for their energy needs and appear to be able to grow as readily as the glycolysis-competent parents. One of the mutants

(Pouysségur *et al.*, 1980) has been shown to be just as capable of forming tumors in experimental animals as the wild-type cells, implying that there is no necessary connection between the rate of glycolysis and malignancy.

3.1. Pyruvate Metabolism

The role of glutamine oxidation in the economy of the cell is established (Chapters 3 and 4) but little attention has been given to the role that the oxidation of pyruvate may play.

Pyruvate is a common constituent of many cell culture media and has been reported to be essential for growth on substrates such as ribose and galactose (Eagle *et al.*, 1958), fructose, galactose, xylose, trehalose, maltotriose, dextrin, amylose, glucose-6-phosphate, and galacturonic acid (Burns *et al.*, 1976).

The effects of pyruvate on the growth of normal and transformed hamster embryo fibroblasts has recently been studied (Sens *et al.*, 1982) and the potential role of pyruvate in cell metabolism discussed. Two effects of pyruvate were noted: a sparing of glucose utilization and an effect on glutamine utilization with a concomitant increase in lactate production. McKeehan (McKeehan *et al.*, 1982; McKeehan, 1984) showed that pyruvate was a key nutrient for the proliferation of normal fibroblasts in defined culture conditions. The role of pyruvate (and other 2-oxocarboxylic acids) could be mediated through an increased ratio of oxidized to reduced pyridine nucleotides, that would result from their reduction. DNA synthesis in rat liver parenchymal cells in serum-free medium is strikingly sensitive to the presence of pyruvate (McGowan *et al.*, 1984) whereas both alanine and glutamine were inhibitory under similar conditions. Glucose acted synergistically with pyruvate to enhance DNA synthesis in a complex mixture (Gunn *et al.*, 1975).

The glycolysis-deficient mutant of CHO cells (Morgan and Faik, 1980) has been shown to obtain essentially no energy from glycolysis but to be absolutely dependent on respiration for its energy needs (Morgan *et al.*, 1981). The rate of glutamine utilization was shown to be insufficient to provide the energy necessary for the growth of the mutant and it was postulated that other substrates present in the medium must also be oxidized and provide additional energy (Morgan *et al.*, 1981).

The oxidation of a variety of substrates was examined by measuring oxygen uptake in washed cells. Glutamine was most readily oxidized followed by pyruvate, glutamate, alanine, and malate. Succinate, citrate, aspartate, arginine, and oxaloacetate were without effect (Morgan *et al.*, 1983a). As the role of glutamine oxidation was already established, the possible place of pyruvate oxidation in the economy of this cell line was investigated.

In the presence of glucose, pyruvate was utilized only slowly by the wild-type cells, but very rapidly by the mutant cells. By the use of specifically labeled

Table IV. ATP Derived from the Utilization of Glucose, Pyruvate, or Glutamine[a]

Product	CHO-K1 (wild type)		R1.1.7 (PGI$^-$,PGK$^-$)	
	Product formed[b]	ATP[b]	Product formed[b]	ATP[b]
Lactate	38	38	<1	<1
CO_2-Ex-glutamine	1.5	18[c]	1.6	19^2.2[c]
CO_2-Ex-C1-pyruvate	2.8	8.4[d]	9.3	27^2.9[d]
CO_2-Ex-C2-pyruvate	—	13.4[e]	3.7	44^2.4[e]
Total ATP yield		77.8		91.5

[a]The ratio of 2-$^{14}CO_2$ evolved/1-$^{14}CO_2$ evolved for R1.1.7 cells has been used to calculate the ATP yield from the operation of the TCA cycle in CHO-K5cells.

[b]nmoles/min per mg cell protein.

[c]1 mole of $^{14}CO_2$ from incomplete oxidation of glutamines equivalent to 12 moles of ATP synthesized.

[d]1 mole of 1-$^{14}CO_2$ equivalent to 3 moles of ATP synthesized.

[e]1 mole of 2-$^{14}CO_2$ equivalent to 12 moles of ATP synthesized.

pyruvate, it could be shown that the decarboxylation of pyruvate was threefold higher than the rate of oxidation of pyruvate (TCA cycle) and it was suggested that the acetyl CoA not oxidized, was probably converted to lipid (Morgan *et al.*, 1983a). Similar observations and conclusions have been made for ascites cells (Lazo, 1981). When the yield of ATP from pyruvate oxidation was calculated, it was shown that pyruvate oxidation in the glycolysis-deficient mutant provided approximately three times as much energy as pyruvate oxidation in the wild-type cells. The total yield of ATP from the combined operation of glycolysis, glutamine oxidation, and pyruvate oxidation in both cell types was essentially similar (Morgan *et al.*, 1983a; Table IV). It has further been shown (Faik, Walker, and Morgan, unpublished data) that the growth of wild-type CHO cells on fructose and galactose and the growth of the glycolysis-deficient cell on glucose, fructose, galactose, and ribose, are accompanied by a rapid utilization of pyruvate. The specific role of pyruvate (and other oxocarboxylic acids) clearly deserves further attention.

4. PENTOSE PHOSPHATE PATHWAY

The two major products of this pathway are NADPH and ribose-5-phosphate; the former is of major importance for reductive synthesis and the latter for nucleic acid synthesis. In liver the generation of NADPH may be of greater importance and the activity of the pentose phosphate pathway may be regulated in concert with the requirements of lipogenesis (Nepokroeff *et al.*, 1974). However, in cultured cells the supply of ribose-5-phosphate and NADPH for nucleic acid synthesis may be of more importance than the supply of NADPH for lipid synthesis in actively growing cells.

Smith and Buchanan (1979) have shown that one of the earliest measurable responses in serum-stimulated fibroblasts is an increase in the synthesis of phosphoribosyl pyrophosphate following an increase in the flux from glucose to ribose-5-phosphate. Thus, the regulation of the pentose phosphate pathway may be an important factor in growth control and deserves further investigation. Singh *et al.* (1974a) showed that transformation of chick embryo cells by RSV resulted in a 70% increase in the activity of glucose-6-phosphate dehydrogenase but there was no change in the activity of 6-phosphogluconate dehydrogenase.

Wagner *et al.* (1978) showed an eightfold increase in activity of both glucose-6-phosphate dehydrogenase and 6-phosphogluconate dehydrogenase within the first 24 hr of regeneration of skeletal muscle. The increase in activity of the nonoxidative enzymes of the pathway (transaldolase, transketolase, ribose-5-phosphate isomerase, and ribulose-5-phosphate 3-epimerase) was more limited.

Reitzer *et al.* (1980) showed that at least one-third of the ribose-5-phosphate synthesized in HeLa cells growing on glucose and almost all of the ribose-5-phosphate generated during growth on fructose were required for nucleic acid synthesis. They suggested that the supply of ribose-5-phosphate was growth-limiting in fructose-grown cells and that the only essential function of sugar metabolism was to provide carbon in the pentose phosphate cycle. The flux of hexose carbon through the oxidative arm of the pentose phosphate pathway was determined by measuring the release of $^{14}CO_2$ from [1-^{14}C]hexose. The velocities of both glucose-6-phosphate and 6-phosphogluconate dehydrogenases were severely limited at the maximum cytoplasmic $NADP^+$ concentration and $NADPH/NADP^+$ ratio found but the coenzyme levels did not change significantly during growth on different hexoses. In contrast, the concentrations of the substrates, glucose-6-phosphate and 6-phosphogluconate, did change severalfold. The changes in the levels of the substrates could account for the observed changes in the flux through the oxidative arm. It was concluded that NADPH production was not an essential function of the oxidative arm in HeLa cells.

Reitzer *et al.* (1980) pointed out that NADPH would be generated by the metabolism of glutamine (Chapter 4) in sufficient amounts and drew attention to a mutant of *Drosophila* (lacking glucose-6-phosphate dehydrogenase activity) that was able to grow and reproduce, presumably obtaining the ribose-5-phosphate for nucleic acid synthesis via the nonoxidative arm (Vozdev, 1976). Such a conclusion is in accord with observations made in a mutant of CHO cells that lacks both PGI and PGK activity, but grows on fructose as readily as the wild-type cells (Faik and Morgan, 1977b). However, it is perhaps an extreme view to suggest that the only essential function of sugar metabolism is to provide carbon in the pentose cycle. This appears to ignore the role of sugar metabolism in providing amino sugars such as glucosamine for glycoprotein synthesis, inositol for phospholipid synthesis (see below), and other precursors for glycoprotein and glycolipid synthesis. Nevertheless, these data support the concept that the essential role of carbohydrates is to provide carbon for biosynthesis and not for energy (see below and Chapter 3 and 4).

It has been claimed (Williams, 1980) that the reaction sequence of the pentose phosphate pathway, as usually represented, requires modifying. In the usual scheme (Fig. 2), glucose-6-phosphate generated from glucose is converted to ribulose-5-phosphate and then is transformed, by the combined operation of an isomerase, epimerase, transaldolase, and transketolase, to give fructose-6-phosphate and glyceraldehyde-3-phosphate (which may be converted to fructose-6-phosphate). The cycle is completed by the isomerization of fructose-6-phosphate to glucose-6-phosphate catalyzed by PGI. In the alternative scheme (Williams, 1980) the final products are the same as the classical scheme, but some of the intermediates (which include octulose mono- and bisphosphates) are different and there are additional enzymatic steps catalyzed by aldolase, a new epimerase, and a new phosphotransferase; there is no role for transaldolase. One crucial difference in the two alternative schemes is the role of the enzyme PGI: in the classical scheme the formation of glucose-6-phosphate in the rearrangement reactions is entirely dependent on the activity of PGI. In the revised scheme (Williams, 1980) glucose-6-phosphate is formed directly (along with fructose-6-phosphate) by the action of transaldolase on erythrose-4-phosphate and oc-tulose-8-phosphate. The formation of glucose-6-phosphate in this scheme is thus not dependent on the activity of PGI.

Mutants of Chinese hamster cells lacking PGI activity (Morgan and Faik, 1980; Pouysségur et al., 1980) have been used to investigate the reaction sequence in these cells. If the Williams pathway operates, then extracts of PGI-deficient cells should still be able to synthesize glucose-6-phosphate from added ribose-5-phosphate. This was shown not to be the case: such extracts convert ribose-5-phosphate only to fructose-6-phosphate and it has therefore been concluded that the modified pentose phosphate pathway does not operate in these cells (Morgan, 1981). Similar findings were made in extracts of PGI-deficient mutants of *Escherichia coli* (R. A. Cooper, personal communication). These findings are still controversial, however, and continue to generate active discussion (Landau and Wood, 1983; Williams et al., 1983).

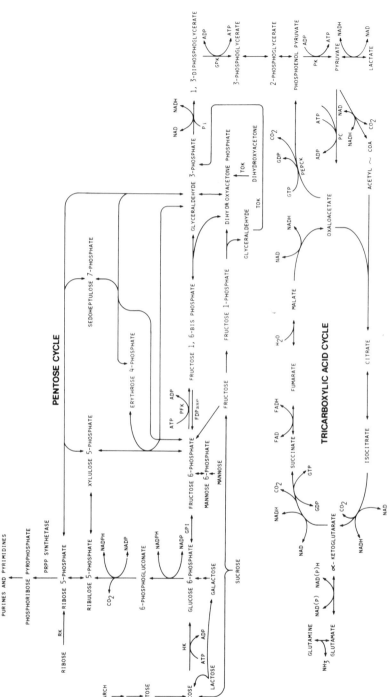

Figure 2. Metabolic scheme of reactions relevant to this review. All reactions depicted have not necessarily been demonstrated in mammalian cells in culture, but are based on metabolic routes known to be present in mammalian tissue. (From Morgan and Faik, 1981.)

Abbreviations: RK, ribokinase; HK, hexokinase; GPI, phosphoglucose isomerase; PFK, phosphofructokinase; GPK, phosphoglycerate kinase; PK, pyruvate kinase; PC, pyruvate carboxylase; PEPCK, phosphoenolpyruvate carboxykinase; TOK, triokinase; FDPase, fructose bisphosphatase; PRPP synthetase, phosphoribosyl pyrophosphate synthetase.

4.1. Glucose-6-phosphate Dehydrogenase

Glucose-6-phosphate dehydrogenase (D-glucose-6-phosphate: NADP$^+$ 1-oxidoreductase, EC 1.1.1.49) catalyzes the oxidation of glucose-6-phosphate to 6-phosphogluconic acid with the generation of NADPH. It is the first step in the pentose phosphate pathway and may be subject to regulation (see above). Absence of this enzyme in humans is responsible for the commonest form of nonspherocytic hemolytic anemia. The gene has been shown to be X-linked. Cell lines lacking glucose-6-phosphate dehydrogenase activity have been isolated from such patients and mutants lacking activity have been isolated in Chinese hamster cells (Rosenstraus and Chasin, 1975). The hamster mutant has been fused with human cells and the resultant hybrid shown to express the human form of the enzyme, but not the hamster form (D'Urso et al., 1983). The mutation in the hamster line is thus probably at the structural gene locus.

A novel activation of the glucose-6-phosphate dehydrogenase gene appears to take place in the malarial parasite, Plasmodium falciparum, when it infects the host red blood cell. Transcription of the parasite's glucose-6-phosphate dehydrogenase gene appears to be switched on when the parasite infects glucose-6-phosphate-deficient red blood cells (Usanga and Luzzatto, 1985). Expression of the gene appears only to occur in glucose-6-phosphate-deficient cells and cannot be detected in normal cells. This may explain why hemizygotic glucose-6-phosphate-deficient males are not protected against the infection, whereas heterozygotes (females) are. In the latter the P. falciparum merozoites emerging from normal red blood cells have, on the average, an even chance of infecting a normal or deficient cell. In the latter instance, the chance to complete successfully the next schizogonic cycle is reduced by 50%. It will be of great interest to study how the induction of this gene in the parasite occurs.

5. DIFFERENTIATION

It has been appreciated for some time that the composition of culture media affects the expression of differentiated functions and indeed the ability of differentiated cells to grow in vitro. In many instances this has been shown to be due to serum factors, especially hormones and growth factors. The carbohydrate source can also have a selective influence: some tissues are characterized by their metabolic versatility; the liver, for example, can metabolize lactate and oxaloacetate since it carries out gluconeogenesis. Medium containing such substrates would thus be selective for liverlike cells (see below). It should also be possible by altering the carbohydrate source to select cell lines, from differentiated tissue and tumors, that are able to metabolize these carbohydrates (Rheinwald and Green, 1974; Avner *et al.*, 1977).

The differentiation of the PCC3/A/1 teratocarcinoma into cartilage, keratin, lipid, muscle, nerve cells, and so on is dependent on the presence of either glucose or mannose (Avner *et al.*, 1977). In the absence of glucose, mainly endoderm is formed, and a number of variant differentiated cell lines were obtained by culturing the teratocarcinoma cells in medium in which other carbohydrates were substituted for glucose. Thus, by varying the carbon source it may be possible to isolate cell lines that correspond to those occurring at different stages of the normal *in vitro* differentiation sequence.

Pinto *et al.* (1982) showed that replacement of glucose by galactose in the medium of the human colon carcinoma cell line HT-29 results in the occurrence of a brush border, and in the appearance of the brush border-associated enzymes alkaline phosphatase, aminopeptidase, and disaccharidases. When the differentiated cells were seeded back into a medium with high concentrations of glucose, they immediately reverted to the undifferentiated phenotype (Pinto *et al.*, 1982). The pattern of glucose metabolism has also been shown to be characteristic of the degree of differentiated function expressed in cultured mouse mammary cells (Emerman *et al.*, 1981).

It thus seems reasonable to assume that the failure of many cell lines to grow in medium other than that containing glucose is due to their inability to express various differentiated functions that would otherwise allow utilization of the carbon source. For example, cells lining the small intestine synthesize a lactase (usually only in young animals) and are able to metabolize lactose. Since cell lines should contain the same genetic information as the animal from which they were derived (even if rearrangements have occurred), and if inactive genes can be reactivated, it should be possible to isolate variants that reexpress enzymes necessary for growth on what are usually considered as nonutilizable carbohydrates. Using this reasoning, a number of variants have been isolated and studied and are described below.

The basis for the widely recognized "dedifferentiation" of primary cultures, such as liver cells, is not well understood. Evidence has recently been obtained that, although there was little change in the concentration of liver-specific mRNA concentrations in the first 24 hr following primary culture of mouse hepatocytes, the synthesis of liver-specific mRNA was greatly diminished within 24 hr. Transcription of "common" genes and transcription of tRNA and rRNA did not decline (Clayton and Darnell, 1983). The basis for this change in transcription has been investigated and it has been shown that mixtures of hormones and growth factors are unable to restore transcription. The possibility that cell position was important and that the lack of transcription was due to disaggregation was also considered (Clayton and Darnell, 1983). This approach appears likely to produce very interesting information on the regulation of differentiated functions.

5.1. Alternative Carbon Sources

Reference has already been made to the restricted ability of cultured animal cells to grow in medium in which glucose is absent and another carbon source is substituted. In this section we propose to discuss the growth of cells on a number of alternative carbon sources (Fig. 2), and in particular will discuss "variants" or "mutants" that have the ability to utilize additional alternative carbon sources.

5.1.1. Fructose

Most cultured animal cells will grow in medium in which fructose is substituted for glucose provided there is an alternative energy source such as glutamine (Chapter 4) or pyruvate (see Section 3). It is not clear what essential role fructose plays in media in which it replaces glucose, but it is clear that it is not the source of energy (Section 4). Schwartz and Johnson (1976) reported that an SV40-transformed BALB mouse 3T3 line (SVT2) grew on glucose and mannose but was unable to grow on galactose or fructose. A variant of this cell (SVT2-SUG$^+$) was isolated that was capable of growing on all four sugars. The relative activity of hexokinase was investigated to determine if the rate of growth on the sugars was related to the relative activity of this enzyme, but no difference between SVT2 and SVT2-SUG$^+$ cells was found.

Wolfrom et al. (1983) examined the effects of glucose versus fructose on the growth and morphology of cultured human skin fibroblasts, emphasizing that their work differed from earlier studies (Eagle et al., 1958; Reitzer et al., 1980; Schwartz and Johnson, 1976) since their studies extended over 20 days of culture. Fructose utilization and lactate production were very low (as had been observed by other workers for a variety of cells) and the pH was higher on fructose-grown cells than on glucose-grown cells, presumably reflecting the low rate of production of lactic acid. Growth on fructose not only affected cell growth, longevity, and morphology but was accompanied by the accumulation of lipid. However, the metabolism of other compounds present in the medium (e.g., glutamine, 2 mM), which could also affect these properties, was not investigated.

The activity of pyruvate dehydrogenase in cells grown on glucose versus cells grown on fructose was determined and a two- to threefold increase in activity was found in fructose-grown cells (Wolfrom et al., 1983). This increase in pyruvate dehydrogenase activity, if not coupled with an increase in activity of citrate synthase, could explain the increased lipid content of the fructose-grown cells since excess acetyl CoA would be channeled into lipid synthesis (Lazo, 1981).

5.1.2. Galactose

The first step in galactose metabolism following its uptake into the cell (Chapter 1) is phosphorylation by a specific hexokinase, galactokinase, the first enzyme of the Leloir pathway, to give galactose-1-phosphate. This is followed by conversion of galactose-1-phosphate to glucose-1-phosphate catalyzed by galactose-1-phosphate uridyltransferase. The glucose-1-phosphate is interconvertible with glucose-6-phosphate via the phosphoglucomutase reaction and thus carbon from galactose enters the glycolytic pathway.

Stern and Krooth (1975) examined the regulation of the three enzymes of the Leloir pathway in human and rat cells by growing them in medium containing either glucose or galactose. The substitution of galactose for glucose had no effect on the activity of the three enzymes of the Leloir pathway. The possibility was considered that the enzymes are expressed constitutively in all cells, but it was pointed out (Stern and Krooth, 1975) that the activity of the pathway in liver is considerably higher than in fibroblasts (i.e., it is inducible). However, since the activity of the pathway would only need to be high if galactose were the sole energy source, and since the medium contained glutamine, the level of the enzymes was high enough to explain how the cells would continue to proliferate in medium containing galactose but not in the same medium in the absence of added galactose (or glucose). Stern and Krooth noted that galactose was essential and that the cells thus had a mechanism for perceiving that something had changed when galactose was substituted for glucose as the sole hexose present; however, they were unable to explain its role. It is likely that galactose was serving as the precursor for cell constituents that cannot be derived from other medium components, and that the cells' energy was derived from the utilization of glutamine and pyruvate (Section 3).

Evidence that growth on galactose is dependent on respiration was obtained by studying chloramphenicol resistance in cultured hamster cells (Ziegler and Davidson, 1979). Chloramphenicol selectively inhibits mitochondrial protein synthesis in eukaryotic cells without affecting cytoplasmic protein synthesis. Chloramphenicol-resistant Chinese hamster cells have been isolated, and grow and survive in the presence of chloramphenicol provided glucose is present in the medium (Ziegler and Davidson, 1979). Substitution of galactose for glucose leads to rapid cell death and it has not been possible to isolate chloramphenicol-resistant mutants when the hexose is galactose (Ziegler and Davidson, 1979). Presumably, such variants would be required to have had a number of alterations in the enzymes of galactose metabolism such that the flux from galactose was almost equal to the flux from glucose.

There is an age-related change in the metabolism of galactose that can be reproduced in cultured hepatocytes. In the newborn mammal, about half of the dietary carbohydrate is contributed by galactose from the metabolism of lactose. This is metabolized almost entirely in the liver and during late fetal life there is an increase in the specific activities of the enzymes of the Leloir pathway. Hepatocytes isolated from rats of different ages show changes in hexose utilization that reflect this alteration: a threefold higher rate of galactose utilization and a higher rate of conversion of galactose to glucose were found in the cells of weaning rats compared to those from adult rats (Fukushima *et al.*, 1981).

The ability of cells to grow in medium in which galactose is substituted for glucose has been utilized in a number of studies on the effect of glucose deprivation on sugar transport (Kalckar and Ullrey, this volume) and will not be discussed here.

5.1.2a. Galactosemia. A major incentive to study galactose metabolism is to develop a greater understanding of the human inherited metabolic disorder, galactosemia, due to a deficiency of galactose-1-phosphate uridyltransferase activity. The enzyme deficiency is expressed in skin fibroblasts from homozygous sufferers of the disease and such galactosemic cells will not grow in medium containing galactose as the sole hexose source under circumstances where normal and heterozygous cells will grow (Krooth and Weinberg, 1961). However, Russell and De Mars (1967) showed that cultured human fibroblasts from galactosemic patients could still metabolize galactose even if they could not grow on it. Hill (1976) investigated the effect of pH on the incorporation of galactose in a normal human cell line and in a galactosemic cell line and concluded that there was an alternative galactose-metabolizing pathway. Kuchka *et al.* (1981) used galactose-containing medium to select for galactose-positive cells from a population of diploid fibroblasts derived from a patient with galactosemia. The galactosemic cells died rapidly in mass culture in galactose medium and three lines were cloned from the surviving cells (Kuchka *et al.*, 1981). Two of the lines appeared to be revertants since they contained normal levels of galactose-1-phosphate uridyltransferase activity, but the other line was still enzyme deficient. There is thus a question to be resolved regarding possible alternative routes for galactose metabolism.

5.1.2b. Galactose Toxicity. Untreated galactosemic patients become mentally retarded and show signs of neuropathy possibly due to the accumulation of high concentrations of galactose or galactose-1-phosphate in the brain. Interestingly, when chickens are maintained on high-galactose diets, neurotoxicity develops. Giovanni *et al.* (1981) showed that treatment of human neuroblastoma cells with galactose caused an inhibition of active sodium channels possibly through an inhibition of excitoporin, a glycoprotein of molecular weight 200,000. Human galactosemia has also been shown to be associated with premature ovarian failure and a striking decrease in oocyte number has been observed in rats exposed, prenatally, to high levels of galactose (Chen *et al.*, 1981). These and other toxic effects of galactose (and other sugars) deserve more attention and may be most readily investigated in studies with cells *in vitro*.

5.1.2c. Isolation of Mutants. Whitfield *et al.* (1978) isolated variants of CHO cells resistant to 2-deoxygalactose in an attempt to obtain cells with a defect in galactose transport or metabolism. Twelve independent clones of resistant cells were isolated and were shown to contain decreased levels of galactokinase activity. The level of enzyme was related to resistance: cells with high resistance to 2-deoxygalactose had 1% or less of the enzyme activity observed in the parental cells, while cells with low resistance had 10 to 30% galactokinase activity. The "mutant" enzyme had a similar K_m for galactose, K_m for ATP, K_i for deoxygalactose, and thermomobility as the native enzyme, suggesting the possibility that the "mutant" was not affected at the structural gene locus, but rather at a regulatory site that determined the number of galactokinase molecules per cell.

Benn *et al.* (1981) showed that the galactose-1-phosphate uridyltransferase revertants isolated by Kuchka *et al.* (1981) contained enzymes that had kinetic properties slightly different from those of the native enzymes and concluded that structural gene changes were probably responsible for the reversion.

Sun *et al.* (1975) isolated mutants of Chinese hamster lung cells by selection for an inability to grow in galactose medium. The mutants were pleiotropically affected in carbohydrate metabolism and could not utilize exogenous galactose, mannose, fructose, galactose-1-phosphate, glucose-1-phosphate, or glucose-6-phosphate. No alteration could be demonstrated in galactose uptake, nor in any of the enzymes of the Leloir pathway. A significant reduction in the activities of phosphoglucomutase and NADP-dependent isocitrate dehydrogenase was shown. However, these changes did not appear to explain the inability of the cells to utilize carbohydrates.

Using a similar procedure to that described by Sun *et al.* (1975), other galactose-negative mutants have been isolated: Maiti *et al.* (1981) and Scheffler's group (Chapter 3) showed that respiration-deficient mutants had a phenotype that was also galactose-negative. These mutants, obtained in three separate laboratories, all appear to be related in the sense of being respiration-deficient mutants rather than galactose-metabolism mutants (Chapter 3). Their isolation emphasizes the essential role of oxidation in the growth of cells on nonglycolytic carbohydrates such as galactose. For further discussion see Chapter 3.

5.1.3. Mannose

The first step in mannose metabolism after its uptake is phosphorylation by hexokinase to mannose-6-phosphate and then a specific phosphomannose isomerase (EC 5.3.1.8) catalyzes the interconversion of mannose-6-phosphate to fructose-6-phosphate. The majority of reports suggest that mannose may be substituted for glucose without any detrimental effects and that it is as good a substrate as glucose.

Mannose uptake may be accomplished by a system different from that for glucose uptake (Faik and Morgan, 1984), but it is likely that mannose can also be transported on the glucose uptake system. A glycolysis-deficient mutant to which mannose is toxic appears to contain a second mannose uptake system, with a higher affinity for mannose than the normal glucose system (Faik and Morgan, 1984). That mannose is phosphorylated by hexokinase is suggested by the kinetic data, since mannose is essentially as good a substrate as glucose, and by the isolation of a hexokinase-deficient mutant of CHO cells (Faik, unpublished data). In this mutant, both the phosphorylation of mannose and of glucose are similarly reduced. Although glucose-positive, mannose-negative, naturally occurring variants have been reported (Dahl et al., 1976), the basis for this inability to metabolize mannose has not been adequately resolved.

Mannose toxicity has been reported in the honeybee where it is claimed to be due to the absence of phosphomannose isomerase activity and the consequent accumulation, to toxic levels, of mannose-6-phosphate (Sols et al., 1960). Mannose, but not glucose, is toxic to the PGI-deficient, PGK-deficient mutant of CHO cells (Section 3) where it is associated with the accumulation of fructose-1,6-bisphosphate (Morgan and Faik, 1981). Whether or not the toxicity is due to the accumulation of this hexose bisphosphate, to the concomitant depletion of ATP, or to some other reason, is not established. However, the glycolysis-deficient mutant (Faik and Morgan, 1977b) and the PGI-deficient mutant (Pouysségur et al., 1980) both accumulate glucose-6-phosphate while growing on glucose without any apparent detriment to the cells (Morgan and Faik, 1981).

5.1.4. Xylose

The pentose sugar xylose is generally regarded as a nonutilizable substitute for glucose. However, growth of chick cells on xylose has been reported (Demetrakopoulos and Amos, 1976; Demetrakopoulos et al., 1977). Xylose uptake may be accomplished via the glucose carrier (Romano and Connell, 1982a) or by a second, specific transport system that does not compete with glucose transport (Demetrakopoulos *et al.*, 1977). Indeed, there may be a major difference in the way in which avian and mammalian cell cultures handle xylose, the latter being unable to utilize it (Eagle *et al.*, 1958; Faik and Morgan, 1977a; Romano and Connell, 1982a; Kajstura and Korohoda, 1983). Pentose-utilizing variants of CHO-K1 cells (Faik and Morgan, 1977b) and of Novikoff hepatoma cells (Hoffee *et al.*, 1977) have been described that also grow on xylose and these are discussed below.

5.1.5. Ribose

The first, and essentially only unique step in ribose metabolism after its transport into the cell would be phosphorylation by a ribokinase (EC 2.7.1.15) to give ribose-5-phosphate, an intermediate of the pentose phosphate pathway (Fig. 2). It is thus somewhat surprising that cells, in general, do not grow in medium in which ribose is substituted for glucose; indeed, attention has been drawn to the toxicity of ribose (Demetrakopoulos and Amos, 1976; Morgan and Faik, 1981). Two laboratories have reported the isolation of pentose-utilizing variants (Faik and Morgan, 1976, 1977a; Hoffee *et al.*, 1977).

Faik and Morgan (1977b) isolated mutants of CHO cells that were able to grow in medium in which ribose replaced glucose (Fig. 3); the variants grew on xylose in addition to ribose, but did not grow on arabinose. The variants could be separated into two classes: one class differed from its wild-type parent only in its ability to utilize ribose and xylose while the other in addition no longer secreted lactic acid while growing on glucose (Fig. 4). This "lactic acid-negative" variant was later shown to be deficient in both PGI and PGK activity (Morgan and Faik, 1980). Both classes of ribose-positive cells were able to accumulate ribose from the medium, contained ribokinase activities, and contained all the enzymes of the classical pentose phosphate pathway. The activities of the pentose phosphate pathway enzymes are elevated about twofold in the ribose-positive variants (Faik and Morgan, unpublished data). Hybrids of wild-type and ribose-positive cells retain the ribose-positive phenotype (Morgan *et al.*, 1980, 1983b) and thus in these cells the ability to grow on ribose is a dominant characteristic.

Pentose-utilizing variants of Novikoff hepatoma cells have been isolated in medium in which glucose was replaced by either ribose, xylose, arabinose, or deoxyribose (Hoffee *et al.*, 1977; Jargiello, 1978). Two major classes of variants were found, distinguished with regard to their specificity for pentoses: one class could grow on ribose, xylose, or arabinose, while the other class grew only on ribose (Jargiello, 1978). Jargiello (1980) investigated ribokinase activity in 15 ribose-positive variants and found alterations in ribokinase activity in 13 of the variants. In 7 of the 10 nonspecific variants, which grew on ribose, xylose, or arabinose, the ribokinase appeared to be a cytoplasmic enzyme in contrast to the

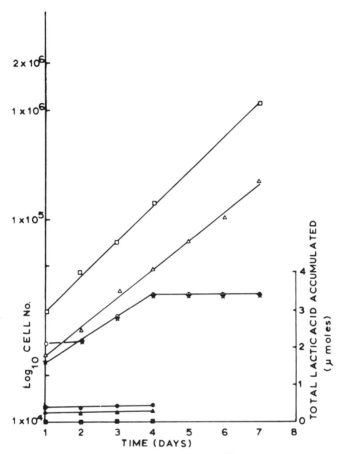

Figure 3. Growth of CHO-K1 (○), R12.19 (△), and R1.1.7 (□) on ribose (5 mM) and of CHO-K1 in the absence of an added energy source (☆). Accumulation of lactic acid: CHO-K1 (●), R12.19 (▲), and R1.1.7 (■). (From Faik and Morgan, 1977b.)

parental cell line, which contained mostly membrane-associated ribokinase activity. This change was associated with other changes in membrane-related functions. In the other ribose variants, the ribokinase activity was membrane-associated as in the parental cells, but the level was approximately sevenfold higher than in the parental cells. This alteration in ribokinase acitivity has been investigated further and shown to vary throughout the cell cycle, achieving high levels at approximately the midpoints of the S, G2, and M phases. The lowest levels of ribokinase were found between these phases (Jargiello, 1982). Ribokinase activity was not inducible by exposure of parental or variant cells to ribose.

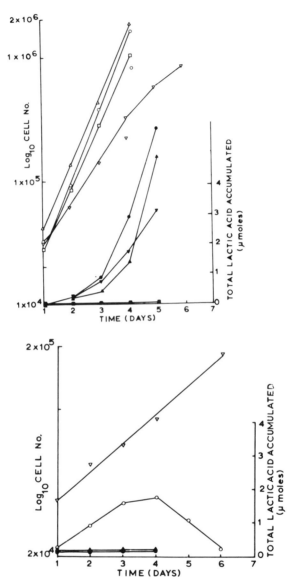

Figure 4. (Top) Growth of CHO-K1 (○), R12.19 (△), R1.1.7 (□), and L52 (▽) on glucose (5.0 mM). Accumulation of lactic acid: CHO-K1 (●), R12.19 (▲), R1.1.7 (■), and L52 (▼). (Bottom) Growth of CHO-K1 (○) and L52 (▽) on lactose (5.0 mM). Accumulation of lactic acid: CHO-K1 (●) and L52 (▲). (From Faik and Morgan, 1977b.)

Genetic analysis (Silnutzer and Jargiello, 1981) of hybrids obtained by fusion of various ribose-positive variants with pentose-negative parental hepatoma cells, showed an initial extinction of ribose utilization as the primary event with eventual reexpression of the ribose-positive phenotype. Karyological analysis did not allow any conclusions with respect to functions on specific chromosomes being involved. However, it was suggested that a differential rate in chromosomal segregation might explain the phenomonon (Silnutzer and Jargiello, 1981).

Although the specific role of ribose (and other pentoses) in the growth of cells in medium in which a pentose substitutes for glucose is not established, it is clear that the pentose only supplies carbon and is not a major energy source: the Chinese hamster, ribose-positive, mutants (Morgan and Faik, 1980) have been shown to require both pyruvate and glutamine when growing on ribose (Faik and Morgan, unpublished data).

5.1.6. Maltose

The ability of cells to grow on maltose should depend on the activity of enzymes capable of hydrolyzing it to glucose (Fig. 2). Serum contains abundant quantities of various enzymes including maltase (Fig. 5) and thus most cells grow in medium in which maltose is the sole carbohydrate source (Rheinwald and Green, 1974; Burns et al., 1976; Faik and Morgan, 1976; Bloch et al., 1977; Scannell and Morgan, 1980, 1982).

A variety of mouse and Chinese hamster cell lines were examined for their ability to grow in medium containing maltose or starch in place of glucose (Rheinwald and Green, 1974). Heat treatment of the serum caused inactivation of amylase and maltase and under these circumstances the cells were unable to grow. Such a system provides selection for variants able to express either amylase or maltase but attempts to isolate such variants were unsuccessful (Rheinwald and Green, 1974).

Burns et al. (1976) used a combination of glucose oxidase and catalase in order to convert glucose, formed by the hydrolysis of various compounds including maltose, to the nonutilizable gluconic acid. Under these conditions, CHO cells would grow on maltose at almost the same rate as on glucose, but would not grow on sucrose or trehalose; no attempt was made to isolate variants able to grow on sucrose or trehalose under these particular conditions (Burns et al., 1976).

In contrast, Scannell and Morgan (1980, 1982) showed that CHO cells would grow in glucose-free medium supplemented with maltose, trehalose, or starch only in the presence of active serum enzymes. Glucose-free medium containing glucose-free serum, heated to inactivate amylase, trehalase, and maltase, was unable to support the growth of the cells even in the presence of

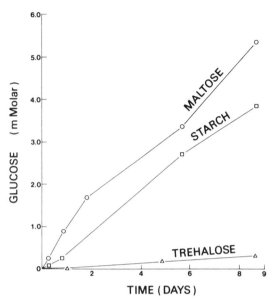

TIME (DAYS)

Figure 5. Hydrolysis of saccharides by serum enzymes. Culture dishes containing Ham's F12 medium supplemented with macroserum (10%) and either starch (1 mg/ml), maltose (5 mM), or trehalose (5 mM) were incubated at 37°C and the glucose content determined by glucostat. (From Scannell and Morgan, 1982.)

glucose. Cell growth was greatly decreased and cell morphology was altered, the cells being either rounded up or extremely stretched over the substratum. The alteration in cell morphology was suggested to be due to the destruction of serum factors necessary for the attachment and spreading of cells (Scannell and Morgan, 1982). Methods were devised to precoat tissue culture dishes such that the cells would attach. Under these conditions, the cells grew in medium containing heat-inactivated serum supplemented with glucose, but not in the equivalent medium supplemented with either maltose, trehalose, or starch (Scannell and Morgan, 1980).

The inability of cells to grow in medium containing heat-treated serum is in contrast to the findings of Bloch *et al.* (1977) who found that mouse 3T3 cells would grow in such medium without any difficulty. The basis for this difference could be in the different ways in which the workers treated the serum. Bloch *et al.* (1977) heated whole serum at 62°C for 1 to 2 hr whereas Scannell and Morgan (1980) first processed the serum to eliminate small molecules such as glucose by passage through a column of Sephadex G-50 (Faik and Morgan, 1977a) and then heated the resultant macroserum at 70°C for 30 min. The cell types investigated may also have contributed to such differences.

Using media containing heat-inactivated serum in which glucose was replaced with either starch or maltose, it was possible, after treatment with chemical mutagens, to isolate variants capable of utilizing starch or maltose in addition to glucose (Scannell and Morgan, 1982). Preliminary tests revealed high maltase activity in one maltose-positive clone and starch-gel electrophoresis of the enzyme showed that it had a similar mobility to maltase present in extracts of Chinese hamster gut. However, a similar band, albeit of much lower intensity, appeared in extracts of wild-type cells and might therefore represent the activity of a lysosomal enzyme. It remains to be seen if this higher activity is due to some type of gene amplification.

5.1.7. Lactose

Since there is no permeability barrier to lactose (Faik and Morgan, 1977a), cells should be able to grow on lactose provided they are capable of hydrolyzing it to glucose and galactose (Fig. 2). However, when examined, cells appear to be unable to grow on lactose (Burns et al., 1976; Faik and Morgan, 1977a). This is rather surprising since cells contain lysosomal acid, β-galactosidase activity, which should enable them to hydrolyze lactose; indeed, they also contain other disaccharidases, which should, for example, allow them to grow on disaccharides such as sucrose. However, the ability to grow on lactose (and sucrose) appears to require the expression of a specific enzyme, lactase, and is thus a differentiated function. Variants of CHO cells able to grow on lactose in addition to glucose have been isolated (Faik and Morgan, 1977b; Fig. 1). The biochemical analysis of these variants is incomplete but two facts have emerged. First, lactose does not substitute for glucose as the sole carbon and energy source: lactose is not converted to lactic acid (Fig. 4), and the lactose-positive cells are dependent on glutamine oxidation for their energy supply while growing on lactose (Faik, unpublished data). Second, there appears to be a three- to fourfold increase in lysosomal β-galactosidase activity in the lactose-positive cells and thus the ability of the cells to grow on lactose may be due to the overproduction of the lysosomal enzymes and possible leakage into the cytoplasm (Faik, unpublished data).

5.2. Gluconeogenesis

5.2.1. Lactate

For growth on lactate, cells would have to be capable of carrying out gluconeogenesis in order to form C_5 and C_6 compounds from the supplied C_3 compound. The only cells reported to grow on lactate are those isolated from CHO cells (Faik and Morgan, 1980). Lactate-positive cells were isolated after treating the wildtype cells with N-methyl-N'-nitrosoguanidine and selecting for cells able to survive and grow in medium in which lactate was substituted for glucose (Fig. 1). The lactate-positive cells grew as well as the wild-type cells on fructose and galactose, less well on glucose and mannose, and not at all (like the wild-type cells) on pyruvate or glycerol. The doubling time of the cells in glucose medium was 19 hr (wild-type cells, 12 hr) and in lactate was 37 hr. The lactate-positive cells utilized lactate from the medium and converted it into protein, nucleic acid, and lipid. However, it was not possible to show a consistent increase in the levels of gluconeogenic enzymes other than a two- to threefold increase in glucose-6-phosphatase activity in lactate-grown cells (Faik and Morgan, 1980).

5.2.2. Oxaloacetic Acid

Gluconeogenesis would also be an essential function in cells growing on the TCA cycle intermediate, oxaloacetate. Using a similar reasoning to that used by Faik and Morgan (1977a,b), Bertolotti (1977a) isolated variants of differentiated rat hepatoma cells (Faza 967) that proliferate in medium in which oxaloacetic acid is substituted for glucose. The variants express the gluconeogenic enzymes fructose bisphosphatase and phosphoenolpyruvate carboxykinase (PEPCK) and in addition express other liver-specific functions such as albumin synthesis. Differentiated variants were also isolated from dedifferentiated rat hepatoma cells by selection in glucose-free medium supplemented with oxaloacetic acid (Deschatrette *et al.*, 1980), and were shown to contain a number of liver-specific functions including PEPCK, fructose bisphosphatase, tyrosine aminotransferase, liver alcohol dehydrogenase, aldolase B, albumin, and alanine aminotransferase. On the assumption that only the two gluconeogenic enzymes were necessary for survival in glucose-free medium containing oxaloacetic acid, it was concluded that the change involved modification in activity of genes responsible for regulation of the entire group of liver functions rather than individual mutations.

Somatic cell hybrids of differentiated rat hepatoma cells and mouse lymphoblastoma cells (Bertolotti, 1977b) were unable to grow in glucose-free medium containing oxaloacetic acid when first isolated, i.e., the differentiated functions were extinguished. However, it was possible to isolate differentiated segregants after selection in glucose-free medium containing either oxaloacetic acid or dihydroxyacetone. The revertants selected in oxaloacetic acid medium expressed fructose bisphosphatase and PEPCK activity, whereas those selected in dihydroxyacetone medium contained fructose bisphosphatase and triokinase activity. The conclusion was drawn that reexpression of these enzymes is not necessarily under coordinate control.

These cells have proved an interesting system for studying the regulation of expression of differentiated functions (Moore and Weiss, 1982; Levilliers and

Weiss, 1983). Selection for cells no longer able to proliferate in glucose-free medium resulted in the isolation of clones that had lost all the differentiated functions; it was not possible, for example, to isolate clones that had only lost fructose bisphosphatase activity (Moore and Weiss, 1982). Complementation analysis of dedifferentiated variants isolated independently from hepatoma cells (Levilliers and Weiss, 1983) was attempted by fusing lines with one another and examining the hybrids for reexpression of the differentiated function. In no instance was reexpression found and it was concluded that complementation analysis, although suitable for the analysis of simple genetic lesions, was unsuitable for the analysis of a problem as complex as cell differentiation where simple one-step deficiencies might not occur or might be inherently unstable (Levilliers and Weiss, 1983). However, success has recently been reported and reexpression of hepatic functions (gluconeogenic enzymes, tyrosine aminotransferase activity, and α-fetoprotein) was found following fusion of mouse hepatoma and rat hepatoma cells (Cassio, 1984). Difficulty was encountered in obtaining such reexpression hybrids, the only ones isolated coming from hybrids that contained a greater than 1S complement of rat chromosomes.

These cells have also been used as a model system to study the effects that glucocorticoids have on differentiation. A series of dexamethasone-resistant variants of differentiated clones of the Faza 967 rat hepatoma cell line (Deschatrette and Weiss, 1974) have been studied to see if there was an associated decrease in the expression of those liver (differentiated) functions that are normally regulated by glucocorticoid hormones (Venetianer and Bosze, 1983). A variety of effects was noted, the most significant of which was that dexamethasone resistance could occur before the loss of any liver-specific function, i.e., dedifferentiation of the hepatoma cells was not a prerequisite for hormone resistance. Thus, growth sensitivity (resistance to dexamethasone) and the expression of differentiated functions were not controlled in a coordinated fashion.

The expression of liver-specific enzymes has also been studied in hybrids constructed between crosses of Faza 967 cells with non-liver-derived human and Chinese hamster cells (Kielty et al., 1981, 1982). The expression of some 13 liver-specific enzymes was demonstrated in the hepatoma cells and the expression of the enzymes in the hybrids was determined. None of the hepatoma–Chinese hamster hybrids expressed any liver-specific functions (Kielty et al., 1981). Some 21 independent rat hepatoma–human hybrids were isolated and examined. Seventeen hybrids continued to express some of the rat liver-specific enzymes and in some hybrids there was expression of human-specific enzymes. The only human enzymes of relevance to the present discussion that were definitely expressed were α-glycerophosphate dehydrogenase and pyruvate kinase. Some evidence was also obtained for the expression of human fructose-1,6-bisphosphatase.

It was not possible to determine which factors controlled the expression of these enzymes in any particular fusion, but it appeared that expression of a particular function might depend on the relative contribution of chromosomes by the different parents. The general picture that emerged was that in human–rodent crosses, in which rodent differentiated functions were expressed, human differentiated functions might also be expressed if the appropriate structural genes were retained. Hybrids that contained a relatively large number of human chromosomes (12 or more) and that also had a rat contribution approaching 2S expressed the most human liver-specific enzymes.

It is interesting to note that in these studies (Kielty *et al.*, 1981, 1982) the medium in which the hybrids were isolated was nonselective and contained glucose; it would be of considerable interest to know if the results would have been any different if the hybrids had been selected in glucose-free medium containing, for example, oxaloacetic acid in place of glucose.

5.2.3. *Phosphoenolpyruvate Carboxykinase*

PEPCK [GTP:oxaloacetate carboxylase (transphosphorylating), EC 4.1.1.32] is one of the key enzymes in gluconeogenesis and is essential for growth on compounds such as lactate or oxaloacetate (see above). Two apparently distinct forms of the enzyme are found in a variety of tissues; one is found predominantly in the cytosol and the other in the mitochondrion. The activity of the cytosolic enzyme appears to be altered by dietary and hormonal manipulation while the mitochondrial form appears to be noninducible. It has, however, been shown (Arinze *et al.*, 1978) that mitochondrial PEPCK in cultured human fibroblasts is induced by the addition of AMP or dexamethasone. The increases in enzyme activity were inhibited by cycloheximide and actinomycin D, suggesting that the increase was dependent on *de novo* protein synthesis. The activity of PEPCK has also been shown to be dependent on the constitution of the cell culture medium (Gunn *et al.*, 1975) but the nature of the active component in the medium that was essential was not determined. The potential role of PEPCK in glutaminolysis is discussed in Chapter 4.

6. OTHER EFFECTS OF CARBOHYDRATES

6.1. *Effects on Morphology*

Changes in the morphology of cells have been noted when cells are switched from glucose to other carbohydrates (Cox and Gesner, 1965; Amos *et al.*, 1976; Johnson and Schwartz, 1976; Jargiello, 1980; Morgan and Faik, 1981; Wolfrom *et al.*, 1983). The biochemical basis of these changes is not established, but it seems likely that they follow from an alteration in the glycoproteins and/or glycolipids of the cell membrane resulting from changes in the pool of sugar-derived precursors (Amos *et al.*, 1976). The ability reversibly, but stably, to alter cell morphology may provide a useful system for studying the regulation of glycoprotein and glycolipid biosynthesis and/or studying the effects of altering these processes on, for example, cell–cell interactions. Further consideration is, however, outside the scope of the present discussion.

6.2. Glucose-Regulated Proteins

A number of laboratories have reported the accumulation of two proteins of apparent molecular weight 90,000–95,000 and 70,000–73,000 in a variety of cells starved for glucose, defective in glycoprotein synthesis or following transformation by tumor viruses (Stone *et al.*, 1974; Isaka *et al.*, 1975; Pouysségur *et al.*, 1977; Lee, 1981). These proteins do not appear to be related to the so-called heat shock proteins (Lanks, 1983; Lin and Lee, 1984).

Glucose starvation appears to result in a specific induction of these two genes against a background of general growth inhibition (Lin and Lee, 1984), and restoration of the glucose concentration in the medium leads to a repression of transcription of the two genes. Expression of the genes is not sensitive to the removal of pyruvate or glutamine and appears to be a specific response to glucose deprivation. The possibility that they might encode proteins that catalyze functions essential to overcome the loss of glucose as a carbon and energy source was investigated by examining the induction of the PEPCK gene since its mRNA was of a similar size to the mRNAs encoded by a cDNA clone of a hamster glucose-regulated protein (Lin and Lee, 1984), but the expression of PEPCK was found to follow a very different pattern. The function of these proteins is still highly speculative, but is naturally of considerable interest.

Partial amino acid sequence data have been obtained from the two glucose-regulated proteins overproduced by a temperature-sensitive mutant of hamster cells (Lee *et al.*, 1984). The amino-terminal sequences of both proteins are relatively abundant in acidic amino acids, particularly aspartic and glutamic acid. The human and *Drosophila* heat shock proteins and the calcium-binding proteins are also acidic in nature and it has been suggested that a basic relationship may exist between these same sets of proteins under stress conditions.

Analysis of the gene for the glucose-regulated protein of molecular weight 75,000, its sensitivity to glucose concentration, together with the temperature-sensitive mutant gene, provides a model system for studying the regulation of gene expression (Attenello and Lee, 1984). This analysis has begun and interesting data have been reported on the regulation of a hybrid gene constructed from a rat glucose-regulated gene (molecular weight 78,000) and the bacterial phosphotransferase gene (*neo*) that confers resistance to the neomycin–kanamycin antibiotic G418. After transfection into hamster fibroblasts, the *neo*-transcripts can be induced to high levels by the absence of glucose and the hybrid gene can be regulated by temperature shifts (Attenello and Lee, 1984). Sequence analysis, deletion analysis, and site-specific mutagenesis of the genes should allow the control sequences of this fascinating system to be defined in detail.

6.3. Hypergravity

Perhaps one of the most unusual, but potentially far-reaching, reports on glucose metabolism emanates from the Space-Lab program. Analysis of lymphocytes from crew members of Soviet and American space missions showed a significant reduction of their reactivity after flight. Studies are now under way to explore the role of gravity in regulating cell proliferation (Tschopp and Cogoli, 1983). Glucose consumption was measured in a variety of cells including lymphocytes, HeLa cells, and chicken embryo fibroblasts. Tschopp and Cogoli found that cell growth was faster at $10g$ than at $1g$, but glucose consumption was similar. Thus, under g-stress conditions the cell appears to be capable of shifting to other metabolic pathways (perhaps glutamine oxidation). Under microgravity conditions, lymphocytes show a dramatic reduction in proliferation rate, reduced glucose consumption, and a strong increase in interferon secretion (Cogoli *et al.*, 1984). It remains to be seen what bearing these studies will have on future Space-Lab missions and whether or not, as suggested by Tschopp and Cogoli (1983), cells contain a ''gravity sensor'' that plays a role in cell proliferation and carbohydrate metabolism.

7. CONCLUDING REMARKS

We hope that we have demonstrated that the study of carbohydrate metabolism in animal cells in culture is still a fruitful field of research.

Fresh insights into the regulation of carbohydrate metabolism and its role in the control of cell proliferation will, we are convinced, come from further studies utilizing carbon sources other than glucose. The scope to penetrate the mysteries of differentiation by manipulating components of cell culture media with the aim of selecting for cell type or expression (or extinction) of cell functions is in its infancy. The ability to isolate well-characterized mutants together with the burgeoning power of molecular genetics, holds a promise of rich dividends to come. The analogy, with the progress of prokaryotic biochemical genetics, described elsewhere in this book, is striking and it is tempting to predict that an analysis of animal cell mutants will provide a deeper understanding of the regulation of carbohydrate metabolism and gene expression in animals.

ACKNOWLEDGMENTS. The authors are pleased to acknowledge the support of the Medical Research Council, the Nuffield Foundation, and the Wellcome Trust.

REFERENCES

Aithal, H. N., Walsh-Reitz, M. M., and Toback, F. G., 1983, Appearance of a cytosolic protein that stimulates glyceraldehyde-3-phosphate dehydrogenase activity during initiation of renal epithelial cell growth, *Proc. Natl. Acad. Sci. USA* **80:**2941–2945.

Amos, H., Leventhal, M., Chu, L., and Karnovsky, M. J., 1976, Modifications of mammalian cell surfaces induced by sugars: Scanning electron microscopy, *Cell* **7:**97–103.

Ardawi, M. S. M., and Newsholme, E. A., 1982, Maximum activities of some enzymes of glycolysis, the tricarboxylic acid cycle and ketone-body and glutamine utilisation pathways in lymphocytes of the rat, *Biochem. J.* **208:**743–748.

Arinze, I. J., Raghunathan, R., and Russell, J. D., 1978, Induction of mitochondrial phosphoenolpyruvate carboxykinase in cultured human fibroblasts, *Biochim. Biophys. Acta* **531:**792–804.

Attenello, J. W., and Lee, A. S., 1984, Regulation of a hybrid gene by glucose and temperature in hamster fibroblasts, *Science* **226:**187–190.

Avner, P., Dubois, P., Nicolas, J. F., Jakob, H., Gaillard, J., and Jacob, F., 1977, Mouse teratocarcinoma: Carbon source utilisation patterns for growth and in vitro differentiation, *Exp. Cell Res.* **105:**39–50.

Bailey, J. M., Gey, G. O., and Gey, M. K., 1959, The carbohydrate nutrition and metabolism of a strain of mammalian cells (MB III strain of mouse lymphoblasts) growing *in vitro*, *J. Biol. Chem.* **234:**1042–1047.

Benn, P. A., Kelley, R. I., Mellman, W. J., Amer, L., Boches, F. S., Markus, H. B., Nichols, W., and Hoffman, B., 1981, Reversion from deficiency of galactose 1-phosphate uridyltransferase (GALT) in an SV40-transformed human fibroblast line, *Somat. Cell Genet.* **7:**667–682.

Bertolotti, R., 1977a, A selective system for hepatoma cells producing gluconeogenic enzymes, *Somat. Cell Genet.* **3:**365–380.

Bertolotti, R., 1977b, Expression of differentiated functions in hepatoma cell hybrids: Selection in glucose-free media of segregated hybrid cells which re-express gluconeogenic enzymes, *Somat. Cell Genet.* **3:**579–602.

Bishayee, S., and Das, M., 1981, Aberrant energy metabolism in a variant epidermal growth factor receptor-negative fibroblastic cell line, *FEBS Lett.* **127:**237–240.

Bloch, R., Betschart, B., and Burger, M. M., 1977, Cell culture in serum depleted of glycosidases by heating, *Exp. Cell Res.* **104:**143–152.

Blomquist, C. H., Gregg, C. T., and Tobey, R. A., 1971, Enzyme and co-enzyme levels, oxygen uptake and lactate production in synchronised cultures of Chinese hamster cells, *Exp. Cell Res.* **66:**75–80.

Bruni, P., Faranraro, M., Vasta, V., and D'Alessandro, A., 1983, Increase of the glycolytic rate in human resting fibroblasts following serum stimulation: The possible role of the fructose 2,6-bisphosphate, *FEBS Lett.* **159:**39–42.

Burns, R. L., Rossenberger, P. G., and Klebe, R. J., 1976, Carbohydrate preferences of mammalian cells, *J. Cell. Physiol.* **88:**307–316.

Bustamante, E., and Pedersen, P. L., 1977, High aerobic glycolysis of rat hepatoma cells in culture: Role of mitochondrial hexokinase, *Proc. Natl. Acad. Sci. USA* **74:**3735–3739.

Bustamante, E., Morris, H. P., and Pedersen, P. L., 1981, Energy metabolism of tumour cells: Requirement for a hexokinase with a propensity for mitochondrial binding, *J. Biol. Chem.* **256:**8699–8704.

Cassio, D., 1984, Re-expression of hepatic functions in mouse hepatoma × rat hepatoma hybrids, *Differentiation* **26:**77–82.

Chen, Y. T., Mattison, D. R., Feigenbaum, L., Fukui, H., and Schulman, J. D., 1981, Reduction in oocyte number following prenatal exposure to a diet high in galactose, *Science* **214:**1145–1147.

Clayton, D. F., and Darnell, J. E., 1983, Changes in liver-specific compared to common gene transcription during primary culture of mouse hepatocytes, *Mol. Cell. Biol.* **3:**1552–1561.

Cogoli, A., Tschopp, A., and Fuchs-Bislin, P., 1984, Cell sensitivity to gravity, *Science* **225:**228–230.

Colby, C., and Romano, A. H., 1974, Phosphorylation but not transport of sugars is enhanced in virus-transformed mouse 3T3 cells, *J. Cell. Physiol.* **85:**15–24.

Cooper, J. A., Reiss, N. A., Schwartz, R. J., and Hunter, T., 1983, Three glycolytic enzymes are phosphorylated at tyrosine in cells transformed by Rous sarcoma virus, *Nature* **302:**218–223.

Cox, R. P., and Gesner, B. M., 1965, Effect of simple sugars on the morphology and growth pattern of mammalian cell cultures, *Proc. Natl. Acad. Sci. USA* **54:**1571–1579.

Dahl, R. H., and Morse, M. L., 1979, Differential metabolism of mannose by Chinese hamster cell lines, *Exp. Cell Res.* **121:**277–282.

Dahl, R. H., Morrissey, A., Puck, T. T., and Morse, M. L., 1976, Carbohydrate energy sources for Chinese hamster cells in culture, *Proc. Soc. Exp. Biol. Med.* **153:**251–253.

Demetrakopoulos, G. E., and Amos, H., 1976, D-Xylose and xylitol: Previously unrecognised sole carbon and energy sources for chick and mammalian cells, *Biochem. Biophys. Res. Commun.* **72:**1169–1176.

Demetrakopoulos, G. E., Gonzalez, F., Colofiore, J., and Amos, H., 1977, Growth of chick and mammalian cells on D-xylose, *Exp. Cell Res.* **106:**167–173.

Deschatrette, J., and Weiss, M. C., 1974, Characterisation of differentiated and dedifferentiated clones from a rat hepatoma, *Biochimie* **56:**1603–1611.

Deschatrette, J., Moore, E. E., Dubois, M., and Weiss, M. C., 1980, Dedifferentiated variants of a rat hepatoma: Reversion analysis, *Cell* **19:**1043–1051.

Diamond, I., Legg, A., Schneider, J. A., and Rozengurt, E., 1978, Glycolysis in quiescent cultures of 3T3 cells: Stimulation by serum, epidermal growth factor, and insulin in intact cells and persistence of the stimulation after cell homogenization, *J. Biol. Chem.* **253:**866–871.

D'Urso, M., Mareni, C., Toniolo, D., Piscopo, M., Schlessinger, D., and Luzzatto, L., 1983, Regulation of glucose 6-phosphate dehydrogenase expression in CHO–human fibroblast somatic cell hybrids, *Somat. Cell Genet.* **9:**429–443.

Eagle, H., Barban, S., Levy, M., and Schulze, H. O., 1958, The utilisation of carbohydrates by human cell cultures, *J. Biol. Chem.* **233:**551–558.

Emerman, J. T., Bartley, J. C., and Bissell, M. J., 1981, Glucose metabolite patterns as markers of functional differentiation in freshly isolated and cultured mouse mammary epithelial cells, *Exp. Cell Res.* **134:**241–250.

Faik, P., and Morgan, M. J., 1976, Carbohydrate metabolism in Chinese hamster cells, *Biochem. Soc. Trans.* **4:**1043–1045.

Faik, P., and Morgan, M. J., 1977a, A method of isolation of Chinese hamster cell variants with an altered ability to utilise carbohydrates, *Cell Biol. Int. Rep.* **1:**555–562.

Faik, P., and Morgan, M. J., 1977b, Properties of carbohydrate utilising variants of Chinese hamster cells, *Cell Biol. Int. Rep.* **1:**563–570.

Faik, P., and Morgan, M. J., 1980, The regulation of carbohydrate metabolism in animal cells: Isolation of variants able to utilise lactate, *Biochem. Soc. Trans.* **8:**632–633.

Faik, P., and Morgan, M. J., 1984, Regulation of hexose uptake in Chinese hamster ovary cells, *Biochem. Soc. Trans.* **12**:10.

Faik, P., Rawson, S., Walker, J. H., and Morgan, M. J., 1986, Introduction of human phosphoglycerate kinase (PGK) cDNA into a PGK-deficient line of Chinese hamster ovary cells, *Genet. Res.*, in press.

Fodge, D. W., and Rubin, H., 1973, Activation of phosphofructokinase by stimulants of cell multiplication, *Nature* **246**:181–183.

Fukushima, N., Cohen-Khallas, M., and Kalant, N., 1981, Galactose and glucose metabolism by cultured hepatocytes: Responsiveness to insulin and the effect of age, *Dev. Biol.* **84**:359–363.

Giovanni, M. Y., Kessel, D., and Gluck, M. C., 1981, Specific monosaccharide inhibition of active sodium channels in neuroblastoma cells, *Proc. Natl. Acad. Sci. USA* **78**:1250–1254.

Gregory, S. H., and Bose, S. K., 1977, Density-dependent changes in hexose transport, glycolytic enzyme levels and glycolytic rates, in uninfected and murine sarcoma virus-transformed rat kidney cells, *Exp. Cell Res.* **110**:387–397.

Gregory, S. H., and Bose, S. K., 1979, Glycolytic enzyme activities in malignant cells grown *in vitro* and *in vivo*, *Cancer Lett.* **7**:319–324.

Gunn, J. M., Shinozuka, H., and Williams, G. M., 1975, Enhancement of phenotypic expression in cultured malignant liver epithelial cells by a complex medium, *J. Cell. Physiol.* **87**:79–89.

Halban, P. A., Praz, G. A., and Wollheim, C. B., 1983, Abnormal glucose metabolism accompanies failure of glucose to stimulate insulin release from a rat pancreatic cell line (RINm5F), *Biochem. J.* **212**:439–443.

Harris, M., and Kutsky, P. B., 1953, Utilisation of added sugars by chick heart fibroblasts in dialysed media, *J. Cell. Comp. Physiol.* **42**:449–466.

Hers, H. G., and Van Schaftingen, E., 1982, Fructose 2,6-bisphosphate 2 years after its discovery, *Biochem. J.* **206**:1–12.

Hill, H. Z., 1976, The effect of pH on incorporation of galactose by a normal human cell line and cell lines from patients with defective galactose metabolism, *J. Cell. Physiol.* **87**:313–320.

Hoffee, P., Jargiello, P., Zaner, L., and Martin, J., 1977, Pentose utilising variants of Novikoff hepatoma cells: Modification of growth and morphological properties, *J. Cell. Physiol.* **91**:39–50.

Hutz, M. H., Michelson, A. M., Antonarakis, S. E., Orkin, S. H., and Kazazian, H. H., 1984, Restriction site polymorphism in the phosphoglycerate kinase gene on the X chromosome, *Hum. Genet.* **66**:217–219.

Isaka, T., Yoshida, M., Owada, M., and Toyoshima, K., 1975, Alterations in membrane polypeptides of chick embryo fibroblasts induced by transformation with avian sarcoma viruses, *Virology* **65**:226–237.

Jargiello, P., 1978, Pentose utilising variants of Novikoff hepatoma cells: Phenotypic characterisation, *Somat. Cell Genet.* **4**:647–660.

Jargiello, P., 1980, Multiple genetic changes determine ribose utilisation by Novikoff hepatoma cell variants, *Biochim. Biophys. Acta* **632**:507–516.

Jargiello, P., 1982, Altered expression of ribokinase activity in Novikoff hepatoma variants, *Biochim. Biophys. Acta* **698**:78–85.

Johnson, G. S., and Schwartz, J. P., 1976, Effects of sugars on the physiology of cultured fibroblasts, *Exp. Cell Res.* **97**:281–290.

Kajstura, J., and Korohoda, W., 1983, Significance of energy metabolism pathways for stimulation of DNA synthesis by cell migration and serum, *Eur. J. Cell Biol.* **31**:9–14.

Kielty, C. M., Povey, S., and Hopkinson, D. A., 1981, Regulation of expression of liver-specific enzymes. 1. Detection in mammalian tissues and cultured cells, *Ann. Hum. Genet.* **45**:341–356.

Kielty, C. M., Povey, S., and Hopkinson, D. A., 1982, Regulation of expression of liver-specific enzymes. 3. Further analysis of a series of rat hepatoma × human somatic cell hybrids, *Ann. Hum. Genet.* **46**:307–327.

Kletzien, R. F., and Perdue, J. F., 1974, Sugar transport in chick embryo fibroblasts, III. Evidence for host-transcriptional and host-translational regulation of transport following serum addition, *J. Biol. Chem.* **249**:3383–3387.

Krooth, R. S., and Weinberg, A. N., 1961, Studies on cell lines developed from the tissues of patients with galactosemia, *J. Exp. Med.* **113**:1155–1171.

Kuchka, M., Markus, H. B., and Mellman, W. J., 1981, Influence of hexose conditions on glutamine oxidation of SV40-transformed and diploid fibroblast human cell lines, *Biochem. Med.* **26**:356–364.

Landau, B. R., and Wood, H. G., 1983, The pentose cycle in animal tissues: Evidence for the classical and against the 'L-type' pathway, *Trends Biochem. Sci.* **8**:292–296.

Lanks, K. W., 1983, Metabolite regulation of heat shock protein levels, *Proc. Natl. Acad. Sci. USA* **80**:5325–5329.

Lazo, P. A., 1981, Amino acids and glucose utilisation by different metabolic pathways in ascites-tumour cells, *Eur. J. Biochem.* **117**:19–25.

Lee, A. S., 1981, The accumulation of three specific proteins related to glucose-regulated proteins in a temperature-sensitive hamster mutant cell line K12, *J. Cell. Physiol.* **106**:119–125.

Lee, A. S., Bell, J., and Ting, J., 1984, Biochemical characterization of the 94- and 78-kilodalton glucose-regulated proteins in hamster fibroblasts, *J. Biol. Chem.* **259**:4616–4621.

Levilliers, J., and Weiss, M. C., 1983, Differentiation is not restored in hybrids between independent variants of a rat hepatoma, *Somat. Cell Genet.* **9**:407–413.

Lin, A. Y., and Lee, A. S., 1984, Induction of two genes by glucose starvation in hamster fibroblasts, *Proc. Natl. Acad. Sci. USA* **81**:988–992.

McGowan, J. A., Russell, W. E., and Bucher, L. R., 1984, Hepatocyte DNA replication: Effect of nutrients and intermediary metabolites, *Fed. Proc.* **43**:131–133.

McKeehan, W. L., 1984, Control of normal and transformed cell proliferation by growth factor–nutrient interactions, *Fed. Proc.* **43**:113–115.

McKeehan, W. L., McKeehan, K. A., and Calkins, D., 1982, Epidermal growth factor modifies Ca^{2+}, Mg^{2+} and 2-oxocarboxylic acid, but not K^+ and phosphate ion requirement for multiplication of human fibroblasts, *Exp. Cell Res.* **140**:25–30.

Maiti, I. B., Comlan de Souza, A., and Thirion, J. P., 1981, Biochemical and genetic characterization of respiration-deficient mutants of Chinese hamster cells with a Gal^- phenotype, *Somat. Cell Genet.* **7**:567–582.

Malaisse, W. J., Malaisse-Lagae, F., Sener, A., Van Schaftingen, E., and Hers, H. G., 1981, Is the glucose-induced stimulation of glycolysis in pancreatic islets attributable to activation of phosphofructokinase by fructose 2,6-bisphosphate?, *FEBS Lett.* **125**:217–219.

Meglasson, M. D., and Matschinsky, F. M., 1984, New perspectives on pancreatic islet glucokinase, *Am. J. Physiol.* **246**:E1–E13.

Melnykovych, G. and Bishop, C. F., 1972, Utilisation of hexoses and synthesis of glycogen in two strains of HeLa cells, *In Vitro* **7**:397–405.

Michelson, A. M., Markham, A. F., and Orkin, S. H., 1983, Isolation and DNA sequence of a full-length cDNA clone for human X chromosome-encoded phosphoglycerate kinase, *Proc. Natl. Acad. Sci. USA* **80**:427–476.

Miwa, S., Nakashima, K., Oda, S., Ogawa, H., Nakafuji, H., Aríma, M., Okuna, T., Nakashima, T., 1972, Phosphoglycerate kinase (PGK) deficiency hereditary nonspherocytic hemolytic anemia: Report of a case found in a Japanese family, *Acta Haematol. Jpn.* **35**:570–574.

Moore, E. E., and Weiss, M. C., 1982, Selective isolation of stable and unstable dedifferentiated variants from a rat hepatoma cell line, *J. Cell. Physiol.* **111**:1–8.

Morgan, J., and Morton, H., 1960, Carbohydrate utilisation by chick embryonic heart cultures, *Can. J. Biochem. Physiol.* **35**:69–78.

Morgan, M. J., 1981, The pentose phosphate pathway: Evidence for the indispensable role of glucose-phosphate isomerase, *FEBS Lett.* **130**:124–126.

Morgan, M. J., and Faik, P., 1980, The regulation of carbohydrate metabolism in animal cells: Isolation of a glycolytic variant of Chinese hamster ovary cells, *Cell Biol. Int. Rep.* **4**:121–127.

Morgan, M. J., and Faik, P., 1981, Carbohydrate metabolism in cultured animal cells, *Biosci. Rep.* **1**:669–686.

Morgan, M. J., Faik, P., and Walker, S. W., 1980, The regulation of carbohydrate metabolism in animal cells: Isolation of a glycolytic variant, *Biochem. Soc. Trans.* **8**:631–632.

Morgan, M. J., Bowness, K. M., and Faik, P., 1981, Regulation of carbohydrate metabolism in cultured mammalian cells: Energy provision in a glycolytic mutant, *Biosci. Rep.* **1**:811–817.

Morgan, M. J., Bowness, K. M., and Faik, P., 1983a, Energy provision in Chinese hamster ovary cells, *Biochem. Soc. Trans.* **11**:725–726.

Morgan, M. J., Faik, P., and Calvert, J., 1983b, Genetics of carbohydrate metabolism in animal cells, *Genet. Res.* **41**:307.

Nepokroeff, C. M., Lakshmann, M. R., Ness, G. C., Muesing, R. A., Kleinsek, D. A., and Porter, J. W., 1974, Co-ordinate control of rat liver lipogenic enzymes by insulin. *Arch. Biochem. Biophys.* **162**:340–344.

Ovadi, J., and Keleti, T., 1978, Kinetic evidence for interaction between aldolase and D-glyceraldehyde-3-phosphate dehydrogenase, *Eur. J. Biochem.* **85**:157–161.

Pauwels, P. J., Opperdoes, F. R., and Trouet, A., 1984, Effect of oxygen and glucose availability on the glycolytic rate in neuroblastoma cells under different conditions of culture, *Neurochem. Int.* **6**:467–473.

Piechaczyk, M., Blanchard, J. M., Riaad-El Sabouty, S., Dani, C., Marty, L, and Jeanteur, P., 1984, Unusual abundance of vertebrate 3-phosphate dehydrogenase pseudogenes, *Nature* **312**:469–471.

Pinto, M., Appay, M. D., Simon-Assman, P., Chevalier, G., Dracopoli, N., Fogh, J., and Zweibaum, A., 1982, Enterocytic differentiation of cultured human colon cancer cells by replacement of glucose by galactose in the medium, *Biol. Cell.* **44**:193–196.

Pouysségur, J., Shiu, R. P. C., and Pastan, I., 1977, Induction of two transformation-sensitive membrane polypeptides in normal fibroblasts by a block in glycoprotein synthesis or glucose deprivation, *Cell* **11**:941–947.

Pouysségur, J., Franchi, A., Salomon, J.-C., and Silvestre, P., 1980, Isolation of a Chinese hamster fibroblast mutant defective in hexose transport and aerobic glycolysis: Its use to dissect the malignant phenotype, *Proc. Natl. Acad. Sci. USA* **77**:2698–2701.

Racker, E., 1976, Why do tumour cells have a high aerobic glycolysis?, *J. Cell. Physiol.* **89**:697–700.

Racker, E., 1984, Resolution and reconstitution of biological pathways from 1919 to 1984, *Fed. Proc.* **42**:2899–2909.

Racker, E., Johnson, J. H., and Blackwell, M. D., 1983, The role of ATPase in glycolysis of Ehrlich ascites tumour cells, *J. Biol. Chem.* **258**:3702–3705.

Reitzer, L. J., Wise, B. M., and Kennel, D., 1980, The pentose cycle: Control and essential function in HeLa cell nucleic acid synthesis, *J. Biol. Chem.* **255**:5616–5626.

Rheinwald, J. G., and Green, H., 1974, Growth of cultured mammalian cells on secondary glucose sources, *Cell* **2**:287–293.

Romano, A. H., and Connell, N. D., 1982a, 6-Deoxy-D-glucose and D-xylose, analogs for the study of D-glucose transport by mouse 3T3 cells, *J. Cell. Physiol.* **111**:77–82.

Romano, A. H., and Connell, N. D., 1982b, Effect of glucose uptake on growth rate of mouse 3T3 cells, *J. Cell. Physiol.* **111**:195–200.

Rosenstraus, M., and Chasin, L. A., 1975, Isolation of mammalian cell mutants deficient in glucose 6-phosphate dehydrogenase activity: Linkage to hypoxanthine phosphoribosyltransferase, *Proc. Natl. Acad. Sci. USA* **72**:493–497.

Russell, J. D., and De Mars, R., 1967, UDP glucose: α-D-galactose-1-phosphate uridyl transferase activity in cultured human fibroblasts, *Biochem. Genet.* **1**:11–24.

Scannell, J., and Morgan, M. J., 1980, The regulation of carbohydrate metabolism in animal cells: Growth on starch and maltose, *Biochem. Soc. Trans.* **8**:633–634.

Scannell, J., and Morgan, M. J., 1982, The regulation of carbohydrate metabolism in animal cells: Isolation of starch- and maltose-utilising variants, *Biosci. Rep.* **2**:99–106.

Schneider, J. A., Diamond, I., and Rozengurt, E., 1978, Glycolysis in quiescent cultures of 3T3 cells: Addition of serum, epidermal growth factor, and insulin increases the activity of phospho-fructokinase in a protein synthesis-indepedent manner, *J. Biol. Chem.* **253**:872–877.

Schwartz, J. P., and Johnson, G. S., 1976, Metabolic effects of glucose deprivation and of various sugars in normal and transformed fibroblast cell lines, *Arch. Biochem. Biophys.* **173**:237–245.

Sens, D. A., Hochstadt, B., and Amos, H., 1982, Effects of pyruvate on the growth of normal and transformed hamster embryo fibroblasts, *J. Cell. Physiol.* **110**:329–335.

Silnutzer, J., and Jargiello, P., 1981, Extinction and expression of the ribose-positive phenotype in hybrid Novikoff hepatoma cells, *Somat. Cell Genet.* **7**:119–131.

Singer-Sam, J., Simmer, R. L., Keith, D. M., Shirley, L., Teplitz, M., Itakura, K., Gartler, S. M., and Riggs, A. D., 1983, Isolation of a cDNA clone for human X-linked 3-phosphoglycerate kinase by use of a mixture of synthetic oligodeoxyribonucleotides as a detection probe, *Proc. Natl. Acad. Sci. USA* **80**:802–806.

Singh, M., Singh, V. N., August, G. T., and Horecker, B. L., 1974a, Alterations in glucose metabolism in chick embryo cells transformed by Rous sarcoma virus: Transformation-specific changes in the activities of key enzymes of the glycolytic and hexose monophosphate shunt pathways, *Arch. Biochem. Biophys.* **165**:240–246.

Singh, M., Singh, V. N., August, G. T., and Horecker, B. L., 1974b, Alterations in glucose metabolism in chick embryo cells transformed by Rous sarcoma virus: Intracellular levels of glycolytic intermediates, *Proc. Natl. Acad. Sci. USA* **71**:4129–4132.

Smith, M. L., and Buchanan, J. M., 1979, Nucleotide and pentose synthesis after serum-stimulation of resting 3T6 fibroblasts, *J. Cell. Physiol.* **101**:293–310.

Sols, A., Cadenas, E., and Alvarado, F., 1960, Enzymatic basis of mannose toxicity in honey bees, *Science* **131**:297–298.

Stern, E. S., and Krooth, R. S., 1975, Studies on the regulation of the three enzymes of the Leloir pathway in cultured mammalian cells: Effect of substitution of galactose for glucose as the sole hexose in the medium in human diploid cell strains and in a rat hepatoma line, *J. Cell. Physiol.* **86**:91–103.

Stone, E. M., Rothblum, K. N., and Schwartz, R. J., 1985, Intron-dependent evolution of chicken glyceraldehyde phosphate dehydrogenase gene, *Nature* **313**:498–500.

Stone, K. R., Smith, R. E., and Joklik, W. K., 1974, Changes in membrane polypeptides that occur when chick embryo fibroblasts and NRK cells are transformed with avian sarcoma viruses, *Virology* **58**:86–100.

Sun, N. C., Chang, C. C., and Chu, E. H. Y., 1975, Mutant hamster cells exhibiting a pleiotropic effect on carbohydrate metabolism, *Proc. Natl. Acad. Sci. USA* **72**:469–473.

Tschopp, A., and Cogoli, A., 1983, Hypergravity promotes cell proliferation, *Experientia* **12**:1323–1329.

Usanga, E. A., and Luzzatto, L., 1985, Adaption of *Plasmodium falciparum* to glucose 6-phosphate dehydrogenase-deficient host red cells by production of parasite-encoded enzyme, *Nature* **313**:793–795.

Venetianer, A., and Bosze, Z., 1983, Expression of differentiated functions in dexamethasone-resistant hepatoma cells, *Differentiation* **25**:70–78.

Vozdev, V. A., 1976, Role of the pentose phosphate pathway in metabolism of *D. melanogaster* elucidated by mutations affecting glucose 6-phosphate and 6-phosphate gluconate dehydrogenase, *FEBS Lett.* **64**:85–88.

Wagner, K. R., Kauffman, F. C., and Max, S. R., 1978, The pentose phosphate pathway in regenerating skeletal muscle, *Biochem. J.* **170**:17–22.

Walker, D. G., 1966, The nature and function of hexokinases in animal tissues, in: *Essays in Biochemistry, Vol. 2* (P. N. Campbell and G. D. Greville, eds.), pp. 33–67, Academic Press, New York.

Wang, T., Marquardt, C., and Foker, J., 1976, Aerobic glycolysis during lymphocyte proliferation, *Nature* **261**:702–705.

Wang, T., Foker, J. E., and Tsai, M. Y., 1980, The shift of an increase in phosphofructokinase activity from protein synthesis-dependent to -independent mode during concanavalin A induced lymphocyte proliferation, *Biochem. Biophys. Res. Commun.* **95**:13–19.

Webber, M. J., Evans, P. K., Johnson, M. A., McNair, T. S., Nakamura, K. D., and Salter, D. W., 1984, Transport of potassium, amino acids, and glucose in cells transformed by Rous sarcoma virus, *Fed. Proc.* **43**:107–112.

Whitfield, C. D., Buchsbaum, B., Bostedor, R., and Chu, E. H. Y., 1978, Inverse relationship between galactokinase activity and 2-deoxygalactose resistance in Chinese hamster ovary cells, *Somat. Cell Genet.* **4**:699–713.

Williams, J. F., 1980, A critical examination of the evidence for the reactions of the pentose pathway in animal tissue, *Trends Biochem. Sci.* **5**:315–320.

Williams, J. F., Arora, K. K., and Longenecker, J. F., 1983, The F-pentose cycle doesn't have the answers for liver tissue, *Trends Biochem. Sci.* **8**:275–277.

Wohlhueter, R. M., and Plagemann, P. G. W., 1981, Hexose transport and phosphorylation by Novikoff rat hepatoma cells as a function of extracellular pH, *J. Biol. Chem.* **256**:869–875.

Wolfrom, C., Loriette, C., Polini, G., Delhotal, B., Lemonnier, F., and Gautier, M., 1983, Comparative effects of glucose and fructose on growth and morphological aspects of cultured skin fibroblasts, *Exp. Cell Res.* **149**:535–546.

Yoshida, A., and Miwa, S., 1974, Characterisation of a phosphoglycerate kinase variant associated with haemolytic anaemia, *Am. J. Hum. Genet.* **26**:378–384.

Ziegler, M. L., and Davidson, R. L., 1979, The effect of hexose on chloramphenicol sensitivity and resistance in Chinese hamster cells, *J. Cell. Physiol.* **98**:627–637.

3

Biochemical Genetics of Respiration-Deficient Mutants of Animal Cells

IMMO E. SCHEFFLER

1. INTRODUCTION

Given an abundant supply of glucose and oxygen, mammalian cells have a capacity for energy production (ATP synthesis) that far exceeds the need of the cell. The control of energy metabolism in normal cells is exquisite, and it has presented a fascinating challenge to several generations of biochemists. On the one hand, ATP is generated only when it is needed, but a viable cell also maintains its energy charge at a constant level (Atkinson, 1968, 1977).

Glycolysis and oxidative phosphorylation have long been known to contribute to the overall energy metabolism, and experiments by Pasteur in 1861 already hinted strongly at a regulatory coupling of the two processes. Explaining and hypothesizing on the Pasteur effect became a favorite preoccupation of leading biochemists (see Racker, 1976). The issue increased in importance by the discovery in Warburg's laboratory in 1926 that tumor cells appeared to differ from normal cells in the relative contribution of glycolysis and oxidative phosphorylation to the overall energy demands of the cells. This Warburg effect more specifically referred to the increased conversion of glucose to lactic acid under aerobic conditions, and it became associated with the notion that tumor cells revert to the more "primitive" pathway of energy generation. While the observation has been repeated in many laboratories, its interpretation has remained controversial to this day (Gregg, 1972; Racker, 1972; Pedersen, 1978; Racker *et*

IMMO E. SCHEFFLER • Department of Biology, University of California at San Diego, La Jolla, California 92093.

al., 1983). There is first of all a problem that can be stated in biochemical terms: Are the mitochondria in tumor cells defective? What is the defect? Which are the rate-controlling enzymes in glycolysis and how is their activity regulated by allosteric mechanisms and small molecules? Much more controversial has been the discussion of the relationship of the Warburg effect to the cancer problem: What is the cause, and what is simply a secondary effect, of transformation? Are genetic mechanisms (mutations in structural genes) involved, or are we dealing with epigenetic phenomena (changes in the level of gene expression)? A speculation in terms of some of the most recent developments would probably include changes in the expression of oncogenes, altered levels of activity of oncogene products such as specific kinases, and shifts in the balance of glycolysis and respiration through the modulation of key enzyme activities by phosphorylation and perhaps other mechanisms. Explorations in this direction are already under way (Cooper *et al.*, 1983).

Thus, while the transformed state has been linked for some time with altered or defective mitochondrial activity, proof for specific mutations or specific biochemical defects has not been found. On the other hand, one may ask the following question: What would be the properties of a cell in which oxidative phosphorylation has been deliberately blocked by a mutation?

After the lessons learned from the study of prokaryotes and lower eukaryotes, such as yeast, it was clear that the availability of mammalian cell mutants with defects in mitochondrial functions would be highly desirable. Although relatively specific inhibitors could be employed, single mutations would be a more elegant and less ambiguous means of blocking certain functions or pathways. The existence of mutants affected in mitochondrial functions would help to focus attention on the genes, on the regulation of their expression in the assembly of functional complexes of the electron transport chain, and on their localization on chromosomes in the nucleus or on the mitochondrial genome. The study of such mutants would therefore address not only questions related to bioenergetics and metabolism, but ultimately it would deal with the very interesting problem of the interaction of the nuclear and mitochondrial genomes and the mechanism of assembly of mitochondria.

It is easy, particularly with the benefit of hindsight, to formulate such sweeping and profound questions. However, at the beginning we were faced with an uncertainty and a practical problem. Yeast cells had been known to grow under anaerobic conditions in the presence of the appropriate carbon source, and yeast mutants had been selected with severe defects in respiration and even without any functional mitochondria (e.g., Dujon *et al.*, 1977). It appeared, in contrast, that mammalian cells in tissue culture could not be grown under anaerobic conditions (McLimans, 1972), and this observation suggested that oxidative phosphorylation might not be indispensable in these cells.

A second set of problems one might have anticipated in the search for respiration-deficient mammalian cell mutants is the following: (1) the cells are at least pseudodiploid, and the frequency of homozygous recessive mutants could be predicted to be extremely low; (2) protocols for selection of such mutants were not available, and would at any rate have been expected to be relatively inefficient, making the isolation of very rare mutuants ($<$ per 10^7) extremely difficult.

It is evident that in 1971 the proposal to isolate respiration-deficient Chinese hamster fibroblasts was far from our minds, and it would probably have been met with justifiable skepticism if it had been made. Serendipity played a large part in our acquisition of the first such mutant, and its characterization led to the development of an enrichment scheme that yielded additional mutants with this general phenotype.

2. THE SELECTION OF RESPIRATION-DEFICIENT MAMMALIAN CELL MUTANTS

2.1. Characterization of the First Mutant

In 1971 the selection for temperature-sensitive mutants of Chinese hamster lung fibroblasts by a "BUdR suicide" technique led to the isolation of a clone* that appeared unable to proliferate at the nonpermissive temperature of 40°C (Scheffler, 1974). During an attempted experiment employing L-15 medium to maintain better pH control over the temperature range of interest (34–40°C), the cells were found to be unable to grow even at the permissive temperature (34°C). Since L-15 medium (Leibovitz, 1963) is made free of bicarbonate to maintain physiological pH in the laboratory atmosphere, our attention became focused on the role of $CO_2/H_2CO_3/HCO_3^-$ in tissue culture medium. A series of experiments with various media with CO_2/bicarbonate buffering or with the bicarbonate replaced by NaCl led us to the following conclusions: (1) Normal Chinese hamster fibroblasts could grow in media with only traces of CO_2/HCO_3^- from the laboratory atmosphere, as long as a purine and a pyrimidine were included in the medium. Undialyzed serum may obscure this requirement for uridine and hypoxanthine, as in the case of L-15 medium, but when dialyzed serum was used, the requirement became evident. (2) The mutant Chinese hamster cell line

---

*In the earlier publications this clone was designated auxCO$_2$. In later publications the name was changed to CCL16-B2 (res⁻); this and another mutant [CCL16-B9 (SDH⁻)] were derived from Chinese hamster lung fibroblasts CCL16 of the American Type Culture Collection. Other mutants, e.g., V79-G7 (res⁻), were derived from the Chinese hamster cell line V79.

(CCL16-B2) was absolutely dependent on exogenous CO_2/HCO_3^-, even in the presence of the purine and pyrimidine (Fig. 1). It appeared that we had rediscovered an observation by Chang *et al.* (1961) and Geyer and Neimark (1958) that greater than 95% of the CO_2 fixed by mammalian cells in culture appears in

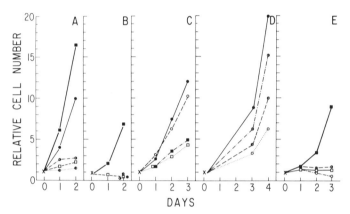

Figure 1. (A) Proliferation of CCL39 cells in MEM without bicarbonate. 10^5 cells were plated on day -1 in normal MEM in 5-cm NUNC plates. On day 0 cells in duplicate plates were counted, and media changes were made in all other plates. Cells in duplicate plates were counted on the two following days; ■, control; □, MEM without bicarbonate; ◑, as for □ but with 10 μ/ml uridine added; ◐, as for □ but with 13.6 μg/ml hypoxanthine added; ●, as for □ but with both uridine and hypoxanthine added. All media contained 10% fetal calf serum and 0.02 M Hepes, pH 7.5.
(B) Proliferation of auxCO₂ in MEM without bicarbonate. The experiment was identical to that described for CCL39 in (A). The points for □, ◑, ◐, and ● all fall on the same curve within experimental uncertainty.
(C) Multiplication of CCL39 and CCL16 cells in L-15 medium, with 10% dialyzed fetal calf serum, 10 μg/ml uridine, 13.6 μg/ml hypoxanthine, 0.02 M Hepes, pH ~ 7.2, 10^5 cells of each cell line were plated in 5 ml of medium in 5-cm NUNC plates, and at the indicated times cells in one plate were counted with the Coulter counter. To half of the plates glucose was added at a concentration equal to that of glucose in MEM. ○ and ●, CCL39 cells without and with glucose, respectively; □ and ■, CCL16 cells without and with glucose.
(D) Growth of CCL16 cells in L-15 medium at different pHs. The medium was buffered with 0.02 M Tricine and the pH was adjusted to the values indicated below by means of 0.3 M NaOH. The medium also contained 10 μ/ml uridine and 13.6 μg/ml hypoxanthine. 10^5 cells were plated in 5-cm Falcon plates, and the total cell number per plate was determined after 3 and 4 days with a Coulter counter. These numbers were then normalized with respect to the original input. ○, pH 6.7; ◐, pH 7.1; ◑, pH 7.5; ●, pH 7.9.
(E) Multiplication of auxCO₂ at pH 8.0 in the presence and absence of bicarbonate. 10^5 cells were plated in MEM in 5-cm NUNC plates and the medium was changed the following day. ■, control with 0.02 M Hepes; □, MEM with NaHCO₃ replaced by NaCl, 0.02 M Hepes, 10 μg/ml uridine, 13.6 μg/ml hypoxanthine, pH adjusted to 8.0 with NaOH; ○, L-15 medium, 0.02 M Hepes, 10 μg/ml uridine, 13.6 μg/ml hypoxanthine, pH 8.0; ◑, L-15 medium, as for ○ but with 2 μg/ml glucose added.

the nucleic acids. In the absence of high levels of $CO_2/HCO_3{}^-$ in the medium, but in the presence of uridine and hypoxanthine, wild-type cells were self-sufficient, presumably due to the intracellular generation of CO_2 from respiration, but it came to be suspected that the mutant was defective in CO_2 production from the citric acid cycle. This conclusion was confirmed by experiments showing that various ^{14}C-labeled sugars, pyruvate, and lactate were not oxidized to $^{14}CO_2$ by the mutant cells, in contrast to wild-type cells. The difference was striking and of several orders of magnitude (Fig. 2).

A follow-up of these studies investigated the fate of several additional labeled carbon compounds, that were expected to yield $^{14}CO_2$ from reactions involving the citric acid cycle (Fig. 3; DeFrancesco et al., 1975). The results could be interpreted in terms of a fairly specific and effective block between α-ketoglutarate and succinyl CoA in the cycle: carbons from succinate, aspartate, or asparagine were evolved as CO_2 at easily measurable rates in both mutant and wild-type cells, but neither glutamate, glutamine, nor compounds entering the cycle via acetyl CoA yielded CO_2 in the mutant cells. Aspartate could be converted to glutamate, but glutamate could not be converted to aspartate in these mutants (DeFrancesco et al., 1975). This latter observation became very important in the subsequent development of an enrichment scheme for selecting additional mutants. For normal cells in tissue culture, aspartate and asparagine are among the so-called nonessential amino acids, because the cells can synthesize these from other precursors provided in the medium. In fibroblasts, it should be noted, oxaloacetate cannot be derived from a direct carboxylation reaction involving pyruvate or phosphoenolpyruvate, and a functioning citric acid cycle is

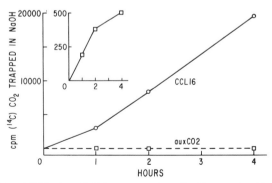

Figure 2. Formation of $^{14}CO_2$ from [2-^{14}C]pyruvate. 3×10^5 cells were plated in each of a series of 25-ml Erlenmeyer flasks in 3 ml of MEM. After 1 day there were 5.1×10^5 cells of CCL16 per flask (0.36 ± 0.03 mg total protein), and 4.2×10^5 cells of auxCO2 (0.34 ± 0.03 mg total protein). The cells were incubated in TD buffer containing 1 μCi/ml [2-^{14}C]pyruvate (specific activity 8.2 mCi/mmole). The inset shows the result for auxCO2 on a different scale.

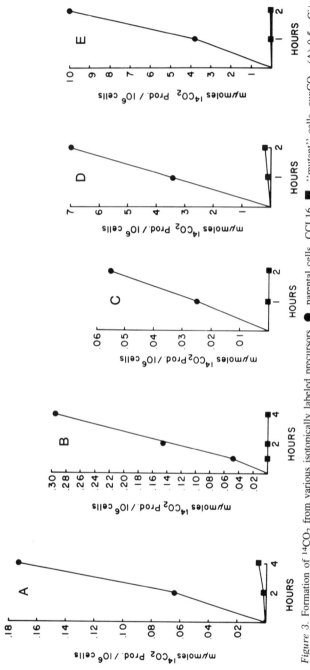

Figure 3. Formation of $^{14}CO_2$ from various isotopically labeled precursors. ●, parental cells, CCL16, ■, "mutant" cells, auxCO₂. (A) 0.5 μCi/ml [1-^{14}C]acetate (55 mCi/mmole); (B) 0.1 μCi/ml [U-^{14}C]palmitic acid (720 mCi/mmole); (C) 0.1 μCi/ml [3-^{14}C]-β-hydroxybutyrate (5.05 mCi/mmole); (D) 0.5 μCi/ml [1-^{14}C]-DL-glutamate (19.2 mCi/mmole); (E) 0.5 μCi/ml [5-^{14}C]-DL-glutamate (4.17 mCi/mmole). Absolute amounts of $^{14}CO_2$ were calculated assuming the specific activity of the CO_2 to be identical to the specific activity of a single carbon of the substrate; for acetate, glutamate, and β-hydroxybutyrate, this is the same as the specific activity of the precursor, but for palmitic acid it is 1/16th of the specific activity of the whole molecule. The amount of $^{14}CO_2$ was then normalized with respect to 10^6 cells.

necessary for aspartate synthesis. *A priori,* one would also expect glutamate and glutamine to be among the nonessential amino acids of cells in culture, but as discussed briefly below and extensively elsewhere in this volume (see also McKeehan, 1982), these five-carbon amino acids have long been known to be required in tissue culture medium. The striking observation made by us at that time was that our mutant required asparagine in the medium. For reasons not very clear to us, this requirement was met less well by aspartate (DeFrancesco *et al.,* 1975).

As a next step in the characterization of the phenotype of the mutant, we showed that the rate of oxygen consumption by whole cells was at least one order of magnitude less in the mutant cells (DeFrancesco *et al.,* 1975). Thus, the mutant cells were characterized as the first respiration-deficient mammalian cells in culture, with an inactive citric acid cycle, and as a consequence a requirement for asparagine and HCO_3^- in the medium. Glycolysis had become the major source of metabolic energy as indicated by the facts that (1) these cells acidified the medium at a dramatic rate and (2) the cells were very sensitive to glucose concentrations and unable to survive in medium with glucose replaced by galactose (see below).

It remained to establish and verify the precise step of the block in the mutant. To our initial surprise, a direct assay of α-ketoglutarate dehydrogenase in detergent-disrupted, isolated mitochondria showed it to be present in normal amounts in mutant mitochrondria (DeFrancesco *et al.,* 1976), and assays of several other citric acid cycle enzymes and pyruvate dehydrogenase also showed normal levels. An explanation of the results obtained with intact cells therefore required the postulate that the observed inhibition was due to feedback regulation rather than mutation. A block in the electron transport chain leading to a "back-up" of the system, and specifically to a high concentration of NADH in the mitochondria, was hypothesized to feedback inhibit several of the dehydrogenases, particularly α-ketoglutarate dehydrogenase.

Reduced minus oxidized difference spectra of oxidized and reduced cytochromes in whole cells revealed that the mutant cells had normal levels of functioning cytochromes. Furthermore, measurements of oxygen consumption by isolated, intact mitochondria in the presence of succinate revealed no difference between mutant and parental cells, and we concluded that complexes II, III, and IV of the electron transport chains were intact and functional in both cell types. On the other hand, substrates (α-ketoglutarate, malate, glutamate) that required an active complex I (NADH-CoQ reductase) stimulated oxygen consumption to less than 10% of wild-type rates (Fig. 4; DeFrancesco *et al.,* 1976).

Finally, a direct measurement of the rotenone-sensitive NADH oxidase activity of detergent (Lubrol)-treated mitochondria revealed that the mutants had less than 10% of this activity, which is associated with complex I of the electron transport chain (DeFrancesco *et al.,* 1976).

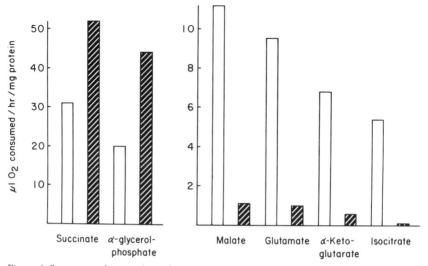

Figure 4. Summary and comparison of oxygen consumption by wild-type and mutant mitochondria with different substrates. Left, non-NADH-linked substrates, right, NADH-linked substrates (note the change in scale). Open bars, CCL16 mitochondria; crosshatched bars, res⁻ mutant mitochondria.

2.2. Protocol for the Isolation of Additional Mutants

As described in the previous section, a block in the electron transport chain in our first mutant caused feedback inhibition of the citric acid cycle and was ultimately responsible for the observed auxotrophies for CO_2/HCO_3^- and asparagine. The pioneering studies by Puck and Kao (1967; Kao and Puck, 1968) with mammalian cells had shown the feasibility of the "BUdR suicide" technique for the enrichment of cell populations in mutants with conditional-lethal mutations, and we made an effort to adapt the protocol to the selection of more mutants with a phenotype similar to that of the first. A successful outcome of such selections was expected to be governed by two major factors. (1) Under nonpermissive conditions (here starvation for CO_2/HCO_3^- and asparagine) the mutant cells must stop progressing through the cell cycle (i.e., DNA synthesis), but must remain viable for a sufficiently long period to allow all wild-type cells to incorporate a lethal amount of the analogue. Our preliminary experiments showed that the mutant cells lost viability under these conditions, but a substantial fraction of viable cells remained after 48 hr. (2) The killing of wild-type cells in the presence of the analogue must be greater than 99.9% efficient. Reconstitution experiments with artificially mixed cultures indicated that a 30-fold enrichment of a particular mutant could be achieved by our protocol (Ditta *et al.*, 1976).

In dealing with mutant selections from mammalian cell populations in tissue culture, a consideration of the expected frequencies of recessive phenotypes is in order. Many cell lines are poly/heteroploid, and at best pseudodiploid, such as the Chinese hamster fibroblasts used by us. Even after mutagenesis the incidence of homozygous recessive mutants is expected to be 10^{-8} or less, unless the locus of interest is (1) on the X chromosome, (2) in a region on an autosome for which the cell happens to be haploid after prolonged subculture *in vitro*, or (3) already heterozygous. Such considerations may seem discouraging, but examples of success because of either condition being fulfilled are now quite numerous.

Our search for mutants included two phases. Mutagenized cultures were subjected to several cycles of enrichment, after which individual surviving clones were picked, grown up to a few thousand cells, and split into two cultures. One was maintained in normal (permissive) medium, the other was switched to medium with galactose substituted for glucose (DME-Gal). Respiration-deficient cells of the phenotype desired by us could be recognized within 24 hr by their disintegration in DME-Gal, because the reduced rate of glycolysis in this medium was not able to sustain the viability of such mutants. An initial experiment with the Chinese hamster lung fibroblasts, CCL16 (DON), yielded two new mutant clones (B9, B10) out of 1500 surviving clones screened. Additional selections were made with Chinese hamster lung fibroblasts, V79, to take advantage of other potential partial haploidies or heterozygosities. Three separate experiments yielded almost 50 new mutants. All of the mutants were shown to require a high concentration of glucose; all but one (see below) had very low rates of oxygen consumption; all were auxotrophs for asparagine and for CO_2/HCO_3^- (Ditta *et al.*, 1976).

While these experiments were in progress, another group of investigators was engaged in a search for Chinese hamster cell (V79) mutants with defects in galactose utilization similar to those in human patients suffering from galactosemia (Chu *et al.*, 1972; Sun *et al.*, 1975). A large number of such Gal$^-$ mutants were isolated and characterized genetically (Chu, 1974) and biochemically (Whitfield *et al.*, 1981; Malczewski and Whitfield, 1982). While there was no specific selection for the phenotype, it turned out that many of these mutants also proved to be defective in the electron transport chain rather than in the Leloir pathway, which is most immediately concerned with galactose utilization.

Since then, another independent search for Chinese hamster cell mutants unable to utilize galactose in Thirion's laboratory (Maiti *et al.*, 1981) has also uncovered several respiration-deficient mutants with a phenotype very similar to that described by us (see below).

3. GLYCOLYSIS AND RESPIRATION IN WILD-TYPE PARENTS AND res⁻ MUTANTS

The discovery of mammalian cell mutants with a very low rate of respiration and therefore almost entirely dependent on glycolysis raised several interesting questions. In particular, one could ask about the relative contributions of glycolysis and respiration to the energy metabolism in the parental cells to determine how much the rate of glycolysis must be increased to compensate for the loss of oxidative phosphorylation. It also occurred to us that with a mutant cell almost entirely dependent on glycolysis, it should be possible to make a measurement of the absolute, total energy requirement for cell division.

We therefore measured glucose consumption and lactate production in wild-type (parental) and our original mutant cells (Fig. 5). To compare the two cell types, our data were, however, not normalized with respect to time, but with respect to the increase in total protein in the cultures. The measurements were made under normal conditions of cell proliferation over periods of up to 10 hr. By eliminating the time from this comparison, we avoided having to consider differences in the growth rates of the two cell types, although these were at any rate rather small. One of the striking observations was that even in wild-type cells, more than 95% of the glucose consumed was converted to and secreted as lactate. This conclusion was based not only on the separate measurements of glucose consumed and lactate produced, but also on the use of isotopically labeled glucose to measure the relatively low rates of utilization of glucose in the pentose shunt and of entry into the citric acid cycle (Table I). Thus, it was straightforward to calculate the amount of ATP generated by glycolysis and to relate such measurements to the increments in total protein in the two cell cultures (Donnelly and Scheffler, 1976).

Two major conclusions could be drawn: The rate of glycolysis in mutant cells was about 50% greater than in wild-type cells. While not unexpected, this result implied that even in wild-type cells under normal conditions for cell growth, about 65% of the total energy requirement is satisfied by glycolysis. With this assumption, namely that mutant cells derive all of their ATP from glycolysis, our measurements established for the first time an absolute energy requirement for mammalian fibroblast growth and proliferation in tissue culture: 110 ± 10 μmoles ATP/mg increment in total protein.

This experimental value is interesting in comparison to some theoretical considerations elaborated by Paul (1965). On the assumption that essentially all low-molecular-weight "building blocks" are provided, the energy requirement for the biosynthetic reactions (protein and nucleic acid synthesis) can be estimated. Protein synthesis represents a major component. A comparison with our value confirms and provides some quantitation for the expectation that cells require energy considerably in excess of that required for biosynthesis. We

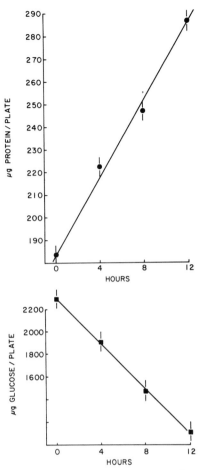

Figure 5. Glucose consumption and growth of CCL16-B2 (res⁻) cells (measured as total protein) at pH 7.3. From these measurements we calculated that 63.5 μmoles of glucose were consumed per milligram increment of protein.

Table I. The Fate of Glucose in Wild-Type and Mutant Cells

	μmoles glucose/Δ mg protein	
	CCL16	res⁻
[1-^{14}C]-Glucose → $^{14}CO_2$	0.88 ± 0.3	1.1 ± 0.2
[6-^{14}C]-Glucose → $^{14}CO_2$	0.064	0.009
Glucose → lactate	31.5 ± 5.0	52.5 ± 7.6

estimate that 60–80% of the total energy requirement is for activities such as active transport and ion pumps, motility and cell division, and others.

The fact that wild-type cells have the same total energy requirement as mutant cells could be confirmed by an experiment in which rotenone was used as a specific inhibitor of complex I to convert wild-type cells into a phenocopy of the mutant CCL16-B2 (Donnelly and Scheffler, 1976). A parallel series of experiments showed that under these conditions (inhibition of respiration by rotenone), glucose consumption per milligram increment of protein was perfectly comparable to that of the mutant cells. Thus, as long as there is an adequate supply of glucose to sustain a high rate of glycolysis, respiration in these fibroblasts can be inhibited quite drastically by either mutation or specific inhibitors without any effect on the proliferation of these cells. This is not to say that mammalian cells can grow under anaerobic conditions, although we have not performed any rigorous experiments to test this possibility. It is not yet clear whether a very low level of electron transport chain activity and oxidative phosphorylation is still essential, or whether oxygen is needed for oxidation involving other reactions (e.g., steroid biosynthesis) in these cells.

A few comments should be made here about our observations that more than 95% of the glucose consumed by wild-type cells can be accounted for from the production and secretion of lactate, even though these cells do exploit oxidative phosphorylation to satisfy their energy needs. Another carbon source must therefore be available and utilized in significant amounts. These considerations led us (Donnelly and Scheffler, 1976) to attempt to identify other oxidizable components in tissue culture medium, and we have provided evidence that glutamine is used far in excess of its requirement for protein synthesis. In wild-type cells a large proportion of the glutamine enters the citric acid cycle and is ultimately oxidized to carbon dioxide. The role of glutamine in cellular energy metabolism will be covered more extensively in another chapter of this book.

So far we have dealt with the total energy requirements for these cells and the sources of this energy. Another question is concerned with the high-energy phosphate state of the cell as defined by the energy charge (Atkinson, 1968):

$$\text{energy charge} = \frac{[ATP] + \frac{1}{2}[ADP]}{[ATP] + [ADP] + [AMP]}$$

Many experiments (Atkinson, 1977) have shown that the energy charge is maintained within a rather narrow range of 0.8–0.9 in all viable, proliferating prokaryotic and eukaryotic cells. Our measurements in both mutant and wild-type cells also yielded a value of ~ 0.9 for this parameter, but we found that while this value was maintained in wild-type cells during various manipulations in the preparation of cell extracts, it dropped significantly in mutant cells within minutes of their exposure to physiological saline or trypsin solutions lacking glucose.

Another significant result was that the total adenylate pool appeared to be 20–40% lower in the respiration-deficient mutants (Soderberg *et al.*, 1980).

4. BIOCHEMICAL CHARACTERIZATION OF MUTANTS

4.1. Defect in NADH-CoQ Reductase

Our first mutant had been characterized as having a defect in NADH-coenzyme Q reductase (complex I) of the electron transport chain. The isolation of additional mutants based on a selection devised with the help of the properties of this mutant immediately raised the question whether very similar defects could be found, or whether a very broad class of respiration-deficient mutants could be selected by this protocol. An answer can be obtained from biochemical and genetic experiments.

Genetic experiments, and more specifically a complementation analysis (to be described in more detail in the next section), showed that our mutants fell into seven complementation groups, i.e., that mutations in at least seven genes had been uncovered by our selection. Single mutants representing two complementation groups had biochemical characteristics that were totally different from those of the original mutant (see below), but all the other mutants representing five complementation groups had properties very similar to those of the first (Breen and Scheffler, 1979). The oxidation of substrates requiring a complete turn of the citric acid cycle was very low, while the entry of aspartate into the cycle was much less affected. Measurements of oxygen consumption by isolated mitochondria and by submitochondrial preparations are shown in Fig. 6A and B. In every case, succinate oxidase activity was comparable in the mitochondria from parental and mutant cells, while the activity of complex I was severely reduced. In these experiments with submitochondrial particles, we used NADH directly as the substrate, and it was necessary to distinguish at least two NADH oxidase activities. About 70% of the total NADH oxidase activity measured in wild-type mitochondria was inhibited by rotenone, a specific inhibitor of complex I of the electron transport chain. None of the NADH-dependent oxygen consumption of mutant mitochondria was inhibited by rotenone, and we concluded that in these cells NADH-CoQ reductase was severely affected by mutation while an "external" pathway, which bypassed the electron transport chain and which therefore was also not sensitive to antimycin, was still intact (Ernster, 1956).

These studies were performed most thoroughly with mutants representing complementation groups I, II, and VII (Breen and Scheffler, 1979), and in a more preliminary fashion with mutants from groups III and VI (Lakin-Thomas, Breen, Scheffler, unpublished). Thus, five genes have been mutated in different mutants and these affect the functioning of complex I of the electron transport

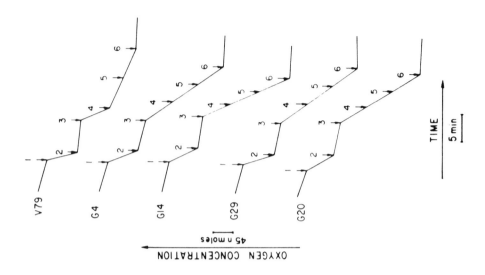

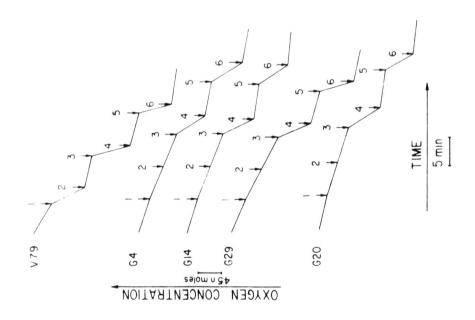

Figure 6. (A) Oxygen consumption by isolated mitochondria from wild-type and mutant cells. The amount of oxygen remaining in solution was measured with a Clark oxygen electrode following the sequential addition (indicated by the arrows) of the following substrates or inhibitors: (1) malate and pyruvate (2.5 mM each); (2) rotenone (5×10^{-7} M); (3) succinate (2.5 mM); (4) malonate (2.5 mM); (5) α-glycerophosphate (1 mM); and (6) antimycin (2.5 μM). Wild-type and mutant mitochondria (mg protein) were present in the following amounts in the 2-ml electrode chamber: V79, 1.9 mg; V79-G4, 2.0 mg; V79-G14, 1.8 mg; V79-G29, 1.7 mg; V79-G20, 1.7 mg.
(B) NADH oxidase and succinate oxidase activities in wild-type and mutant mitochondria. The amount of oxygen remaining in solution was measured with an oxygen electrode after the sequential addition (indicated by arrows) of the following substrates or inhibitors: (1) succinate (2.5 mM); (2) malonate (2.5 mM); (3) NADH (0.5 mM); (4) rotenone (5×10^{-7} M); (5) antimycin (2.5 μM); and (6) cyanide (1 mM). The mitochondria (mg protein) were present in the electrode chamber in the following amounts: V79, 20.9 mg; V79-G4, 1.8 mg; V79-G14, 1.9 mg; V79-G29, 1.8 mg; V79-G20, 2.1 mg.

chain. Moreover, it appears that mutants from at least two additional complementation groups have been characterized by Thirion's laboratory (Maiti *et al.*, 1981) as having a very similar defect in complex I.

The multitude of mutations affecting complex I is not particularly surprising when considered in the light of the biochemical characterization of this complex in the inner mitochondrial membrane (Ragan, 1980; Ragan *et al.*, 1982). Heron *et al.* (1979) have described the electrophoretic separation of at least 25 different polypeptides associated with this complex in beef heart mitochondria, and it is obvious that a large number of genes must be involved. It remains to be established whether each of these peptides is coded for by an individual gene, or whether larger peptides made in the cytoplasm from a single message become incorporated and then proteolytically processed (see below).

4.2. Defect in Succinate Dehydrogenase (SDH)

The mutant CCL16-B9 representing complementation group IV was shown very early to be quite different from all the others, and attention became focused very quickly on measurement of SDH activity (Soderberg *et al.*, 1977). The yield of $^{14}CO_2$ from [1-^{14}C]glutamate and [U-^{14}C]malate was almost comparable in wild-type and mutant cells, but [5-^{14}C]glutamate and [1,4-^{14}C]succinate yielded two orders of magnitude less $^{14}CO_2$ in the mutant.

Direct assays of SDH were carried out by several methods. Oxygen consumption by whole, isolated mitochondria was stimulated by succinate only in wild-type mitochondria, while α-glycerophosphate stimulated oxygen consumption in both wild-type and mutant mitochondria at comparable rates (Fig. 7). When the mitochondria were "activated" by treatment with phospholipase and Ca^{2+} at 38°C (Gutman *et al.*, 1971), succinate-dependent reduction of artificial electron acceptors (phenazine methosulfate and 2,6-dichlorophenol-indophenol) could be measured spectrophotometrically (Fig. 8). An alternative electron acceptor used in a spectrophotometric assay was ferricyanide [$Fe(CN)_6^{3-}$] (Veeger *et al.*, 1969). Again, only wild-type mitochondria showed an appreciable reaction, while both wild-type and mutant mitochondria reduced these substrates with α-glycerophosphate as the electron donor. A third assay, less quantitative but carried out with whole cells permeabilized by detergent, was similar to the spectrophotometric assay but used nitro-blue tetrazolium as the electron acceptor. This turned out to be a convenient histochemical assay for SDH (Soderberg *et al.*, 1977).

All of these assays revealed that in the mutant cells, SDH activity was severely depressed—to 1–3% of wild-type levels. The stimulation of O_2 uptake by mutant mitochondria in the presence of α-glycerophosphate showed that the

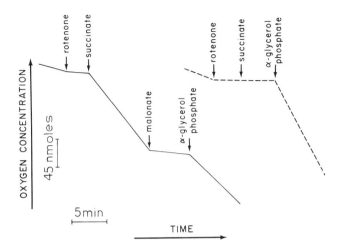

Figure 7. Respiration of wild-type and mutant (CCL16-B9) mitochondria with succinate and α-glycerophosphate as substrates. Mitochondria were isolated and resuspended at 1–2 mg/ml in TD buffer. Oxygen concentrations were determined at 30°C with a Clark electrode. At the times indicated by the arrows, rotenone (5×10^{-8} M), succinate (5×10^{-3} M), malonate (5×10^{-3} M), and α-glycerophosphate (5×10^{-3} M) were added to the mitochondrial suspension in the electrode chamber. Solid traces, wild-type mitochondria; dashed traces, mutant mitochondria.

electron transport chain from CoQ to oxygen was unaffected. Furthermore, the assays of SDH with the artificial electron acceptors showed that the defect was most likely in one of the two subunits of SDH (3OK and 7OK) (Davis and Hatefi, 1971; Hatefi, 1976) and not in one of the smaller peptides associated with complex II, which serve to couple SDH to CoQ (Capaldi *et al.*, 1977; Merli *et al.*, 1979; Ackrell *et al.*, 1982). Whether the defect is in the 70K flavoprotein or in the 30K iron–sulfur protein remains to be established.

4.3. Defect in Mitochondrial Protein Synthesis

An analysis of difference spectra of cytochromes of some of the mutants led to the discovery of one mutant derived from V79 cells in which cytochromes aa_3 appeared to be completely absent (Fig. 9; Ditta *et al.*, 1977). There was also an indication of some deficiency in the region of absorption of cytochrome *b*. This

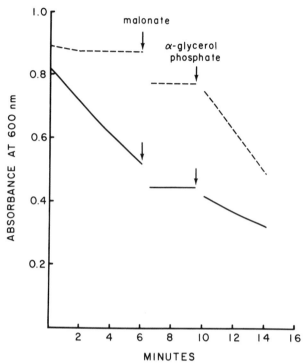

Figure 8. Spectrophotometric assay of succinate dehydrogenase with PMS and DCIP. After cooling the activated mitochondria to 25°C, KCN (1 mM), PMS (0.2 mM), and DCIP (52 μM) were added, and the succinate-driven reduction of DCIP was monitored spectrophotometrically at 600 nm. Arrows show the time of addition of malonate (15 mM) to inhibit SDH, and α-glycerophosphate (15 mM) to measure α-glycerophosphate dehydrogenase activity. The solid trace was obtained with 36 μg of wild-type mitochondria, the dashed trace with 88 μg of mutant mitochondria.

mutant, V79-G7, in complementation group V (Soderberg *et al.*, 1977), was shown to have no cytochrome *c* oxidase activity in assays with reduced cytochrome *c* from beef heart and detergent (Lubrol)-treated mitochondria. Complex IV, responsible for this activity, is made up of seven subunits, three of which were known (from work with yeast mutants) to be made inside the mitochondira and coded for by the mitochondrial genome. This prompted us to examine mitochondrial protein synthesis in these mutant cells. The first experiments were carried out with intact cells in which cytoplasmic protein synthesis had been inhibited with cycloheximide (Fig. 10) and it became clear that mitochondrial protein synthesis was severely inhibited in this mutant. This work was extended in a later study (Burnett and Scheffler, 1981), in which the defect was

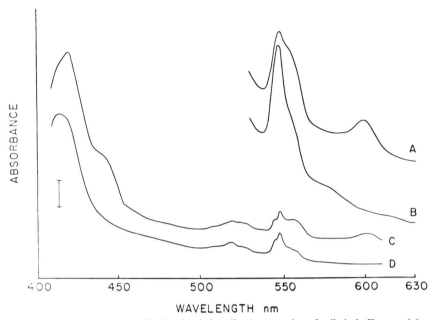

Figure 9. Absorption spectra of reduced, whole cells. A suspension of cells in buffer containing NaCl (137 mM), KCl (5 mM), Na$_2$HPO$_4$ (0.7 mM), Tris (25 mM), pH 7.3, was reduced by the addition of a few grains of dithionide and then frozen in quartz spectrophotometer cuvettes at a density of 10^7 cells/ml. Absorption spectra were recorded at liquid nitrogen temperature in a Cary 14 spectrophotometer, with the output from the photomultiplier fed into a PDP 8/1 computer for the summation of multiple scans. Approximately equal numbers of wild-type and mutant cells were compared in each experiment. (A,C) Wild-type cells; (B,D) mutant cells. The bar represents an absorbance change of 0.1 (curves C and D), and the scale is changed by a factor of 4 for curves A and B in which the scan was from 530 to 630 nm.

also demonstrated with isolated mitochondria by measurements of the incorporation of [^{35}S]methionine into mitochondrial proteins. Inhibition of mitochondrial protein synthesis could be estimated to be greater than 95%.

Again based on information from studies with yeast, one would predict, in addition to the observed cytochrome *b* deficiency, a deficiency or absence of the mitochondrial, oligomycin-sensitive ATPase (complex V). This complex is made up of a collection of ten peptides, two or three of which are coded for by the mitochondrial genome and synthesized inside the mitochondria. Measurements of this activity in isolated mitochondria showed a residual level of less than 10% (Ditta *et al.*, 1977).

The existence of a mammalian cell mutant with obvious similarities to yeast mutants with the ''petite'' phenotype raised several additional questions, some

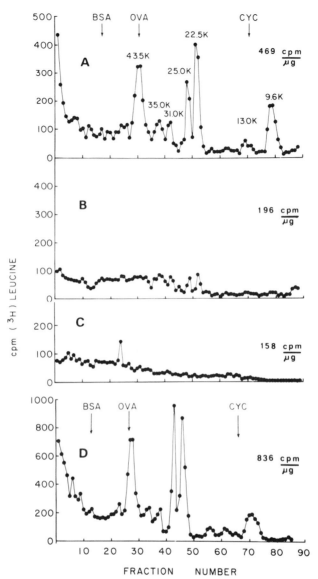

Figure 10. Protein synthesis in wild-type and mutant mitochondria. Intact cells were preincubated for 30 min with 10 μg/ml cycloheximide, and then labeled for 2 hr with [³H]leucine, followed by a chase (30 min) with an excess of cold leucine. Purified mitochondria were dissolved in a buffer containing β-mercaptoethanol and SDS and then fractionated by SDS–polyacrylamide gel electrophoresis. The gels were sliced into 1-mm thin sections, which were solubilized with NCS (Amersham/Searle) and then counted. Bovine serum albumin (BSA), ovalbumin (OVA), and cytochrome *c* (CYC) were run on the same gels in parallel tracks as molecular weight markers. (A) V79 wild-type mitochondria; (B) V79-G7 res⁻ mitochondria; (C) V79 cells pretreated with chloramphenicol at 100 μg/ml for 5 hr before the addition of label; (D) CCL16-B2 res⁻ mitochondria, i.e., from the mutant defective in complex I.

of which were answered (Burnett and Scheffler, 1981). The number of mitochondria in the mutant cells was comparable to wild type, but these mitochondria
were morphologically abnormal. Most noticeable were the disorganized and
apparently tubular and inflated cristae, when compared to the more typical,
lamellar cristae in wild-type mitochondria (Fig. 11). In view of this altered
morphology, it was perhaps a little surprising that two-dimensional polyacrylamide gel electrophoresis of mitochondrial membrane and matrix proteins
revealed very few differences. It was well known that more than 90% of all
mitochondrial proteins are made in the cytoplasm, and their synthesis was not
expected to be directly affected. However, the almost total absence of mitochondrial protein synthesis and oxidative phosphorylation initially led us to anticipate
a more pronounced effect on mitochondrial biogenesis. In the case of complex
IV, we were also unable to detect any accumulation of ^{35}S-methionine-labeled
peptides or precursors in the cytoplasm, using a specific antiserum that was
shown to precipitate seven peptides from wild-type mitochondria.

In an effort to define more closely the defective step responsible for the
inhibition of mitochrondrial protein synthesis, we examined the transcription of
the mitochondrial rRNA genes and the assembly of the rRNA into mitochondrial
ribosomes. No evidence for any defect was obtained, although the ribosomes
made could only be examined for their hydrodynamic properties (sedimentation
coefficient) and not for their biological activity. The nature of the specific defect
is still completely unknown, and it would require the *in vitro* reconstitution of a
mitochondrial protein-synthesizing system and *in vitro* complementation assays
to identify the missing factor. Such an experiment is logistically very difficult to
perform at this time.

There has been one other report on mammalian cell mutants with a defect in
mitochondrial protein synthesis (Wiseman and Attardi, 1979). A fairly substantial number of reports have been made by now on mammalian cells resistant to
specific inhibitors of mitochondrial functions such as chloramphenicol, oligomycin, and antimycin. In a number of cases, the drug-resistance phenotype has been
shown to be due to a mutation in the mitochondrial genome, and in most instances ability to plate in the presence of the drug is the only phenotype. However, a set of mutants isolated by Wiseman and Attardi from the human cell line
VA_2-B is not only resistant to chloramphenicol, but defective in mitochondrial
protein synthesis. As a result, these mutants have very similar properties when
compared to our V79-G7 mutant: dependence on abundant glucose, and reduction in cytochrome *c* oxidase and rutamycin-sensitive ATPase activities. It
should be emphasized that other chloramphenicol-resistant cell lines do not show
this behavior, and it is not yet clear whether a single mutation in the mitochondrial large rRNA gene is responsible for all the phenotypic alterations. Fusions with
enucleated cells have shown that drug resistance and glucose dependence are
cotransferable with the cytoplasm from these mutants.

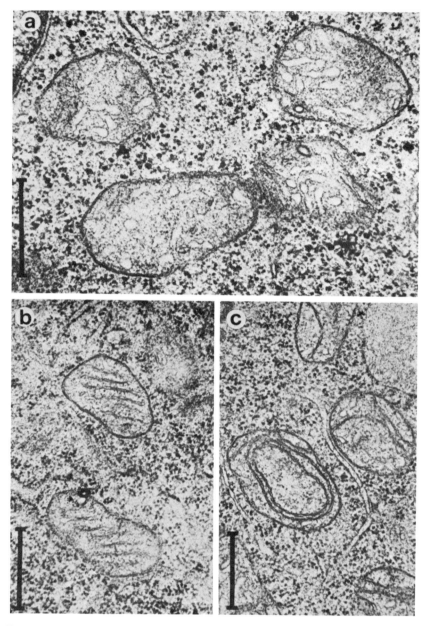

Figure 11. Electron micrographs of representative mitochondria from Chinese hamster lung fibroblasts. (A) G7 mitochondria ×63,000. (B) Mitochondria of V79 cells maintained in high glucose,
×57,000. (C) Mitochondria of V79 cells treated for 3 days with 100 µg/ml chloramphenicol,
×50,000. Bars = 0.5 µm.

5. GENETIC CHARACTERIZATION OF MUTANTS

Are these respiration-deficient cells real mutants? The possibility that an altered phenotype of a mammalian cell in culture is due to an epigenetic change rather than due to a structural gene mutation has been suggested at various times, and it is an issue that should not be ignored (Harris, 1982). The answer to this question is not always easy to obtain.

All of our mutants, except the mutant V79-G7, have been very stable over many years, and even mutagen-induced reversions have been very rare. More recently, azacytidine has been found to induce gene reactivation in somatic cells in culture most likely by a mechanism that includes hypomethylation (Harris, 1982), but we have not yet carried out similar experiments.

Spontaneous revertants were observed in populations of the mutant defective in mitochondrial protein synthesis, and they were shown to arise at a rate of 8.6×10^{-7}/cell per generation, which was only slightly higher than the mutation rate of thioguanine or ouabain resistance (Soderberg et al., 1979). Although rates for epigenetic events are not known, our data on stability do not argue against a true mutation.

All the mutations were shown to be recessive in intraspecies hybridization (Soderberg et al., 1977), indicative of a missing function rather than of a dominant regulatory or suppressor mutation. There is no evidence that any of our mutants are defective in a cytoplasmic (mitochondrial) gene. In some cases this has been shown explicitly by the complementation with a particular chromosome (see below), while in one initially suspect case (V79-G7) the conclusion was based on hybridizations with enucleated cells (cytoplasts) carrying a cytoplasmically inherited, selectable marker (Soderberg et al., 1977).

A complementation analysis (Fig. 12) involving pairwise fusions of mutant cells has allowed us to sort our mutants into seven complementation groups, and additional complementation groups are defined by the mutants of others (Chu, 1974; Soderberg et al.,1977; Maiti et al., 1981). One may conclude, therefore, that our enrichment scheme yields mutants of a general class of respiration-deficient mutants, of which mutants with a defect in the citric acid cycle (SDH), a defect in the electron transport chain (complex I), or defective mitochondrial protein synthesis are representative examples.

On the other hand there were some surprises: (1) Several independent experiments yielded mutants in the same complementation group, mutants in group I being particularly numerous (2) Independent selections with two different Chinese hamster cell lines (CCL16 or DON, and V79) yielded mutants in the same two complementation groups (3) As observed previously by Chu and colleagues with their Gal⁻ mutants (Sun et al., 1975), we also had a case of one mutant failing to complement mutants from two different complementation groups (Table II). The repeated isolation of the same mutation should also be

B 2   G4   B9   B10   G7

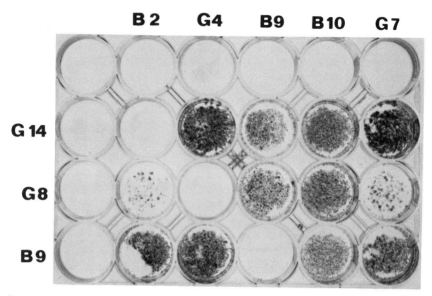

G 14

G 8

B 9

Figure 12. Complementation testing of Chinese hamster mutants with defects in respiration. The first column and first row contain only cells from the indicated cell lines, while mixtures of cells were plated in the other wells. All wells were exposed to PEG to induce cell fusion, and the cells were incubated for 2 days in complete medium, before selective conditions (DME-Gal) were imposed. A week later the plates were stained with crystal violet. Complementation is indicated by the presence of viable, adherent cells. Mutant cells and hybrids in which there is no complementation detach and disintegrate quickly in DME-Gal.

discussed in conjunction with the relatively high frequency at which these re-cessive mutants could be isolated from pseudodiploid mammalian cells (Ditta *et al.*, 1976). Such a result was reasonable only if one postulated the cells to be haploid at this locus (X-linked, or due to a partial deletion of a chromosome), or to be functionally haploid due to a previous mutation or gene inactivation. For the two largest complementation groups, which include the mutants in NADH-CoQ reductase, the problem has most likely been resolved by mapping studies that located the affected genes on the X chromosomes of the hamster and the mouse (Day and Scheffler, 1982). Mutants in these two genes therefore form a linkage group, and it remains to be seen whether these genes are closely linked, which would be very interesting in view of their functional relationship. The mutation in the mutant overlapping these two complementation groups has also been mapped to the X chromosome, leaving open the possibility of a deletion of two closely linked genes, but further evidence will be required to clarify this situation. Preliminary data with a mutant in NADH-CoQ reductase and in a separate complementation group suggest an autosomal gene, but mapping of this and of the other complex I mutants is incomplete.

Table II. Complementation Groups[a]

I	II	III	IV	V	VI	VII	VIII
	CCL16-B2						gal13-3
		CCL16-B10	CCL16-B9				
V79-G4		V79-G18					
G5							
G19							
V79-G8	V79-G24			V79-G7			
G9	G42						
G10							
G21							
G22							
G23							
G37							
G38							
G43							
	V79-G20						
V79-G12	V79-G14	V79-G35				V79-G11	V79-G29
G13	G31						
G26							
G28							
G32							
G49							
G50							
G52							

[a]The mutants are grouped vertically according to the experiment in which they were isolated.

About the mutant defective in mitochondrial protein synthesis we can only say that it carries a nuclear mutation, as indicated earlier. A gene complementing a defective SDH gene of the hamster has been mapped to human chromosome 1 (Mascarello *et al.*, 1980), and it remains to be established whether this is the gene for the 70K flavoprotein or the 3OK iron–sulfur protein of SDH. Studies based on this question are in progress. We are particularly interested to learn whether the two genes are linked, as they are in prokaryotes (Hederstedt and Rutberg, 1981).

(4) The mutants isolated by Chu's laboratory based on their inability to grow on galactose substituted for glucose (Chu *et al.*, 1972; Sun *et al.*, 1975) have been characterized further by Whitfield and co-workers (1981; Malczewski and Whitfield, 1982). Mutants from several different complementation groups were examined and shown to have a spectrum of pleiotropic alterations in the following: complex I activity, complex III activity, CoQ content, and coupling of respiration and phosphorylation of ADP. Some were most drastically affected in complex I activity, others were primarily affected in complex III, some ap-

peared to be seriously deficient in CoQ, and therefore all activities requiring this cofactor and electron carrier were significantly reduced.

Since the experiments and the experimental protocols have been somewhat different between our laboratory and that of Whitfield, it is probably not easy to make precise comparisons between our complex I mutants and those characterized by her laboratory. We also made comparisons of difference spectra between wild-type cells and some of our mutants, and no cytochrome peaks were found missing except in the mutant with a defect in mitochondrial protein synthesis. In the case of all of our complex I mutants, succinate- and α-glycerophosphate-driven oxygen consumption in isolated mitochondria were comparable to the rate in wild-type mitochondria. On the other hand, we did not make any direct measurements of CoQ levels in our mutants.

There is at this point a very fundamental difference in our interpretation of the biochemical asepcts characteristic of the observed mutations, and it is proba-bly fair to say that considerably more biochemical analysis is required to resolve the difference. While we favor an interpretation in terms of mutations in struc-tural genes for peptides of complex I, Whitfield et al. (1981) have interpreted their data in terms of different mutations within the same gene (cistron) affecting a protein that is (1) either common to complexes I and III, or (2) needed for incorporation of CoQ, or (3) required for the assembly of the complexes in the inner mitochondrial membrane, for example, a protease needed for processing of precursors. Mapping studies by us suggest that both X-linked and autosomal loci are mutated (see above), arguing against a single cistron being responsible. It should also be stressed that the two laboratories may be looking at related but different sets of mutants. The Gal-13 mutant of Chu forms a different comple-mentation group when tested by hybridization with all of ours (Soderberg et al., 1977); the others have not yet been compared.

Another independently derived set of mutants was obtained by Thirion's group (Maiti et al., 1981). The biochemical assays employed in their charac-terization were very similar to ours, and they resulted in the identification of several mutants defective in NADH-CoQ reductase, and another with a possible defect in SDH. Interestingly, the former mutants defined two new complementa-tion groups in addition to ours, and one mutant failed to complement mutants in several of our complementation groups.

6. WORK IN PROGRESS AND FUTURE PROSPECTS

It should be apparent from the previous discussion that a considerable number of mammalian cell mutants are now available with defects in mitochon-drial functions and respiration. In every case, many specific questions remain to be answered. We still do not know the identity of the polypeptide that has been altered by mutation in any of the mutants, although the answer should be forthcoming in the case of the SDH⁻ mutant in the not too distant future. In general, the biochemical analysis is made difficult by the fact that we are dealing with integral membrane proteins and protein complexes from a source that is somewhat limited: a kilogram of beef heart is much more readily available than even just tens of grams of mutant mammalian cells grown in tissue culture. A heroic effort will be necessary to isolate sufficient amounts to be able to examine the constituents of complex I, and depending on the nature of the defect, the block may be localizable to one of the various intermediate reactions. The exploitation of antisera and even monoclonal antibodies made against individual peptides of complex I may be possible, but frequently these reagents are made against the protein from beef heart mitochondria, and cross-reactivity with the hamster proteins may be a problem (Ragan and Scheffler, unpublished observations).

It remains to be seen whether the lack of activity is due to a defective polypeptide in the assembled complex or due to a block in the processing and assembly of the various polypeptides in the inner mitochondrial membrane.

A less formidable problem with complex I mutants is presented by the further genetic analysis, and particularly the mapping of the various genes on human, mouse, or hamster chromosomes. Such information may ultimately be very useful in understanding the coordinate expression of these genes. In principle, the isolation and cloning of these genes is not out of the question, since a powerful selection system for respiration-competent cells is available. Some of these genes have already been mapped on the X chromosome, and one would like to have such information about the other genes defined by the complementation groups. The approach, using interspecies hybrids, appears straightforward, but we have had some indications that a certain degree of incompatibility may exist between hamster mitochondrial protein complexes and human gene products that are to substitute for a defective hamster protein. The extent or generality of this incompatibility has to be further defined, but it is already quite clear that human mitochondria (i.e., with a human mitochondrial genome) cannot exist in hamster cells. Thus, complementation analysis and mapping with interspecies hybrids may be frustrating and complicated in the case of mitochondrial membrane complexes.

It is now fairly clear that cytoplasmic precursors of mitochondrial proteins are further processed upon and after binding and entry into mitochondria (for recent reviews see Neupert and Schatz, 1981; Reid and Schatz, 1982; Schatz and Butow, 1983). An early idea on a single large precursor for all of the cytoplasmically made peptides of complex III (Poyton and McKemmie, 1979) has been experimentally disproved (Lewin et al., 1980), and each of the peptides of complex V (the F_0F_1 ATPase) also appears to have its own precursor (Maccechini et al., 1979; Lewin et al., 1980). Complex I, with 26 different polypeptides, may be a candidate with which to resurrect the hypothesis of a single precursor for at least some of the peptides.

A considerably less complicated situation exists in the case of SDH (complex II). A 70K flavoprotein and a 30K iron–sulfur protein form a complex (SDH) that is thought to be anchored to the inner mitochondrial membrane with the help of two rather small, very hydrophobic peptides. Interestingly, the genes for these four proteins form an operon in *B. subtilis* (Hederstedt and Rutberg, 1981) and likely in *E. coli* (Cole and Guest, 1982), and there is an analogous operon for the closely related complex, fumarate reductase, which is induced under anaerobic conditions (Cole *et al.*, 1982; Lemire *et al.*, 1982). It will be most interesting to find out if this linkage group has been preserved in higher organisms and specifically mammalian cells.

We have mapped an SDH gene on human chromosome 1 (Mascarello *et al.*, 1980). Some hybrids were obtained in which only a very small fragment ($\leq 1\%$ of the total human genome) of a human chromosome was retained in the hamster cells under selective conditions, and we presume it to contain an SDH gene. Libraries have been constructed from two such independent clones, using the lambda phage ends for cloning, and a variety of strategies are now under consideration for identifying and cloning an SDH gene. Further studies limited only by experimental obstacles and not by theoretical problems will lead to information about nucleotide and amino acid sequences, and information about the peptide precursor that is synthesized in the cytoplasm and then imported into the inner mitochondrial membrane. Of obvious interest will be to obtain information about the nature of the ''sorting'' signal on this precursor that presumably interacts with some receptor on the outer mitochondrial membrane. It is also most likely that there will be proteolytic processing of this precursor. A more complicated problem will be to understand how the four peptides of complex II are assembled: (1) Is the assembly sequential, with the two small hydrophobic peptides incorporated first? (2) Are the two other peptides first imported into the mitochondrial matrix as soluble precursors? (3) How is complex II integrated into the rest of the electron transport chain? (4) If separate genes are involved rather than an operon as in bacteria, the coordination of their expression becomes another issue to be addressed in connection with mapping data, when such data become available.

The mutant defective in mitochondrial protein synthesis also poses questions of interest for the future. It appears on the one hand that oxidative phosphorylation can be replaced almost entirely by an increase in the rate of glycolysis, and one may ask whether the mitochondria in mammalian cells perform essential functions unrelated to energy metabolism. In fact, a small residual rate of oxidative phosphorylation may still be necessary, because it appears that these mutant cells are exceptionally sensitive to efrapeptin, a specific inhibitor of the F_1 complex of the mitochondrial ATPase (Burnett and Scheffler, unpublished observations). Unfortunately, attempts to demonstrate the existence of a membrane potential in these mitochondria in intact cells have been inconclusive so far, but such an energized membrane would have to be present to explain the energy-dependent transport of proteins into the inner mitochondrial membrane and matrix (Nelson and Schatz, 1979; Schatz and Butow, 1983). Furthermore, the transcription of the mitochondrial genome appears to occur at a normal rate, requiring nucleoside triphosphates. These could presumably be imported from the cytosol, a reversal of the usual export of ATP.

Thus, a fundamental question remains: Is the mutation leaky, permitting a trace of electron transport and oxidative phosphorylation sufficient for mitochondrial biogenesis and other unknown functions, or is there indeed an efficient mechanism to "import" into the mitochondria energy (ATP) derived from glycolysis?

7. SUMMARY

One of the most significant early conclusions from our studies was that mammalian cells in tissue culture can satisfy essentially all their energy needs from glycolysis. Thus, the feasibility of isolating mammalian cell mutants with defects in mitochondrial functions associated with energy metabolism became established, and the properties of the first mutant even suggested to us an experimental protocol for enriching populations of Chinese hamster cells in culture in respiration-deficient mutants. A variety of mutants were isolated that represent the following general types of defects: (1) a block in the electron transport chain (in complex I, or NADH-CoQ reductase); (2) a block in the citric acid cycle (SDH); and (3) a block in mitochondrial protein synthesis (causing multiple deficiencies in mitochondrial functions). Similar or related sets of mutants have since also been described by other laboratories.

Energy metabolism (glycolysis versus respiration) will probably be less of an issue in the future, although it may be of interest to establish whether a small, residual flow of electrons through the electron transport chain is absolutely essential, not so much to contribute to the energy supply of the cell, but to maintain mitochondria and other indispensable mitochondrial functions. Mammalian cell mutants absolutely equivalent to the rho mutants of yeast have not yet been found.

A more precise biochemical characterization of the defects and the identification of the defective or missing peptide in each of the mutants will be a problem for the future, and in many cases this represents a considerable challenge from the experimental and practical point of view. Only the SDH-deficient mutant promises to be resolvable in the near future. The complex I mutants are not only difficult to analyze because of the multitude of peptides involved, but it remains to be clarified whether there are different, specific, defective complex I peptides in each of the mutants, or whether a mutation affects in a more pleiotropic fashion the assembly of the electron transport chain in the inner mitochondrial membrane, as suggested by Whitfield's group (Whitfield *et al.*, 1981).

Some progress has been made in the mapping of the mutations on mammalian chromosomes, and future efforts promise to reveal whether other complementation groups represent X-linked genes, as shown for our most abundant mutants in two complementation groups. A gene for SDH has been mapped on human chromosome 1.

Since these mutants are quite stable and the wild-type phenotype can be selected for very efficiently, prospects for gene transfers and gene isolation are promising, and it is hoped that such studies will provide interesting information on the very general problem of mitochondrial biogenesis.

ACKNOWLEDGMENTS.   Support for this research over the past decade has been provided by the United States Public Health Service, the American Cancer Society, and the National Science Foundation.

None of this work would have been completed without the stimulation, ideas, and criticisms, perseverance, and the competence at the lab bench of several collaborators to whom I wish to express my sincere thanks: Drs. L. DeFrancesco, G. Ditta, K. Soderberg, G. Breen, J. Mascarello, K. Burnett, C. Day, and Mr. M. Donnelly. E. Gabriel and E. Waltzer provided valuable technical assistance.

REFERENCES

Ackrell, B. A. C., Ramsay, R. R., Kearney, E. B., Singer, T. P., White, G. A., and Thorn, G. D., 1982, Two small polypeptides from complex II and their role in the reconstruction of Q-reductase activity and in the binding of TTF, in: *Function of Quinones in Energy Conserving Systems* (B. L. Trumpower, ed.), pp. 319–332, Academic Press, New York.

Atkinson, D. E., 1968, The energy charge of the adenylate pool as a regulatory parameter: Interaction with feedback modifiers. *Biochemistry* **7**:4030–4034.

Atkinson, D. E., 1977, *Cellular Energy Metabolism and Its Regulation,* Academic Press, New York.

Breen, G. A. M., and Scheffler, I. E., 1979. Respiration-deficient Chinese hamster cell mutants: Biochemical characterization, *Somat. Cell Genet.* **5**:441–451.

Burnett, K. G., and Scheffler, I. E., 1981, Integrity of mitochondria in a mammalian cell mutant defective in mitochondrial protein synthesis, *J. Cell Biol.* **90**:108–115.

Capaldi, R. A., Sweetland, J., and Merli, A., 1977, Polypeptides in the succinate-coenzyme Q segment of the electron transport chain, *Biochemistry,* **16**:5707–5710.

Chang, R. S., Liepins, H., and Margolish, M., 1961, Carbon dioxide requirement and nucleic acid metabolism of HeLa and conjunctival cells, *Proc. Soc. Exp. Biol. Med.* **106**:149–152.

Chu, E. H. Y., 1974, Induction and analysis of gene mutations in cultured mammalian somatic cells, *Genetics* **78**:115–132.

Chu, E. H. Y., Sun, N. C., and Chang, C. C., 1972, Induction of auxotrophic mutations by treatment of Chinese hamster cells with 5-bromodeoxyuridine and black light, Proc. Natl. Acad. USA **69**:3459–3463.

Cole, S. T., and Guest, J. R., 1982, Molecular genetic aspects of the succinate–fumarate oxidoreductases of *E. coli*, *Biochem. Soc. Trans.* **10**:473–475.

Cole, S. T., Grundstrom, T., Jaurin, B., Robinson, J. J., and Weiner, J. H., 1982, Location and nucleotide sequence of frdB, the gene coding for the iron–sulphur protein subunit of the fumarate reductase of *E. coli*, *Eur. J. Biochem.* **126**:211–216.

Cooper, J. A., Russ, N. A., Schwartz, R. J., and Hunter, T., 1983, Three glycolytic enzymes are phosphorylated at tyrosine in cells transformed by Rous sarcoma virus, *Nature* **302**:218–223.

Davis, K., and Hatefi, Y., 1971, Succinate dehydrogenase. I. Purification, molecular properties, and substructure, *Biochemistry* **10**:2509–2516.

Day, C., and Scheffler, I. E., 1982, Mapping of the genes for some components of complex I of the electron transport chain on the X chromosome of mammals, *Somat. Cell Genet.* **8**:691–707.

DeFrancesco, L., Werntz, D., and Scheffler, I. E., 1975, Conditionally lethal mutations in Chinese hamster cells: Characterization of a cell line with a possible defect in the Krebs cycle, *J. Cell. Physiol.* **85**:293–306.

DeFrancesco, L., Scheffler, I. E., and Bissell, M. J., 1976, A respiration deficient Chinese hamster cell line with a defect in NADH-coenzyme Q reductase, *J. Biol. Chem.* **251**:4588–4595.

Ditta, G., Soderberg, K., Landy, F., and Scheffler, I. E., 1976, The selection of Chinese hamster cells deficient in oxidative energy metabolism, *Somat. Cell Genet.* **2**:331–344.

Ditta, G., Soderberg, K., and Scheffler, I. E., 1977, Chinese hamster cell mutant with defective mitochondrial protein synthesis, *Nature* **268**:64–66.

Donnelly, M., and Scheffler, I. E., 1976, Energy metabolism in respiration-deficient and wild type Chinese hamster fibroblasts in culture, *J. Cell. Physiol.* **89**:39–52.

Dujon, B., Colson, A. M., and Sloninski, P. P., 1977, The mitochondrial genetic map of Saccharomyces cerevisiae: compilation of mutations, genes, genetic and physical maps, in: *Mitochondria 1977, Genetics and Biogenesis of Mitochondria* (W. Bandlow, R. J. Schwegen, K. Wolf, and F. Kandewitz, eds.), pp. 579–669, W. de Gruyter, Berlin.

Ernster, L., 1956, Organization of mitochondrial DPN-linked systems. II. Regulation of alternate electron transfer pathways, *Exp. Cell Res.* **10**:721–732.

Geyer, R. P., and Neimark, J. M., 1958, Response of CO_2 deficient human cells *in vitro* to normal cell extracts, *Proc. Soc. Exp. Biol. Med.* **99**:599–601.

Gregg, C. T., 1972, Some aspects of energy metabolism of mammalian cells, in: *Growth, Nutrition, and Metabolism of Cells in Culture* (G. H. Rothblat and V. J. Cristofalo, eds.), pp. 83–136, Academic Press, New York.

Gutman, M., Kearney, E. B., and Singer, T. P., 1971, Control of succinate dehydrogenase in mitochondria, *Biochemistry* **10**:4763–4770.

Harris, M., 1982, Induction of thymidine kinase in enzyme deficient Chinese hamster cells, *Cell* **29**:483–492.

Hatefi, Y., 1976, The enzymes and enzyme complexes of the mitochondrial oxidative phosphorylation system, in: *The Enzymes of Biological Membranes* (A. Martonosi, ed.), pp. 3–42, Plenum Press, New York.

Hederstedt, L., and Rutberg, L., 1981, Succinate dehydrogenase—A comparative review, *Microbiol. Rev.* **45**:542–555.

Heron, C., Smith, S., and Ragan, C. I., 1979, An analysis of the polypeptide composition of bovine heart mitochondrial NADH ubiquinone oxidoreductase by two-dimensional gel electrophoresis, *Biochem. J.* **181**:435–443.

Kao, F.-T., and Puck, T. T., 1968, Genetics of somatic mammalian cells. VII. Induction and isolation of nutritional mutants in Chinese hamster cells, *Proc. Natl. Acad. Sci. USA* **60**:1275–1281.

Leibovitz, A., 1963, The growth and maintenance of tissue cell culture in free gas exchange with the atmosphere, *Am. J. Hyg.* **78**:173–180.

Lemire, B. C., Robinson, J. J., and Weiner, J. H., 1982, Identification of membrane anchor polypeptides of *E. coli* fumarate reductase, *J. Bacteriol.* **152**:1126–1131.

Lewin, A. S., Gregor, I., Mason, T. L., Nelson, N., and Schatz, G., 1980, Cytoplasmically made subunits of yeast mitochondrial F_1-ATPase and cytochrome c oxidase are synthesized as individual precursors, not as polyproteins, *Proc. Natl. Acad. Sci. USA* **77**:3998–4002.

Maccechini, M. L., Rudin, Y., Blobel, G., and Schatz, G., 1979, Import of proteins into mitochondria: Precursor forms of the extramitochondrially made F_1 ATPase subunits in yeast, *Proc. Natl. Acad. Sci. USA* **76**:343–347.

McKeehan, W. L., 1982, Glycolysis, glutaminolysis, and cell proliferation, *Cell Biol. Int. Rep.* **6**:635–650.

McLimans, W. F., 1972, The gaseous environment of the mammalian cell in culture, in: *Growth, Nutrition and Metabolism of the Mammalian Cell in Culture* (G. H. Rothblat and V. J. Cristofalo, eds.), pp. 137–170, Academic Press, New York.

Maiti, I. B., Comlan de Souza, A., and Thirion, J. P., 1981, Biochemical and genetic characterization of respiration-deficient mutants of Chinese hamster cells with a Gal⁻ phenotype, *Somat. Cell Genet.* **7**:567–582.

Malczewski, R. M., and Whitfield, C. D., 1982, Respiration defective Chinese hamster cell mutants containing low levels of NADH-ubiquinone reductase and cytochrome c oxidase, *J. Biol. Chem.* **257**:8137–8142.

Mascarello, J. T., Soderberg, K., and Scheffler, I. E., 1980, Assignment of a gene for succinate dehydrogenase to human chromosome 1 by somatic cell hybridization, *Cytogenet. Cell Genet.* **28**:121–135.

Merli, A., Capaldi, R. A., Ackrell, B. A. C., and Kearney, E. G., 1979, Arrangement of complex II (succinate–ubiquinone reductase) in the mitochondrial inner membrane, *Biochemistry* **18**:1393–1400.

Nelson, N., and Schatz, G., 1979, Energy-dependent processing of cytoplasmically made precursors to mitochondrial proteins, *Proc. Natl. Acad. Sci. USA* **76**:4365–4369.

Neupert, W., and Schatz, G., 1981, How proteins are transported into mitochondria, *Trends Biochem. Sci.* **6**:1–4.

Paul, J., 1965, Carbohydrate and energy metabolism, in: *Cells and Tissues in Culture* (E. N. Wittmer, ed.), pp. 239–276, Academic Press, New York.

Pedersen, P. L., 1978, Tumor mitochondria and the bioenergetics of cancer cells, *Prog. Exp. Tumor Res.* **22**:190–274.

Poyton, R. O., and McKemmie, E., 1979, A polyprotein precursor to all four cytoplasmically translated subunits of cytochrome c oxidase from *S. cerevisiae*, *J. Biol. Chem.* **254**:6763–6771.

Puck, T. T., and Kao, F.-T., 1967, Genetics of mammalian calls. V. Treatment with 5' BUdR and visible light for isolation of nutritionally deficient mutants, *Proc. Natl. Acad. Sci. USA* **58**:1227–1234.

Racker, E., 1972, Bioenergetics and the problem of tumor growth, *Am. Sci.* **60**:56–63.

Racker, E., 1976, *A New Look at Mechanisms in Bioenergetics*, pp. 153–175, Academic Press, New York.

Racker, E., Johnson, J. H., and Blackwell, M. T., 1983, The role of ATPase in glycolysis of Ehrlich ascites tumor cells, *J. Biol. Chem* **258**:3702–3705.

Ragan, C. I., 1980, The molecular organization of NADH dehydrogenase, in: *Subcellular Biochemistry, Vol. 7* (D. B. Roodyn, ed.), pp. 267–307, Plenum Press, New York.

Ragan, C. I., Galante, Y. M., Hatefi, Y., and Ohnishi, T., 1982, Resolution of mitochondrial NADH dehydrogenase and isolation of two iron–sulfur proteins, *Biochemistry* **21**:590–594.

Reid, G. A., and Schatz, G., 1982, Biogenesis of mitochondrial membrane proteins, in: *Membranes in Growth and Development: Progress in Clinical and Biological Research* (J. F. Hoffman, G. H. Giebisch, and L. Bolis, eds.), pp. 49, Alan R. Liss, New York.

Schatz, G., and Butow, R. A., 1983, How are proteins imported into mitochondria?, *Cell* **32:**39–52.

Scheffler, I. E., 1974, Conditional lethal mutants of Chinese hamster cells: Mutants requiring exogenous carbon dioxide for growth, *J. Cell. Physiol.* **83:**219–230.

Soderberg, K., Mascarello, J. T., Breen, G. A. M., and Scheffler, I. E., 1979, Respiration-deficient Chinese hamster cell mutants: Genetic characterization, *Somat. Cell Genet.* **5:**225–240.

Soderberg, K., Ditta, G., and Scheffler, I. E., 1977, Mammalian cells with defective mitochondrial functions: A Chinese hamster cell line lacking succinate dehydrogenase activity, *Cell* **10:**697–702.

Soderberg, K., Nissinen, E., Bakay, B., and Scheffler, I. E., 1980, The energy charge in wild-type and respiration-deficient Chinese hamster cell mutants, *J. Cell. Physiol.* **103:**169–172.

Sun, N. C., Chang, C. C., and Chu, E. H. Y., 1975, Mutant hamster cells exhibiting a pleiotropic effect on carbohydrate metabolism, *Proc. Natl. Acad. Sci. USA* **72:**469–473.

Veeger, C., Der Vartanian, D. V., and Zeylemaker, W. F., 1969, Succinate dehydrogenase, in: *Methods in Enzymology* (J. M. Lowenstein, ed.), pp. 81–90, Academic Press, New York.

Whitfield, C. D., Bostedor, R., Goodrum, D., Haak, M., and Chu, E. H. Y., 1981, Hamster cell mutants unable to grow on galactose and exhibiting an overlapping complementation pattern are defective in the electron transport chain, *J. Biol. Chem.* **256:**6651–6656.

Wiseman, A., and Attardi, G., 1979, Cytoplasmically inherited mutations of a human cell line resulting in deficient mitochondrial protein synthesis, *Somat. Cell Genet.* **5:**241–262.

4

Glutaminolysis in Animal Cells

WALLACE L. McKEEHAN

1. GLUTAMINE METABOLISM IN MAMMALS

Glutamine is the most abundant amino acid in plasma and most tissues (Van Slyke *et al.*, 1943). Because of both empirical reasoning and cellular require- ments determined experimentally, it is the most abundant amino acid in most cell culture media (Ham and McKeehan, 1979). Although other amino acids have metabolic functions in addition to protein and peptide synthesis, glutamine is the most versatile (Krebs, 1980). It is the major source of urinary nitrogen and a key factor in acid–base balance in mammals. The carbon skeleton of glutamine is an important precursor of glucose in kidney cortex and thus contributes to renal gluconeogenesis (Krebs, 1963; Goodman *et al.*, 1966). Glutamine is a vehicle for transporting nitrogen among tissues. Skeletal muscle is the principal site of glutamine production. Release of glutamine from muscle is nearly four times that that can be accounted for by direct protein breakdown (Blackshear *et al.*, 1975; Pardridge and Casenello-Ertl, 1979; Garber, 1980). The principal site of net glutamine metabolism appears to be the gut (Windmueller and Spaeth, 1974; Hanson and Parsons, 1977) followed by the liver (Blackshear *et al.*, 1975). Glutamine is a key metabolite for elimination of toxic ammonia in nerve tissue and may be an important precursor of glutamate and α-aminobutyrate, a synaptic transmitter (Waelsch, 1960; Takagaki *et al.*, 1961). In addition to its specific roles in multiple tissues, glutamine is the primary amino group donor in synthesis of purines and pyrimidines, amino sugars, pyridine nucleotides, and asparagine in mammalian cells. The reader is referred to the following books and reviews for an in-depth picture of the role of glutamine (and glutamate) in mammals:

WALLACE L. McKEEHAN • W. Alton Jones Cell Science Center, Lake Placid, New York 12946.

Meister (1956, 1965), Lund *et al.* (1970), Prusiner and Stadtman (1973), Shepartz (1973), Meister (1978), Munro (1978), Mora and Palacios (1980), Kovacevic and McGivan (1983).

In recent years, the importance of glutamine as a respiratory fuel and source of anabolic metabolites in several tissues has become increasingly apparent. In this respect, glutamine metabolism is intimately related to carbohydrate metabolism and an appropriate subject for inclusion in a synopsis on carbohydrate metabolism in cultured cells. A strong correlation has emerged between tissues that proliferate or have proliferative potential and those that extensively metabolize glutamine. It is this extensive metabolism of glutamine that I refer to as glutaminolysis and is the focus of this chapter.

2. GLUTAMINOLYSIS IN TISSUES

2.1. Liver and Kidney

Glutamine metabolism in the liver and kidney occupies special domains of overall glutamine metabolism in animals that will not be considered in detail here. Intact liver and isolated hepatocytes can both extract and deliver glutamine from and to the blood, depending on blood glutamine levels (McMenamy *et al.*, 1962; Lund, 1971; Saheki and Katunuma, 1975; Lund and Watford, 1976; Deaciuc and Petrescu, 1980). Under conditions of metabolic acidosis, the kidney extracts large quantities of glutamine for ammonia production (Van Slyke *et al.*, 1943; Pitts *et al.*, 1972). The magnitude and significance of renal glutamine metabolism under normal conditions are unclear. The carbon skeleton of glutamine is converted to both glucose and carbon dioxide, but the relative importance of the two pathways and the nature of the pathway that produces carbon dioxide are unresolved (Watford *et al.*, 1980). Both liver and kidney cells have a high capacity for compensatory growth, but the role of glutamine metabolism in proliferation of liver and renal epithelial cells is not understood.

2.2. Brain

Arteriovenous differences across the brain suggest a net glutamine synthesis (Lund, 1971; Hills *et al.*, 1972). However, the metabolism of glutamine is compartmented among the neurons, nerve endings, and glial cells (astrocytes) (for reviews, see Berl and Clarke, 1969; Berl *et al.*, 1975; Tapia, 1980; Shank and Aprison, 1981). Glutamine metabolism consists of degradation to glutamate and γ-aminobutyrate in neurons. After release of glutamate at nerve terminals, it is taken up by glial cells, which resynthesize glutamine and then release it into the extracellular space. The quantitative importance of this ''glutamine cycle'' to pools of glutamate and γ-aminobutyrate and neurotransmission as well as to the energy and ammonia metabolism of nerve cells is unclear (Bradford and Ward, 1975, 1976; Tapia, 1980; Hertz et al., 1980; Shank and Aprison, 1981; Kovacevic and McGivan, 1983). Glutamine as a possible energy source in the brain has recently been reviewed (Tildon, 1983).

2.3. Pancreas

Differences in arterial and venous substrates across a section of dog pancreas revealed a large net uptake of glutamine (-300%) and glucose (-900%) and a net production of ammonia ($+310\%$), glutamate ($+145\%$), and glycine ($+140\%$) (Pinkus and Berkowitz, 1980). [$1\text{-}^{14}C$]-Glutamine was metabolized to labeled carbon dioxide (25–30%), glutamate (40–50%), amino/organic acids (10–20%), and acid-insoluble material (5–10%). The large release of glutamate into the blood by the pancreas differed markedly from intestine and other tissues and cells, but was similar to the fetus (Section 2.6).

2.4. Mammary Gland

Arteriovenous differences across the perfused mammary gland of lactating rats indicated a selective removal of glutamine and alanine (Vina and Williamson, 1981). The amount of glutamate extracted from the perfusion medium was negligible. The results were interpreted to indicate the increased demand for carbon for lipogenesis in the differentiated mammary gland.

2.5. The Intestine

Early animal studies revealed a large net uptake of glutamine by organs drained by the hepatic portal vein that was subsequently traced to the small intestine (reviewed by Windmueller, 1982). Arteriovenous perfusion studies found that small intestinal cells in man and most animals removed 20 to 30% of plasma glutamine from a single pass of blood (Windmueller and Spaeth, 1974). Over 50% of the removed glutamine was oxidized to carbon dioxide, which constituted 35% of the total carbon dioxide produced by the tissue. Only about 15% of glutamine carbons appeared in acid-insoluble material in the tissue. The remainder distributed among lactate (8%), citrate (5%), citrulline (6%), proline (5%), alanine (1%), other organic acids (4%), and amino acids (0.1%). A net release of nitrogen in ammonia (38%), citrulline (28%), alanine (24%), proline (7%), glutamate (2%), and ornithine (1%) accounted for the net uptake of glutamine nitrogen during perfusion (Windmueller and Spaeth, 1980). Glutamine utilization was localized in the villus and crypt cells of the mucosal epithelium. These elegant studies clearly established the small intestine as a major site of glutaminolysis in normal animals and humans. Currently, it is unclear what the relative contribution of glutaminolysis is to the extensive cell proliferation observed in the intestinal crypts and to normal intestinal cell function. It is possible that specific steps of glutaminolysis differ among different cell types in the intestine, especially between proliferating and resting cells.

2.6. Embryonic and Placental Tissue

Measurements across the umbilical artery and vein indicated a large utilization of plasma glutamine by the fetus (Ishikawa, 1976). There was no uptake of acidic amino acids and a net flux of glutamate from the fetus (Lemons et al., 1976; Schneider et al., 1979). The fetus-derived glutamate was almost totally taken up by the placenta with negligible release to the maternal circulation (Schneider et al., 1979). In addition, very little maternal glutamate reached the fetal circulation except at very high levels in the perfusion medium (Steginck et al., 1975). The amount of glutamate retained in the placenta accounted for little of the glutamate uptake, which suggested significant metabolism of the glutamate by the trophoblast (Schneider et al., 1979). Placental homogenates and extracts have the capability to deamidate glutamine to glutamate and ammonia (Luchinsky, 1951; Adachi, 1967). Glutamine, but not glutamate, caused an increase in ammonia production by placental tissue slices in vitro (Holzman et al., 1979). Glutamine and glutamate also caused a small increase in lactate production (Holzman et al., 1979). Glutamate supported oxidative phosphorylation in isolated mitochondria from human placenta (Olivera and Meigs, 1975). Glutamine was the most efficient substrate for ammonia production in isolated placental mitochondria although glutamate also increased ammonia release in the presence of ADP and NADP (Makarewicz and Swierczynski, 1982). These experiments indicated an extensive cooperative metabolism of glutamine and glutamate between fetus and placenta. They indicated that glutaminolysis may be an important pathway for disposal of fetus-produced glutamate as well as a source of energy for placental function.

2.7. Tumors

Glutamine is taken up in high quantities relative to other amino acids by a variety of neoplasms (Roberts and Borges, 1955; Rabinovitz et al., 1956; Roberts et al., 1956; Roberts and Simonsen, 1960). Plasma glutamine levels were significantly decreased in animals bearing breast tumors, Hodgkin's disease, reticular sarcomas, and leukemias (Roberts and Simonsen, 1960). This was largely interpreted as a requirement for glutamine for protein synthesis and an amino group donor for nucleic acid biosynthesis. However, tracer work showed a substantial metabolism of the resulting glutamate to other acid-soluble metabolites as well. Robert *et al.* (1956) showed that [^{14}C]glutamate, injected into Yoshida sarcoma-bearing rats, appeared in glutamate, succinate, and aspartate. Tracer studies of Nyhan and Busch (1958a), which involved injection of [U-^{14}C]glutamate into rats bearing Walker 256 carcinosarcomas, indicated that the isotope appeared in lactate in the tumor at all times tested. At 3 to 8 min after injection, 25% of the isotope from [^{14}C]glutamate that was utilized by the tumor was in lactate. The remainder of the isotope appeared in protein, carbon dioxide, glutamine, succinate, and aspartate. In contrast, significant quantities of isotope in nontumor host tissues were not found in lactate at 30 to 60 sec after injection of isotope. Isotope from [2-^{14}C]succinate also appeared significantly in lactate in tumor tissue accounting for over 51% of the isotope utilized by the tumor (Nyhan and Busch, 1958b). This was in contrast to normal tissues where the isotope appeared in organic acids other than lactate. These experiments suggested an active pathway in tumors for oxidation of glutamate to carbon dioxide, four- and three-carbon intermediates, pyruvate, and lactate relative to normal tissues.

Arterial–iliac vein differences across rat thigh muscle implanted with a Walker carcinosarcoma indicated a high extraction rate of glutamine by the tumor (Ishikawa, 1976). Sauer *et al.* (1982) and Sauer and Dauchy (1983) employed a method that restricted the blood supply and drainage of an implanted tumor to a single artery and vein in order to sample blood levels of various metabolites before and after passing through the tumor. Arteriovenous differences across several hepatomas and a sarcocarcinoma indicated that glutamine was the amino acid most extensively utilized. The utilization of glutamine, glucose, lactate, and ketone bodies directly correlated with arterial blood concentrations of each metabolite. Lactate was either produced or utilized depending on arterial blood concentrations, but was surprisingly independent of arterial glucose concentration or the glucose utilization rate. Tumors growing in fasted rats extracted and utilized glutamine at an even higher rate than fed rats (Sauer and Dauchy, 1983). Although the pathway and enzymology of glutamine metabolism in various tumors *in vivo* have not been elucidated, the experiments described in this section implicate glutamine and glutaminolysis as a potentially important fuel and amino group and carbon source for maintenance and growth of tumors *in vivo*.

3. GLUTAMINOLYSIS IN ISOLATED TISSUES AND PRIMARY CELL SUSPENSIONS

3.1. Enterocytes

Probably the earliest report of intestinal glutaminolysis was from an experimental system using isolated ileac tissue *in vitro* (Neptune, 1965). Studies from rats, hamsters, guinea pigs, rabbits, and monkeys established that high rates of conversion of [^{14}C]glutamine to $^{14}CO_2$ occurred in the intestine. Subsequent studies showed that glutamine stimulated oxygen uptake in rabbit intestinal mucosa (Frizzell *et al.*, 1974) and in dissociated cell suspensions (Towler *et al.*, 1978; Watford *et al.*, 1979a; Porteous, 1980). Glutamine carbon and nitrogen generally appeared in metabolites reported for the perfusion studies of intact intestine (Windmueller, 1982). However, in contrast to the perfusion studies, a large amount of glutamine-derived glutamate appeared in the medium of isolated rat enterocytes while glutamine carbon could not be detected in citrulline or proline (Watford *et al.*, 1979a). It was unclear whether these differences represented damaged cells that leaked glutamate, loss of key enzyme activities, or the selection for a specific cell population *in vitro*.

3.2. Blood Cells

Mammalian reticulocytes extensively consume oxygen even after external substrates are removed (Warburg *et al.*, 1931). The endogenous substrate was found to be amino acids from the breakdown of stromal proteins (Schweiger *et al.*, 1956). Subsequent study revealed that glutamine was the most extensively oxidized of several substrates tested (Ababei *et al.*, 1962). Glutamine was a 4- to 16-fold better source for carbon dioxide than glutamate (Rapoport *et al.*, 1971). Carbon dioxide formed from glutamine or glutamate accounted for up to 80% of oxygen consumed. At the average glutamine concentration of plasma, it was estimated that up to 50% of respiratory carbon dioxide produced in the reticulocyte was due to glutamine. The contribution of glutamine to oxidative metabolism of reticulocytes may be even higher in the bone marrow, since glutamine concentrations up to 20 times that of plasma have been reported (Rapoport *et al.*, 1971).

Maximal enzyme activities, oxygen consumption, and metabolite distributions suggest that glutamine may be an important respiratory fuel and source of anabolic substrates during activation of primary suspensions of lymphocytes from rat mesenteric nodes (Ardawi and Newsholme, 1982a,b). The major acid-soluble products of glutamine were glutamate, aspartate, and ammonia. Exposure of the lymphocytes to the mitogen concanavalin A stimulated glutamine metabolism and accumulation of ammonia by 50 to 80% (Ardawi and Newsholme, 1982b).

3.3. Lens

The presence of oxygen is required for maintenance of the clarity, salt balance, labile organic phosphate and protein synthesis of calf lens in the absence of glucose. This suggested the presence of an important aerobic substrate other than glucose for lens *in vitro* (Trayhurn and Van Heyningen, 1971). The substrate was subsequently identified as amino acids (Trayhurn and Van Heyningen, 1973a). Of the amino acids tested, glutamate was the most extensively oxidized. Subsequent studies indicated that glutamine was probably the source of glutamate under physiological conditions since glutamate was poorly transported and little glutamate was synthesized from glucose in lens tissue (Trayhurn and Van Heyningen, 1973b). The uptake of glutamine into lens tissue *in vitro* was three times that of glutamate. Tracer glutamine was readily converted to carbon dioxide, protein, and other metabolites. In contrast to other glutamine-metabolizing tissues, over 30% of isotope from glutamine appeared in glutathione (Trayhurn and Van Heyningen, 1973a,b). It was concluded that glutaminolysis likely occurred in the epithelial cells in the outermost layers of the lens.

3.4. Germ Cells

In the presence of only vitamins A, E, C and glutamine, gonocytes differentiated into pachytene spermatocytes in short-term organ cultures of rat testes (Steinberger and Steinberger, 1966). The fate and role of glutamine and derived glutamate were not elucidated. Hamster oocytes developed *in vitro* in the presence of only glutamine, isoleucine, phenylalanine, methionine, and a 286 mOsm salt solution (Gwatkin and Haidri, 1973). Glutamine and proline, but not other amino acids, stimulated rabbit follicular oocytes to develop to the prophase and metaphase stages in the absence of carbohydrate (Bae and Foote, 1975a,b). Significant amounts of tracer glutamine carbons appeared in carbon dioxide. Carbon dioxide production by associated cumulus cells was about threefold that of oocytes in the primary cell suspensions.

3.5. Calvaria

Calvaria from newborn rats converted significant amounts of glutamine to ammonia, glutamate, alanine, aspartate, proline, ornithine, and carbon dioxide (Blitz *et al.*, 1982). Glutamine did not give rise to lactate although addition of glutamine at 0.5mM increased and at 5mM decreased lactate production from glucose.

3.6. Astrocytes

Glutamate can substitute for glucose as a metabolic substrate for astrocytes in primary culture (Yu *et al.*, 1982; Tildon, 1983). Substantial amounts of glutamate are oxidized to carbon dioxide by the cultures. The conversion was not sensitive to the aminotransferase inhibitor, aminooxyacetate.

4. GLUTAMINOLYSIS IN NORMAL AND TUMOR CELLS IN CULTURE

4.1. Tumor-Derived Cells and Transformed Cell Lines

Both tumor cells and normal cells that proliferate in culture generally exhibit a high glutamine requirement (Eagle *et al.*, 1956; Ham *et al.*, 1977; Ham and McKeehan, 1979). This has largely been explained as a requirement for protein and nucleic acid biosynthesis (Levintow *et al.*, 1957; Salzman *et al.*, 1957) as well as an instability of glutamine under cell culture conditions (Gilbert *et al.*, 1949; Tritsch and Moore, 1962; Griffiths and Pirt, 1967; Wein and Goetz, 1973).

4.1.1. Ehrlich Ascites Cells

Examination of the fate of [U-^{14}C]glutamine in Ehrlich ascites cells revealed that the major end products of glutamine metabolism were carbon dioxide, glutamate, and aspartate (Coles and Johnstone, 1962).

4.1.2. Mouse L Cells

Kitos *et al.* (1962) demonstrated that mouse L fibroblasts converted [^{14}C]glutamine to carbon dioxide and a significant amount of acid-soluble products relative to acid-insoluble products, but concluded that the rate of conversion was not sufficient for consideration of glutamine as a principal energy source. Significant glutamine carbons appeared in the medium as glutamate and proline, but not alanine and aspartate. Proliferating mouse L-M fibroblasts utilized glutamine in excess of the requirements for new cellular components (Stoner and Merchant, 1972). About 55% of glutamine carbons appeared as carbon dioxide and it was estimated that glutamine provided up to 35% of cellular energy requirements in the presence of high glucose.

4.1.3. Hepatomas

The rate of glutamine utilization was directly correlated with the degree of malignancy in a series of rat hepatomas (Knox *et al.*, 1969; Linder-Horowitz *et al.*, 1969). A comparison of the rate of oxygen consumption, glutaminase activity, and carbon dioxide production in the presence of glutamine, glucose, glutamate, and pyruvate suggested that glutamine was potentially the most important substrate for oxidative metabolism in rapidly proliferating, malignant hepatoma cells (Kovacevic, 1971; Kovacevic and Morris, 1972). Fluoropyruvate and vinblastine inhibited oxygen uptake in sarcoma 180 and three Morris hepatoma cell lines (Regan *et al.*, 1973). The inhibition could be reduced by increasing the level of glutamate in the culture medium.

4.1.4. Lymphoma

Lavietes *et al.* (1974) also demonstrated the relative contributions of labeled glucose, glutamine, and glutamate to $^{14}CO_2$ production and oxygen uptake in 6C3HED lymphoma cells. Glutamine oxidation accounted for 70 to 80% of oxygen uptake while glucose supported less than 10%. Both glutamine and glutamate supported respiration, but glutamine was oxidized much more rapidly.

4.1.5. Chinese Hamster Ovary Cells

Using a genetic approach, Scheffler and co-workers (see Scheffler, this volume) generated mutants from wild-type Chinese hamster cell lines with defects in respiration that were highly dependent on glycolysis for energy. By comparison of glycolysis and glutaminolysis in the mutant and wild-type cells, they concluded that glutaminolysis supplies about 40% of the energy for cell proliferation in the wild-type ovary cell line (Donnelly and Scheffler, 1976).

4.1.6. HeLa Cells

Reitzer *et al.* (1979) demonstrated that HeLa cells from a human cervical carcinoma grew at a similar rate when the sugar source was glucose, galactose, or fructose. With fructose as the carbohydrate, glycolytic activity was reduced to 0.1% of that in the presence of glucose, yet cell proliferation continued. Irrespective of the sugar present, glutamine was metabolized rapidly to carbon dioxide (35%), lactate (13%), and macromolecules (18–25%). Only 2% of glutamine carbons appeared in protein. It was calculated that when glucose was present, glutaminolysis provided over 50% of the cell's energy and near 98% when fructose or galactose was substituted for glucose.

4.2. Untransformed Normal Cells in Culture

4.2.1. Fibroblasts

Zielke *et al.* (1976) demonstrated that normal human diploid fibroblasts doubled one to two times without glucose utilization provided the medium was supplemented with hypoxanthine, thymidine, and uridine. In the absence of glucose, cells continued to produce lactate, but at a reduced rate. Glutamine utilization was extensive and calculated to supply at least 30% of the cellular energy requirement in the presence of glucose (Zielke *et al.*, 1978). A significant amount of glutamine carbons appeared in the culture medium as lactate (10%) and glutamate (15%).

4.2.2. Skeletal Muscle

Pardridge et al. (1978) and Pardridge and Casenello-Ertl (1979) have examined aspects of glutamine metabolism in a myogenic cell line, L6, that displayed some properties of skeletal muscle tissue. In contrast to tumor cells and proliferating normal fibroblasts, which converted nearly stoichiometric amounts of glucose to lactate, only 35 to 40% of metabolized glucose could be accounted for as lactate in L6 myotubes. However, many aspects of amino acid and glucose metabolism in myotube cultures were dissimilar to that in skeletal tissue (Pardridge et al., 1978). Although alanine production was similar in degree to adult skeletal muscle, the L6 myotubes exhibited a net utilization of glutamine (Pardridge et al., 1978) and production of ammonia much in contrast to normal muscle, which is a site of net glutamine production (Garber et al., 1976). Depletion of glutamine in the medium resulted in a marked reduction of cellular aspartate, and a lesser reduction in cellular glutamate, alanine, and malate (Pardridge and Casenello-Ertl, 1979). Surprisingly, there was no reduction in ammonia production. Subsequent experiments revealed that primary cultures of neonatal rat skeletal muscle cells that fused into myotubes did not exhibit accelerated rates of glutamine utilization (Pardridge et al., 1980). This property more closely resembled muscle in vivo. Stanisz et al. (1983) compared the oxidation of glucose, glutamine, and fatty acids in proliferating cultures of chick heart muscle cells to HeLa cells (Reitzer et al., 1979). Both the cultured heart muscle cells and HeLa cells exhibited similar patterns of glucose and glutamine utilization. However, muscle cells were capable of producing at least 100 times as much carbon dioxide from external fatty acids as were HeLa cells (Stanisz et al., 1983). The differences in fatty acid oxidation were not absolute, but depended on the type and concentration of serum in the test medium.

4.2.3. Adrenal Cells

Hornsby and Gill (1981) compared the relative production of carbon dioxide from glutamine and glucose in bovine adrenal cortical cells in culture. The results were expressed as a ratio of carbon dioxide formed from [U-^{14}C] glutamine to $^{14}CO_2$ formed from [2-^{14}C]pyruvate. The ratio rose as the number of population doublings of the culture increased. As the ratio rose, cell growth became more sensitive to the aminotransferase inhibitor aminooxyacetate and lipid peroxidation (Hornsby and Gill, 1981). The authors interpreted these results to indicate a progressive decrease in the capacity of cells to control lipid peroxidation, which inhibited normal pyruvate oxidation and caused the shift toward glutamine/glutamate oxidation.

5. GLUTAMINOLYSIS—THE PATHWAY OF GLUTAMINE OXIDATION

5.1. Enzymology

Figure 1 shows a generalized model of a complete pathway of metabolism of glutamine to pyruvate. If viewed as a linear pathway of glutamine to pyruvate, glutaminolysis has some similarities to glycolysis or, more accurately, the Embden–Meyerhof pathway of glucose to pyruvate (McKeehan, 1982). Both glucose and glutamine are readily permeable to plasma membranes and probably enter cells by active, controlled transport mechanisms. Subsequent metabolites, the dicarboxylic acids of glutaminolysis and the phosphorylated intermediates of glycolysis, are far less permeable and essentially trapped within the cell (Eagle *et al.*, 1956; Hems *et al.*, 1968; W. L. McKeehan, unpublished results). Glutaminolysis is perceived to proceed by an irreversible deamidation of glutamine to glutamate (Fig. 1, step 1) followed by a reversible, near-equilibrium second step that removes the 2-amino group from glutamate (step 2). This is followed by an irreversible third step that involves oxidation of 2-oxoglutarate to succinyl CoA (step 3). The pathway then proceeds by two reversible, energy-yielding steps (steps 4 and 5) followed by a reversible hydratase reaction to yield malate (step 6). The pathway ends in a potentially irreversible conversion of malate to pyruvate (step 7). Both glycolysis and glutaminolysis are potentially characterized by essentially irreversible, rate-limiting reactions at the first, third, and last steps of the pathways. Similar to glycolysis, the first, third, and last steps of glutaminolysis are also potentially catalyzed by complicated allosteric enzymes that respond to a variety of effectors.

5.1.1. Glutamine to Glutamate (Fig. 1, Step 1)

The removal of the amide amino group of glutamine to yield glutamate involves a large free energy drop and is an essentially irreversible reaction (Meister, 1962, 1965). Thus, this step potentially constitutes the rate-limiting reaction that initiates glutaminolysis and is a good candidate for overall regulation of the pathway.

5.1.1a. Glutaminase (Fig. 1, Step 1a). The presence and level of simple glutaminase activity, glutamine $\rightarrow$ glutamate $+ NH_4^+$, are strongly correlated with tissues and cells that exhibit high rates of glutamine oxidation (Williams and Manson, 1958; Knox *et al.*, 1967, 1969; Linder-Horowitz *et al.*, 1969; Kovacevic, 1974; Pinkus and Windmueller, 1977; Sevdalian *et al.*, 1980; Blitz *et al.*, 1982). Detailed studies on the reaction kinetics and physicochemical properties

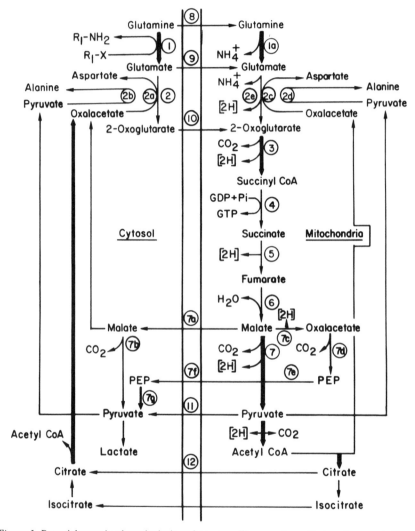

Figure 1. Potential steps in glutaminolysis and some ancillary reactions. Darkened arrows indicate possible one-way reactions. Some cofactors are omitted for clarity. Reactions with carbon dioxide and reducing equivalents ([2H]) as products are indicated. Circled numbers indicate reaction steps described in the text.

of purified glutaminase activities from different glutaminolytic tissues and cells are scarce. Glutaminases exist in two general classes first recognized by Krebs (1935): "brain-type" and "liver-type." "Brain-type" was called "kidney-type" in later studies (Horowitz and Knox, 1968). The two isozymes differ by requirement for P_i and other anions, pH optima, K_m for glutamine, and response to various effectors (Horowitz and Knox, 1968; Katunuma et al., 1967, 1973). The requirement of the solubilized liver-type glutaminase for P_i ($K_m = 28$ mM) is lower and the requirement for glutamine (K_m 28 to 42 mM) is higher (Horowitz and Knox, 1968; Huang and Knox, 1976) than the kidney-type. Generally, glutaminase activity in crude extracts of glutaminolytic cells and tissues appears to be the phosphate-activated, kidney-type isozyme of the two classes of isozymes in rat tissues reviewed by Katunuma et al. (1967, 1973). In small intestine, glutaminase was located in the mucosal epithelial cells (Windmueller and Spaeth, 1974), was membrane-bound, and sedimented with the mitochondria on density gradients (Pinkus and Windmueller, 1977). The apparent K_m for glutamine at pH 8.1 was 2.2 mM, much lower than the K_m reported for glutaminase activity from liver (Katunuma et al., 1967). Glutaminase activity was activated by phosphate at a $K_{0.50}$ of 22 mM (Pinkus and Windmueller, 1977). Immunochemical tests indicated that extracts of intestinal mitochondria reacted with antiserum to purified kidney glutaminase. Liver mitochondrial extracts containing glutaminase activity did not react with the same antiserum (Windmueller, 1982). Based on examination of the possible glutamine-degrading enzyme activities in crude extracts of small intestine, Windmueller (1982) concluded that phosphate-dependent glutaminase activity was the most active enzyme capable of conversion of glutamine to glutamate and the single enzyme with sufficient activity in vitro to account for the high rates of glutamine carbon metabolism and NH_4^+ formation exhibited by intestine in vivo.

The glutaminase activities of tumors and isolated, normal proliferating cells have not been extensively purified and have been studied in far less detail than those of small intestine, brain, liver, and kidney tissue (Katunuma et al., 1973). The apparent K_m for glutamine for glutaminase activity in extracts of Ehrlich ascites cells was about 4.5 mM, the activity was stimulated maximally by 50 mM phosphate (K_a for phosphate was 15.6 mM), and the activity was located in the mitochondria (Kovacevic, 1974). HeLa cell glutaminase activity was also activated by anions and localized in the mitochondria (Williams and Manson, 1958). Glutaminase activity in cell extracts increased during the growth cycle of human diploid fibroblasts (Sevdalian et al., 1980).

The extensive appearance of ammonia during glutamate production from glutamine (Levintow et al., 1957; Kvamme and Svenneby, 1961; Ardawi and Newsholme, 1982a,b) makes phosphate-dependent glutaminase a strong candidate for the significant enzyme responsible for the glutamine-to-glutamate step of glutaminolysis during cell proliferation. Glutaminase activity correlated positively with growth rates and degree of malignancy within the Morris hepatomas (Knox *et al.*, 1967, 1969; Linder-Horowitz *et al.*, 1969). Although the phosphate-dependent glutaminase activity was high in normal resting kidney, the isozyme increased tenfold in kidney tumors (Katunuma *et al.*, 1972). Mammary tumors expressed the kidney-type glutaminase although it was absent in the normal resting gland (Knox, 1976). It is noteworthy that fetal liver contained the kidney-type glutaminase, which became repressed in adult liver where the liver-type isozyme predominated (Knox, 1976). The liver-type isozyme was repressed in regenerating liver (Knox, 1976). As described above, the kidney-type isozyme of glutaminase predominated in small intestine, which exhibits extensive normal cell proliferation (Katunuma *et al.*, 1972). Despite these strong correlations, several other possibilities for the glutamine-to-glutamate step should be noted in the absence of purified and characterized glutaminases and quantitative data on the relationship between external glutamine and ammoniagenesis in various glutaminolytic cell types. First, although glutaminase activity fluctuates with growth rates, total activity in hepatomas is actually lower than in many normal tissues (Haruno, 1956; Raina and Ramakrishnan, 1964; Knox *et al.*, 1967; Lamar, 1968). Second, all the ammonia liberated may not be from glutamine. Other routes of ammonia production, independent of or indirectly from glutamine amino groups (Lowenstein, 1972; Wanders *et al.*, 1980), have not been extensively studied in proliferating cells. As noted earlier, the reduction of glutamine did not cause a proportional decrease in NH_4^+ production by cultured myotubes (Pardridge and Casenello-Ertl, 1979). Third, many glutamine amidotransferases exhibit significant simple glutaminase activity in the absence of the appropriate amino group acceptor (Meister, 1962, 1965; Buchanan, 1973).

5.1.1b. Glutamine Amidotransferases (Fig. 1, Step 1). Regardless of the amino group acceptor (R_1-X, Fig. 1, step 1), glutamate is a common product of glutamine amidotransferase reactions. The contribution of amidotransferase activity to the glutamate pool has not been carefully dissected from the contribution of simple glutaminase activity in proliferating, glutaminolytic cells. Most glutamine amidotransferases are labile upon homogenization and, therefore, it is difficult to accurately quantitate activities in crude cell extracts. Most glutamine amidotransferases are found in the cytosol in contrast to the simple glutaminases, which are localized in the mitochondrial matrix (Buchanan, 1973). Since mammalian cells cannot utilize free NH_4^+ in the biosynthesis of asparagine, purines, pyrimidines, and amino sugars (Levintow *et al.*, 1957; Salzman *et al.*, 1957; Meister, 1962), it can be expected that increased demand for these anabolic substrates during cell proliferation might increase production of glutamate from glutamine. Some tumor cells have constitutively elevated levels of asparagine synthase (Horowitz *et al.*, 1968), carbamoyl phosphate synthase II (glutamine-hydrolyzing) (Aoki *et al.*, 1982), glutamine-phosphoribosyl pyrophosphate amidotransferase (EC 2.4.2.14) (Prajda *et al.*, 1975), and glutamine:fructose-6-phosphate amidotransferase (Tsuiki *et al.*, 1972).

The properties of glutamine amidotransferases have been reviewed in detail by Meister (1962, 1965) and Buchanan (1973).

5.1.2. *Glutamate:2-Oxocarboxylate Aminotransferases and Glutamate Dehydrogenase*

Removal of the 1-amino group of glutamate is the second step of glutaminolysis (Fig. 1, step 2). This step is catalyzed by one or more glutamate:2-oxocarboxylate aminotransferases (steps 2a–d) or NAD(P)$^+$-dependent glutamate dehydrogenase (EC 1.4.1.3) (step 2e). Because of their abundance in most cells and tissues, aspartate aminotransferase (EC 2.6.1.1) (steps 2a, 2c) and alanine aminotransferase (EC 2.6.1.2) (steps 2b, 2d) are most likely the aminotransferases involved. The relative contribution of single aminotransferases to the conversion of glutamate to 2-oxoglutarate in various glutaminolytic cell types has not been established.

In isolated rat enterocytes, molar amounts of ammonia appeared in the medium that sometimes exceeded glutamine uptake (Baverel and Lund, 1979; Watford *et al.*, 1979a). The aminotransferase inhibitor aminooxyacetate reduced alanine formation, but had no effect on carbon dioxide produced from glutamine (Baverel and Lund, 1979). The authors interpreted this to suggest an important role for glutamate dehydrogenase (step 2e) relative to aminotransferases. However, perfusion studies in rat and dog intestine always resulted in equimolar ammonia produced to glutamine utilized (Windmueller, 1982). Alanine was a more significant fate than aspartate of both carbons and nitrogen from glutamine in both perfused and cultured intestinal cells (Hanson and Parsons, 1977; Watford *et al.*, 1979a; Windmueller, 1982). Therefore, in the absence of a significant involvement of glutamate dehydrogenase (Hanson and Parsons, 1980), this implies an important role of alanine aminotransferase (steps 2b, 2d) in the glutamate-to-2-oxoglutarate conversion in intestinal cells. The maximum activities of several enzymes of glutaminolysis in activated lymphocytes suggested that the aminotransferases were likely more important than glutamate dehydrogenase for conversion of glutamate to 2-oxoglutarate (Ardawi and Newsholme, 1982a). Glutamate dehydrogenase activity was also low in lymphocytes and tumor tissue (Glazer *et al.*, 1974). In lymphocytes, the activity of aspartate aminotransferase was much higher than that of alanine aminotransferase. Kinetic studies suggested that alanine aminotransferase may be the major path of glutamate to 2-oxoglutarate (Moreadith and Lehninger, 1984a). Under their test conditions, which may be similar to tumor cell conditions in vivo, glutamate dehydrogenase would be strongly inhibited by negative metabolite effectors.

No clear positive correlation between either aspartate or alanine aminotransferase activity and the growth rate of tumor cells has been established (Morris, 1972; Shaffer and Felder, 1983). Aspartate was a significant product of glutamine metabolism in some tumor cells and normal fibroblasts under some conditions (Coles and Johnstone, 1962; Zielke *et al.*, 1981). Kovacevic (1971) demonstrated in isolated kidney, liver, and tumor cell mitochondria that inhibitors of aspartate production had little effect on glutamine oxidation, but inhibited carbon dioxide production from glutamate by more than 50%. He reasoned that, under physiological conditions, glutamate dehydrogenase was the most important enzyme in conversion of glutamate to 2-oxoglutarate in glutaminolytic tumor cells. This study and others with isolated mammalian mitochondria generally suggested two pathways for glutamate metabolism, dependent on the origin of the glutamate (Borst, 1962; Kovacevic, 1971; Schoolwerth and LaNoue, 1980). External glutamate is primarily transaminated to 2-oxoglutarate whereas glutamate generated from glutamine within the mitochondria is oxidatively deaminated by glutamate dehydrogenase. Therefore, more knowledge on the relative contribution of the cytosol and mitochondria to the glutamine-to-glutamate step of glutaminolysis may shed more light on the second step of the pathway (Fig. 1).

5.1.3. 2-Oxoglutarate Dehydrogenase Complex, Succinyl CoA Synthase, Succinate Dehydrogenase, Fumarase

The conversion of 2-oxoglutarate to malate likely occurs in four well-characterized steps, the enzymes of which, except for possibly fumarase (Janski and Cornell, 1980), are confined to the mitochondria (Lehninger, 1975) (Fig. 1, steps 3–6). Each of these enzymes is part of the TCA cycle in cells that extensively oxidize glucose and has been reviewed elsewhere in that role (Lowenstein, 1969; Singer *et al.*, 1973; Whittaker and Danks, 1978; Hansford, 1980; Ottaway *et al.*, 1981). Succinyl CoA:acetoacetate CoA transferase may also contribute to the succinyl CoA-to-succinate step (Fig. 1, step 4) of glutaminolysis in some cells (Fenselau *et al.*, 1975, 1976).

5.1.4. NAD(P)$^+$-Linked Malic Enzymes

Malic enzymes are likely involved in the malate-to-pyruvate step of glutaminolysis (Fig. 1, step 7). Three distinct isozymes of malic enzyme have been described in mammals. Strictly NADP$^+$-dependent (EC 1.1.1.40) forms are found in cytosol and mitochondria (Brdiczka and Pette, 1971; Tsoncheva, 1974). Mitochondria of some tissues also contain a malic enzyme that works with either NADP$^+$ or NAD$^+$ (Linn and Davis, 1974; Mandella and Sauer, 1975). Based on electrophoretic mobility toward the anode, the NADP$^+$-linked mitochondrial, the NAD(P)$^+$-linked mitochondrial, and the NADP$^+$-linked cytosolic malic enzymes have been designated isozymes 1, 2, and 3, respectively (Sauer and Dauchy, 1978). Malic enzymes 1 and 3 catalyzed the decarboxylation of malate to pyruvate in the presence of $NADP^+$ as well as the carboxylation of pyruvate to malate in the presence of NADPH and bicarbonate (Simpson and Estabrook, 1969; Linn and Davis, 1974). Both are thought to play major roles in the production of NADPH for reductive biosynthesis (Wise and Ball, 1964; Simpson and Estabrook, 1969). In contrast to malic isozymes 1 and 3, isozyme 2 catalyzed the irreversible decarboxylation of malate to pyruvate in the presence of either NAD^+ or $NADP^+$ (Linn and Davis, 1974; Mandella and Sauer, 1975). Only isozyme 2 was present in the mitochondria of ascites tumors and hepatomas (Hansford and Lehninger, 1973; Sauer and Dauchy, 1978; Sauer et al., 1980). No strictly $NADP^+$-dependent mitochondrial malic enzyme activity (isozyme 1) was evident in either type of tumor mitochondria. The amount of isozyme 2 activity increased with increasing growth rates of hepatomas. A survey for the presence of the $NAD(P)^+$-dependent and $NADP^+$-dependent malic enzyme activities in the mitochondria of tissues of rats revealed the presence of the $NAD(P)^+$-linked activity (isozyme 2) in tissues that undergo extensive rates of cell renewal (Nagel et al., 1980). The mitochondria of small intestinal mucosa, spleen, thymus, lung, and testis had significant $NAD(P)^+$-dependent malic enzyme. Brain, heart, kidney, and skeletal muscle mitochondria contained only the $NADP^+$-linked enzyme (isozyme 1). Small intestinal mucosa, spleen, and thymus mitochondria contained only the $NAD(P)^+$-dependent enzyme (Nagel et al., 1980). $NAD(P)^+$-dependent malic enzyme activity was present in the mitochondria of both human term placenta (Swierczynski et al., 1982) and cultured human fibroblasts (McKeehan and McKeehan, 1982). Both of these normal tissues exhibited high levels of glutaminolysis as discussed earlier.

$NAD(P)^+$-linked malic enzyme exhibited several properties that suggested it is a regulatory enzyme and thus well suited to end the proposed pathway shown in Fig. 1 (step 7). In contrast to the $NADP^+$-dependent malic isozymes, it catalyzes the irreversible decarboxylation of malate to pyruvate, but does not catalyze the conversion of oxalacetate to pyruvate (Linn and Davis, 1974). Succinate, fumarate, and isocitrate have been identified as positive activators and ATP and ADP were competitive inhibitors with respect to malate (Sauer, 1973; Linn and Davis, 1974; Mandella and Sauer, 1975). $NAD(P)^+$ malic enzyme has been purified from mitochondria of canine small intestine (Nagel and Sauer, 1982) and, more recently, from Ehrlich ascites tumor mitochondria (Moreadith and Lehninger, 1984b).

5.1.5. Alternate Pathways of Malate to Pyruvate

Several studies have considered more complicated pathways to account for the appearance of glutamine carbons in pyruvate, alanine, and lactate in glutaminolytic cells and tissues. Small intestinal cells are equipped to convert malate to oxaloacetate via malate dehydrogenase (Fig. 1, step 7c) and the resultant oxaloacetate to pyruvate via oxaloacetate decarboxylase (EC 4.1.1.3) (not shown in Fig. 1) or oxaloacetate to phosphoenolpyruvate via phosphoenolpyruvate carboxykinase (PEPCK) (step 7d) for conversion to pyruvate by pyruvate kinase in the cytosol (step 7g) (Windmueller, 1982; Hanson and Parsons, 1977; Watford *et al.*, 1979b). A specific inhibitor of PEPCK, mercaptopicolinate, had no effect on alanine formation from glutamine in perfused intestinal tissue or intact enterocytes, which argued against involvement of PEPCK (Hanson and Parsons, 1977, 1980, Watford *et al.*, 1979b). Watford *et al.* (1979a) concluded that cytosolic NADP$^+$-linked malic enzyme (EC 1.1.1.40) was the most attractive candidate for the conversion of glutamine-derived malate to pyruvate (step 7b). Hanson and Parsons (1980) argued against a significant role of cytosolic NADP$^+$-dependent malic enzyme on the grounds that while the activity of the enzyme dropped in half in rats during food deprivation (Tyrrell and Anderson, 1971), the metabolism of glutamine to pyruvate increased under similar conditions. The same authors also pointed out that although oxaloacetate decarboxylase had high activity in small intestinal cells (Dean and Bartley, 1973; Watford *et al.*, 1979a), it is doubtful that decarboxylation of oxaloacetate is a physiological activity of the enzyme (Dean and Bartley, 1973). Hanson and Parsons (1980), therefore, considered the mitochondrial NAD(P)$^+$-dependent malic enzyme (Fig. 1, step 7) as the strongest candidate for conversion of malate to pyruvate in small intestine (Sauer *et al.*, 1979). Reitzer *et al.* (1979) assumed that glutamine-derived malate exited the mitochondria and was converted to pyruvate by cytosolic NADP$^+$-dependent malic enzyme (EC 1.1.1.40) in HeLa cells (step 7b). Sumbilla *et al.* (1981) showed that the specific activities of mitochondrial PEPCK and cytosolic NADP$^+$-dependent malic enzyme, malate dehydrogenase, and aspartate aminotransferase increased during growth of human fibroblasts, but the authors offered no conclusion as to which enzyme was most important for conversion of malate to pyruvate. Based on the maximal activities of relevant enzymes, Ardawi and Newsholme (1982a,b) suggested that the mitochondrial NAD$^+$-dependent malic enzyme (step 7) might be important in conversion of glutamine-derived malate to pyruvate in lymphocytes.

Recent kinetic study of effects of substrates on glutamine oxidation in isolated tumor mitochondria by Moreadith and Lehninger (1984a) disagree with a simple linear pathway of glutaminolysis (McKeehan, 1982; Fig. 1). They observed nearly quantitative release of citrate and alanine into the medium of mitochondria during oxidation of glutamine. In contrast to normal tissues, the tumor mitochondrial pathway of malate oxidation depended on the origin of the malate. External malate was almost exclusively oxidized to pyruvate by NAD(P)$^+$ malic enzyme, whereas intramitochondrial malate was oxidized to oxalacetate and its products by malate dehydrogenase.

5.2. Compartmentation

Glutaminolysis may occur completely within the mitochondria or may involve translocation of intermediates between cytosol and mitochondria (Fig. 1). For general reviews on mitochondrial transport of relevant intermediates, see Whittaker and Danks (1978) and LaNoue and Schoolwerth (1979). As pointed out above, knowledge of the significance of the activity and location of glutaminolytic enzymes in different glutaminolytic cell types is scarce. It is nearly certain that the intermediate conversion of 2-oxoglutarate to malate occurs in the mitochondria via a linear segment of the TCA cycle as described for glucose-oxidizing cells (Fig. 1). However, the location of the glutamine-to-glutamate-to-2-oxoglutarate steps as well as the malate-to-pyruvate step (or steps) has not been firmly established. Mouse ascites tumor and rat hepatoma cells are equipped to oxidize glutamine carbons to acetyl CoA and carbon dioxide completely within the mitochondria (Sauer and Dauchy, 1978; Sauer *et al.*, 1980). Both cell types contained significant levels of both phosphate-dependent glutaminase and $NAD(P)^+$-dependent malic enzyme within the mitochondria. Although the activity and location of glutamine amidotransferase and glutaminase activities have not been clarified, the mitochondria of cultured human diploid fibroblasts appear to also be able to completely oxidize glutamate (Zielke *et al.*, 1978; Sumbilla *et al.*, 1981; McKeehan and McKeehan, 1982; W. L. McKeehan, unpublished results).

Cytosolic enzymes may contribute to one or both steps in the two-step conversion of glutamine to glutamate to 2-oxoglutarate (Fig. 1, steps 1, 2, 2a, 2b). Glutamine amidotransferases are predominantly cytosolic (Meister, 1962, 1965; Buchanan, 1973) while phosphate-dependent glutaminases are mitochondrial. Glutaminolytic fibroblasts (Sumbilla *et al.*, 1981; W. L. McKeehan, unpublished results) and hepatomas (Morris, 1972) are equipped with high levels of cytosolic aspartate aminotransferase that could account for a cytosolic conversion of glutamate to 2-oxoglutarate in the presence of permissive ratios of cytosolic oxaloacetate and aspartate (step 2a). In contrast, cytosolic alanine aminotransferase is low or absent in normal fibroblasts, normal, proliferating prostate epithelial cells, and HeLa cells (W. L. McKeehan, unpublished results). Total alanine aminotransferase activity is generally reduced in hepatomas relative to normal liver (Morris, 1972). A relatively high amount of glutamine-derived ammonia (presumably α-amino groups) and a low percentage of glutamine-derived carbons appeared in alanine during glutaminolysis in small intestinal cells and most alanine carbons were derived from lactate and glucose (Windmueller and Spaeth, 1974). From these data, Hanson and Parsons (1980) proposed a significant cytosolic conversion of glutamate to 2-oxoglutarate and pyruvate to alanine (step 2b). Significant levels of particle-free alanine aminotransferase activity have been reported in small intestine (Volman-Mitchell and Parsons, 1974; Herzfeld and Roper, 1979). The differences in cytosolic alanine aminotransferase activity between small intestine and cultured fibroblasts may reflect the mixture of cell types present in intact intestinal tissue relative to cultured cells or represent a fundamental property of intestinal epithelial cells where glutaminolysis plays a specialized function in overall glutamine metabolism in the animal.

If the glutamine-to-glutamate conversion (Fig. 1, step 1) occurs significantly in the cytosol, then the transport of glutamate across mitochondrial membranes (step 9) may be an important step in glutaminolysis (Azzi *et al.*, 1967; Williamson, 1976; Hoek and Njogu, 1976). If the conversion of glutamate to 2-oxoglutarate occurs in the cytosol, then the transport of 2-oxoglutarate across the mitochondrial membrane on the specific 2-oxoglutarate carrier or less specific dicarboxylic acid transporter (Klingenberg, 1971; Palmieri *et al.*, 1972) may be an important step (step 10). As discussed earlier, the prime candidate for the malate-to-pyruvate conversion is NAD(P) $^+$-linked malic enzyme, which is strictly mitochondrial (step 7). Alternatively, the conversion may occur by exit of glutamine-derived malate from mitochondria to cytosol on the dicarboxylate carriers or the tricarboxylate carrier (step 7a) (Chappell, 1968; Klingenberg, 1970) where it can be converted to pyruvate by cytosolic NADP $^+$-linked malic enzyme (step 7b). If malate is converted to oxaloacetate (step 7c) and then to PEP (step 7d) in the mitochondria, then the tricarboxylate carrier may also be involved in exit of PEP to the cytosol (step 7f) where it can be converted to pyruvate via cytosolic pyruvate kinase (step 7g). The tricarboxylate carrier appears to be the means of PEP exit in liver and heart mitochondria (Robinson, 1971a,b).

The compartmentation of glutaminolytic reactions has especially important implications for regulation of the pathway in the scheme proposed for tumor cells by Moreadith and Lehninger (1984a).

5.3. Regulation of Glutaminolysis

The paucity of study of glutaminolysis as a pathway means little is known of how the pathway is regulated in a variety of glutaminolytic cell types. From the known properties of the enzymes shown in the scheme in Fig. 1, there are three strategic sites that are good candidates for regulatory control. The first is the glutamine-to-glutamate step (steps 1, 1a). All known mammalian glutaminase or glutamine 5-amidotransferase reactions are essentially irreversible (Meister, 1962, 1965; Katunuma *et al.*, 1972; Buchanan, 1973; Curthoys *et al.*, 1973). Therefore, the glutamine-to-glutamate step may be the first flux-generating step of glutaminolysis. Glutamine amidotransferases catalyze rate-limiting steps of other pathways and their control has been extensively studied (Buchanan, 1973). The mitochondrial phosphate-dependent glutaminase in the rat kidney is a complicated allosteric enzyme. In the presence of phosphate and other anions, it undergoes a reversible association–reassociation, which affects enzyme activity (Godfrey et al., 1977; Morehouse and Curthoys, 1981). Beside phosphate, glutaminase activity in situ is stimulated by NH_4^+, HCO_3^-, glucagon, dibutyryl cAMP, leucine, isoleucine, valine, and ATP (Lund, 1980). The solubilized enzyme is inhibited by glutamate and activated by phosphate, riboflavin-5-phosphate, other phosphate esters, ATP, GTP, ITP, acetate, and di- and tricarboxylic acids. cAMP and cGMP inhibit (Lund, 1980).

The second likely site for regulatory control of the pathway is the 2-oxoglutarate-to-succinyl CoA step catalyzed by the 2-oxoglutarate dehydrogenase complex (Fig. 1, step 3). This step is considered a primary site of control for the segment of the TCA cycle (2-oxoglutarate to malate) that is relevant to glutaminolysis (Denton and McCormack, 1980; Williamson and Cooper, 1980; Hansford, 1980). Ca^{2+} is a strong direct activator of 2-oxoglutarate dehydrogenase from a variety of sources (McCormack and Denton, 1981). The activity of 2-oxoglutarate dehydrogenase is also affected by ADP, ATP, H^+, and the ratios [NAD$^+$]/[NADH] and [succinyl CoA]/[CoASH].

The third strong candidate for regulatory control of glutaminolysis is the final step—the malate-to-pyruvate conversion (Fig. 1, step 7). Preliminary evidence suggests that mitochondrial NAD(P)$^+$-dependent malic enzyme has the properties that are expected of a pacemaker enzyme at the end of a pathway. NAD(P)$^+$-linked malic enzyme catalyzes a nonequilibrium reaction in the malate-to-pyruvate direction, is activated by succinate, fumarate, and isocitrate, and is inhibited by ATP and ADP.

If the cytosol participates in glutaminolysis, the flux of intermediates across mitochondrial membranes may also be a site for regulatory control of the pathway. Mitochondrial transport of glutamine has not been rigorously studied in isolated glutaminolytic cell types. In kidney mitochondria where glutamine entry is usually accompanied by the exit of glutamate, the exact mechanism is controversial (LaNoue and Schoolwerth, 1979). One scheme favors a glutamine/glutamate exchange (Crompton and Chappell, 1973) and the other favors glutamine entry by a uniport mechanism accompanied by exit of glutamate on a separate carrier (LaNoue and Schoolwerth, 1979). Experimental evidence for both electroneutral (Curthoys and Shapiro, 1978) and electrogenic (Kovacevic, 1975) uniport of glutamine have been reported. LaNoue and Schoolwerth (1979) reasoned that the uptake of glutamine in rat kidney mitochondria is unlikely a limiting factor in glutamine metabolism because of its rapid rate relative to glutaminase activity. 2-Oxoglutarate was a competitive inhibitor (K_i = 300 μM) of the glutamine carrier, which had a K_m of 2.7 mM for glutamine (Goldstein and Boylan, 1978).

The known mitochondrial membrane carriers for glutamate, 2-oxoglutarate, and malate are largely reversible and the direction and degree of their activity depend on substrate, redox ratio, and ionic gradients between cytosol and mitochondria (LaNoue and Schoolwerth, 1979). Few studies are available on the properties and regulation of the activity of relevant mitochondrial carriers in glutaminolytic cells. Noteworthy are two recent studies. Kaplan *et al.* (1982) demonstrated that the tricarboxylate carrier (Fig. 1, step 12) in the mitochondria of rat hepatoma cells has an apparent V_{max} at physiological temperatures that was 109% higher than that in normal liver mitochondria. The scheme of Moreadith and Lehninger (1984a) for glutaminolysis in tumor cell mitochondria implies a significant role of malate–citrate influx–efflux which probably involves the tricarboxylate carrier. The V_{max} of mitochondrial pyruvate uptake in hepatomas was depressed relative to that of normal rat liver mitochondria (Paradies *et al.*, 1983).

Plasma membrane uptake of glutamine may limit cellular glutaminolysis, yet little is known of the mechanism and regulation in glutaminolytic cells. Kilberg *et al.* (1980) described a Na^+-dependent transport system (system N) in liver that exhibited specificity for glutamine, asparagine, histidine, and related analogues, insensitivity to glucagon and insulin, and increased activity during amino acid starvation. System N was also present in a hepatoma cell line, but absent in cultured fibroblasts and rat intestinal segments (Shotwell *et al.*, 1983). Glutamine is a substrate for transport systems ASC, A, and L in Ehrlich ascites tumor cells (Koser and Christensen, 1971; Thomas and Christensen, 1971). System A appears the most hormone-responsive system in a wide variety of cell types studied (Guidotti *et al.*, 1978). However, whether glutamine transport via system A can be rate-limiting to glutaminolysis remains to be demonstrated.

An additional control of glutaminolysis and defect of control in tumor cells and other pathologies may lie in multienzyme complexes of sequential enzymes in the pathway especially in mitochondria (Moreadith and Lehninger, 1984a).

6. GLUTAMINOLYSIS AND GLYCOLYSIS IN CELL GROWTH AND FUNCTION

There is a striking correlation between cells and tissues that exhibit high rates of glutaminolysis and the presence of high rates of aerobic glycolysis when both glutamine and glucose are available. To some extent, the availability of glucose reduces the rate of glutamine oxidation in normal and tumor cells (Kvamme and Svenneby, 1961; Donnelly and Scheffler, 1976; Zielke *et al.*, 1978; Reitzer *et al.*, 1979; Lazo, 1981). In contrast, addition of either glucose or glutamine to lymphocytes increased rates of utilization of both nutrients (Ardawi and Newsholme, 1982b). When availability of glucose was unlimited in fibroblasts and HeLa cells, glutaminolysis continued at a significant rate and could contribute up to 40% of cellular energy (Stoner and Merchant, 1972; Donnelly and Scheffler, 1976; Zielke *et al.*, 1976, 1978; Reitzer *et al.*, 1979). Zielke *et al.*

(1981) demonstrated that extracellular aspartate was a key metabolite that fluctuated with the amount of glucose in the medium and the proliferative state of glutaminolytic human fibroblasts. Glutamate biosynthesis was unaffected by external glucose concentration. These results have been confirmed for intracellular aspartate from the author's laboratory (unpublished results). The basis of the changes in cellular aspartate due to glucose and proliferative state is currently unclear, and elucidation of the cause may shed some further light on the relationship between the two pathways during fibroblast proliferation.

6.1. The Role of Glycolysis in Cell Proliferation

A high rate of glycolysis is generally characteristic of tumor cells (Warburg, 1926) and normal cells with proliferative potential (Gregg, 1972; Roos and Loos, 1973; Weinhouse, 1976; Donnelly and Scheffler, 1976; Pederson, 1978; Hume et al., 1978). Much effort has been directed toward implication of the high flux of glucose to lactate in processes essential to cell proliferation (Warburg, 1926; Papaconstantinou et al., 1963; Gregg, 1972; Singh et al., 1974a,b; Wenner, 1975; Racker, 1976; Sols, 1976; Wang et al., 1976; Weinhouse, 1976; Fagan and Racker, 1978; Hume et al., 1978; Lazo and Sols, 1980; Eigenbrodt and Glossmann, 1980). Despite this effort, the cause of the high rate of glucose metabolism and whether glucose metabolism is a cause or consequence of cell growth are still debated. A number of studies indicate that a high rate of glucose uptake and flux of glucose to lactate through the complete glycolytic pathway are unessential to normal cell proliferation and unregulated cell growth. An increase in membrane transport and the intracellular pool size of glucose have been dissociated from the rate of increase in DNA synthesis and cell growth (Romano, 1976; Barsh and Cunningham, 1977; Romano and Cornell, 1982). Mutations have been introduced that exhibited a reduced rate of glucose transport (Pouysségur et al., 1980) arid a deficiency of phosphoglucose isomerase and phosphoglycerate kinase activities (Morgan and Faik, this volume). Lactate production in the mutants was reduced to as low as 10% of wild-type cells, yet the mutants survived, proliferated, and exhibited malignant properties. These results suggested that increased glucose uptake, the conversion of glucose-6-phosphate to fructose-6-phosphate, and the ATP-yielding steps of glycolysis (1,3-diphosphoglycerate to lactate) were unessential to cell proliferation under the test conditions. Reduction of glucose in the medium of cultured cells, which normally exhibited high rates of glycolysis when the supply of glucose was high, resulted in a marked reduction in lactate production without a proportional decrease in rate of cell proliferation (Eagle et al., 1958; Graff et al., 1965; Renner et al., 1972; Rheinwald and Green, 1974; Hume et al., 1978). At low glucose concentrations or in the presence of other hexoses, the flux of hexose carbons into the pentose phosphate pathway was significant (Renner et al., 1972; Reitzer *et al.*, 1979, 1980). Normal human fibroblasts in dense culture were able to undergo limited cell divisions in a medium without glucose containing hypoxanthine, thymidine, uridine, and glycine (Zielke *et al.*, 1976). High levels of uridine or cytidine (>100 μM) completely displaced the carbohydrate required for continuous proliferation of HeLa cells (Wice *et al.*, 1981). High uridine supported clonal growth of human diploid fibroblasts in the absence of other carbohydrate in serum-free medium (W. L. McKeehan, unpublished results). However, a maximal rate of clonal growth still required a low level of glucose (> 100 μM). In HeLa cells, about 90% of the metabolized uridine in carbohydrate-deficient medium appeared in uracil and 23% of the ribose carbon appeared in nucleic acids (Wice *et al.*, 1981). Significant amounts of ribose carbons appeared in carbon dioxide (13–24%) by recycling through the nonoxidative arm of the pentose cycle to fructose-6-phosphate, then glucose-6-phosphate, and then via the oxidative arm of the pentose phosphate pathway (Wice *et al.*, 1981). Appearance of ribose carbons in pyruvate and lactate was negligible. Thus, the essential function of glucose in cell proliferation appears to be to provide anabolic substrates rather than fuel. Hume *et al.* (1978) and Eigenbrodt and Glossmann (1980) have proposed that the apparent high flux of glucose through the intact, complete glycolytic pathway is a consequence of the requirement for high cytosolic levels of essential carbohydrate-derived anabolic precursors (especially ribose phosphate). At a minimum, the essential requirement for carbohydrate may be to supply fructose-6-phosphate and glyceraldehyde-3-phosphate. The latter are essential to the nonoxidative production of ribose-5-phosphate, fructose-6-phosphate is essential for glucosamine-6-phosphate, and glyceraldehyde-3-phosphate is essential for triose phosphate, the precursor to serine, glycine, and glycerol, all of which are essential for macromolecule and lipid biosynthesis during cell proliferation.

6.2. Can Cells Proliferate in the Absence of Glutaminolysis?

Glutamine is an essential requirement for optimal proliferation of cells under conditions so far studied (Ham and McKeehan, 1979). Only specialized normal cells and mutants can synthesize glutamine at high rates from carbohydrate or other amino acids (Tate and Meister, 1973; Goetz *et al.*, 1973; Tiemeier *et al.*, 1973; Miller and Carrino, 1981). The anabolic role of glutamine as an amino group donor in purine, pyrimidine, amino sugar, and asparagine biosynthesis and translational protein synthesis is essential to cell proliferation. Free ammonium ions cannot substitute for the 5-amido group of glutamine (Levintow *et al.*, 1957; Salzman *et al.*, 1957; Meister, 1962).

There is extensive evidence that tumor cell proliferation depends on oxygen supply (for key references, see Olivotto and Paoletti, 1981). However, mutants of wild-type glutaminolytic fibroblasts have been isolated that exhibited reduced oxygen uptake and reduced oxidation of certain TCA cycle intermediates (see Scheffler, this volume; Franchi *et al.*, 1981). Mutants with single specific lesions in NADH-coenzyme Q reductase (complex I) and succinate dehydrogenase have been demonstrated (Scheffler, this volume). The mutant cells still proliferated *in vitro* and were highly dependent on external glucose and glycolysis. Transformed cell mutants that exhibited reduced rates of oxygen uptake also formed tumors in immunosuppressed hosts (Franchi *et al.*, 1981). These results suggest that, similar to the results discussed earlier for glycolysis, a high rate of flux of glutamine carbons to pyruvate through the complete pathway of glutaminolysis (Fig. 1) is unessential to cell proliferation and the malignant phenotype in mutant cells. However, the single-site mutants do not eliminate the remaining anabolic and energy-producing segments of glutaminolysis as essential or important to proliferation. Although a lesion at NADH-coenzyme Q reductase (site I) would severely reduce energy production from oxidation of NADH derived during glutaminolysis (Fig. 1, steps 3, 7, and possibly 2e), substrate level phosphorylation from step 4 and reducing equivalents from step 5 oxidized downstream of NADH-reductase may still contribute significantly to cell proliferation. The single mutation at succinate dehydrogenase should impair neither energy production from steps upstream of the lesion (Fig. 1, steps 1–4) nor oxidation of NADH produced from step 7 (Fig. 1) if alternate sources of fumarate and malate are available. Respiration-deficient mutants exhibit an auxotrophy for CO_2/HCO_3^- and aspartate or asparagine (Scheffler, this volume). This suggests, in agreement with earlier reports (Salzman *et al.*, 1957; Levintow *et al.*, 1957; Coles and Johnstone, 1962; Zielke *et al.*, 1980), that glutaminolysis may be an important contributor of aspartate for anabolic reactions important to cell proliferation (e.g., asparagine) and purine biosynthesis in proliferating cells. The partial substitution of CO_2/HCO_3^- for the aspartate/asparagine requirement may reflect the requirement for CO_2/HCO_3^- in carboxylation reactions that produce oxaloacetate or malate from glucose-derived pyruvate or PEP in the absence of sufficient glutamine-derived malate or oxaloacetate (DeFrancesco *et al.*, 1975; Zielke *et al.*, 1981).

In separate studies aimed at elucidating the role of the electron transport chain in tumor cell proliferation, Olivotto and co-workers (Olivotto and Paoletti, 1981; Olivotto *et al.*, 1983) examined the contribution of the electron transport chain to the entry of Yoshida AH130 hepatoma cells from a noncycling state (G1) to the cycling state (S). They found that anaerobiosis, excess pyruvate, oxaloacetate, or antimycin A inhibited entry into the cell cycle, while 2,4-dinitrophenol did not. The addition of purine bases or folate to the culture medium in part overcame the inhibition. They concluded that the respiration-dependent step of the G1-to-S transition in the hepatoma cells was not dependent on respiratory ATP, but somehow essential in the interconversion of folate derivatives for *de novo* purine biosynthesis (Olivotto *et al.*, 1983).

6.3. The Role of Glutaminolysis in Cell Specialization

In this chapter, I have emphasized the correlation of the presence of glutaminolysis with cells that proliferate or have proliferative potential and the potential importance of the pathway as a source of energy and anabolic precursors for cell proliferation. Although most glutaminolytic tissues exhibit or are capable of significant cell proliferation, glutaminolysis in the proliferating and resting, differentiated fractions of cells from a single tissue has not been dissected. Small intestine is clearly the best example of a normal glutaminolytic tissue that quantitatively plays a specialized role in normal mammalian glutamine metabolism and that exhibits an extensive turnover of normal cells. Glutaminolysis in small intestine has been examined with both cell growth and function in mind (Hanson and Parsons, 1977, 1980; Sauer et al., 1979; Watford et al., 1979a; Nagel et al., 1980; Windmueller, 1982). It has been considered that the major role of intestinal glutamine metabolism is to convert muscle-derived glutamine into alanine for subsequent gluconeogenesis in liver (Hanson and Parsons, 1980; Windmueller, 1982). Watford et al. (1979a) reconciled the concurrent high rate of aerobic glycolysis and glutaminolysis in small intestinal mucosa. They emphasized the need for pyruvate for the potentially important role of the mucosa in conversion of glutamine-derived glutamate carbon and nitrogen to alanine as a detoxification mechanism protecting the circulation from glutamate overload. Liver, in turn, is equipped to metabolize the nontoxic alanine product to glucose, while poorly equipped to handle glutamate. Hanson and Parsons (1980) argued that, since extracellular glucose did not affect glutamine utilization and alanine production in perfused rat jejunum, muscle-derived glutamine is probably oxidized to provide energy for intestinal epithelial cell functions and pyruvate for alanine biosynthesis. This spares circulating glucose from complete oxidation in the intestine and delivers alanine to the circulation for subsequent conversion to glucose in the liver.

Glutaminolysis may also have a specialized function in placental tissue, e.g., disposal of fetus-derived glutamate and provision of energy for other placental functions (Makarewicz and Swierczynski, 1982). Glutaminolysis may also play a specialized role in individual cell types or under certain abnormal physiological conditions in liver, kidney, brain, pancreas, and mammary gland, but the nature and significance of the pathway are unclear as described earlier.

6.4. Potential Impact of Tumor Glutaminolysis on Host Glutamine and Glucose Metabolism

Cachexia is a major cause of death in tumor-bearing individuals (Robins, 1957). The condition is characterized by progressive weight loss and muscle weakness, altered host metabolism, anemia, and anorexia (Calman, 1982). Host muscle and adipose mass are depleted while liver, kidney, adrenals, and spleen are largely unaffected or even enlarged. The cause of cachexia is thought, in part, to have a metabolic basis (Gold, 1974; Costa, 1977; Lawson *et al.,* 1982). Because of the high rate of aerobic glycolysis that is generally associated with tumors (Warburg, 1926), the impact of tumor on host carbohydrate metabolism has been most studied. Figure 2 summarizes overall glutamine and glucose metabolism in mammals. High aerobic glycolysis, which is a property of most neoplasms, is thought to disrupt the normal, regulated cycling of glucose to muscle to lactate (Cori, 1981) and alanine (Felig, 1973), which then is converted back to glucose via gluconeogenesis in liver and kidney (Gold, 1974). Thermodynamic considerations have been described whereby a progressive energy loss could occur in tumor-bearing hosts by the futile cycling of glucose and lactate as well as alanine between tumor and liver and kidney (Gold, 1966). For example, one complete Cori cycle would use 2 ATPs from glucose in the tumor cell and 12 ATPs in host gluconeogenic tissue to resynthesize glucose from 2 lactates (Gold, 1974). In its glycolytic role, the tumor acts like a constantly exercising muscle, competing with host muscle, adipose tissue, and intestine for glucose while wastefully metabolizing glucose to lactate at a rate far beyond its needs for energy and anabolic substrates. Normal glutamine metabolism can be viewed as a glutamine–glucose cycle of glucose from gluconeogenic organs (liver and kidney) to both ''glutaminogenic'' tissues (muscle and adipose) and glutaminolytic intestine, a cycle that produces lactate and alanine from both glucose and glutamine for reconversion to glucose (Fig. 2). Highly glycolytic and glutaminolytic tumors may disrupt this cycle by competing with intestine for both glucose and glutamine and excessively metabolizing both glucose and glutamine to NH_4^+, lactate, alanine, and other products (Fig. 2). Thus, tumor-derived glutaminolytic products, especially lactate and alanine, may further con-

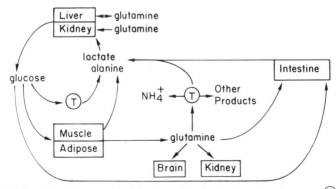

Figure 2. Schematic of glucose and glutamine metabolism in tumor-bearing hosts. Ⓣ, tumors.

tribute substrates or energy to the futile cycling of glucose and lactate between tumor and gluconeogenic tissue. Liver appears to utilize or synthesize glutamine according to supply and demand (Lund, 1971; Lund and Watford, 1976; Deaciuc and Petrescu, 1980). In an individual bearing a highly glutaminolytic and glycolytic tumor, the liver may be faced with demand for circulating glutamine due to tumor utilization as well as its own demand for energy to maintain gluconeogenesis from tumor-derived lactate and alanine. Whether liver produces or consumes glutamine in the tumor-bearing host likely depends on availability of alternatives to glutamine such as fatty acids or branched-chain amino acids to fuel liver gluconeogenesis. Whether glutamine, fatty acids, or branched-chain amino acids are consumed to fuel gluconeogenesis or glutamine is produced in the liver to further cycle between tumor and liver via lactate and alanine, muscle and adipose tissue are the net losers since they are the primary source of glutamine, branched-chain amino acids, and fatty acids. These combined events may underlie the progressive wasting of muscle and adipose mass in the tumor-bearing host.

7. CONCLUSIONS

A common property of normal cells with proliferative potential and tumor cells is a high rate of glucose metabolism to pyruvate (glycolysis) with little oxidation of pyruvate and a high rate of metabolism of glutamine carbons to carbon dioxide and pyruvate through oxidative pathways (glutaminolysis). Studies on the role of glycolysis in cell proliferation and function have focused on its role in energy production with less emphasis on its role in supplying anabolic substrates. The initial observations by Warburg (1926, 1956) that very little glucose-derived pyruvate is oxidized in tumor cells diverted attention from the presence of concurrent oxidative metabolism of other substrates in tumors. Studies on the role of glutamine in cell proliferation and function have centered on its anabolic role as an amido group donor with less emphasis on its role as a respiratory fuel. In this chapter, I have viewed glutaminolysis as a linear pathway that runs concurrently with glycolysis to maintain maximum cellular levels of anabolic precursors and energy, a situation that is favorable to cell proliferation and specialized function of some cell types under anaerobic or aerobic conditions and conditions of carbohydrate deficiency. Nutritional, inhibitor, and genetic studies indicate that a high simultaneous flux of glucose and glutamine through the complete pathways of glycolysis and glutaminolysis is unessential to cell proliferation and malignancy. Dependent on conditions, cells can rely on segments of both pathways to acquire adequate anabolic substrates and energy. This constitutive flexibility may be a key factor in the growth advantage that tumor cells possess over normal tissues. When both glucose and glutamine are in abundant supply, tumors appear to compete with host tissues for and metabolize both glucose and glutamine in excess of energy and anabolic needs to increase tumor mass. The combined unregulated metabolism of host glucose and glutamine may underlie, in part, the cachexia that is the cause of death in many tumor-bearing hosts.

Many questions are in need of resolution:

1. What are the enzymes of glutaminolysis, their cofactors, activators, inhibitors, and subcellular location, in different normal cell types?
2. Does the pathway of glutaminolysis differ with proliferative activity or malignant state in a single cell type?
3. Which steps of glutaminolysis are under hormonal control?
4. Do glutaminolysis and glycolysis sense the activity of the other? If so, what are the sensors and are they altered by proliferative state, malignancy, or hormones?

8. ADDENDUM

While this review was in press, two relevant studies appeared. Wasilenko and Marchok (1984) confirmed that, although isolated normal rat tracheal epithelial cells required pyruvate for growth, the pyruvate requirement was reduced in counterpart tumor cells, which had markedly higher levels of particulate NAD(P^+) malic enzyme activities (Wasilenko and Marchok, 1985). Particulate-bound malate dehydrogenase activity was higher in both normal and preneoplastic epithelial cells than in tumor cells.

Newsholme et al. (1985) presented a strong explanation for the concurrent high rates of both glycolysis and glutaminolysis in rapidly proliferating normal and tumor cells based on quantitative principles of metabolic control (Crabtree and Newsholme, 1985). It was suggested that high rates of glycolysis and glutaminolysis are a consequence, not of energy requirements or anabolic substrates, but maintenance of high sensitivity of both pathways to specific regulators that stimulate cell proliferation when needed, e.g., lymphocytes in response to massive infection and specialized cells involved in wound healing. Although not always necessary, the high rates of flux ensure high rates of proliferation as a "fail-safe" mechanism (Newsholme et al., 1985). How these high rates of flux become fixed in tumor cells remains to be elucidated.

REFERENCES

Ababei, L., Sarkar, S. R., and Rapoport, S., 1962, Deamination as a locus of action of glucose inhibition of respiration in reticulocytes, Acta Biol. Med. Ger. 8:266.
Adachi, H., 1967, The placenta and hormones, J. Jpn. Obstet. Gynecol. Soc. 19:665.

Aoki, T., Morris, H. P., and Weber, G., 1982, Regulatory properties and behavior of carbamoyl phosphate synthetase II (glutamine-hydrolyzing) in normal and proliferating tissues, *J. Biol. Chem.* **257**:432.

Ardawi, M. S. M., and Newsholme, E. A., 1982a, Maximum activities of some enzymes of glycolysis, the tricarboxylic acid cycle and ketone-body and glutamine utilization pathways in lymphocytes of the rat, *Biochem. J.* **208**:743.

Ardawi, M. S. M., and Newsholme, E. A., 1982b, Glutamine metabolism in lymphocytes of the rat, *Biochem. J.* **212**:835.

Azzi, A., Chappell, J. B., and Robinson, B. H., 1967, Penetration of mitochondrial membrane by glutamate and aspartate, *Biochem. Biophys. Res. Commun.* **29**:148.

Bae, I. H., and Foote, R. H., 1975a, Utilization of glutamine for energy and protein synthesis by cultured rabbit follicular oocytes, *Exp. Cell Res.* **90**:432.

Bae, I. H., and Foote, R. H., 1975b, Carbohydrate and amino acid requirements and ammonia production of rabbit follicular oocytes matured in vitro, *Exp. Cell Res.* **91**:113.

Barsh, G. S., and Cunningham, D. D., 1977, Nutrient uptake and control of animal cell proliferation, *J. Supramol. Struct.* **7**:61.

Baverel, G., and Lund, P., 1979, A role for bicarbonate in the regulation of mammalian glutamine metabolism, *Biochem. J.* **184**:599.

Berl, S., and Clarke, D. D., 1969, Compartmentation of amino acid metabolism, in: *Handbook of Neurochemistry* (A. Lajtha, ed.), Vol. 2, pp. 447–450, Plenum Press, New York.

Berl, S., Clarke, D. D., and Schneider, D. (eds.), 1975, *Metabolic Compartmentation and Neurotransmission,* Plenum Press, New York.

Blackshear, P. J., Holloway, P. A. H., and Alberti, K. G. M., 1975, Factors regulating amino-acid release from extrasplanchnic tissues in rat—Interactions of alanine and glutamine, *Biochem. J.* **150**:379.

Blitz, R. M., Letteri, J. M., Pellegrino, E. D., and Pinkus, L., 1982, Glutamine: A new metabolic substrate, *Adv. Exp. Med. Biol.* **157**:423.

Borst, P., 1962, The pathway of glutamate oxidation by mitochondri isolated from different tissues, *Biochim. Biophys. Acta* **57**:256.

Bradford, H. F., and Ward, H. K., 1975, Glutamine as a metabolic substrate for isolated nerve-endings: Inhibition by ammonium ions, *Biochem. Soc. Trans.* **3**:1223.

Bradford, H. F., and Ward, H. K., 1976, On glutaminase activity in mammalian synaptosomes, *Brain Res.* **110**:115.

Brdiczka, D., and Pette, D., 1971, Intra- and extramitochondrial isozymes of (NADP) malate dehydrogenase, *Eur. J. Biochem.* **19**:546.

Buchanan, J. M., 1973, The amidotransferases, *Adv. Enzymol.* **39**:91.

Calman, K. C., 1982, Cancer cachexia, *Br. J. Hosp. Med.* **27**:28.

Chappell, J. B., 1968, Systems used for transport of substrates into mitochondria, *Br. Med. Bull.* **24**:150.

Coles, N. W., and Johnstone, R. M., 1962, Glutamine metabolism in Ehrlich ascites-carcinoma cells, *Biochem. J.* **83**:284.

Cori, C. F., 1981, The glucose-lactic acid cycle and gluconeogenesis, *Curr. Top. Cell. Regul.* **18**:2237.

Costa, G., 1977, Cachexia, the metabolic component of neoplastic diseases, *Cancer Res.* **37**:2237.

Crabtree, B., and Newsholme, E. A., 1985, A quantitative approach to metabolic control, *Curr. Topics Cell. Regul.* **25**:21.

Crompton, M., and Chappell, J. B., 1973, Transport of glutamine and glutamate in kidney mitochondria in relation to glutamine deamidation, *Biochem. J.* **132**:35.

Curthoys, N. P., and Shapiro, R. A., 1978, Effect of metabolic acidosis and of phosphate on presence of glutamine within matrix space of rat renal mitochondria during glutamine transport, *J. Biol. Chem.* **253**:63.

Curthoys, N. P., Sindel, R. W., and Lowry, O. H., 1973, Rat kidney glutaminase isozymes, in: *The*

Enzymes of Glutamine Metabolism (S. Prusiner and E. R. Stadtman, eds.), pp. 259–276, Academic Press, New York.

Deaciuc, I. V., and Petrescu, I., 1980, Regulation of glutamine catabolism in the perfused guinea-pig liver in relation to ureogenesis and gluconeogenesis, *Int. J. Biochem.* **12**:605.

Dean, B., and Bartley, W., 1973, Oxaloacetate decarboxylases of rat liver, *Biochem. J.* **135**:667.

DeFrancesco, L., Wentz, D., and Scheffler, I. E., 1975, Conditionally lethal mutations in Chinese hamster cells: Characterization of a cell line with a possible defect in the Krebs cycle, *J. Cell. Physiol.* **85**:293.

Denton, R. M., and McCormack, J. G., 1980, On the role of the calcium-transport cycle in heart and other mammalian mitochondria, *FEBS Lett.* **119**:1.

Donnelly, M., and Scheffler, I. E., 1976, Energy metabolism in respiration-deficient and wild-type Chinese hamster fibroblasts in culture, *J. Cell. Physiol.* **89**:39.

Eagle, H., Oyama, V. I., Levy, M., Horton, C. L., and Fleischman, R., 1956, The growth response of mammalian cells in tissue culture to L-glutamine and L-glutamic acid, *J. Biol. Chem.* **218**:607.

Eagle, H., Barban, S., Levy, M., and Schulze, H. O., 1958, The utilization of carbohydrates by human cell cultures, *J. Biol. Chem.* **233**:551.

Eigenbrodt, E., and Glossmann, H., 1980, Glycolysis—One of the keys to cancer? *Trends Pharmacol. Sci.* **May**:240.

Fagan, J. B., and Racker, E., 1978, Determinants of glycolytic rate in normal and transformed chick embryo fibroblasts, *Cancer Res.* **38**:749.

Felig, P., 1973, The glucose–alanine cycle, *Metabolism* **22**:179.

Fenselau, A., Wallis, K., and Morris, H. P., 1975, Acetoacetate coenzyme A transferase activity in rat hepatomas, *Cancer Res.* **35**:2315.

Fenselau, A., Wallis, K., and Morris, H. P., 1976, Subcellular localization of acetoacetate co-enzyme A transferase in rat hepatomas, *Cancer Res.* **36**:4429.

Franchi, A., Silvestre, P., and Pouysségur, J., 1981, A genetic approach to the role of energy metabolism in the growth of tumor cells: Tumorigenicity of fibroblast mutants deficient either in glycolysis or in respiration, *Int. J. Cancer* **27**:819.

Frizzel, R. A., Markscheid-Kaspi, L., and Schultz, S. G., 1974, Oxidative metabolism of rabbit ileal mucosa, *Am. J. Physiol.* **226**:1142.

Garber, A. J., 1980, Glutamine metabolism in skeletal muscle, in: *Glutamine: Metabolism, Enzymology and Regulation* (J. Mora and R. Palacios, eds.), pp. 259–284, Academic Press, New York.

Garber, A. J., Karl, I. E., and Kipnis, D. M., 1976, Alanine and glutamine synthesis and release from skeletal muscle. I. Glycolysis and amino acid release, *J. Biol. Chem.* **251**:826.

Gilbert, J. B., Price, V. E., and Greenstein, J. P., 1949, Effect of anions on the non-enzymatic desamidation of glutamine, *J. Biol. Chem.* **180**:209.

Glazer, R. I., Vogel, C. L., Potel, I. R., and Anthony, P. P., 1974, Glutamate dehydrogenase activity related to histopathological grade of hepatocellular carcinoma in man, *Cancer Res.* **34**:2975.

Godfrey, S., Kuhlenschmidt, T., and Curthoys, N. P., 1977, Correlation between activation and dimer formation of rat renal phosphate-dependent glutaminase, *J. Biol. Chem.* **252**:1927.

Goetz, I. E., Weinstein, C., and Roberts, E., 1973, Properties of a hamster tumor cell line grown in a glutamine-free medium, *In Vitro* **9**:46.

Gold, J., 1966, Metabolic profiles in human tumors. I. A new technic for the utilization of human solid tumors in cancer research and its application to the anaerobic glycolysis of isologous benign and malignant colon tissues, *Cancer Res.* **26**:695.

Gold, J., 1974, Cancer cachexia and gluconeogenesis, *Ann. N.Y. Acad. Sci.* **230**:103.

Goldstein, L., and Boylan, J. M., 1978, Renal mitochondrial glutamine transport and metabolism—Studies with a rapid-mixing, rapid-filtration technique, *Am. J. Physiol.* **234**:F514.

142   WALLACE L. McKEEHAN

Goodman, A. D., Fuisz, R. E., and Cahill, G. F., 1966, Renal gluconeogenesis in acidosis, alkalosis, and potassium deficiency: Its possible role in regulation of renal ammonia production, *J. Clin. Invest.* **45**:612.

Graff, S., Moser, H., Kastner, O., Graff, A. M., and Tannenbaum, M., 1965, The significance of glycolysis, *J. Natl. Cancer Inst.* **34**:511.

Gregg, C. T., 1972, Some aspects of the energy metabolism of mammalian cells, in: *Growth Nutrition and Metabolism of Cells in Culture* (G. H. Rothblat and V. J. Cristofalo, eds.), pp. 83–129, Academic Press, New York.

Griffiths, J. B., and Pirt, S. J., 1967, The uptake of amino acids by mouse cells (strain LS) during growth in batch culture and chemostat culture, *Proc. R. Soc. London Ser. B* **168**:421.

Guidotti, G. G., Borghetti, A. F., and Gozzola, G. C., 1978, The regulation of amino acid transport in animal cells, *Biochim. Biophys. Acta* **515**:329.

Gwatkin, R. B. L., and Haidri, A. A., 1973, Requirements for the maturation of hamster oocytes in vitro, *Exp. Cell Res.* **76**:1.

Ham, R. G., and McKeehan, W. L., 1979, Media for growth of cells in culture, *Methods Enzymol.* **5**:44.

Ham, R. G., Hammond, S. L., and Miller, L. L., 1977, Critical adjustment of cysteine and glutamine concentrations for improved clonal growth of WI-38 cells, *In Vitro* **13**:1.

Hansford, R. G., 1980, Control of mitochondrial substrate oxidation, *Curr. Top. Bioenerg.* **10**:217.

Hansford, R. G., and Lehninger, A. L., 1973, Active oxidative decarboxylation of malate by mitochondria isolated from L-1210 ascites tumor cells, *Biochem. Biophys. Res. Commun.* **51**:480.

Hanson, P. J., and Parsons, D. S., 1977, Metabolism and transport of glutamine and glucose in vascularly perfused small intestine of rat, *Biochem. J.* **166**:509.

Hanson, P. J., and Parsons, D. S., 1980, The interrelationship between glutamine and alanine in the intestine, *Biochem. Soc. Trans.* **8**:506.

Haruno, K., 1956, Changes in glutaminase activity of liver tissue from rats during the development of hepatic tumors by carcinogen feeding, *Gann* **47**:231.

Hems, R., Stubbs, M., and Krebs, H. A., 1968, Restricted permeability of rat liver for glutamate and succinate, *Biochem. J.* **107**:807.

Hertz, L., Yu, A., Svenneby, G., Kvamme, E., Fosmark, H., and Shousboe, A., 1980, Absence of preferential glutamine uptake into neurons—an indication of a net transfer of TCA constituents from nerve endings to astrocytes, *Neurosci. Lett.* **16**:103.

Herzfeld, A., and Roper, S. M., 1979, Effects of cortisol or starvation on the activities of four enzymes in small intestine and liver of the rat during development, *J. Dev. Physiol.* **1**:315.

Hills, A. G., Reid, E. L., and Kerr, W. D., 1972, Circulatory transport of L-glutamine in fasted mammals: Cellular sources of urine ammonia, *Am. J. Physiol.* **223**:1470.

Hoek, J. B., and Njogu, R. M., 1976, Glutamate transport and trans-membrane pH gradient in isolated rat liver mitochondria, *FEBS Lett.* **71**:341.

Holzman, I. R., Phillips, A. F., and Battaglia, F. C., 1979, Glucose metabolism, lactate, and ammonia production by the human placenta in vitro, *Pediatr. Res.* **13**:117.

Hornsby, P. J., and Gill, G. N., 1981, Regulation of glutamine and pyruvate oxidation in cultured adrenocortical cells by cortisol, antioxidants, and oxygen: Effects on cell proliferation, *J. Cell. Physiol.* **109**:111.

Horowitz, B., Madras, B. K., Meister, A., and Stockert, E., 1968, Asparagine synthetase activity of mouse leukemias, *Science* **160**:533.

Horowitz, M. L., and Knox, W. E., 1968, A phosphate activated glutaminase in rat liver different from that in kidney and other tissues, *Enzymol. Biol. Clin.* **9**:241.

Huang, Y. Z., and Knox, W. E., 1976, A comparative study of glutaminase isozymes in rat tissues, *Enzyme* **21**:408.

Hume, D. A., Radik, J. L., Ferber, E., and Weidemann, M. J., 1978, Aerobic glycolysis and lymphocyte transformation, *Biochem. J.* **174**:703.

Ishikawa, E., 1976, Regulation of uptake and output of amino-acids by rat tissues, *Adv. Enzyme Regul.* **14**:117.

Janski, A. M., and Cornell, N. W., 1980, Subcellular distribution of enzymes determined by rapid digitonin fractionation of isolated hepatocytes, *Biochem. J.* **186**:423.

Kaplan, R. S., Morris, H. P., and Coleman, P. S., 1982, Kinetic characteristics of citrate influx and efflux with mitochondria from Morris hepatomas 3924A and 16, *Cancer Res.* **42**:4399.

Katunuma, N., Huzino, A., and Tomino, I., 1967, Organ specific control of glutamine metabolism, *Adv. Enzyme Regul.* **5**:55.

Katunuma, N., Kuroda, Y., Yoshida, T., Sanada, Y., and Morris, H. P., 1972, Relationship between degree of differentiation and growth rate of minimal deviation hepatomas and kidney cortex tumors studied with glutaminase isozymes, *Gann* **13**:143.

Katunuma, N., Katsunuma, T., Towatari, T., and Tomino, I., 1973, Regulatory mechanisms of glutamine catabolism, in: *The Enzymes of Glutamine Metabolism* (S. Prusiner and E. R. Stadt-man, eds.), pp. 227–258, Academic Press, New York.

Kilberg, M. S., Handlogten, M. E., and Christensen, H. N., 1980, Characteristics of an amino acid transport system in rat liver for glutamine, asparagine, histidine, and closely related analogs, *J. Biol. Chem.* **255**:4011.

Kitos, P. A., Sinclair, R., and Waymouth, C., 1962, Glutamine metabolism by animal cells growing in a synthetic medium, *Exp. Cell Res.* **27**:307.

Klingenberg, M., 1970, Metabolite transport in mitochondria: An example for intracellular mem-brane function, in: *Essays in Biochemistry* (P. N. Campbell and F. Dickens, eds.), Vol. 6, pp. 119–159, Academic Press, New York.

Klingenberg, M., 1971, Kinetic study of the dicarboxylic carrier in rat liver mitochondria, *Eur. J. Biochem.* **22**:66.

Knox, W. E., 1976, *Enzyme Patterns in Fetal, Adult and Neoplastic Rat Tissues,* 2nd ed., Karger, Basel.

Knox, W. E., Tremblay, G. C., Spanier, B. B., and Friedell, G. H., 1967, Glutaminase activities in normal and neoplastic tissues of the rat, *Cancer Res.* **27**:1456.

Knox, W. E., Horowitz, M. L., and Friedell, G. H., 1969, The proportionality of glutaminase content to growth rate and morphology of rat neoplasmas, *Cancer Res.* **29**:669.

Koser, B. H., and Christensen, H. N., 1971, Effect of substrate structure on coupling ratio for Na$^+$-dependent transport of amino-acids, *Biochim. Biophys. Acta* **241**:9.

Kovacevic, Z., 1971, The pathway of glutamine and glutamate oxidation in isolated mitochondria from mammalian cells, *Biochem. J.* **125**:757.

Kovacevic, Z., 1974, Properties and intracellular localization of Ehrlich ascites tumor cell glu-taminase, *Cancer Res.* **34**:3403.

Kovacevic, Z., 1975, Possible mechanisms of efflux of glutamate from kidney mitochondria gener-ated by activity of mitochondrial glutaminase, *Biochim. Biophys. Acta* **396**:325.

Kovacevic, Z., and McGivan, J. D., 1983, Mitochondrial metabolism of glutamine and glutamate and its physiological significance, *Phys. Rev.* **63**:547.

Kovacevic, Z., and Morris, H. P., 1972, The role of glutamine in the oxidative metabolism of malignant cells, *Cancer Res.* **32**:326.

Krebs, H. A., 1935, Metabolism of amino-acids: The synthesis of glutamine from glutamic acid and ammonia, and the enzymic hydrolysis of glutamine in animal tissues, *Biochem. J.* **29**:1951.

Krebs, H. A., 1963, Renal gluconeogenesis, *Adv. Enzyme Regul.* **1**:385.

Krebs, H. A., 1980, Glutamine metabolism in the animal body, in: *Glutamine: Metabolism, En-zymology, and Regulation* (J. Mora and R. Palacios, eds.), pp. 319–329, Academic Press, New York.

144   WALLACE L. McKEEHAN

Kvamme, E., and Svenneby, G., 1961, The effect of glucose on glutamine utilization by Ehrlich ascites tumor cells, *Cancer Res.* **21**:92.

Lamar, C., 1968, Studies on two glutaminase systems from rat kidney, *Biochim. Biophys. Acta* **151**:188.

LaNoue, K. F., and Schoolwerth, A. C., 1979, Metabolite transport in mitochondria, *Annu. Rev. Biochem.* **48**:871.

Lavietes, B. B., Regan, D. H., and Demopoulos, H. B., 1974, Glutamate oxidation of 6C3HED lymphoma: Effects of L-asparaginase on sensitive and resistant lines, *Proc. Natl. Acad. Sci. USA* **71**:3993.

Lawson, D. H., Richmond, A., Nixon, D. W., and Rudman, D., 1982, Metabolic approaches to cancer cachexia, *Annu Rev. Nutr.* **2**:277.

Lazo, P., 1981, Amino acids and glucose utilization by different metabolic pathways in ascites-tumour cells, *Eur. J. Biochem.* **117**:19.

Lazo, P. A., and Sols, A., 1980, Energetics of tumour cells: Enzyme basis of aerobic glycolysis, *Biochem. Soc. Trans.* **8**:579.

Lehninger, A. L., 1975, *Biochemistry: The Molecular Basis of Cell Structure and Function,* Worth, New York.

Lemons, L. H., Adkock, E. W., Jones, M. D., Naughton, M. A., Meschia, G., and Battaglia, F. C., 1976, Umbilical uptake of amino acids in the unstressed fetal lamb, *J. Clin. Invest.* **58**:1428.

Levintow, L., Eagle, H., and Piez, K. A., 1957, The role of glutamine in protein biosynthesis in tissue culture, *J. Biol. Chem.* **227**:929.

Linder-Horowitz, M., Knox, W. E., and Morris, H. P., 1969, Glutaminase activities and growth rates of rat hepatomas, *Cancer Res.* **29**:1195.

Linn, R. C., and Davis, E. J., 1974, Malic enzymes of rabbit heart mitochondria: Separation and comparison of some characteristics of a nicotinamide adenine dinucleotide-preferring and a nicotinamide adenine dinucleotide phosphate-specific enzyme, *J. Biol. Chem.* **249**:3867.

Lowenstein, J. M., 1969, *Citric Acid Cycle, Control and Compartmentation,* Dekker, New York.

Lowenstein, J. M., 1972, Ammonia production in muscle and other tissue: The purine nucleotide cycle, *Phys. Rev.* **52**:382.

Luchinsky, H. L., 1951, The activity of glutaminase in the human placenta, *Arch. Biochem. Biophys.* **31**:132.

Lund, P., 1971, Control of glutamine synthesis in rat liver, *Biochem. J.* **124**:653.

Lund, P., 1980, Glutamine metabolism in the rat, *Biochem. J.* **117**:K86.

Lund, P., and Watford, M., 1976, Glutamine as a precursor of urea, in: *The Urea Cycle* (S. Grisolia, R. Bagenna, and F. Mayor, eds.), pp. 479–485, Wiley, New York.

Lund, P., Bresnan, J. T., and Eggleston, L. V., 1970, The regulation of ammonia metabolism in mammalian tissue, in: *Essays in Cell Metabolism* (W. Bartley, H. L. Kornberg, and J. R. Quayle, eds.), pp. 167–180, Wiley, New York.

McCormack, J. G., and Denton, R. M., 1981, A comparative study of the regulation by Ca^{2+} of the activities of the 2-oxoglutarate dehydrogenase complex and NAD^+-isocitrate dehydrogenase from a variety of sources, *Biochem. J.* **196**:619.

McKeehan, W. L., 1982, Glycolysis, glutaminolysis and cell proliferation, *Cell Biol. Int. Rep.* **6**:635.

McKeehan, W. L., and McKeehan, K. A., 1982, Changes in $NAD(P)^+$-dependent malic enzyme and malate dehydrogenase activities during fibroblast proliferation, *J. Cell. Physiol.* **110**:142.

McMenamy, R. H., Shoemaker, W. C., Richmond, J. E., and Elwyn, D., 1962, Uptake and metabolism of amino acids by the dog liver perfused in situ, *Am. J. Physiol.* **202**:407.

Makarewicz, W., and Swierczynski, J., 1982, Ammonia formation from some amino acids by human term placental mitochondria, *Biochem. Med.* **28**:135.

Mandella, R. D., and Sauer, L. A., 1975, The mitochondrial malic enzymes: Submitochondrial localization and purification and properties of the NAD(P)$^+$-dependent enzyme from adrenal cortex, *J. Biol. Chem.* **250**:5877.

Meister, A., 1956, Metabolism of glutamine, *Physiol. Rev.* **36**:103.

Meister, A., 1962, Amide nitrogen transfer (survey), *The Enzymes* **16**:247.

Meister, A., 1965, Glutamic acid and glutamine, in: *Biochemistry of the Amino Acids*, Vol. 2, pp. 617–635, Academic Press, New York.

Meister, A., 1978, Biochemistry of glutamate, glutamine and glutathione, in: *Glutamic Acid: Advances in Biochemistry and Physiology* (L. J. Filer, Jr., S. Garattini, M. R. Kare, W. A. Reynolds, and W. J. Wartman, eds.), pp. 369–385, Raven Press, New York.

Miller, R. E., and Carrino, D. A., 1981, An association between glutamine synthetase activity and adipocyte differentiation in cultured 3T3-L1 cells, *Arch. Biochem. Biophys.* **209**:486.

Mora, J., and Palacios, R. (eds.), 1980, *Glutamine: Metabolism, Enzymology and Regulation*, Academic Press, New York.

Moreadith, R. W., and Lehninger, A. L., 1984a, The pathways of glutamate and glutamine oxidation by tumor cell mitochondria. Role of mitochondrial NAD(P)$^+$-dependent malic enzyme, *J. Biol. Chem.* **259**:6215.

Moreadith, R. W., and Lehninger, A. L., 1984b, Purification, kinetic behavior, and regulation of NAD(P)$^+$ malic enzyme of tumor mitochondria, *J. Biol. Chem.* **259**:6222.

Morehouse, R. F., and Curthoys, N. P., 1981, Properties of rat renal phosphate-dependent glutaminase coupled to Sepharose, *Biochem. J.* **193**:709.

Morris, H. P., 1972, Isozymes in selected hepatomas and some biological characteristics of a spectrum of transplantable hepatomas, *Gann* **13**:95.

Munro, H. M., 1978, Factors in the regulation of glutamate metabolism, in: *Glutamic Acid: Advances in Biochemistry and Physiology* (L. J. Filer, Jr., S. Garattini, M. R. Kare, W. A. Reynolds, and R. J. Wartman, eds.), pp. 55–65, Raven Press, New York.

Nagel, W. O., Dauchy, R. T., and Sauer, L., 1980, Mitochondrial malic enzymes: An association between NAD(P)$^+$-dependent malic enzyme and cell renewal in Sprague–Dawley rat tissues, *J. Biol. Chem.* **255**:3849.

Nagel, W. O., and Sauer, L. A., 1982, Mitochondrial malic enzymes. Purification and properties of the NAD(P)$^+$-dependent malic enzyme from canine intestinal mucosa.

Neptune, E. M., Jr., 1965, Respiration and oxidation of various substrates by ileum in vitro, *Am. J. Physiol.* **209**:329.

Newsholme, E. A., Crabtree, B., and Ardawi, M. S. M., The role of high rates of glycolysis and glutamine utilization in rapidly dividing cells, *Bioscience Rep.* **5**:393.

Nyhan, W. L., and Busch, H., 1958a, Metabolic patterns for L-glutamate-U-C^{14} in tissues of tumor-bearing rats, *Cancer Res.* **18**:385.

Nyhan, W. L., and Busch, H., 1958b, Metabolic patterns for succinate-2-C^{14} in tissues of tumor-bearing rats, *Cancer Res.* **18**:1203.

Olivera, A., and Meigs, R., 1975, Mitochondria from human term placenta. I. Isolation and assay conditions for oxidative phosphorylation, *Biochim. Biophys. Acta* **376**:426.

Olivotto, M., and Paoletti, F., 1981, The role of respiration in tumor cell transition from the noncycling to the cycling state, *J. Cell. Physiol.* **107**:243.

Olivotto, M., Caldini, R., Chevanne, M., and Apolleschi, M. G., 1983, The respiration-linked limiting step of tumor cell transition from the noncycling to the cycling state: Its inhibition by oxidizable substrates and its relation to purine metabolism, *J. Cell. Physiol.* **116**:149.

Ottaway, J. H., McClellan, J. A., and Saunderson, C. L., 1981, Succinic thiokinase and metabolic control, *Int. J. Biochem.* **13**:401.

Palmieri, F., Quagliariello, E., and Klingenberg, M., 1972, Kinetics and specificity of the oxoglutarate carrier in rat-liver mitochondria, *Eur. J. Biochem.* **29**:408.

Papaconstantinou, J., Goldberg, E. B., and Colowick, S. P., 1963, The role of glycolysis in the growth of tumor cells, in: *Control Mechanisms in Respiration and Fermentation* (B. Wright, ed.), pp. 243–251, Ronald Press, New York.

Paradies, G., Capuano, F., Palombi, G., Galeotti, T., and Papa, S., 1983, Transport of pyruvate in mitochondria from different tumor cells, *Cancer Res.* **43**:5068.

Pardridge, W. M., and Casenello-Ertl, D., 1979, Effects of glutamine deprivation on glucose and amino-acid metabolism in tissue culture, *Am. J. Physiol.* **236**:E234.

Pardridge, W. M., Davidson, M. B., and Casenello-Ertl, D., 1978, Glucose and amino acid metabolism in an established line of skeletal muscle cells, *J. Cell. Physiol.* **96**:309.

Pardridge, W. M., Duducgian-Vartavarian, L., Casenello-Ertl, D., Jones, M. R., and Kopple, J. D., 1980, Glucose and amino acid metabolism in neonatal rat skeletal muscle in tissue culture, *J. Cell. Physiol.* **102**:91.

Pederson, P. L., 1978, Tumor mitochondria and the bioenergetics of cancer cells, *Prog. Exp. Tumor Res.* **22**:190.

Pinkus, L. M., and Berkowitz, J. M., 1980, Utilization of glutamine by canine pancreas in vivo and acinar cells in vitro, *Fed. Proc.* **39**:1902.

Pinkus, L. M., and Windmueller, H. G., 1977, Phosphate-dependent glutaminase of small intestine: Localization and role in intestinal glutamine metabolism, *Arch. Biochem. Biophys.* **182**:506.

Pitts, R. F., Pilkington, L. A., MacLeod, M. B., and Leal-Pinto, E., 1972, Metabolism of glutamine by intact functioning kidney of dog—Studies in metabolic-acidosis and alkalosis, *J. Clin. Invest.* **51**:557.

Porteous, J. W., 1980, Glutamate, glutamine, aspartate, asparagine, glucose and ketone body metabolism in chick intestinal brush-border cells, *Biochem. J.* **188**:619.

Pouysségur, J., Franchi, A., Salomon, J.-C., and Silvestre, P., 1980, Isolation of a Chinese hamster fibroblast mutant defective in hexose transport and aerobic glycolysis: Its use to dissect the malignant phenotype, *Proc. Natl. Acad. Sci. USA* **77**:2698.

Prajda, N., Katunuma, N., Morris, H. P., and Weber, G., 1975, Imbalance of purine metabolism in hepatomas of different growth rates as expressed in behavior of glutamine-phosphoribosyl-pyrophosphate amidotransferase (amidophosphoribosyltransferase, EC 2.4.2.14), *J. Biol. Chem.* **250**:432.

Prusiner, S., and Stadtman, E. R. (eds.), 1973, *The Enzymes of Glutamine Metabolism*, Academic Press, New York.

Rabinovitz, M., Olson, M. E., and Greenburg, D. M., 1956, Role of glutamine in protein synthesis by the Ehrlich ascites cells, *J. Biol. Chem.* **231**:879.

Racker, E., 1976, Why do tumor cells have a high aerobic glycolysis?, *J. Cell. Physiol.* **89**:697.

Raina, P. N., and Ramakrishnan, C. V., 1964, Glutaminase activity in rat tissues, *Oncologia* **18**:14.

Rapoport, S., Rost, J., and Schultze, M., 1971, Glutamine and glutamate as respiratory substrates of rabbit reticulocytes, *Eur. J. Biochem.* **23**:1966.

Regan, D. H., Lavietes, B. B., Regan, M. G., Demopoulos, H. B., and Morris, H. P., 1973, Glutamate-mediated respiration in tumors, *J. Natl. Cancer Inst.* **51**:1013.

Reitzer, L. J., Wice, B. M., and Kennell, D., 1979, Evidence that glutamate, not sugar, is the major energy source for cultured HeLa cells, *J. Biol. Chem.* **254**:2669.

Reitzer, L. J., Wice, B. M., and Kennell, D., 1980, The pentose cycle: Control and essential function in HeLa cell nucleic acid synthesis, *J. Biol. Chem.* **255**:5616.

Renner, E. D., Plagemann, P. G. W., and Bernlohr, R. W., 1972, Permeation of glucose by simple and facilitated diffusion by Novikoff rat hepatoma cells in suspension culture and its relation to glucose metabolism, *J. Biol. Chem.* **247**:5765.

Rheinwald, J. G., and Green, H., 1974, Growth of cultured mammalian cells on secondary glucose sources, *Cell* **2**:287.

Roberts, E., and Borges, P. R. F., 1955, Patterns of free amino acids in growing and regressing tumors, *Cancer Res.* **15**:697.

Roberts, E., and Simonsen, D. G., 1960, Free amino acids and similar substances in normal and neoplastic tissue, in: *Amino Acids, Proteins and Cancer Biochemistry* (J. T. Edsell, ed.), pp. 127–135, Academic Press, New York.

Roberts, E., Tanaka, K. K., Tanaka, T., and Simonsen, D. G., 1956, Free amino acids in growing and regressing ascites cell tumors: Host resistance and chemical agents, *Cancer Res.* **16**:970.

Robins, S., 1957, *Textbook of Pathology*, Saunders, Philadelphia.

Robinson, B. H., 1971a, Transport of phosphoenolpyruvate by the tricarboxylate transporting system in mammalian mitochondria, *FEBS Lett.* **14**:309.

Robinson, B. H., 1971b, The role of the tricarboxylate transporting system in the production of phosphoenolpyruvate by ox liver mitochondria, *FEBS Lett.* **16**:267.

Romano, A. H., 1976, Is glucose transport enhanced in virus-transformed mammalian cells? A dissenting view, *J. Cell. Physiol.* **89**:737.

Romano, A. H., and Cornell, N. D., 1982, Transport of 6-deoxy-D-glucose and D-xylose by untransformed and SV40-transformed 3T3 cells, *J. Cell. Physiol.* **111**:83.

Roos, D., and Loos, J. A., 1973, Changes in the carbohydrate metabolism of mitogenically stimulated human peripheral lymphocytes. II. Relative importance of glycolysis and oxidative phosphorylation on phytohaemagglutinin stimulation, *Exp. Cell Res.* **77**:127.

Saheki, T., and Katunuma, N., 1975, Analysis of regulatory factors for urea synthesis by isolated perfused rat-liver. 1. Urea synthesis with ammonia and glutamine as nitrogen sources, *J. Biochem.* **77**:659.

Salzman, N. P., Eagle, H., and Sebring, E. D., 1957, The utilization of glutamine, glutamic acid, and ammonia for the biosynthesis of nucleic acid bases in mammalian cell cultures, *J. Biol. Chem.* **227**:1001.

Sauer, L. A., 1973, An NAD- and NADP-dependent malic enzyme with regulatory properties in rat liver and adrenal cortex mitochondrial fractions, *Biochem. Biophys. Res. Commun.* **50**:524.

Sauer, L. A., and Dauchy, R. T., 1978, Identification and properties of the nicotinamide adenine dinucleotide (phosphate) $^+$-dependent malic enzyme in mouse ascites tumor mitochondria, *Cancer Res.* **38**:1751.

Sauer, L. A., and Dauchy, R. T., 1983, Ketone body, glucose, lactic acid, and amino acid utilization by tumors and in vivo in fasted rats, *Cancer Res.* **43**:3497.

Sauer, L. A., Dauchy, R. T., and Nagel, W. O., 1979, Identification of an NAD(P)$^+$-dependent "malic" enzyme in small-intestinal-mucosal mitochondria, *Biochem. J.* **184**:185.

Sauer, L. A., Dauchy, R. T., Nagel, W. O., and Morris, H. P., 1980, Mitochondrial malic enzymes: Mitochondrial NAD(P)$^+$-dependent malic enzyme activity and malate-dependent pyruvate formation are progression-linked in Morris hepatomas, *J. Biol. Chem.* **255**:3844.

Sauer, L. A., Stayman, J. W., III, and Dauchy, R. T., 1982, Amino acid, glucose, and lactic acid utilization in vivo by rat tumors, *Cancer Res.* **42**:4090.

Schneider, H., Mohlen, K. H., Challier, J. C., and Dancis, J., 1979, Transfer of glutamic acid across the human placenta perfused in vitro, *Br. J. Obstet. Gynaecol.* **86**:299.

Schoolwerth, A. C., and LaNoue, K. F., 1980, The role of microcompartmentation in the regulation of glutamate metabolism by rat kidney mitochondria, *J. Biol. Chem.* **255**:3403.

Schweiger, H. G., Rapoport, S. and Schozel, F., 1956, Nitrogen metabolism in erythrocyte maturation: Residual nitrogen formation and hemoglobin synthesis, *Hoppe-Seylers Z. Physiol. Chem.* **306**:33.

Sevdalian, D. A., Ozand, P. T., and Zielke, H. R., 1980, Increase in glutaminase activity during the growth cycle of cultured human diploid fibroblasts, *Enzyme* **25**:142.

Shaffer, J. B., and Felder, M. R., 1983, Turnover of cytoplasmic and mitochondrial aspartate aminotransferase isozymes in mouse liver and transplantable hepatomas, *Arch. Biochem. Biophys.* **223**:649.

Shank, R. P., and Aprison, M. H., 1981, Present status and significance of the glutamine cycle in neural tissue, *Life Sci.* **28**:837.

Shepartz, B., 1973, *Regulation of Amino Acid Metabolism in Mammals*, Saunders, Philadelphia.

Shotwell, M. A., Kilberg, M. S., and Oxender, D. L., 1983, The regulation of neutral amino acid transport in mammalian cells, *Biochim. Biophys. Acta* **737**:267.

Simpson, E., and Estabrook, R. W., 1969, Mitochondrial malic enzyme: The source of reduced nicotinamide adenine dinucleotide phosphate for steroid hydroxylation in bovine adrenal cortex mitochondria, *Arch. Biochem. Biophys.* **129**:384.

Singer, T. P., Kearney, E. B., and Kenney, W. C., 1973, Succinate dehydrogenase, *Adv. Enzymol.* **37**:189.

Singh, M., Singh, V. N., August, J. T., and Horecker, B. L., 1974a, Alterations in glucose metabolism in chick embryo cells transformed by Rous sarcoma virus: Transformation-specific changes in the activities of key enzymes of the glycolytic and hexose monophosphate shunt pathways, *Arch. Biochem. Biophys.* **165**:240.

Singh, V. N., Singh, M., August, J. T., and Horecker, B. L., 1974b, Alterations in glucose metabolism in chick embryo cells transformed by Rous sarcoma virus: Intracellular levels of glycolytic intermediates, *Proc. Natl. Acad. Sci. USA* **71**:4129.

Sols, A., 1976, The Pasteur effect in the allosteric era, in: *Reflections on Biochemistry* (A. Kornberg, B. L. Horecker, L. Cornudella, and J. Oró, eds.), pp. 199–206, Pergamon Press, Elmsford, N.Y.

Stanisz, J., Wice, B. R., and Kennell, D. E., 1983, Comparative energy metabolism in cultured heart muscle and HeLa cells, *J. Cell. Physiol.* **115**:320.

Stegink, L. D., Pitkin, R. M., Reynolds, W. A., Filer, L. J., Booz, D. P., and Brummel, M. C., 1975, Placental transfer of glutamate and its metabolites in the primate, *Am. J. Obstet. Gynecol.* **122**:70.

Steinberger, A., and Steinberger, E., 1966, Stimulatory effect of vitamins and glutamine on the differentiation of germ cells in rat testes organ culture grown in chemically defined media, *Exp. Cell Res.* **44**:429.

Stoner, G. D., and Merchant, D. J., 1972, Amino acid utilization of L-M strain mouse cells in a chemically defined medium, *In Vitro* **7**:330.

Sumbilla, C. M., Ozand, P. T., and Zielke, H. R., 1981, Activities of enzymes required for the conversion of 4-carbon TCA cycle compounds to 3-carbon glycolytic compounds in human diploid fibroblasts, *Enzyme* **26**:201.

Swierczynski, J., Scislowski, P., Aleksandrowicz, A., and Zelewski, L., 1982, NAD(P)-dependent malic enzyme activity in human term placental mitochondria, *Biochem. Med.* **28**:247.

Takagaki, G., Berl, S., Clarke, D. D., Purpura, D. P., and Waelsch, H., 1961, Glutamic acid metabolism in brain and liver during infusion with ammonia labelled with nitrogen-15, *Nature* **189**:326.

Tapia, R., 1980, Glutamine metabolism in brain, in: *Glutamine: Metabolism, Enzymology and Regulation* (J. Mora and R. Palacios, eds.), pp. 285–297, Academic Press, New York.

Tate, S. S., and Meister, A., 1973, Glutamine synthetases of mammalian liver and brain, in: *The Enzymes of Glutamine Metabolism* (S. Prusiner, ed.), pp. 77–127, Academic Press, New York.

Thomas, E. L., and Christensen, H. N., 1971, Nature of cosubstrate action of Na$^+$ and neutral amino-acids in a transport system, *J. Biol. Chem.* **246**:1682.

Tiemeier, D. C., Smotkin, D., and Milman, G., 1973, Regulation of glutamine synthetase in Chinese hamster cells, in: *The Enzymes of Glutamine Metabolism* (S. Prusiner, ed.), pp. 145–166, Academic Press, New York.

Tildon, J. T., 1983, Glutamine: a possible energy source for the brain, in: *Glutamine, Glutamate, and GABA in the Central Nervous System* (L. Hertz, E. Kvamme, E. G. McGeer, and A. Schousboe, eds.), pp. 415–429, Alan R. Liss, Inc., New York.

Towler, C. M., Pugh-Humphreys, G. P., and Porteous, J. P., 1978, Characterization of columnar absorptive epithelial cells isolated from rat jejunum, *J. Cell Sci.* **29**:53.

Trayhurn, P., and Van Heyningen, R., 1971, Aerobic metabolism in the bovine lens, *Exp. Eye Res.* **12**:315.

Trayhurn, P., and Van Heyningen, R., 1973a, The metabolism of amino acids in the bovine lens, *Biochem. J.* **136**:67.

Trayhurn, P., and Van Heyningen, R., 1973b, The metabolism of glutamine in the bovine lens: Glutamine as a source of glutamate, *Exp. Eye Res.* **17**:149.

Tritsch, G. L., and Moore, G. E., 1962, Spontaneous decomposition of glutamine in cell culture media, *Exp. Cell Res.* **28**:360.

Tsoncheva, A., 1974, Some properties of isozymes of NADP-malate dehydrogenase from cortical layers of rat kidneys, *Biokhimiya* **39**:1172.

Tsuiki, S., Sato, K., Mijagi, T., and Kikuchi, H., 1972, Isozymes of fructose 1,6-disphosphatase, glycogen synthetase, and glutamine:fructose 6-phosphate amidotransferase, *Gann* **13**:153.

Tyrrell, J. B., and Anderson, J. N., 1971, Glycolytic and pentose phosphate pathway enzymes in jejunal mucosa: Adaptive responses to alloxan-diabetes and fasting in the rat, *Endocrinology* **89**:1178.

Van Slyke, D. D., Phillips, R. A., Hamilton, P. B., Archibald, R. M., Futcher, P. H., and Hiller, A., 1943, Glutamine as source material of urinary ammonia, *J. Biol. Chem.* **150**:481.

Vina, J. R., and Williamson, D. H., 1981, Utilization of L-alanine and L-glutamine by lactating mammary gland of the rat, *Biochem. J.* **196**:757.

Volman-Mitchell, H., and Parsons, D. S., 1974, Distribution and activities of dicarboxylic amino acid transaminases in gastrointestinal mucosa of rat, mouse, hamster, guinea pig, chicken and pigeon, *Biochim. Biophys. Acta* **334**:316.

Waelsch, H., 1960, An attempt at integration of structure and metabolism in the nervous system, in: *Structure and Function of the Cerebral Cortex* (D. B. Tower and J. P. Shade, eds.), pp. 313–326, Elsevier, Amsterdam.

Wanders, R. J. A., Hoek, J. B., and Tager, J. M., 1980, Origin of the ammonia found in protein-free extracts of rat-liver mitochondria and rat hepatocytes, *Eur. J. Biochem.* **110**:197.

Wang, T., Marquardt, C., and Foker, J., 1976, Aerobic glycolysis during lymphocyte proliferation, *Nature* **261**:701.

Warburg, O., 1926, *Über den Stoffwechsel der Tumoren,* Springer-Verlag, Berlin (Translation: *The Metabolism of Tumors,* Arnold Constable, London, 1930).

Warburg, O., 1956, On the origin of cancer cells, *Science* **123**:309.

Warburg, O., Kubowitz, F., and Christian, W., 1931, Uber die wirkung von phenylhydrazin und phenylhydroxylamin auf der stoffwechsel de roten blutzellen, *Biochem. Z.* **242**:170.

Wasilenko, W. J., and Marchok, A. C., 1984, Pyruvate regulation of growth and differentiation in primary cultures of rat tracheal epithelial cells, *Exp. Cell Res.* **155**:507.

Wasilenko, W. J., and Marchok, A. C., 1985, Malic enzyme and malate dehydrogenase activities in rat tracheal epithelial cells during the progression of neoplasia, *Cancer Letters* **28**:35.

Watford, M., Lund, P., and Krebs, H. A., 1979a, Isolation and metabolic characteristics of rat and chicken enterocytes, *Biochem. J.* **178**:589.

Watford, M., Vinay, P., Lemieux, G., and Gougoux, A., 1979b, The formation of pyruvate from citric acid-cycle intermediates in kidney cortex, *Biochem. Soc. Trans.* **7**:753.

Watford, M., Vinay, P., Lemieux, G., and Gougoux, A., 1980, The regulation of glucose and pyruvate formation from glutamine and citric-acid-cycle intermediates in the kidney cortex of rats, dogs, rabbits and guinea pigs, *Biochem. J.* **188**:741.

Wein, J., and Goetz, I. E., 1973, Asparaginase and glutaminase activities in culture media containing dialyzed fetal calf serum, *In Vitro* **9**:186.

Weinhouse, S., 1976, The Warburg hypothesis fifty years later, *Z. Krebsforsch. Klin. Onkol.* **87**:115.

Wenner, C. E., 1975, Regulation of energy metabolism in normal and tumor cells, in: *Cancer: A Comprehensive Treatise* (F. F. Becker, ed.), Vol. 3, pp. 389–403, Plenum Press, New York.

Whittaker, P. A., and Danks, S. M., 1978, *Mitochondria: Structure, Function and Assembly*, Longman, London.

Wice, B. M., Reitzer, L. J., and Kennell, D., 1981, The continuous growth of vertebrate cells in the absence of sugar, *J. Biol. Chem.* **256**:7812.

Williams, W. J., and Manson, C. A., 1958, Glutaminase of the human malignant cell, strain HeLa, *J. Biol. Chem.* **232**:229.

Williamson, J. R., 1976, Role of anion transport in the regulation of metabolism, in: *Gluconeogenesis: Its Regulation in Mammalian Species* (R. W. Hanson and M. A. Mehlman, eds.), pp. 165–181, Wiley, New York.

Williamson, J. R., and Cooper, R. H., 1980, Regulation of the citric-acid cycle in mammalian systems, *FEBS Lett.* **117**:K73.

Windmueller, H. G., 1982, Glutamine utilization by the small intestine, *Adv. Enzymol.* **53**:201.

Windmueller, H. G., and Spaeth, A. E., 1974, Uptake and metabolism of plasma glutamine by the small intestine, *J. Biol. Chem.* **249**:5070.

Windmueller, H. G., and Spaeth, A. E., 1980, Respiratory fuels and nitrogen metabolism in vivo in small intestine of fed rats, *J. Biol. Chem.* **255**:107.

Wise, E. M., and Ball, E. G., 1964, Malic enzyme and lipogenesis, *Proc. Natl. Acad. Sci. USA* **52**:1255.

Yu, A. C., Shousbae, A., and Hertz, L., 1982, Metabolic fate of C-labeled glutamate in astrocytes in primary cultures, *J. Neurochem.*, **39**:958.

Zielke, H. R., Ozand, P. T., Tildon, J. T., Sevdalian, D. A., and Cornblath, M., 1976, Growth of human diploid fibroblasts in the absence of glucose utilization, *Proc. Natl. Acad. Sci. USA* **73**:4110.

Zielke, H. R., Ozand, P. T., Tildon, J. T., Sevdalian, D. A., and Cornblath, M., 1978, Reciprocal regulation of glucose and glutamine utilization by cultured human diploid fibroblasts, *J. Cell. Physiol.* **95**:41.

Zielke, H. R., Sumbilla, C. M., Sevdalian, D. A., Hawkins, R. L., and Ozand, P. T., 1980, Lactate: A major product of glutamine metabolism by human diploid fibroblasts, *J. Cell. Physiol.* **104**:433.

Zielke, H. R., Sumbilla, C. M., and Ozand, P. T., 1981, Effect of glucose on aspartate and glutamate synthesis by human diploid fibroblasts, *J. Cell. Physiol.* **107**:251.

5

The Metabolism and Utilization of Carbohydrates by Suspension Cultures of Plant Cells

GAGIK STEPAN-SARKISSIAN and MICHAEL W. FOWLER

1. INTRODUCTION

The subject of carbohydrate metabolism of plant cell cultures may be approached from two principal directions. The first concerns the use of cell cultures to study the physiology and biochemistry of carbohydrate metabolism in plants but at the level of the cell. The second relates to the utilization of carbohydrates as a primary carbon source and their influence upon biomass and secondary metabolite productivity in cell cultures. Such a division is of course arbitrary and superficial, the two areas often being inextricably linked together. The literature relating to carbohydrate metabolism in plant cell cultures is both disperse and, in many areas, sparse. For this reason and to examine the subject in a structured fashion, we have chosen to discuss the available data through a series of linked topics. Such an approach also has the advantage of highlighting key areas where little information is available alongside those for which a substantial body of information already exists. The topics covered concern the nature and efficiency of various carbon sources tested for their ability to support growth, the mode of utilization and fate of different carbon sources, and the effects of different carbon sources on biomass accumulation and secondary metabolite synthesis.

In this review, we have concentrated on cell suspension cultures rather than on solid or callus (Street, 1974) cultures. The often high degree of heterogeneity

GAGIK STEPAN-SARKISSIAN and MICHAEL W. FOWLER • Wolfson Institute of Biotechnology, University of Sheffield, Sheffield S10 2TN, United Kingdom.

of the latter makes data interpretation often difficult and open to question. Data from callus cultures have only been used in two situations: (1) to reinforce data from suspension cultures and (2) where there is no information available from suspension cultures but where a point needs to be made. We have also made no attempt to cover photosynthetic/autotrophic systems. Few plant cell cultures are able to support themselves autotrophically and this topic has recently been reviewed (Yamada *et al.*, 1982).

2. CARBON SOURCES FOR CULTURE GROWTH

As so few cell cultures are able to maintain themselves autotrophically, the choice of an appropriate exogenous carbon source to support culture growth is of key importance.

2.1. The Range of Carbon Sources Tested

A wide range of carbon sources has now been investigated for the ability to support cell growth and division in culture. The list includes mono-, di,- and trisaccharides, sugar alcohols as well as polysaccharides and their derivatives (Table I). Much of the early work has been reviewed by Gautheret (1955), Maretzki *et al.* (1974), and Fowler *et al.* (1982).

In spite of the wide-ranging nature of investigations into potential carbon sources, sucrose remains the source of choice for most workers, with glucose closely following. Both sugars generally sustain high growth rates in cell cultures and carbon conversion efficiency is also high (see Fowler, 1982). As pointed out by Maretzki *et al.* (1974), the rapid utilization of sucrose and glucose by many cell cultures makes them potentially growth-limiting, and therefore of importance in growth limitation studies. Some cell lines appear to exhibit a specific preference for sucrose (Maretzki *et al.*, 1974) although the data are by no means clear-cut on this point. The initial sucrose level applied to cell cultures ranges from 1 to 4% (w/v), with a substantial amount of evidence to suggest that the optimal concentration lies around 2% (Gamborg, 1966). Care must be exercised when considering data on sucrose utilization. That hydrolysis (to varying degrees) of sucrose occurs during autoclaving is well established (Ball, 1953). Consequently, the apparent preference for sucrose in terms of uptake and subsequent metabolism could be due to the hydrolysis products rather than to sucrose itself. This problem may be circumvented by filter sterilization of the added carbohydrate.

Of the constituent monosaccharides of sucrose, glucose has been more extensively studied than fructose. Glucose nutrition has been investigated in a wide range of cell cultures, including bean (*Phaseolus vulgaris* L.) (Bertola and

Table I. Carbon Sources and Their Ability to Support Growth of Plant Cells in Suspension Cultures

Carbon source	Medium concn (% w/v)	Species	Growth[a]	Reference
Arabinose	2	*Morinda citrifolia*	−	Zenk *et al.* (1977)
	2	*Saccharum officinarum*	−	Nickell and Maretzki (1970)
	3	*Digitalis purpurea*	−	Hagimori *et al.* (1982)
	4	*Lolium multiflorum*	−	Smith and Stone (1973)
Cellobiose	2	*Daucus carota*	−	Verma and Dougall (1977)
	2	*Saccharum officinarum*	+++	Nickell and Maretzki (1970)
Dulcitol	2	*Morinda citrifolia*	−	Zenk *et al.* (1977)
Fructose	1	*Cannabis sativa*	+++	Jones and Veliky (1980)
	1	*Daucus carota*	++	Jones and Veliky (1980)
	1	*Ipomoea* sp.	++	Jones and Veliky (1980)
	1	*Populus*	+++	Matsumoto *et al.* (1973)
	2	*Acer pseudoplatanus*	+++	Simpkins *et al.* (1970)
	2	*Acer pseudoplatanus*	+++	Simpkins and Street (1970)
	2	*Catharanthus roseus*	+	Fowler (1982)
	2	*Daucus carota*	+++	Verma and Dougall (1977)
	2	*Morinda citrifolia*	+++	Zenk *et al.* (1977)
	2	*Saccharum officinarum*	+++	Nickell and Maretzki (1970)
	3	*Coptis japonica*	+++	Sato and Yamada (1984)
	3	*Digitalis purpurea*	++	Hagimori *et al.* (1982)
	4	*Lolium multiflorum*	++	Smith and Stone (1973)
Galactose	1	*Canabis sativa*	+	Jones and Veliky (1980)
	1	*Daucus carota*	++	Jones and Veliky (1980)
	1	*Ipomoea* sp.	+	Jones and Veliky (1980)
	1	*Populus*	+	Matsumoto *et al.* (1973)
	2	*Acer pseudoplatanus*	+++	Simpkins *et al.* (1970)
	2	*Catharanthus roseus*	+	Fowler (1982)
	2	*Daucus carota*	+++	Verma and Dougall (1977)
	2	*Morinda citrifolia*	+++	Zenk *et al.* (1977)
	2	*Saccharum officinarum*	++	Nickell and Maretzki (1970)
	3	*Digitalis purpurea*	++	Hagimori *et al.* (1982)
	4	*Lolium multiflorum*	++	Smith and Stone (1973)
Glucose	0.5	*Haplopappus gracilis*	+	Eriksson (1965)
	1	*Cannabis sativa*	++	Jones and Veliky (1980)
	1	*Daucus carota*	++	Jones and Veliky (1980)
	1	*Haplopappus gracilis*	+	Eriksson (1965)
	1	*Ipomoea* sp.	+	Jones and Veliky (1980)
	1	*Populus*	+++	Matsumoto *et al.* (1973)
	2	*Acer pseudoplatanus*	+++	Grout *et al.* (1976)
	2	*Acer pseudoplatanus*	+++	Simpkins and Street (1970)
	2	*Acer pseudoplatanus*	+++	Simpkins *et al.* (1970)
	2	*Catharanthus roseus*	+	Fowler (1982)

(continued)

Table 1. (Cont.)

Carbon source	Medium concn (% w/v)	Species	Growth[a]	Reference
	2	*Daucus carota*	+++	Verma and Dougall (1977)
	2	*Haplopappus gracilis*	++	Eriksson (1965)
	2	*Morinda citrifolia*	+++	Zenk *et al.* (1977)
	2	*Nicotiana tabacum*	+	Sahai and Shuler (1984)
	2	*Nicotiana tabacum*	+	Ikeda *et al.* (1976)
	2	*Saccharum officinarum*	+++	Nickell and Maretzki (1970)
	3	*Catharanthus roseus*	++	Fowler (1982)
	3	*Coptis japonica*	+++	Sato and Yamada (1984)
	3	*Daucus carota*	+++	Okamura *et al.* (1975)
	3	*Digitalis purpurea*	+++	Hagimori *et al.* (1982)
	3	*Haplopappus gracilis*	+++	Eriksson (1965)
	3	*Nicotiana tabacum*	++	Ikeda *et al.* (1976)
	3	*Nicotiana tabacum*	++	Sahai and Shuler (1984)
	4	*Acer pseudoplatanus*	+++	Simpkins *et al.* (1970)
	4	*Catharanthus roseus*	+++	Fowler (1982)
	4	*Haplopappus gracilis*	++	Eriksson (1965)
	4	*Lolium multiflorum*	+++	Smith and Stone (1973)
	4	*Nicotiana tabacum*	+++	Ikeda *et al.* (1976)
	5	*Nicotiana tabacum*	+++	Ikeda *et al.* (1976)
	6	*Nicotiana tabacum*	+++	Sahai and Shuler (1984)
	8	*Acer pseudoplatanus*	+++	Simpkins *et al.* (1970)
Glucose +	1 + 1	*Acer pseudoplatanus*	+++	Simpkins *et al.* (1970)
fructose	1 + 1	*Acer pseudoplatanus*	+++	Simpkins and Street (1970)
	1 + 1	*Cannabis sativa*	++	Jones and Veliky (1980)
	1 + 1	*Daucus carota*	+++	Jones and Veliky (1980)
	1 + 1	*Ipomoea* sp.	+++	Jones and Veliky (1980)
	2 + 2	*Lolium multiflorum*	+++	Smith and Stone (1973)
Glycerine	2	*Morinda citrifolia*	−	Zenk *et al.* (1977)
Glycerol	1	*Cannabis sativa*	+	Jones and Veliky (1980)
	1	*Daucus carota*	+	Jones and Veliky (1980)
	1	*Ipomoea* sp.	++	Jones and Veliky (1980)
	2	*Acer pseudoplatanus*	+++	Grout *et al.* (1976)
Inositol	3	*Digitalis purpurea*	−	Hagimori *et al.* (1982)
	4	*Lolium multiflorum*	−	Smith and Stone (1973)
Inulin	1	*Populus*	+	Matsumoto *et al.* (1973)
Lactose	1	*Populus*	+	Matsumoto *et al.* (1973)
	2	*Cannabis sativa*	−	Jones and Veliky (1980)
	2	*Catharanthus roseus*	+	Fowler (1982)
	2	*Daucus carota*	+	Jones and Veliky (1980)
	2	*Daucus carota*	−	Verma and Dougall (1977)
	2	*Ipomoea* sp.	+	Jones and Veliky (1980)
	2	*Morinda citrifolia*	++	Zenk *et al.* (1977)
	2	*Saccharum officinarum*	++	Nickell and Maretzki (1970)
	3	*Medicago sativa*	+++	Chaubet and Pareilleux (1982)

Table I. (Cont.)

Carbon source	Medium concn (% w/v)	Species	Growth[a]	Reference
Maltose	1	*Populus*	+	Matsumoto *et al.* (1973)
	2	*Acer pseudoplatanus*	+++	Simpkins *et al.* (1970)
	2	*Cannabis sativa*	+	Jones and Veliky (1980)
	2	*Catharanthus roseus*	+	Fowler (1982)
	2	*Daucus carota*	+++	Jones and Veliky (1980)
	2	*Daucus carota*	+++	Verma and Dougall (1977)
	2	*Glycine max*	+++	Limberg *et al.* (1979)
	2	*Ipomoea* sp.	+	Jones and Veliky (1980)
	2	*Saccharum officinarum*	+	Nickell and Maretzki (1970)
	3	*Coptis japonica*	++	Sato and Yamada (1984)
	3	*Digitalis purpurea*	+++	Hagimori *et al.* (1982)
Mannitol	1	*Cannabis sativa*	−	Jones and Veliky (1980)
	1	*Daucus carota*	−	Jones and Veliky (1980)
	1	*Ipomoea* sp.	−	Jones and Veliky (1980)
	2	*Morinda citrifolia*	−	Zenk *et al.* (1977)
Mannose	1	*Cannabis sativa*	−	Jones and Veliky (1980)
	1	*Daucus carota*	++	Jones and Veliky (1980)
	1	*Ipomoea* sp.	++	Jones and Veliky (1980)
	2	*Daucus carota*	+++	Verma and Dougall (1977)
	2	*Saccharum officinarum*	−	Nickell and Maretzki (1970)
	3	*Digitalis purpurea*	−	Hagimori *et al.* (1982)
	4	*Lolium multiflorum*	−	Smith and Stone (1973)
Melezitose	1	*Populus*	+	Matsumoto *et al.* (1973)
	2	*Saccharum officinarum*	−	Nickell and Maretzki (1970)
Melibiose	2	*Daucus carota*	−	Verma and Dougall (1977)
	2	*Saccharum officinarum*	+++	Nickell and Maretzki (1970)
Raffinose	1	*Populus*	++	Matsumoto *et al.* (1973)
	2	*Daucus carota*	+++	Verma and Dougall (1977)
	2	*Morinda citrifolia*	++	Zenk *et al.* (1977)
	2	*Saccharum officinarum*	+++	Nickell and Maretzki (1970)
	3	*Digitalis purpurea*	+++	Hagimori *et al.* (1982)
Rhamnose	2	*Morinda citrifolia*	−	Zenk *et al.* (1977)
	2	*Saccharum officinarum*	−	Nickell and Maretzki (1970)
	3	*Digitalis purpurea*	−	Hagimori *et al.* (1982)
Ribose	1	*Cannabis sativa*	−	Jones and Veliky (1980)
	1	*Daucus carota*	−	Jones and Veliky (1980)
	1	*Ipomoea* sp.	−	Jones and Veliky (1980)
	2	*Saccharum officinarum*	++	Nickell and Maretzki (1970)
Sorbitol	1	*Cannabis sativa*	−	Jones and Veliky (1980)
	1	*Daucus carota*	−	Jones and Veliky (1980)
	1	*Ipomoea* sp.	−	Jones and Veliky (1980)
	2	*Morinda citrifolia*	−	Zenk *et al.* (1977)

(continued)

156 GAGIK STEPAN-SARKISSIAN and MICHAEL W. FOWLER

Table I. (Cont.)

Carbon source	Medium concn (% w/v)	Species	Growth[a]	Reference
Sorbose	2	*Saccharum officinarum*	−	Nickell and Maretzki (1970)
Stachyose	2	*Daucus carota*	+++	Verma and Dougall (1977)
Starch	1	*Cannabis sativa*	−	Jones and Veliky (1980)
	1	*Daucus carota*	+	Jones and Veliky (1980)
	1	*Ipomoea* sp.	+	Jones and Veliky (1980)
	1	*Populus*	+	Matsumoto *et al.* (1973)
	1	*Saccharum officinarum*	+++	Maretzki *et al.* (1971)
	2	*Acer pseudoplatanus*	+++	Simpkins *et al.* (1970)
	2	*Catharanthus roseus*	+	Fowler (1982)
	2	*Saccharum officinarum*	+++	Nickell and Maretzki (1970)
	3	*Coptis japonica*	++	Sato and Yamada (1984)
	3	*Digitalis purpurea*	−	Hagimori *et al.* (1982)
	3	*Digitalis purpurea*	−	Hagimori *et al.* (1982)
Sucrose	0.3	*Populus*	+++	Matsumoto *et al.* (1973)
	0.7	*Populus*	+++	Matsumoto *et al.* (1973)
	1	*Acer pseudoplatanus*	+	Simpkins *et al.* (1970)
	1	*Coptis japonica*	+++	Sato and Yamada (1984)
	1	*Digitalis purpurea*	−	Hagimori *et al.* (1982)
	1	*Haplopappus gracilis*	+	Eriksson (1965)
	1	*Populus*	+++	Matsumoto *et al.* (1973)
	1	*Saccharum officinarum*	+++	Maretzki *et al.* (1971)
	1.2	*Populus*	+++	Matsumoto *et al.* (1973)
	1.5	*Vitis*	+++	Yamakawa *et al.* (1983)
	2	*Acer pseudoplatanus*	++	Carceller *et al.* (1971)
	2	*Acer pseudoplatanus*	+++	Simpkins and Street (1970)
	2	*Acer pseudoplatanus*	+++	Simpkins *et al.* (1970)
	2	*Cannabis sativa*	+++	Jones and Veliky (1980)
	2	*Catharanthus roseus*	+	Fowler (1982)
	2	*Daucus carota*	+++	Jones and Veliky (1980)
	2	*Daucus carota*	+++	Verma and Dougall (1977)
	2	*Haplopappus gracilis*	++	Eriksson (1965)
	2	*Ipomoea* sp.	+++	Jones and Veliky (1980)
	2	*Morinda citrifolia*	+++	Zenk *et al.* (1977)
	2	*Nicotiana tabacum*	+	Ikeda *et al.* (1976)
	2	*Rosa* sp.	+++	Danks *et al.* (1975)
	3	*Catharanthus roseus*	+++	Stafford and Fowler (1983)
	3	*Coptis japonica*	+++	Sato and Yamada (1984)
	3	*Digitalis purpurea*	+++	Hagimori *et al.* (1982)
	3	*Haplopappus gracilis*	+++	Eriksson (1965)
	3	*Nicotiana tabacum*	++	Ikeda *et al.* (1976)
	3	*Nicotiana tabacum*	++	Gamanetz and Gamburg (1981)
	3	*Vitis*	+++	Yamakawa *et al.* (1983)
	3	*Populus*	+++	Matsumoto *et al.* (1973)
	4	*Acer pseudoplatanus*	+++	Carceller *et al.* (1971)

Table I. (Cont.)

Carbon source	Medium concn (% w/v)	Species	Growth[a]	Reference
	4	*Acer pseudoplatanus*	+++	Simpkins *et al.* (1970)
	4	*Catharanthus roseus*	++	Fowler (1982)
	4	*Haplopappus gracilis*	+++	Eriksson (1965)
	4	*Lolium multiflorum*	+++	Smith and Stone (1973)
	4	*Malus pumila*	+++	Pareilleux and Chaubet (1980)
	4	*Nicotiana tabacum*	+++	Ikeda *et al.* (1976)
	5	*Digitalis purpurea*	+++	Hagimori *et al.* (1982)
	5	*Haplopappus gracilis*	+++	Eriksson (1965)
	5	*Nicotiana tabacum*	+++	Ikeda *et al.* (1976)
	5	*Populus*	+++	Matsumoto *et al.* (1973)
	5	*Vitis*	+++	Yamakawa *et al.* (1983)
	6	*Acer pseudoplatanus*	+++	Carceller *et al.* (1971)
	6	*Acer pseudoplatanus*	+++	Simpkins *et al.* (1970)
	6	*Catharanthus roseus*	+++	Fowler (1982)
	6	*Coptis japonica*	+++	Sato and Yamada (1984)
	6	*Haplopappus gracilis*	++	Eriksson (1965)
	8	*Acer pseudoplatanus*	++	Carceller *et al.* (1971)
	8	*Acer pseudoplatanus*	++	Simpkins *et al.* (1970)
	8	*Catharanthus roseus*	+++	Fowler (1982)
	10	*Acer pseudoplatanus*	++	Carceller *et al.* (1971)
	10	*Acer pseudoplatanus*	++	Simpkins *et al.* (1970)
	10	*Catharanthus roseus*	+++	Fowler (1982)
	10	*Coptis japonica*	+++	Sato and Yamada (1984)
	10	*Digitalis purpurea*	+	Hagimori *et al.* (1982)
	10	*Vitis*	++	Yamakawa *et al.* (1983)
	15	*Acer pseudoplatanus*	+	Carceller *et al.* (1971)
Trehalose	1	*Populus*	+	Matsumoto *et al.* (1973)
	2	*Daucus carota*	−	Verma and Dougall (1977)
	2	*Morinda citrifolia*	−	Zenk *et al.* (1977)
	2	*Saccharum officinarum*	+++	Nickell and Maretzki (1970)
Xylose	2	*Morinda citrifolia*	−	Zenk *et al.* (1977)
	2	*Saccharum officinarum*	−	Nickell and Maretzki (1970)
	3	*Digitalis purpurea*	−	Hagimori *et al.* (1982)
	4	*Lolium multiflorum*	−	Smith and Stone (1973)

[a]Relative growth is expressed as +, ++, and +++ indicating low, moderate, and high growth, respectively. A minus sign signifies no growth.

Klis, 1979), sycamore (*Acer pseudoplatanus* L.) (Simpkins *et al.*, 1970; Simpkins and Street, 1970; King, 1976, 1977), sugarcane (Nickell and Maretzki, 1970), *Lolium multiflorum* endosperm cells (Smith and Stone, 1973), and *Digitalis purpurea* L. (Hagimori *et al.*, 1982). In general, species growing on glucose exhibit high growth rates and biomass yields.

In the case of fructose, more work has been carried out with callus (Gautheret, 1955; Butenko *et al.*, 1972; Klenovska, 1973; Mathes *et al.*, 1973; Chin *et al.*, 1981) than with suspension cultures. Where data are available for suspension cultures, they are equivocal and may indicate species or cell line dependence. Data from work with sugarcane cell suspensions led Nickell and Maretzki (1970) to suggest that with these cell lines at least, fructose was utilized almost as effectively as sucrose. However, the data were obtained from a single 3-week-old subculture, which is an inadequate test of nutritional value of any carbon source. There are, however, a number of other examples where fructose has been shown capable of supporting cell growth and biomass production (Maretzki *et al.*, 1974). In contrast, there are other examples where transfer of cells from a sucrose- or glucose-based medium to fructose, results in a much reduced growth rate (Fig. 1) (Fowler, 1982). To further complicate matters, it has been shown that for shoot-forming cultures of *D. purpurea* L., fructose is able to actively support growth but is ineffectual in supporting natural product synthesis (Hagimori *et al.*, 1982).

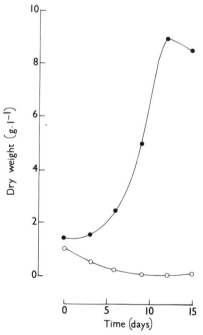

Figure 1. Growth of *Catharanthus roseus* cells with 2% (w/v) fructose as carbon source. Fructose is able to support cell growth during the first passage (●) but this ability does not appear to extend to the second passage (○). (From Fowler, 1982.)

Of the different carbon sources tested for their ability to support growth, none has been found to be as effective as sucrose or glucose. Galactose has been used as sole carbon source in cultures of sugarcane (Nickell and Maretzki, 1970: Maretzki and Thom, 1978), wild carrot (Verma and Dougall, 1979), and *D. purpurea* L. (Hagimori *et al.*, 1982), and in conjunction with other sugars in soybean (*Glycine max* L.) (Sabinski *et al.*, 1982), *Catharanthus roseus* (Fowler *et al.*, 1982), and sycamore (Simpkins *et al.*, 1970). Disaccharides other than sucrose generally result in much reduced growth rates. Such a situation is well documented for maltose, lactose, cellobiose, melibiose, and trehalose, when tested as potential carbon sources for sugarcane suspension cultures (Nickell and Maretzki, 1970). (The lack of good growth on lactose is particularly disappointing as there is an abundance of the disaccharide in milk whey, which exists in large part as a waste product of the dairy industry and is therefore a potential crude substrate material.) Low growth rates on lactose were also reported by Fowler *et al.* (1982) working with *C. roseus*. In contrast, good growth rates were achieved with the monosaccharide components of lactose, i.e., glucose and galactose. When cells grown on these two monosaccharides were transferred to lactose, a much reduced growth rate resulted. Such a situation is generally ascribed to a low level of β-galactosidase activity in plant cells (Fowler *et al.*, 1982). However, there would appear to be exceptions to this. Chaubet *et al.* (1981) reported good growth of *Medicago sativa* cell cultures on lactose, and were able to demonstrate appreciable levels of β-galactosidase activity in both cells and culture medium.

In some cases it has been found possible, over a period of time, to "adapt" cell lines to grow on unusual carbon sources. For instance, both soybean (Limberg *et al.*, 1979) and *A. pseudoplatanus* (Simpkins *et al.*, 1970) cell cultures have been adapted to grow on maltose. Of the other carbon sources tested, none has been found that supports growth to the same degree as sucrose or glucose. Some do, however, support growth to a limited extent; they include raffinose [sugarcane cells (Nickell and Maretzki, 1970)], glycerol [various cell lines (Jones and Veliky, 1980)], and soluble starch [sycamore cells (Simpkins *et al.*, 1970), tobacco cells (Voliva *et al.*, 1982), *C. roseus* cells (Fowler, 1982), *D. purpurea* cells (Hagimori *et al.*, 1982)]. On the whole, sugar alcohols such as inositol (Hagimori *et al.*, 1982) do not appear to support growth, although there are reports that sorbitol can act as sole carbon source for callus initiation and growth of apple cultures (Chong and Taper, 1972, 1974).

2.2. Effects on Growth (Physiology and Biochemistry)

Inextricably linked to the nature of the carbon source used to support culture growth is the relationship of that carbon source with kinetic aspects of culture growth and physiology (Gautheret, 1955: Maretzki *et al.*, 1974; Fowler, 1982).

Growth data are generally presented as packed cell volume, fresh and dry weights, and cell numbers. However, certain factors must be considered in interpreting data obtained from these parameters. First, there may be discriminatory uptake mechanisms for different sugars or, subsequent to uptake, one sugar may undergo a more rapid metabolism than another. Second, certain cell cultures have a tendency to store the incoming sugars in the form of starch, which is then reflected in biomass measurements (Fowler, 1982; Fowler et al., 1982).

Much of the data relating carbon source to culture growth have come from work with batch cultures, and while this has provided much base information (see King and Street, 1973) the transient nature of batch cultures limits their use in definitive quantitative studies relating kinetic parameters to the physiological status of the cells. Continuous (chemostat) culture is a suitable alternative approach, and while this has been explored with plant cells (King et al., 1973; King, 1977; Fowler, 1977; Dougall and Weyrauch, 1980; Wilson, 1980), no thoroughgoing study has as yet been carried out with a carbon source as the limiting nutrient.

As stated above, sucrose and glucose are the most widely used carbon sources for plant cell cultures. However, and although related structurally, the two sources do in some cases have different effects on growth. For instance, in C. roseus cells maximum biomass yields were achieved in a shorter period of time with glucose than with sucrose (Fowler, 1982). At the same time, the efficiency of sucrose utilization declined with increasing initial sugar concentrations. In contrast, it has been shown that sycamore cells grow equally well on sucrose or glucose (Simpkins et al., 1970). The biomass yield of these cultures was enhanced with increased (up to 6%) levels of sucrose. There also appear to be complex interactions between synthetic and natural phytohormones and carbohydrate utilization. For instance, glucose utilization, in contrast to sucrose, was not affected by high kinetin levels supplied to sycamore cell cultures. Kinetin inhibits cell growth by reducing O_2 uptake of the cells (Simpkins and Street, 1970). However, the cell yield and the degree of cell clumping were significantly altered by high auxin [2,4-dichlorophenoxyacetic acid (2,4-D)] and kinetin concentrations (Simpkins and Street, 1970; Carceller et al., 1971). One of the earliest reports of the influence of growth regulators on cell cultures was by Goris (1948) who noted the influence of indoleacetic acid (IAA) on the rate of reserve sugar disappearance in tissue cultures of Jerusalem artichoke. Growth regulators such as naphthaleneacetic acid (NAA), IAA, 2,4-D, and gibberellic acid (GA) have all been shown to influence the uptake of carbohydrate and the level of intracellular sugars in tissue cultures of Zizyphus jujuba (Tandon and Arya, 1979). Also, the interaction between NAA and kinetin has been shown in potato tissue cultures to enhance the participation of the pentose phosphate pathway in glucose catabolism (Kikuta et al., 1977).

The effects of other carbon sources on the growth kinetics of cultured plant cells are varied. In recent years, the availability of high-fructose syrup has raised interest in this sugar as a possible carbon nutrient. Available data on fructose utilization by suspension cultures indicate that its future use will depend on the species used. In *C. roseus* cultures, biomass yield and carbon conversion values were lower with fructose than with glucose. Moreover, the cultures appeared to lose their ability to utilize fructose after successive subculturing (Fig. 1) (Fowler, 1982). In contrast, data of Nickell and Maretzki (1970) from work with sugar-cane cell culture, suggest that for this species at least, fructose is a reasonable alternative carbon source, supporting a level of growth 20% below that achieved with sucrose and 8% with glucose. Unfortunately, experiments were only carried out over a single 3-week culture, with no indication of growth stability in subsequent subculturing. Consequently, their value is limited. With sycamore cell cultures, Simpkins *et al.* (1970) found that fructose could replace glucose even after a second culture passage. In green shoot-forming cultures of *D. purpurea*, which were incapable of growth without an external carbon source, fructose at 3% (w/v) supported a biomass yield only some 20% below that achieved with 3% (w/v) sucrose (Hagimori *et al.*, 1982).

As mentioned earlier, galactose has been shown to support cultures both in the presence (Maretzki and Thom, 1978; Sabinski *et al.*, 1982; Nickell and Maretzki, 1970) and absence (Maretzki *et al.*, 1974; Hagimori *et al.*, 1982; Maretzki and Thom, 1978; Simpkins *et al.*, 1970; Verma and Dougall, 1979) of other carbon sources. Maretzki and Thom (1978) showed that the rate of utilization depended upon the activity of UDP-galactose-4-epimerase (EC 5.1.3.2), a key enzyme in the utilization of galactose. This enzyme had a high activity in galactose-adapted cells. In contrast, when sucrose-adapted cells were transferred to galactose, toxicity symptoms soon became apparent and significant amounts of extracellular polysaccharide were detected in the cultures. Maretzki and Thom (1978) attributed these symptoms to a low activity of UDP-galactose-4-epimerase, which was then unable to cope with excess galactose and subsequent impairment of cell wall formation (see also Hughes and Street, 1974).

We have mentioned the limitations of lactose as a carbon source. However, in cell cultures of *M. sativa* grown on lactose as sole carbon source, average biomass yields of 0.42 g dry wt/g have been reported (Chaubet *et al.*, 1981). It is also of interest to note that unlike sucrose, which is rapidly hydrolyzed to its constituent monosaccharides, lactose breakdown to glucose and galactose proceeds until the end of the exponential phase.

Investigations involving other, less common, carbon sources have for the most part centered on the growth potential of these nutrients rather than on their effects on the physiology and metabolism of cultured cells (see Table I).

For instance, the trisaccharide raffinose is able to sustain growth of sugar-cane cell cultures, at a rate comparable to that obtained with sucrose (Nickell and

Maretzki, 1970). However, no detailed information is available as to the uptake and breakdown of this carbohydrate into its component sugars (glucose, fructose, galactose) nor as to the possible changes in the cellular architecture and enzymological repertoire that its metabolism might entail. Sugarcane cell cultures grow on starch as well as on sucrose. The starch is broken down by an extracellular α-amylase secreted by the cells when starch is supplied in the culture medium (Maretzki et al., 1971, 1974). Similar results were obtained with normal and crown gall tumor tobacco cells in tissue culture (Voliva et al., 1982). Replacement of sucrose by starch resulted in increased intracellular amylolytic activity, which subsequently manifested itself in the nutrient medium. Studies with cycloheximide and actinomycin D indicated that synthesis of mRNA was required for the enzyme induction although other possibilities such as zymogen or cofactor activation were not discounted. Thus far, all the carbon sources discussed have had a hexose structure as the fundamental unit. Little work has been carried out with other smaller molecules except for glycerol. In this latter case, cells of A. pseudoplatanus grown on glycerol had a higher proportion of larger (diameter > 100 μm) cells than cultures grown on glucose (Grout et al., 1976). Oxygen consumption of glycerol-grown cells was considerably lower, although the total dry weight increase and maximum cell number were comparable with glucose-metabolizing cultures.

2.3. Effects on Natural Product Synthesis

The choice of a carbon source for plant cell suspension cultures acquires another dimension when the capacity of cultures for natural product synthesis is considered. Ideally, the carbon source supporting an efficient cell growth would be suitable for product formation and, when more practical applications of cell cultures are envisaged, would also be readily available and inexpensive.

In recent years, a number of workers have reviewed various general aspects of cell culture and secondary metabolite synthesis (Butcher, 1977; Misawa, 1977; Staba, 1977, 1980; Street, 1977; Tabata, 1977; Yeoman et al., 1980; Shuler, 1981; Fowler and Stepan-Sarkissian, 1983). For most of these reviews, the effect of carbon nutrition on the synthetic potential of cultured cells has received peripheral attention, although a few authors have included a separate discussion in their articles (Maretzki et al., 1974; Dougall, 1980).

It is now well established that a medium supporting maximum growth is not necessarily one favoring product formation (Zenk et al., 1977). Although most of the available data suggest that plant growth regulators and selected precursors have a significant influence on the biosynthetic potential of plant cell cultures (Butcher, 1977; Bohm, 1978), the effect of the carbon source on natural product formation is by no means negligible (Table II). In suspension cultures of sycamore cells, for instance, the lignin content per cell increased more than 5-fold as the initial sucrose concentration was increased from 2% to 15% (w/v) (Carceller *et al.,* 1971). At the same time, the amount of lignin in spent medium increased about 10-fold. However, above 6% sucrose concentration, the number of cells (10^6/ml) decreased, indicating a progressive cellular disorganization in media favoring lignin synthesis. Similar results were observed by Constabel (1968) working with callus cultures of *Juniperus communis* L., where tannin content of the cells increased when the medium glucose concentration was raised from 1% to 6%. Both growth and polyphenol synthesis were enhanced when sucrose concentration was increased from 2% to 4% in cell cultures of Paul's Scarlet rose (*Rosa* sp.), with the effect being more pronounced in the case of product synthesis (Davies, 1972). In comparison, glucose, at 2%, was able to sustain product formation at levels comparable to sucrose only at high auxin levels (5 × 10^{-6} M). Matsumoto *et al.* (1973) investigated the effect of a number of carbon sources at 1% concentration on growth and anthocyanin production of *Populus* cells. Sucrose, glucose, fructose, and raffinose were all effective in sustaining cell growth and product formation, with both parameters at a maximum for sucrose and glucose. Growth rate on fructose was comparable to that on sucrose and glucose, but anthocyanin production was 50% lower. All the other carbohydrates tested (galactose, sorbitol, maltose, trehalose, melibiose, and starch) were ineffective in support of pigment formation. Increasing sucrose concentration from 0.3% to 5.0% brought about a marked increase (about 20-fold) in anthocyanin biosynthesis with little effect on cell growth. The same workers also investigated the effects of various (2–5%) sucrose and glucose concentrations on the formation of ubiquinone by tobacco cells (Ikeda *et al.,* 1976). With both sugars, ubiquinone content was highest at 2% concentration while maximum cell yield was obtained at 5% concentration. The production of total phenols and leucoanthocyanins in cell cultures of Paul's Scarlet rose was also promoted when the initial medium glucose concentration was raised (Amorim *et al.,* 1977). In this as in most of the previously cited cases, product accumulation occurred in the stationary phase of growth in general and in the later stages of that phase in particular. An apparent relationship may therefore exist between substrate depletion and product formation. A similar phenomenon in yeasts has been ascribed to catabolite repression (Gordon and Stewart, 1969). In plant cells, a more complex situation probably exists. This is, however, an area in need of detailed investigation. A contrasting picture to that described above is provided by recent work of Hagimori *et al.* (1982) with *D. purpurea* cells. The data presented indicate that for these cells, sucrose, glucose, and raffinose were effective for growth and digitoxin formation, whereas maltose, fructose, and galactose supported cell growth but not product formation. Other sources (mannose, xylose, rhamnose, arabinose, inositol, and starch) sustained neither digitoxin synthesis nor growth. Experiments over a range of initial sucrose concentrations revealed that the optimum concentration for product formation was 3%. Recently, Sato and

Table II. *Effect of Various Carbon Sources on the Formation of Secondary Products in Plant Cell Cultures*

Carbon source	Medium concn (% w/v)	Species	Secondary product[a]	Product formation	Reference
Fructose	1	*Populus*	Anthocyanin	+ +	Matsumoto *et al.* (1973)
	2	*Morinda citrifolia*	Anthraquinone	+	Zenk *et. al* (1977)
	3	*Coptis japonica*	Berberine	+ +	Sato and Yamada (1984)
	3	*Digitalis purpurea*	Digitoxin	+	Hagimori *et al.* (1982
Galactose	1	*Populus*	Anthocyanin	−	Matsumoto *et al.* (1973)
	2	*Morinda citrifolia*	Anthraquinone	+ +	Zenk *et al.* (1977)
	3	*Digitalis purpurea*	Digitoxin	−	Hagimori *et al.* (1982)
Glucose	1	*Populus*	Anthocyanin	+ + +	Matsumoto *et al.* (1973)
	2	*Morinda citrifolia*	Anthraquinone	+	Zenk *et al.* (1977)
	2	*Nicotiana tabacum*	Ubiquinone	+ + +	Ikeda *et al.* (1976)
	2	*Nicotiana tabacum*	Phenolics	+	Sahai and Shuler (1984)
	3	*Coptis japonica*	Berberine	+ +	Sato and Yamada (1984)
	3	*Digitalis purpurea*	Digitoxin	+ + +	Hagimori *et al.* (1982)
	3	*Nicotiana tabacum*	Ubiquinone	+ + +	Ikeda *et al.* (1976)
	3	*Nicotiana tabacum*	Phenolics	+ +	Sahai and Shuler (1984)
	4	*Nicotiana tabacum*	Ubiquinone	+ +	Ikeda *et al.* (1976)
	5	*Nicotiana tabacum*	Ubiquinone	+	Ikeda *et al.* (1976)
	6	*Nicotiana tabacum*	Phenolics	+ + +	Sahai and Shuler (1984)
Lactose	1	*Populus*	Anthocyanin	−	Matsumoto *et al.* (1973)
	2	*Morinda citrifolia*	Anthraquinone	+ +	Zenk *et al.* (1977)
Maltose	1	*Populus*	Anthocyanin	−	Matsumoto *et al.* (1973)
	3	*Coptis japonica*	Berberine	+ +	Sato and Yamada (1984)
	3	*Digitalis purpurea*	Digitoxin	+ +	Hagimori *et al.* (1982)
Mannose	3	*Digitalis purpurea*	Digitoxin	−	Hagimori *et al.* (1982)
Melezitose	1	*Populus*	Anthocyanin	−	Matsumoto *et al.* (1973)
Raffinose	1	*Populus*	Anthocyanin	+	Matsumoto *et al.* (1973)
	2	*Morinda citrifolia*	Anthraquinone	+ +	Zenk *et al.* (1977)
	3	*Digitalis purpurea*	Digitoxin	+ + +	Hagimori *et al.* (1982)
Starch	1	*Populus*	Anthocyanin	−	Matsumoto *et al.* (1973)
	3	*Coptis japonica*	Berberine	+ +	Sato and Yamada (1984)
	3	*Digitalis purpurea*	Digitoxin	−	Hagimori *et al.* (1982)
Sucrose	0.3	*Digitalis purpurea*	Digitoxin	−	Hagimori *et al.* (1982)
	0.3	*Populus*	Anthocyanin	−	Matsumoto *et al.* (1973)
	0.7	*Populus*	Anthocyanin	+ +	Matsumoto *et al.* (1973)
	1	*Coptis japonica*	Berberine	+ +	Sata and Yamada (1984)
	1	*Digitalis purpurea*	Digitoxin	−	Hagimori *et al.* (1982)
	1	*Populus*	Anthocyanin	+ + +	Matsumoto *et al.* (1973)
	1.2	*Populus*	Anthocyanin	+ + +	Matsumoto *et al.* (1973)
	1.5	*Vitis*	Anthocyanin	+	Yamakawa *et al.* (1983)
	2	*Acer pseudoplatanus*	Lignin	+ +	Carceller *et al.* (1971)
	2	*Morinda citrifolia*	Anthraquinone	+ + +	Zenk *et al.* (1977)
	2	*Nicotiana tabacum*	Ubiquinone	+ + +	Ikeda *et al.* (1976)
	2	*Rosa* sp.	Polyphenols	+	Davies (1972)

Table II. (Cont.)

Carbon source	Medium concn (% w/v)	Species	Secondary product[a]	Product formation	Reference
	3	*Coptis japonica*	Berberine	+++	Sato and Yamada (1984)
	3	*Digitalis purpurea*	Digitoxin	+++	Hagimori *et al.* (1982)
	3	*Nicotiana tabacum*	Ubiquinone	+++	Ikeda *et al.* (1976)
	3	*Populus*	Anthocyanin	+++	Matsumoto *et al.* (1973)
	3	*Rosa* sp.	Polyphenols	++	Davies (1972)
	3	*Vitis*	Anthocyanin	++	Yamakawa *et al.* (1983)
	4	*Nicotiana tabacum*	Ubiquinone	++	Ikeda *et al.* (1976)
	4	*Rosa* sp.	Polyphenols	+++	Davies (1972)
	5	*Digitalis purpurea*	Digitoxin	++	Hagimori *et al.* (1982)
	5	*Nicotiana tabacum*	Ubiquinone	+	Ikeda *et al.* (1976)
	5	*Populus*	Anthocyanin	+++	Matsumoto *et al.* (1973)
	5	*Vitis*	Anthocyanin	++	Yamakawa *et al.* (1983)
	6	*Acer pseudoplatanus*	Lignin	+++	Carceller *et al.* (1971)
	6	*Coptis japonica*	Berberine	+++	Sato and Yamada (1984)
	10	*Coptis japonica*	Berberine	++	Sato and Yamada (1984)
	10	*Digitalis purpurea*	Digitoxin	+	Hagimori *et al.* (1982)
	10	*Vitis*	Anthocyanin	+++	Yamakawa *et al.* (1983)
	15	*Acer pseudoplatanus*	Lignin	+++	Carceller *et al.* (1971)
Trehalose	1	*Populus*	Anthocyanin	−	Matsumoto *et al.* (1973)
Xylose	3	*Digitalis purpurea*	Digitoxin	+	Hagimori *et al.* (1982)

[a]+, ++, and +++ indicate low, moderate, and high levels of product formation, respectively. A minus sign denotes no secondary product formation.

Yamada (1984) have studied the effect of various carbon sources on growth and berberine production of *Coptis japonica* cells. Sucrose was the most efficient source for both growth and berberine production. A wide range of sucrose concentrations (1–10%) supported good growth and alkaloid production, although the berberine yield was somewhat lower at 10% concentration. The optimum surcose concentration for both parameters was found to be 3%.

3. UPTAKE MECHANISMS FOR CARBON SOURCES

The availability of relatively homogeneous and finely dispersed cell cultures with a defined cultivation history has provided the biochemist with an important tool with which to study uptake mechanisms. Hart and Filner (1969) were among the first to make use of cell cultures for such studies. They used tobacco cell cultures to study sulfate uptake. This and related work has been reviewed by Dougall (1980).

A variety of mechanisms exist through which nutrient sugars may enter the cell. To reach an understanding of these is an important aspect of investigations into cellular carbohydrate metabolism and physiology.

3.1. Differential Mechanisms

The great majority of information on carbohydrate uptake into plant cell culture relates to sucrose. Thus, the following discussion is centered around this sugar.

Mechanisms of sucrose entry into the cell can be broadly divided into two categories, hydrolytic and nonhydrolytic (Fig. 2). There are many examples of sucrose hydrolysis occurring prior to uptake in a wide range of cell culture systems. This possibly represents the simplest situation, where hydrolysis takes place in the external medium or on the cell surface (cell wall) (Maretzki *et al.*, 1974; Ueda *et al.*, 1974). It would appear that in the majority of these cases, hydrolysis to glucose and fructose is not essential, but that the level of hydrolysis is a variable depending upon such factors as cell age, nutrient status, sucrose availability, and growth rate (Fowler, 1982). The one major exception to this is sugarcane where in both cell cultures (Maretzki *et al.*, 1974) and the whole cane plant (Sacher *et al.*, 1963; Bowen, 1972), hydrolysis is a necessary, if not absolute, prerequisite to uptake.

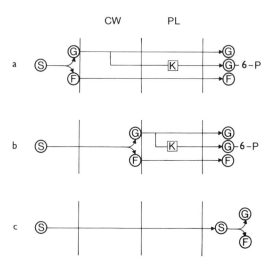

Figure 2. Models for hydrolytic (a, b) and nonhydrolytic (c) modes of sucrose entry into plant cells. Sucrose (S) hydrolysis into glucose (G) and fructose (F) can take place either in the external medium or within the cell barrier [cell wall (CW) + plasmalemma (PL)]. The formation of sugar phosphates by a kinase (K) may be a special case for either route.

A slightly more complicated hydrolytic route for sucrose entry may involve a mechanism whereby the disaccharide is preferentially removed from the nutrient medium and bound in some way to the cell wall or plasmalemma. Hydrolysis would then take place with the disaccharide in a bound state, the constituent monosaccharides then being released into the cell sap. Such a model needs detailed investigation.

Nonhydrolytic uptake of sucrose involves the intact molecule passing the cell wall and plasmalemma with inversion occurring intracellularly. In the case of carrot callus cultures, there is definitive evidence for such a mechanism (Edelman and Hanson, 1971a,b). Information on this mode of uptake in cultured cells is limited, although it has been more thoroughly investigated in whole plants (e.g., Sacher, 1966; Engel and Kholodova, 1969; Jenner, 1974; Chin and Weston, 1975; Giaquinta, 1977).

Sugar uptake may also involve phosphorylation mechanisms. Maretzki and Thom (1972) reported that sugar phosphates appear almost immediately after cells of *Saccharum* sp. are incubated in medium containing hexoses. Although the formation of phosphorylated intermediates may be an important stage in the uptake process, it would not be specifically associated with any of the models in Fig. 2 but would constitute a special case for any one of them.

A number of workers have investigated the uptake of hexoses, and in particular glucose. There are many difficulties associated with such studies, mainly due to the fact that glucose is a rapidly and widely metabolized substrate. The methods suggested to overcome the problems involved are not themselves without drawbacks. For example, use of radioactive glucose must take into account the loss of label in the form of carbon dioxide. Similarly, use of glucose analogues must be based on firm evidence that the uptake system does not differentiate between the analogue and the original substrate.

Data for glucose uptake by plant cells in suspension culture, principally *Saccharum* sp., suggest that it may be an active process (Maretzki and Thom, 1972, 1979; Maretzki *et al.*, 1974). In contrast, observations with carrot (Parr and Edelman, 1976) and tobacco (Opekarova and Kotyk, 1973) callus tissue discount the operation of active uptake. As has been pointed out by Fowler (1978), the type of culture system selected to study the phenomenon of sugar uptake may not represent the *in vivo* situation and therefore caution must be exercised in the interpretation of data from such systems.

In general, information on interrelationships between different sugars and various uptake mechanisms is limited, mainly because studies involving detailed kinetics of uptake phenomena are difficult to perform. Although in the case of cultured plant cells the problem of substrate diffusion through tissues (see Ehward *et al.*, 1979) is very much reduced, the complications of rapid sugar metabolism subsequent to uptake remain. As mentioned earlier, the data available on sugar uptake by plant cells have largely come from work with whole plant tissue and callus cultures, with contributions from suspension cultures being minimal. This is surprising in view of the advantages offered by cell suspension cultures as model systems.

3.2. Cellular Location

Before entering the internal environment of the plant cell, sugars must pass through two physical barriers, the cell wall and the plasmalemma. There are various opinions concerning the role that these limiting structures play in allowing the penetration of carbohydrates. Available evidence, mostly from whole plant tissues, suggests that the cell wall and plasmalemma may behave differently depending on the type of sugar and plant species concerned.

3.2.1. Role of the Cell Wall

It is generally agreed that the cell wall is the major site of sucrose hydrolysis in most plant cells (Copping and Street, 1972; Ueda et al., 1974; Fowler, 1978, 1982). Working with different plant organs and tissue cultures, Straus (1962) demonstrated the presence of a cell wall-located sucrose-hydrolyzing system in at least one strain of tobacco cultures. Later work from the same laboratory (Straus and Campbell, 1963) showed that the cell walls of a variety of tissue cultures were able to release enzymes such as phosphatase, amylase, and peroxidase into the medium. The major metabolic role of the cell wall is, however, considered to be its sucrose-hydrolyzing capacity due to the presence of the enzyme invertase (EC 3.2.1.26). Depending on its pH optimum, this enzyme occurs in acid, alkaline, and neutral forms and has been described in a wide range of plant tissues (e.g., Straus, 1962).

Copping and Street (1972) reported the presence of both "acid" and "neutral" invertase activities from suspension cultures of A. pseudoplatanus cells. An appreciable activity of the neutral invertase appeared in the cell wall fraction, whereas acid invertase showed an equal distribution between the cell wall and soluble fractions. The highest amount of enzyme recoverable from the cell wall fraction was 50% of the total present, indicating a strong binding of the enzyme to cell wall material. A similar situation was found in suspension cultures of Daucus carota by Ueda et al. (1974), who reported 50–60% release of invertase activity after treatment with cell wall-solubilizing enzymes. The enzyme from both cell wall and cell homogenate preparations had an acid pH optimum and no neutral invertase was found.

Further proof for a strong association between invertase activity and the cell wall was presented by Parr and Edelman (1975) working with carrot cell cultures. They reported a release of up to 80% of total enzyme activity by altering the ionic strength or pH of the extraction medium. In further work with carrot cell cultures, Watson and Fowler (see Fowler, 1982) found that the rate of hydrolysis of [U-^{14}C]sucrose by protoplasts was 50% lower than that by intact cells. The presence of hydrolytic activity in protoplasts was explained by the presence of residual cell wall material. In cultures of sugarcane cells (Maretzki *et al.*, 1974), a slightly different invertase profile was observed. The cell wall fraction was found to contain an acid invertase whereas both acid and neutral types were present in the soluble fraction with the neutral form predominating. The cell wall enzyme was found to be tightly bound and could not be removed by repeated washing. Earlier reviews contain detailed discussions of the role of invertase in cultured plant cells (Maretzki *et al.*, 1974; Fowler, 1978).

3.2.2. Role of the Plasmalemma

While there is a great deal of evidence to suggest that cell wall-located enzyme activity has a major impact on sucrose uptake, the role of the plasmalemma should not be underestimated. Indeed, there are indications that the plasmalemma may play a substantial role in mediating sugar uptake into plant cells.

Studies by Maretzki and Thom (1972) with sugarcane suspension cultures suggested that in these cells the uptake of glucose was actively mediated and that the plasmalemma played a role in this. The possibility that a coupled phosphorylation may be involved was also considered, although no evidence of a membrane-bound hexokinase was presented. A particulate hexokinase (EC 2.7.1.1) has been reported in cell cultures of *A. pseudoplatanus* (Fowler and Clifton, 1974, 1975). However, this was subsequently found to be associated with the mitochondrion and it is difficult to see how this enzyme could be involved in a phosphorylative entry of glucose into the cell through the plasmalemma (Fowler, 1978).

In contrast to the views of Maretzki and Thom (1972) from their work with sugarcane suspension cultures, Parr and Edelman (1976) suggested that the passage of sugars across the plasmalemma of carrot callus cells is primarily a passive process. They ascribed the reported active uptake of sugars to the operation of specific sites on the tonoplast membrane surrounding the vacuole. Similar results obtained with tobacco callus tissue led Opekarova and Kotyk (1973) to suggest that uptake of sugars across the plasmalemma may be a simple diffusion through hydrophilic pores or intercellular spaces. Although it is possible that the type of culture system used (suspension versus callus) might have had an effect on the results obtained (Fowler, 1978), the possibility of different uptake systems coexisting in the same tissue should not be overlooked. It has been suggested that in sugar beet root tissue a close contact exists between the plasmalemma and the tonoplast, allowing direct entry of sucrose into the vacuole (Wyse, 1979) without entering the cytoplasm. It is not clear if both membranes are served with the same carrier system or if passage through the plasmalemma is passive and that through the tonoplast active. Further studies using callus and cell suspension cultures are needed to compare the uptake mechanisms of free cells and those of cell groups and tissues.

3.3. Effect of Internal Pools

The influence internal "pools" of carbohydrates may exert on their own uptake by plant cell systems is an area in need of in-depth and comprehensive investigation. References in the literature to this particular phenomenon are scattered and generally are discussed in a peripheral manner in relation to other aspects of cell physiology and biochemistry.

In an early study with suspension cultures of *Ipomoea* sp., Rose *et al.* (1972) reported an increase in total cellular carbohydrate after a lag of some 48 hr. Comparison of the rates of sucrose uptake and accumulation suggested that net accumulation began some 24 hr after subculture and attained a maximum shortly after the peak of sucrose uptake from the medium. In this and the majority of studies relating to internal pools, it is the effect of the extracellular carbon concentration on the internal sugar content of the cell (rather than vice versa) that has been emphasized. Such an example is seen in work with tissue cultures of *Beta vulgaris* L. (Butenko *et al.*, 1972) where the cellular contents of sucrose, glucose, and fructose decreased sharply and remained low in medium containing 0.5% (w/v) sucrose after transfer from 2% sucrose. When the level of medium sucrose was raised to 5%, the intracellular content of sucrose and monosaccharides rose and remained high. Where the external level of sucrose was low, monosaccharides were utilized preferentially to sucrose. Where the external level of sucrose was high, the increase in internal sucrose level was greater than that for monosaccharides. With 20% sucrose in the medium, a seven-fold rise in sugar content was observed over a period when no tissue growth was detectable. The authors reported a negative correlation between the rate of growth and sugar content in the tissue. Furthermore, the sugar content of sucrose-grown cells was higher than that of cells grown on media containing glucose, fructose, or an equimolar mixture of the two.

The data of Maretzki *et al.* (1974) from cultures of *Saccharum* sp. show an initial rise in endogenous reducing sugars (after 2–3 days) before declining to 50% or less of the peak level after 10 days of growth. Such a situation presumably mirrors the uptake of sucrose from the external medium. Working with sugar beet tissue cultures, Angelova *et al.* (1974) observed an inverse correlation between sugar content and enzyme activity. Unfortunately, their data did not include profiles of the nutrient medium sugar levels, which may have provided some indication as to the relationship among invertase activity, sugar uptake, and cellular content. Cells of *Saccharum* sp. suspension cultures grown initially on sucrose and then transferred to galactose for a period of 3 to 7 days accumulated very significant levels of galactose in the pool of free sugars (Maretzki and Thom, 1978). This increase was at the expense of intracellular sucrose content, which dropped to one-third of its level in the overall sugar pool compared with sucrose-grown cultures. Appreciable increases in UDP-galactose and galactose-1-phosphate were also detected. In suspension cultures of *Dioscorea deltoidea*, the level of soluble sugars in the cells increased to 11–18% of the dry mass during the lag phase and remained within this range until the middle of the log phase, after which there was a decrease to 5% (Tarakanova *et al.*, 1979). Toward the later stages of the stationary phase, however, the level of soluble sugars did increase slightly. Such a trend was also observed in the case of sugar beet tissue grown on 5% (w/v) sucrose (Butenko *et al.*, 1972) and with *Ipomoea* cells (Rose *et al.*, 1972) and could be due to the breakdown of internal cell reserves in order to maintain metabolism.

The accumulation of lactose has also been reported. Cells of *M. sativa* accumulated high levels of lactose (around 200 mg/g dry wt) when grown in suspension culture with the disaccharide as sole carbon source (Chaubet *et al.*, 1981). Endogenous concentrations of glucose and galactose were much lower, but remained relatively constant throughout the growth cycle whereas the level of lactose dropped to 20% of its peak. It is difficult to speculate on the nature of this particular situation given the limited amount of data available.

4. INTRACELLULAR FATE OF CARBON SOURCE— BIOCHEMISTRY: OXIDATION, BIOSYNTHESIS, STORAGE

Subsequent to mobilization and uptake into the cell, the carbon source may be channeled in three ways: (1) oxidation to satisfy the energy requirements of the cell, (2) utilization as precursors for biosynthetic pathways, and (3) storage, usually in the form of a polymer. In studies dealing with carbohydrate nutrition of plant cell cultures, comparatively few have focused on the biochemical aspects of carbon utilization. Although there is good evidence for the presence and operation of glycolytic and pentose phosphate pathways of hexose catabolism in cultured systems (Fowler, 1971, 1978; Fowler and Clifton, 1974; Thorpe and Meier, 1974; Komamine and Shimizu, 1975; Stafford and Fowler, 1983), only a limited amount of work has been done concerning the enzymological potential of plant cells in the degradation, mobilization, and subsequent utilization of less traditional carbon sources that have been shown to support the growth of cell cultures.

Many of the early studies on intermediary metabolism of plant cells grown in suspension culture were carried out with cells of *A. pseudoplatanus*. The presence and operation in these cell cultures of the glycolytic and pentose phosphate pathways was demonstrated using classical enzymological and respiratory analyses (see Givan and Collin, 1967; Simpkins and Street, 1970). A more detailed analysis of enzymological activity and carbon flux through the two pathways was carried out by Fowler (1971). In this work, the presence of glycolytic and pentose phosphate pathways was demonstrated by measuring the activities of six enzymes of the former and four enzymes of the latter pathway. The activities of all enzymes assayed were associated with the "soluble" fraction. The relative contribution of each pathway to carbohydrate oxidation was also assessed using classical [^{14}C]glucose labeling techniques. Data from both experimental approaches indicated that during the lag and early exponential phases of growth, both pathways make significant contributions to glucose catabolism, with the capacity of the pentose phosphate pathway being relatively greater than that of glycolysis. In the later phases of growth, glycolysis appears to be the main channel of carbohydrate oxidation. The situation appears to be in contrast to that in tobacco callus cultures where no correlation between the activities of carbohydrate oxidation enzymes and the stage of tissue growth was observed (Thorpe and Meier, 1972).

The work of Fowler (1971) on cytosolic metabolism in sycamore cells was extended by Wilson (1971) to include mitochondrial respiration. The latter was shown to resemble systems from whole plants with somewhat reduced efficiency in energy conservation. Wilson (1971) was also able to demonstrate that the cell cultures possessed a significant capacity for cyanide-insensitive respiration, which appeared to be particularly active during the lag phase. He suggested that the high activity of the cyanide-resistant pathway during the lag phase, when appreciable biosynthetic activity is in progress, may indicate that the TCA cycle is supplying essential substrates as biosynthetic precursors. High activity of the pentose phosphate pathway during the same phase was suggested by Fowler (1971) as a means of providing the necessary reducing power in the form of NADPH to support biosynthesis. In embryogenic cultures of carrot cell suspensions, the operation of the pentose phosphate pathway was considerably enhanced during increased lipid biosynthesis and a similar interaction with respiratory electron transport systems may be operational (Kikuta *et al.*, 1981).

Cell cultures have been used fairly extensively to study interactions between the respiratory pathways and other facets of metabolism. For example, the induction of nitrate assimilation in sycamore cells was accompanied by increased activity through the pentose phosphate pathway, presumably through demand for and supply of NADPH for nitrite reduction (Jessup and Fowler, 1977). A similar correlation has been observed with suspension cultures of *C. roseus* (Stafford and Fowler, 1983). In the latter case, batch cultures of *C. roseus* cells were grown under conditions of either nitrogen or carbon limitation and different parameters of cell growth as well as activities of enzymes of carbohydrate oxidation investigated. Major changes in enzyme activity, but particularly in key control enzymes of both glycolysis [phosphofructokinase (EC 2.7.1.11) and pyruvate kinase (EC 2.7.1.4.0)] and the pentose phosphate pathway [glucose-6-phosphate dehydrogenase (EC 1.1.1.49)], were observed. These changes were closely associated with changes in the nutritional status of the cells.

All of the above work was carried out with batch cultures. As pointed out previously, such cultures are limited in their application by their transient nature and difficulties of growth control. Continuous (chemostat) culture provides an alternative, if somewhat lengthy, approach. However, apart from some studies by Fowler and Clifton (1974) on the stability of respiratory enzyme levels in steady states of an *A. pseudoplatanus* L. chemostat culture, this technique has yet to be applied in depth to plant cell cultures.

Information on the metabolic utilization of carbon substrates other than sucrose, glucose, and fructose is extremely limited. Maretzki and Thom (1978) have proposed a pathway for the entry of galactose into intermediary metabolism in suspension cultures of galactose-adapted *Saccharum* cells (Fig. 3). In the absence of galactose-1-phosphate uridyltransferase (EC 2.7.7.12), they suggested that UDP-galactose was formed from galactose-1-phosphate by the enzyme UDP-galactose pyrophosphorylase (EC 2.7.7.10). A tenfold higher activity of UDP-galactose-4-epimerase (EC 5.1.3.2) in galactose-propagated cultures provided further support for the operation of the proposed pathway in these cultures. No data exist for other potential substrates.

In studies concerned with the biosynthetic and storage fates of exogenous carbon, most have been concerned with cell wall synthesis and starch accumulation, respectively.

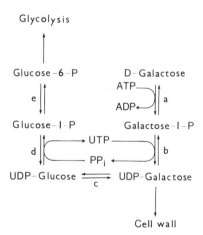

Figure 3. Galactose metabolism in sugarcane cell cultures. The enzymes involved in the pathway are: (a) galactokinase, (b) UDP-galactose pyrophosphorylase, (c) UDP-galactose-4-epimerase, (d) UDP-glucose pyrophosphorylase, and (e) phosphoglucomutase. (Redrawn from Maretzki and Thom, 1978.)

As early as 1953, Lamport used sycamore cells to study oxygen fixation into the hydroxyproline moiety of a primary cell wall-specific protein. Sycamore cells were also used by Becker *et al.* (1964) to investigate the synthesis of extracellular polysaccharides by plant cell suspension cultures. They reported a close similarity of structure between the extracellular polysaccharides and the noncellulosic cell wall polysaccharides of their particular sycamore cells. The course and pattern of synthesis was followed via [^{14}C]glucose. Radioactive glucose fed to the cultures was rapidly taken up and the label was detected in the galacturonic acid moieties of both cell wall and external polysaccharides, the specific activity being higher in the latter. Further studies from the same laboratory (Nevins *et al.*, 1966) indicated that the nature of the carbon source could alter the composition of the cell wall polymer. A variety of carbon sources including sucrose, maltose, glucose, galactose, mannose, and glycerol were tested for such an effect. In all cases, the most abundant sugar in the cell walls was arabinose followed by galactose. When galactose was the carbon source, the incorporation of this sugar in the cell wall was significantly greater than in cells grown on glucose or mannose. Otherwise, the wall sugar ratios remained relatively constant for other carbon sources. These results indicated that in the case of cultured cells, the medium could perhaps exercise catabolite repression in the synthesis of cell wall material, thus overruling strict genetic control. Unfortunately, no growth data were presented for the different carbon sources and thus it is not possible to distinguish any effects on culture performance.

Plant growth substances have also been found to influence polysaccharide synthesis by cultured cells. Street *et al.* (1968) reported that high kinetin levels produced a denser and more granular extracellular hemicellulose in sycamore cultures. Alterations in the chemical composition were observed at reduced galactose and xylose levels and at increased glucose level (Simpkins and Street, 1970). These changes were ascribed to kinetin suppression of phosphoglucose isomerase (EC 5.3.1.9) and other enzymes participating in monosaccharide conversions. The auxin 2,4-D was shown to enhance the incorporation of label from radioactive glucose into the arabinose moieties of sycamore cell wall polysaccharides (Rubery and Northcote, 1970). A detailed analysis of cell wall components of sycamore cells in suspension cultures has been published by Talmadge *et al.* (1973). Studies of cell wall synthesis in carrot cultures by Asamizu and Nishi (1979, 1980) suggested that cells grown in a medium containing sucrose and myoinositol incorporated the glucose moiety of the disaccharide in the neutral sugars of the cell wall, whereas myoinositol served as a precursor of acidic sugars and pentoses. Although myoinositol cannot sustain growth of cell cultures as sole carbon source (e.g., Smith and Stone, 1973; Verma and Dougall, 1977), there is sufficient evidence that its inclusion in the medium promotes the growth of certain culture systems (e.g., Shantz *et al.*, 1967). It is possible that myoinositol acts as the substrate in a glucogenic pathway. Further studies are needed to test this hypothesis and to ascertain whether different mechanisms control the operation of such a pathway in cells responding to myoinositol treatment. Studies into starch metabolism in cell cultures are comparatively few, which is surprising in view of the fact that many cell cultures actively synthesize starch from incoming carbohydrate and that such systems could usefully serve as a highly versatile experimental tool to study starch biosynthesis *in vivo* and complement results obtained from cell-free systems. Some work has been carried out into starch synthesis by callus cultures [e.g., Thorpe and Meier (1972, 1974) with tobacco cells], but little with suspension cultures. Chu and Shannon (1975) found that suspension cultures of corn (*Zea mays* L.) endosperm growing on 2% (w/v) sucrose, exhibited a major increase in starch deposition when the sucrose level was increased to 6% 9 days after subculture. This effect could not be mimicked with glucose or fructose. More recently, Gamanetz and Gamburg (1981) reported that adenine increased the starch content of tobacco cells in suspension cultures. This phenomenon was dependent on sucrose and the auxin NAA, both of which promoted the adenine effect on starch accumulation.

5. SUMMARY COMMENTS

As can be seen from this review, the subject of carbohydrate metabolism in plant cell cultures covers an extremely wide and diverse array of interests. This is reflected in the status of the literature for the subject. Some parts have been reasonably extensively investigated while others are comparatively untouched. Much of the data also suffer from being of a qualitative rather than a quantitative nature. Given its fundamental importance, it is to be hoped that a more concerted effort can be mounted in the area of plant cell carbohydrate metabolism, and we would hope that this review goes some way to stimulate this.

REFERENCES

Amorim, H. V., Dougall, D. K., and Sharp, W. R., 1977, The effect of carbohydrate and nitrogen concentration on phenol synthesis in Paul's Scarlet rose cells grown in tissue culture, *Physiol. Plant.* **39**:91.

Angelova, A. A., Atanasov, A. I., Stambolova, M. A., and Nikolov, T. K., 1974, Invertase activity and sugar content in cultures of sugar beet tissue cultivated on a medium containing chloramphenicol, *Fiziol. Rast.* **21**:1021.

Asamizu, T., and Nishi, A., 1979, Biosynthesis of cell-wall polysaccharides in cultured carrot cells, *Planta* **146**:49.

Asamizu, T., and Nishi, A., 1980, Regenerated cell wall of carrot protoplasts isolated from suspension-cultured cells, *Physiol. Plant.* **48**:207.

Ball, E., 1953, Hydrolysis of sucrose by autoclaving media: A neglected aspect in the tissue culture of plant tissues, *Bull. Torrey Bot. Club* **80**:409.

Becker, G. E., Hui, P. A., and Albersheim, P., 1964, Synthesis of extracellular polysaccharide by suspensions of *Acer pseudoplatanus* cells, *Plant Physiol.* **39**:913.

Bertola, M. A., and Klis, F. M., 1979, Continuous cultivation of glucose-limited bean cells (*Phaseolus vulgaris* L.) in a modified bacterial fermentor, *J. Exp. Bot.* **30**:1223.

Bohm, H., 1978, Regulation of alkaloid production in plant cell cultures, in: *Frontiers of Plant Tissue Culture 1978* (T. A. Thorpe, ed.), pp. 201–211, University of Calgary, Calgary.

Bowen, J. E., 1972, Sugar transport in immature internodal tissue of sugarcane, *Plant Physiol.* **49**:82.

Butcher, D. N., 1977, Secondary production in tissue cultures, in: *Applied and Fundamental Aspects of Plant Cell, Tissue and Organ Culture* (J. Reinert and Y. P. S. Bajaj, eds.), pp. 668–693, Springer-Verlag, Berlin.

Butenko, R. G., Kholodova, V. P., and Urmantseva, V. V., 1972, Regularities of growth and certain correlations between growth and sugar content in cells of sugar beet tissue cultures, *Sov. Plant Physiol.* **19**:786.

Carceller, M., Davey, M. R., Fowler, M. W., and Street, H. E., 1971, The influence of sucrose, 2,4-D and kinetin on the growth, fine structure and lignin content of cultured sycamore cells, *Protoplasma* **73**:367.

Chaubet, N., and Pareilleux, A., 1982, Characterization of β-galactosidases of *Medicago sativa* suspension-cultured cells growing on lactose: Effect of the growth substrates on the activities, *Z. Pflanzenphysiol.* **106**:401.

Chaubet, N., Petiard, V., and Pareilleux, A., 1981, β-Galactosidases of suspension-cultured *Medicago sativa* cells growing on lactose, *Plant Sci. Lett.* **22**:369.

Chin, C. K., and Weston, C. D., 1975, Sucrose absorption and synthesis by excised *Lycopersicon esculentum* roots, *Phytochemistry* **14**:69.

Chin, C. K., Haas, J. C., and Still, C. C., 1981, Growth and sugar uptake of excised root and callus of tomato, *Plant Sci. Lett.* **21**:229.

Chong, C., and Taper, C. D., 1972, *Malus* tissue cultures. I. Sorbitol (D-glucitol) as a carbon source for callus initiation and growth, *Can. J. Bot.* **50**:1399.

Chong, C., and Taper, C. D., 1974, *Malus* tissue cultures. II. Sorbitol metabolism and carbon nutrition, *Can. J. Bot.* **52**:236.

Chu, L. J. C., and Shannon, J. C., 1975, *In vitro* cultures of maize endosperm: A model system for studying *in vivo* starch biosynthesis, *Crop Sci.* **15**:814.

Constabel, F., 1968, Gerbostoffproduktion der calluskulturen von *Juniperus communis* L., *Planta* **79**:58.

Copping, L. G., and Street, H. E., 1972, Properties of the invertases of cultured sycamore cells and changes in their activity during culture growth, *Physiol. Plant.* **26**:346.

Danks, M. L., Fletcher, J. S., and Rice, E. L., 1975, Effects of phenolic inhibitors on growth and metabolism of glucose-UL-^{14}C in Paul's Scarlet rose cell suspension cultures, *Am. J. Bot.* **62**:311.

Davies, M. E., 1972, Polyphenol synthesis in cell suspension cultures of Paul's Scarlet rose, *Planta* **104**:50.

Dougall, D. K., 1980, Nutrition and metabolism, in: *Plant Tissue Culture as a Source of Biochemicals* (E. J. Staba, ed.), pp. 21–58, CRC Press, Boca Raton, Fla.

Dougall, D. K., and Weyrauch, K. W., 1980, Growth and anthocyanin production by carrot suspension cultures grown under chemostat conditions with phosphate as the limiting nutrient, *Biotechnol. Bioeng.* **22**:337.

Edelman, J., and Hanson, A. D., 1971a, Sucrose suppression of chlorophyll synthesis in carrot callus cultures, *Planta* **98**:150.

Edelman, J., and Hanson, A. D., 1971b, Sucrose suppression of chlorophyll synthesis in carrot tissue culture: The role of invertase, *Planta* **101**:122.

Ehward, R., Mescheryakov, A. B., and Kholodova, V. P., 1979, Hexose uptake by storage parenchyma of potato and sugar beet at different concentrations and different thicknesses of tissue slices, *Plant Sci. Lett.* **16**:181.

Engel, O. S., and Kholodova, V. P., 1969, Activity of invertase and accumulation of sucrose in sugar beet roots, *Sov. Plant Physiol.* **16**:973.

Eriksson, T., 1965, Studies on the growth requirements and growth measurements of cell cultures of *Haplopappus gracilis*, *Physiol. Plant.* **18**:976.

Fowler, M. W., 1971, Studies on the growth in culture of plant cells. XIV. Carbohydrate oxidation during the growth of *Acer pseudoplatanus* L. cells in suspension culture, *J. Exp. Bot.* **22**:715.

Fowler, M. W., 1977, Growth of cell cultures under chemostat conditions, in: *Plant Tissue Culture and Its Biotechnological Application* (W. Barz, E. Reinhard, and M. H. Zenk, eds.), pp. 253–263, Springer-Verlag, Berlin.

Fowler, M. W., 1978, Regulation of carbohydrate metabolism in cell suspension cultures, in: *Frontiers of Plant Tissue Culture 1978* (T. A. Thorpe, ed.), pp. 443–452, University of Calgary, Calgary.

Fowler, M. W., 1982, Substrate utilization by plant cell cultures, *J. Chem. Tech. Biotechnol.* **32**:338.

Fowler, M. W., and Clifton, A., 1974, Activities of enzymes of carbohydrate metabolism in cells of *Acer pseudoplatanus* L. maintained in continuous (chemostat) culture, *Eur. J. Biochem.* **45**:445.

Fowler, M. W., and Clifton, A., 1975, Hexokinase activity in cultured sycamore cells, *New Phytol.* **75**:533.

Fowler, M. W., and Stepan-Sarkissian, G., 1983, Chemicals from plant cell fermentation, in: *Advances in Biotechnological Processes* (A. Mizrahi and A. L. van Wezel, eds.), Vol. 2, pp. 135–158, Liss, New York.

Fowler, M. W., Watson, R., and Lyons, I., 1982, Substrate utilization, carbon and nitrogen, by suspension cultured plant cells, in: *Plant Tissue Culture 1982* (A. Fujiwara, ed.), pp. 225–228, Japanese Association for Plant Tissue Culture, Tokyo.

Gamanetz, L. V., and Gamburg, K. Z., 1981, The effect of adenine on growth, starch and ADPG content and ADPG pyrophosphorylase activity in suspension-cultured tobacco cells, *Z. Pflanzenphysiol.* **104**:61.

Gamborg, O. L., 1966, Aromatic metabolism in plants. II. Enzymes of the shikimate pathway in suspension cultures of plant cells, *Can. J. Biochem.* **44**:791.

Gautheret, R. J., 1955, The nutrition of plant tissue cultures, *Annu. Rev. Plant Physiol.* **6**:433.

Giaquinta, R., 1977, Sucrose hydrolysis in relation to phloem translocation in *Beta vulgaris*, *Plant Physiol.* **60**:339.

Givan, C. V., and Collin, H. A., 1967, Studies in the growth in culture of plant cells. II. Changes in respiration rate and nitrogen content associated with the growth of *Acer pseudoplatanus* L. cells in suspension culture, *J. Exp. Bot.* **18**:321.

Gordon, P. A., and Stewart, P. R., 1969, Ubiquinone formation in wild-type and petite yeast: The effect of catabolite repression, *Biochim. Biophys. Acta* **127**:358.

Goris, A., 1948, Epuisement des réserves glucidiques de fragments de tubercules de Topinambour cultivés in vitro sur milieux depourvus de sucres: Influence de l'acide indol-3-acétique, *C. R. Acad. Sci.* **226**:742.

Grout, B. W. W., Chan, K. W., and Simpkins, I., 1976, Aspects of growth and metabolism in a suspension culture of *Acer pseudoplatanus* (L.) grown on a glycerol carbon source, *J. Exp. Bot.* **27**:77.

Hagimori, M., Matsumoto, T., and Obi, Y., 1982, Studies on the production of *Digitalis* cardenolides by plant tissue culture. III. Effects of nutrients on digitoxin formation by shoot-forming cultures of *Digitalis purpurea* L. grown in liquid media, *Plant Cell Physiol.* **23**:1205.

178   GAGIK STEPAN-SARKISSIAN and MICHAEL W. FOWLER

Hart, J. W., and Filner, P., 1969, Regulation of sulfate uptake by amino acids in cultured tobacco cells, *Plant Physiol.* **44:**1253.

Hughes, R., and Street, H. E., 1974, Galactose as an inhibitor of expansion of root cells, *Ann. Bot.* **38:**555.

Ikeda, T., Matsumoto, T., and Noguchi, M., 1976, Effects of nutritional factors on the formation of ubiquinone by tobacco plant cells in suspension culture, *Agric. Biol. Chem.* **40:**1765.

Jenner, C. F., 1974, An investigation of the association between the hydrolysis of sucrose and its absorption by grains of wheat, *Aust. J. Plant Physiol.* **1:**319.

Jessup, W., and Fowler, M. W., 1977, Interrelationship between carbohydrate metabolism and nitrogen assimilation in cultured plant cells. III. Effect of the nitrogen source on the pattern of carbohydrate oxidation in cells of *Acer pseudoplatanus* L. grown in culture, *Planta* **137:**71.

Jones, A., and Veliky, I. A., 1980, Growth of plant cell suspension culture on glycerol as sole source of carbon and energy, *Can. J. Bot.* **58:**648.

Kikuta, Y., Harada, T., Akemine, T., and Tagawa, T., 1977, Role of kinetin in activity of the pentose phosphate pathway in relation to growth of potato tissue cultures, *Plant Cell Physiol.* **18:**361.

Kikuta, Y., Masuda, K., and Okazawa, Y., 1981, Embryogenesis and glucose metabolism in carrot cell suspension cultured *in vitro*, *J. Fac. Agric. Hokkaido Univ.* **60:**250.

King, P. J., 1976, Studies on the growth in culture of plant cells. XX. Utilization of 2,4-dichlorophenoxyacetic acid by steady-state cell cultures of *Acer pseudoplatanus* L., *J. Exp. Bot.* **27:**1053.

King, P. J., 1977, Growth limitation by nitrate and glucose in chemostat cultures of *Acer pseudoplatanus* L., *J. Exp. Bot.* **28:**142.

King, P. J., and Street, H. E., 1973, Growth patterns in cell culture, in: *Plant Tissue and Cell Culture* (H. E. Street, ed.), pp. 269–337, Blackwell, Oxford.

King, P. J., Mansfield, K. J., and Street, H. E., 1973, Control of growth and cell division in plant cell suspension cultures, *Can. J. Bot.* **51:**1807.

Klenovska, S., 1973, Water relations and the dynamics of the sugar content in tobacco callus tissue cultures when using polyethyleneglycol as osmotic agent, *Acta Fac. Rerum Nat. Univ. Comenianae Physiol. Plant* **7:**19–29.

Komamine, A., and Shimizu, T., 1975, Changes in some enzyme activities and respiration in the early stage of callus formation in a carrot root tissue culture, *Physiol. Plant.* **33:**47.

Lamport, D. T. A., 1963, Oxygen fixation into hydroxyproline of plant cell wall protein, *J. Biol. Chem.* **238:**1438.

Limberg, M., Cress, D., and Lark, K. G., 1979, Variants of soybean cells which can grow in suspension with maltose as a carbon energy source, *Plant Physiol.* **63:**718.

Maretzki, A., and Thom, M., 1972, Membrane transport of sugars in cell suspension of sugarcane. I. Evidence for sites and specificity, *Plant Physiol.* **49:**177.

Maretzki, A., and Thom, M., 1978, Characteristics of a galactose-adapted sugarcane cell line grown in suspension culture, *Plant Physiol.* **61:**544.

Maretzki, A., and Thom, M., 1979, Glucose transport in *Saccharum* sp. cell suspensions, *Plant Physiol.* **63**(Suppl.):148.

Maretzki, A., de la Cruz, A., and Nickell, L. G., 1971, Extracellular hydrolysis of starch in sugarcane cell suspensions, *Plant Physiol.* **48:**521.

Maretzki, A., Thom, M., and Nickell, L. G., 1974, Utilization and metabolism of carbohydrates in cell and callus cultures, in: *Tissue Culture and Plant Science 1974* (H. E. Street, ed.), pp. 329–361, Academic Press, New York.

Mathes, M. C., Morselli, M., and Marvin, J. W., 1973, Uses of various carbon sources by isolated maple callus cultures, *Plant Cell Physiol.* **14:**797.

Matsumoto, T., Nishida, K., Noguchi, M., and Tamaki, E., 1973, Some factors affecting the anthocyanin formation by *Populus* cells in suspension cultures, *Agric. Biol. Chem.* **37**:561.

Misawa, M., 1977, Production of natural substances by plant cell cultures described in Japanese patents, in: *Plant Tissue Culture and Its Biotechnological Application* (W. Barz, E. Reinhard, and M. H. Zenk, eds.), pp. 17–26, Springer-Verlag, Berlin.

Nevins, D. J., English, P. D., and Albersheim, P., 1966, The specific nature of plant cell-wall polysaccharides, *Plant Physiol.* **42**:900.

Nickell, L. G., and Maretzki, A., 1970, The utilization of sugars and starch as carbon sources by sugarcane cell suspension cultures, *Plant Cell Physiol.* **11**:183.

Okamura, S., Sueki, K., and Nishi, A., 1975, Physiological changes of carrot cells in suspension culture during growth and senescence, *Physiol. Plant* **33**:251.

Opekarova, M., and Kotyk, A., 1973, Uptake of sugars by tobacco callus tissue, *Biol. Plant.* **15**:312.

Pareilleux, A., and Chaubet, N., 1980, Growth kinetics of apple plant cell cultures, *Biotechnol. Lett.* **2**:291.

Parr, D. R., and Edelman, J., 1975, Release of hydrolytic enzymes from the cell walls of intact and disrupted carrot cell tissue, *Planta* **127**:111.

Parr, D., and Edelman, J., 1976, Passage of sugars across the plasmalemma of carrot callus cells, *Phytochemistry* **15**:619.

Rose, D., Martin, S. M., and Clay, P. P. F., 1972, Metabolic rates for major nutrients in suspension cultures of plant cells, *Can. J. Bot.* **50**:1301.

Rubery, P. H., and Northcote, D. H., 1970, The effect of auxin (2,4-dichlorophenoxyacetic acid) on the synthesis of cell wall polysaccharides in cultured sycamore cells, *Biochim. Biophys. Acta* **222**:95.

Sabinski, F., Barckhaus, R. H., Fromme, H. G., and Spener, F., 1982, Dynamics of galactolipids and plastids in non-photosynthetic cells of *Glycine max* suspension cultures: A morphological and biochemical study, *Plant Physiol.* **70**:610.

Sacher, J. A., 1966, The regulation of sugar uptake and accumulation in bean pod tissue, *Plant Physiol.* **41**:181.

Sacher, J. A., Hatch, M., and Glasziou, K. T., 1963, Sugar accumulation cycle in sugarcane. III. Physical and metabolic aspects of cycle in immature storage tissue, *Plant Physiol.* **38**:348.

Sahai, O. P., and Shuler, M. L., 1984, Environmental parameters influencing phenolics production by batch cultures of *Nicotiana tabacum*, *Biotechnol. Bioeng.* **26**:111.

Sato, F., and Yamada, Y., 1984, High berberine-producing cultures of *Coptis japonica* cells, *Phytochemistry* **23**:281.

Shantz, E. M., Sugii, M., and Steward, F. C., 1967, The interaction of cell division factors with myo-inositol and their effect on cultured carrot tissue, *Ann. N.Y. Acad. Sci.* **144**:335.

Shuler, M. L., 1981, Production of secondary metabolites from plant tissue culture—Problems and prospects, *Ann. N.Y. Acad. Sci.* **369**:65.

Simpkins, I., and Street, H. E., 1970, Studies on the growth in culture of plant cells. VII. Effects of kinetin on the carbohydrate and nitrogen metabolism of *Acer pseudoplatanus* L. cells grown in suspension culture, *J. Exp. Bot.* **21**:170.

Simpkins, I., Collin, H. A., and Street, H. E., 1970, The growth of *Acer pseudoplatanus* cells in a synthetic liquid medium: Response to the carbohydrate, nitrogenous and growth hormone constituents, *Physiol. Plant.* **23**:385.

Smith, M. M., and Stone, B. A., 1973, Studies on *Lolium multiflorum* endosperm in tissue culture. I. Nutrition, *Aust. J. Biol. Sci.* **26**:123.

Staba, E. J., 1977, Tissue culture and pharmacy, in: *Applied and Fundamental Aspects of Plant Cell, Tissue and Organ Culture* (J. Reinert and Y. P. S. Bajaj, eds.), pp. 694–707, Springer-Verlag, Berlin.

Staba, E. J., 1980, Secondary metabolism and biotransformation, in: *Plant Tissue Culture as a Source of Biochemicals* (E. J. Staba, ed.), pp. 59–97, CRC Press, Boca Raton, Fla.

Stafford, A., and Fowler, M. W., 1983, Effect of carbon and nitrogen growth limitation upon nutrient uptake and metabolism in batch cultures of *Catharanthus roseus* (L) G. Don., *Plant Cell Tissue Organ Cult.* **2**:239.

Straus, J., 1962, Invertase in cell walls of plant tissue cultures, *Plant Physiol.* **37**:342.

Straus, J., and Campbell, W. A., 1963, Release of enzymes by plant tissue cultures, *Life Sci.* **2**:50.

Street, H. E. (ed.), 1974, *Tissue Culture and Plant Science 1974*, Academic Press, New York.

Street, H. E., 1977, Applications of cell suspension cultures, in: *Applied and Fundamental Aspects of Plant Cell, Tissue and Organ Culture* (J. Reinert and Y. P. S. Bajaj, eds.), pp. 649–677, Springer-Verlag, Berlin.

Street, H. E., Collin, H. A., Short, K., and Simpkins, I., 1968, Hormonal control of cell division and expansion in suspension cultures of *Acer pseudoplatanus* L.: The action of kinetin, in: *Biochemistry and Physiology of Plant Growth Substances* (F. Wightman and G. Setterfield, eds.), pp. 489–504, Runge Press, Ottawa.

Tabata, M., 1977, Recent advances in the production of medicinal substances by plant cell cultures, in: *Plant Tissue Culture and Its Biotechnological Application* (W. Barz, E. Reinhard, and M. H. Zenk, eds.), pp. 3–16, Springer-Verlag, Berlin.

Talmadge, K. W., Keegstra, K., Bauer, W. D., and Albersheim, P., 1973, The structure of plant cell walls. I. The macromolecular components of the walls of suspension-cultured sycamore cells with a detailed analysis of the pectic polysaccharides, *Plant Physiol.* **51**:158.

Tandon, P., and Arya, H. C., 1979, Effect of growth regulators on carbohydrate metabolism of *Zizyphus jujuba* gall and normal stem tissues in culture, *Biochem. Physiol. Pflanz.* **174**:772.

Tarakanova, G. A., Gudskov, N. L., and Vinnikova, N. V., 1979, Certain characteristics of the primary metabolism and accumulation of diosgenin in culture of *Dioscorea deltoidea* cells, *Fiziol. Rast.* **26**:54.

Thorpe, T. A., and Meier, D., 1972, Starch metabolism, respiration and shoot formation in tobacco callus culture, *Physiol. Plant.* **27**:365.

Thorpe, T. A., and Meier, D., 1974, Enzymes of starch metabolism in *Nicotiana tabacum* callus, *Phytochemistry* **13**:1329.

Ueda, Y., Ishiyama, M., Fukui, M., and Nishi, A., 1974, Invertase in cultured *Daucus carota* cells, *Phytochemistry* **13**:383.

Verma, D. C., and Dougall, D. K., 1977, Influence of carbohydrates on quantitative aspects of growth and embryo formation in wild carrot suspension cultures, *Plant Physiol.* **59**:81.

Verma, D. C., and Dougall, D. K., 1979, Myo-inositol biosynthesis and galactose utilization by wild carrot (*Daucus carota* L. var. *carota*) suspension cultures, *Ann. Bot.* **43**:259.

Voliva, C., Moessen, G. W., and Matthysse, A. G., 1982, Starch-enhanced synthesis and release of amylolytic enzymes from normal and crown gall tumor tobacco tissue culture cells, *Can. J. Bot.* **60**:1474.

Wilson, G., 1980, Continuous culture of plant cells using the chemostat principle, *Adv. Biochem. Eng.* **16**:1.

Wilson, S. B., 1971, Studies of the growth in culture of plant cells. XIII. Properties of mitochondria isolated from batch cultures of *Acer pseudoplatanus* cells, *J. Exp. Bot.* **22**:725.

Wyse, R., 1979, Sucrose uptake by sugar beet tap root tissue, *Plant Physiol.* **64**:837.

Yamada, Y., Sato, F., and Watanbe, K., 1982, Photosynthetic carbon metabolism in cultured photoautotrophic cells, in: *Plant Tissue Culture 1982* (A. Fujiwara, ed.), pp. 249–250, Japanese Association for Plant Tissue Culture, Tokyo.

Yamakawa, T., Kato, S., Ishida, K., Kodama, T., and Minoda, Y., 1983, Production of anthocyanins by *Vitis* cells in suspension culture, *Agric. Biol. Chem.* **47**:2185.

Yeoman, M. M., Miedzybrodzka, M. B., Lindsey, K., and McLauchlan, W. R., 1980, The synthetic potential of cultured plant cells, in: *Plant Cell Cultures: Results and Perspectives* (F. Sala, B. Parisi, R. Cella, and O. Ciferri, eds.), pp. 327–343, Elsevier/North-Holland, Amsterdam.

Zenk, M. H., El-Shagi, H., Arens, H., Stokigt, J., Weiler, E. W., and Deus, B., 1977, Formation of the indole alkaloids serpentine and ajmalicine in cell suspension cultures of *Catharanthus roseus,* in: *Plant Tissue Culture and Its Biotechnologicál Application* (W. Barz, E. Reinhard, and M. H. Zenk, eds.), pp. 27–43, Springer-Verlag, Berlin.

6

Carbohydrate Metabolism in African Trypanosomes, with Special Reference to the Glycosome

ALAN H. FAIRLAMB and FRED R. OPPERDOES

1. INTRODUCTION

1.1. General Background and Scope

Over the last decade, our knowledge of the biochemistry and molecular biology of trypanosomes has expanded so much that the African trypanosome, *Trypanosoma brucei*, is now the equivalent of *E. coli* to the biochemical parasitologist. Trypanosomes are of interest to scientists not only because of their medical and veterinary importance, but also because of several unique features of their biochemistry and molecular biology. Two such features have been reviewed recently: the mitochondrial DNA network (the kinetoplast) and its role in the life cycle (Hajduk, 1978; Borst and Hoeijmakers, 1979; Barker, 1980; Englund *et al.*, 1982), and the variant surface glycoprotein and its role in evading the immune response of the host (Englund *et al.*, 1982; Turner, 1982).

The purpose of this chapter is to review recent advances in our understanding of carbohydrate metabolism in trypanosomes. Particular emphasis will be placed on the recently discovered subcellular organelle, the glycosome, which contains several enzymes of glycolysis as well as some of the pyrimidine synthetic pathway, part of the ether-lipid biosynthetic pathway, and carbon dioxide fixation. The role of the glycosome in subcellular compartmentation of carbohy-

ALAN H. FAIRLAMB • Laboratory of Medical Biochemistry, The Rockefeller University, New York, New York 10021.   FRED R. OPPERDOES • Research Institute for Tropical Diseases, International Institute for Cellular and Molecular Pathology, B-1200 Brussels, Belgium.

drate metabolism will also be discussed. Wherever possible, we have restricted the scope to the salivarian trypanosome, *T. brucei,* drawing supplemental information from the morphologically indistinguishable *T. rhodesiense* where gaps exist in our knowledge of *T. brucei.* The reader is referred to earlier reviews for information on other salivarian trypanosomes (Bowman and Flynn, 1976) or the stercorarian trypanosome, *T. cruzi* (Bowman, 1974; Gutteridge and Rogerson, 1979).

Trypanosomes are flagellated protozoa responsible for several serious parasitic diseases of humans and domestic animals. Today, the trypanosomiases present formidable medical and economic obstacles to development in many African and South American countries and rank among the top six tropical diseases selected for scientific study by the World Health Organisation with a view to developing new or more effective treatments (Trigg, 1979). It is worth noting that the study of African trypanosomes is as old as chemotherapy itself. It is ironic, therefore, that these life-threatening protozoal diseases should have lagged so far behind the enormous advances in chemotherapy of bacterial diseases. To this day, suramin and the organic arsenicals remain the mainstay of chemotherapy, despite their many disadvantages. Worst still, there is no effective treatment for South American trypanosomiasis. Hopefully, the continuing advances in our knowledge of these parasites will lead to better strategies for drug development through the rational approach (Cohen, 1979; Fairlamb, 1982; Opperdoes, 1983a,b; Meshnick, 1984).

1.2. Biology of the Kinetoplastida

Trypanosomes belong to the class Zoomastigophora, order Kinetoplastida and are so named because they possess a kinetoplast—a DNA-containing organelle, visible by light microscopy and situated near the base of the single flagellum. The position of the kinetoplast relative to the nucleus is the distinguishing feature of the trypomastigote (kinetoplast posterior to the nucleus) and epimastigote (anterior to the nucleus) forms of various stages in the life cycle (Fig. 1). Kinetoplastida display a variety of life cycles, from simple free-living forms (the bodonidae) to others that infect birds, mammals, fish, frogs, insects (*Crithidia* spp.), and even plants. Many have complex life cycles involving transmission via an invertebrate blood-sucking vector (flies, bugs, fleas, or leeches) to the vertebrate host. Parasitic trypanosomes of medical importance all belong to the genus *Trypanosoma.* Depending on their mode of transmission by the insect vector, this genus is generally subdivided into the Salivaria (salivary transmission) and Stercoraria (fecal transmission). The remainder of this chapter will be concerned with the salivarian type, as exemplified by *T. brucei.* The reader is referred to other articles for more information on the biology of other species (Hoare, 1972; Vickerman and Preston, 1976; McGhee and Cosgrove, 1980).

IN MAMMALS

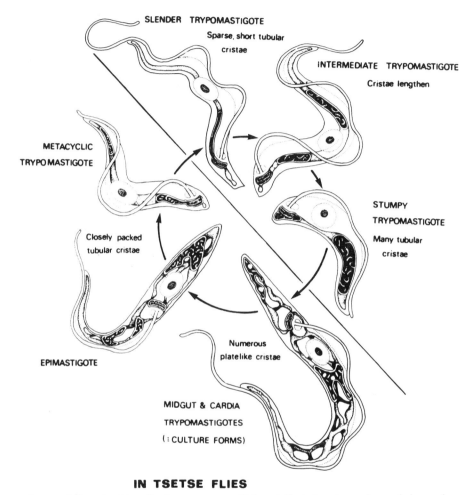

IN TSETSE FLIES

Figure 1. Life cycle of the salivarian trypanosome, *T. brucei,* illustrating changes in morphology and mitochondrial development. (From Vickerman, 1970.)

1.3. Medical and Economic Importance of African Trypanosomiasis

In humans, salivarian trypanosomes (*T. rhodesiense* and *T. gambiense*) cause the fatal disease sleeping sickness. Some 35 million Africans are at risk of contracting sleeping sickness and about 10,000 new cases are reported each year (De Raadt, 1976). However, difficulties in monitoring the incidence of the disease in rural areas of Africa suggest that this represents a gross underestimate.

Moreover, when medical surveillance and methods of control break down, the disease can reach epidemic proportions, such as occurred recently during the political upheavals in Uganda (Gashumba, 1981). If left untreated, human sleeping sickness is inevitably fatal, due to invasion and destruction of the central nervous system. The duration of the disease ranges from weeks to months in *T. rhodesiense* infections and from months to years in *T. gambiense* infections. Suramin and pentamidine are used in treatment of early stages of the disease, but the more toxic aromatic arsenicals or nitrofurazone are used when there is CNS involvement, as the former drugs do not cross the "blood–brain barrier" (Apted, 1970).

In animals, salivarian trypanosomes cause a variety of diseases including nagana in cattle (*T. brucei, T. vivax, T. congolense*), surra in horses and camels (*T. evansi*), and dourine, a venereal disease of horses and donkeys (*T. equiperdum*). Animal trypanosomiasis is a serious obstacle to human welfare, due to the severe nutritional and economic problems it causes. Over 3 million cattle die each year and the rearing of high-meat- and milk-producing domestic cattle, sheep, and goats is impossible in 10 million km^2 of Africa (WHO, 1979). Control measures such as eradication of the insect vector, and eliminating the reservoir of infection in game animals have largely proved unsuccessful. Furthermore, the widespread incidence of drug resistance has further exacerbated efforts to control the disease.

As a note of caution, it is worth emphasizing that the human-infective species, *T. rhodesiense* and *T. gambiense,* are morphologically and biochemically identical to the animal-infective *T. brucei*. An accidental laboratory infection with a stock designated as *T. brucei* has led to its reclassification as *T. rhodesiense* (Robertson *et al.*, 1980). The ability of human serum to induce lysis of *T. brucei*, but not *T. rhodesiense* and *T. gambiense,* forms the basis of the *in vitro* and *in vivo* infectivity tests (Rickman and Robson, 1970; Hawking, 1973). The factor in human serum responsible for selective lysis of *T. brucei* has been identified as high-density lipoprotein (Rifkin, 1978).

2. LIFE CYCLE

T. brucei and other salivarian trypanosomes are transmitted from animal to animal by the tsetse fly. When an infected fly bites an uninfected animal to obtain a blood meal, it injects metacyclic forms along with its salivary secretions into the skin. Metacyclics multiply into trypomastigote forms at the site of the bite forming a chancre. Subsequently, parasites migrate via the lymphatics to the lymph nodes and then to the bloodstream. Here, they multiply in large numbers as well as invade the intercellular spaces of other tissues.

In a chronic, relapsing infection, bloodstream and tissue forms of *T. brucei* display considerable heterogeneity in their morphology (pleomorphism), ranging from long-slender to short-stumpy forms (see Fig. 1). It is generally accepted that long-slender forms, which are numerous in rising parasitemia, are capable of division in the mammalian host, whereas the short-stumpy forms, which are generally seen during the remission phase of the infection, are not. As discussed later, the biochemical differences found between long-slender and short-stumpy organisms would support the suggestion that stumpy forms are a preadaptation for survival in the insect midgut.

After an uninfected tsetse feeds on an infected animal, the ingested stumpy forms transform in the midgut of the fly into procyclic trypomastigote forms. After multiplication, the procyclic trypomastigotes migrate to the salivary glands, where they undergo transformation first into epimastigote forms and then into the infective metacyclic forms (Fig. 1).

As discussed in detail in later sections, the life cycle of the salivarian trypanosome is associated with a series of remarkable changes in mitochondrial morphology and oxidative metabolism. In the slender bloodstream trypomastigote, the single mitochondrion is a simple unbranched tube with a few tubular cristae. There are no cytochromes and respiration is not inhibited by cyanide. Glucose is the sole source of energy in the mammal and the major end product of aerobic metabolism is pyruvate, as the enzymes of the tricarboxylic acid cycle are absent. The transition from slender to stumpy form is associated with the synthesis of TCA cycle enzymes including the pyruvate and α-oxoglutarate decarboxylases (Flynn and Bowman, 1973). These enzymes can be used to stain the mitochondrion histochemically with tetrazolium salts (''NADH diaphorase'') and have been used to differentiate long-slender and short-stumpy forms (Vickerman, 1965). However, due to the low levels of citrate synthetase and succinate dehydrogenase, the TCA cycle is thought to be inoperative *in vivo* (Flynn and Bowman, 1973).

Upon transfer to axenic culture at 25°C, the bloodstream trypomastigote transforms into an organism indistinguishable from the procyclic trypomastigote found in the midgut lumen of the tsetse. The mitochondrion proliferates from a simple tube into a complex branched network with numerous platelike cristae (Vickerman, 1965; Bohringer and Hecker, 1974, 1975). Synthesis of cytochromes takes place and, depending on culture conditions, respiration becomes sensitive to cyanide. Proline can replace glucose as an energy source in culture procyclics; this is probably a reflection of the environment that the trypanosome would encounter in the insect host as proline is the major component in the insect's hemolymph, where glucose is scarce. Little is known about the biochemistry of the other insect stages. However, morphological studies indicate that the mitochondrion of the epimastigote is similar to that of the procyclic form. The transformation of the epimastigote into the metacyclic form completes the cycle and the mitochondrion undergoes involution to a simple tube once more.

3. METHODS FOR CULTIVATION

All stages of the life cycle can now be cultivated either *in vivo* or *in vitro*, although the numbers of organisms obtained by some of the *in vitro* methods are rather limited. Furthermore, with the exception of the procyclic trypomastigote form, cultivation *in vitro* of all other stages requires the presence of either mammalian cell lines (for mammalian trypomastigote forms) or salivary gland explants (for epimastigote and metacyclic trypomastigote forms). As these cultures generally yield small amounts of material, most of the biochemical studies on the model organism *T. brucei* have relied on obtaining parasites from infected rodents.

3.1. Mammalian Forms of T. brucei

The "bloodstream" trypomastigote forms of *T. brucei* in a natural or a relapsing infection often show considerable pleomorphism, ranging from long-slender to short-stumpy forms. No successful method for separating these morphologically and biochemically differing types is available at present. However, laboratory-adapted strains, obtained by repeated sequential passage in rodents, frequently lose their pleomorphism and become "monomorphic" showing the morphology of the long-slender trypomastigote form. Such monomorphic lines can also be cultivated *in vitro*. Pure population of short-stumpy forms cannot be grown either *in vivo* or *in vitro*. Thus, study of the biochemistry of these forms has been carried out using mixed populations of cell types and by subsequent extrapolation.

3.1.1. Growth of Trypomastigotes in Vivo

Good yields of bloodstream forms can be obtained from laboratory infections in rats or mice. With virulent stocks of *T. brucei* (e.g., S427), fulminating parasitemias of 10^9 parasites/ml can be obtained 3 to 4 days after intraperitoneal inoculation of 10^6 parasites/20-g mouse or 10^7 parasites/200-g rat. Trypanosomes are purified free of erythrocytes, leukocytes, and platelets by passage through a column of DEAE-cellulose (Lanham, 1968; Lanham and Godfrey, 1970). With the appropriate strain, a yield of 0.5–1.0 g wet weight of cells can be obtained per rat.

Short-stumpy trypomastigotes are difficult to obtain in such large quantities as they appear in the remission phase of the parasitemia. Laboratory rodents are infected with an appropriate relapsing strain and organisms harvested from animals showing a suitably high parasitemia and high proportion of stumpy forms and purified as above.

3.1.2. Cultivation of Trypomastigotes in Vitro

The first successful long-term cultivation *in vitro* of animal-infective try-pomastigotes was achieved by Hirumi *et al.* (1977a,b). These authors were able to continuously maintain growth of *T. brucei* at 37°C in the presence of bovine fibroblastlike cells in Hepes-buffered RPMI 1640 supplemented with 20% fetal bovine serum. The organisms retained the histochemical and morphological appearance of long-slender bloodstream forms and were fully infective to animals. Hill *et al.* (1978a) were able to substitute a defined cell line of Chinese hamster lung cells for the undefined bovine fibroblastlike cells employed by Hirumi and co-workers to successfully cultivate the human-infective form *T. rhodesiense*. The cultured trypanosomes were morphologically identical to the mammalian bloodstream form, infective for rodents, and possessed the characteristic cyanide-insensitive terminal oxidase of mammalian forms.

Subsequently, it has been demonstrated that several mammalian cell lines can act as a feeder cell layer for *in vitro* cultivation, including fibroblastlike cells derived from buffalo lung (Hill *et al.*, 1978a,b), embryonic tissue from rabbits and mountain voles (Brun *et al.*, 1981), fetal rat lung (Stuart, 1980), and rabbit diaphragm (Zweygarth *et al.*, 1983) as well as primary murine bone marrow cultures (Balber, 1983). Studies by Tanner (1980) have shown that short-term intercellular contact between the feeder cell layer and the trypanosomes is essential for growth. Growth was sustained neither in fibroblast-conditioned media, nor if the trypanosomes and fibroblastlike cells were separated by means of microporous membranes. The reason for this requirement is not understood*, but the preference for the trypanosomes to localize within the intercellular spaces of the feeder cells has been noted by many workers (Hirumi *et al.*, 1977a,b; Hill *et al.*, 1978a,b; Brun *et al.*, 1979, 1981; Balber, 1983).

Successful cultivation of many members of the *Trypanozoon* subgenus have been reported, including *T. brucei* (Hirumi *et al.*, 1977a,b; Hill *et al.*, 1978b; Brun *et al.*, 1979, 1981), *T. rhodesiense* (Hill *et al.*, 1978a; Brun *et al.*, 1981), *T. gambiense* (Brun *et al.*, 1981; Balber, 1983), and *T. evansi* (Zweygarth *et al.*, 1983). Cultivation of dyskinetoplastic *T. brucei*—a mutant that lacks a Giemsa-staining kinetoplast—has also been reported (Stuart, 1980).

Cultivation *in vitro* has a number of disadvantages over growing the parasites in laboratory rodents. These methods are time-consuming, expensive, and have an obligatory requirement for a mammalian feeder cell layer. Moreover, the maximum parasite cell densities achieved are about 2×10^6 cells/ml, which compares unfavorably to parasitemias of 10^8 to 10^9/ml from an infected animal. Thus, most biochemical studies on carbohydrate metabolism have relied on the *in vivo* method of growth and results obtained in this way have been entirely confirmed to date by the limited biochemical studies performed on trypanosomes cultured *in vitro*.

---

*See addendum, page 217.

3.2. Insect Forms of T. brucei

3.2.1. Cultivation of Procyclic Trypomastigotes in Vitro

Procyclic forms can be cultivated in blood-containing, biphasic medium of Tobie *et al.* (1950) or Weinman (1960) and the monophasic medium of Pittam (1970). Cross and Manning (1973) developed the first nearly defined medium HX25 for cultivation of procyclics. Brun and Jenni (1977) found that this medium was unsuitable for growth of a large number of trypanosomes or for direct adaptation of bloodstream forms to procyclic forms, and have developed a semi-defined medium based on HO-MEM medium (Berens *et al.*, 1976) and HX25 (Cross and Manning, 1973). Brun and Schonenberger (1979) have made further modifications to this medium so that good growth of most *T. brucei* stocks can be achieved. Trypanosomes from either a stumpy or intermediate population of blood forms or midgut forms from an infected fly can be used to initiate such cultures. Culture procyclic trypomastigotes are morphologically identical to those found in the midgut of the tsetse fly and are generally not infective to the mammalian host (Mendez and Honigberg, 1972; Cunningham, 1973; Brown *et al.*, 1973; Ghiotto *et al.*, 1979).

3.2.2. Cultivation of Epimastigote and Metacyclic Trypomastigotes in Vitro

A proportion of procyclic forms can be stimulated to transform into meta-cyclic trypomastigote forms by cultivation at 28°C in a liquid medium (Cunningham, 1977) containing tsetse fly salivary gland organ cultures (Cunningham and Honigberg, 1977). After 8 to 10 days of cultivation, some noninfective procyclic forms transform into metacyclic forms infective for mice (Cunningham *et al.*, 1981). Metacyclic forms can be separated from the procyclics using DEAE-cellulose column chromatography (Lanham and Godfrey, 1970). The morphology of these organisms was similar to metacyclic stages formed in the salivary glands of the tsetse, including the characteristic surface coat. The total number of infective organisms is small—from a single 4-ml culture containing 5 $\times$ 10^7 trypanosomes, a maximum of 5 $\times$ 10^4 metacyclics were recovered from the DEAE column (Gardiner *et al.*, 1980). No biochemical studies have been reported on these forms.

4. SUBSTRATES AND END PRODUCTS OF METABOLISM

Little in the way of new information has been added since oxidative metabolism was extensively reviewed by Bowman and Flynn (1976). The essential features of carbohydrate metabolism in *T. brucei* and *T. rhodesiense* are summarized in Tables I and II using a "plus minus" notation. Unfortunately, there has been no comprehensive study of carbohydrate metabolism through each stage in the life cycle of one species, so the data have been compiled from a number of sources, using different strains and conditions for cultivation. Moreover, no information is available on the epimastigote and metacyclic forms of *T. brucei*. Now that methods for cultivation of each stage in the life cycle are available, we hope this information will be forthcoming.

4.1. Mammalian Forms

4.1.1. Aerobic Metabolism

All morphological bloodstream forms respire actively on glucose, fructose, mannose, or glycerol (Ryley, 1956, 1962). In addition, short-stumpy trypomastigotes will respire with α-oxoglutarate at about half the rate found with glucose (Flynn and Bowman, 1973). Neither form is able to respire using amino acids or fatty acids (Bowman and Flynn, 1976) and do not possess any energy reserves as their ATP levels are rapidly depleted in the absence of an external source of carbohydrate (Opperdoes *et al.*, 1976). Thus, in the mammalian host, long-slender and short-stumpy trypomastigotes are dependent on a continuous supply of glucose from plasma or other extracellular fluids. Respiratory activity is extremely high—about 60–90 nmoles O_2/min per 10^8 cells at 37°C or 120–180 nmoles/min per mg protein, compared to 1–10 nmoles/min per mg protein for mammalian cells (Table I). Reoxidation of NADH produced during glycolysis in the glycosome is mediated through the mitochondrial *sn*-glycerol-3-phosphate oxidase (glycerophosphate oxidase) (Grant and Sargent, 1960; Opperdoes *et al.*, 1977a,b; Fairlamb and Bowman, 1977a). This terminal respiratory system is unique in that respiration is insensitive to inhibition by cyanide and does not contain cytochromes (Grant and Sargent, 1960; Flynn and Bowman, 1973; Fairlamb and Bowman, 1977a). Further discussion of this respiratory system can be found in Section 5.1.

In long-slender forms, glucose is metabolized aerobically almost exclusively to pyruvate (> 98%), together with trace amounts of CO_2 and glycerol (Table II). Earlier researchers reported accumulation of significant amounts of glycerol (10–20% of glucose carbon used), but these findings are likely to be due to partial anaerobiosis in the incubation media, since later studies have not confirmed these reports (Opperdoes *et al.*, 1976; Brohn and Clarkson, 1978, 1980; Fairlamb and Bowman, 1980a,b). Pyruvate is the major end product of aerobic glucose metabolism and is not further metabolized by long-slender forms as they are completely lacking in lactate dehydrogenase (Dixon, 1966) and pyruvate decarboxylase (Flynn and Bowman, 1973).

Table 1. Respiratory Activity of Trypanosomes of the brucei Subgroup with Various Substrates[a]

Morphological type	Substrate						
	None	Glucose	Glycerol	α-Oxoglutarate	Proline	Palmitate	Ref[b]
Mammal							
Long-slender trypomastigote	–	++++	++++	–	–	–	1, 2
Short-stumpy trypomastigote	–	++++	++++	+++	±	–	1
Insect							
Procyclic trypomastigote	±	++	++	++	++	–	3
Epimastigote	0	0	0	0	0	0	
Metacyclic trypomastigote	0	0	0	0	0	0	

[a] Data are expressed as nmoles O_2/min per 10^8 organisms, where ++++ represents 51–100 nmoles/min per 10^8 cells; +++, 26–50; ++, 6–25; +, 1–5; and ±, 1. Some data have been recalculated assuming 2×10^{10} cells is approximately equal to 1 g wet wt of cells, 100 mg dry wt, or 14 mg total nitrogen estimated by Kjeldahl digestion.

[b] 1, Flynn and Bowman (1973); 2, Brohn and Clarkson (1980); 3, Ryley (1962), Ford and Bowman (1973), Srivastava and Bowman (1971).

Table II. End Products of Aerobic Glucose Metabolism in Trypanosomes of the brucei Subgroup

Morphological type	Product[a]						
	Pyruvate	Glycerol	Acetate	Lactate	Succinate	CO_2	Ref.[b]
Mammal							
Long-slender trypomastigote	+ + + +	±[c]	−	−	−	±	1–5
Short-stumpy trypomastigote	+ + + +	±[c]	+ +	−	±	+	3
Insect							
Procyclic trypomastigote	−	−	+	0	+	+ + +	2, 6
Epimastigote	0	0	0	0	0	0	
Metacyclic trypomastigote	0	0	0	0	0	0	

[a]The amount of carbon recovered in each product is expressed as a percentage of the carbon of glucose utilized, where + + + + represents 51%; + + +, 26–50%; + +, 6–25%; +, 1–5%; ±, 1%; 0, no available data.
[b]1, Ryley (1956); 2, Ryley (1962); 3, Flynn and Bowman (1973); 4, Opperdoes *et al.* (1976); 5, Brohn and Clarkson (1980); 6, Cross *et al.* (1975).
[c]Variable amounts of glycerol have been reported by various authors, which is probably due to partial anaerobiosis.

In the short-stumpy forms, some pyruvate is further metabolized to acetate, CO_2, and succinate (Flynn and Bowman, 1973). The yields of CO_2 and pyruvate from glucose were found to increase in a time-dependent manner, suggesting that the accumulation of pyruvate in the extracellular medium (up to 7 mM) may have the effect of initiating or driving metabolic pathways not acting at a significant rate *in vivo*. Extrapolation to zero time indicates that in pleomorphic *T. rhodesiense*, the major end product is pyruvate (75% of glucose carbon used) with glycerol (5%), acetate (9%), succinate (1%), and CO_2 (3%) accounting for most of the remainder. Short-stumpy forms also differ biochemically from long-slender types in their ability to use α-oxoglutarate as an energy source. In the absence of other exogenous carbohydrate, α-oxoglutarate preferentially confers survival to stumpy forms. Slender forms rapidly lose motility and disintegrate, whereas stumpy forms remain motile for at least 3 hr *in vitro* (Vickerman, 1965). Respiration with proline, a major energy source in the insect procyclic form, is less than 3% of that with glucose (Table I).

4.1.2. Anaerobic Metabolism

In the absence of oxygen, or when the glycerophosphate oxidase is inhibited by hydroxamic acids such as salicylhydroxamic acid (SHAM), long-slender bloodstream forms continue to utilize glucose at about the same rate as found under aerobic conditions (Opperdoes *et al.*, 1976). Because the glycerophosphate oxidase is inoperative, glucose is metabolized into equimolar amounts of pyruvate and glycerol (Ryley, 1956; Opperdoes *et al.*, 1976). Trace amounts of alanine and dihydroxypropionate have also been found in a recent ^{13}C NMR study (Mackenzie et al., 1983). Glycerol is not used under anaerobic conditions, and, as first noted by Ryley (1962), millimolar concentrations of glycerol will completely inhibit utilization of glucose. This is important with respect to the mechanism by which SHAM–glycerol mixtures kill long-slender T. brucei or T. rhodesiense and will be discussed in Sections 7.1.2 and 8.4.

No studies of anaerobic metabolism in short-stumpy trypomastigotes have been reported.

4.2. Insect Forms of the brucei Subgroup

4.2.1. Aerobic Metabolism

Established procyclic culture forms of T. rhodesiense will actively respire on glucose, mannose, fructose, or glycerol, but not with other mono- or disaccharides (Ryley, 1962). A number of TCA cycle intermediates, notably α-oxoglutarate and succinate, will also support respiration, particularly under acid conditions (Table I). This is in contrast to the short-stumpy bloodstream form, which contains low levels of succinate dehydrogenase but does not respire on succinate at pH 7.4 or 5.6 (Flynn and Bowman, 1973).

In contrast to bloodstream forms, established culture procyclic forms of T. brucei and T. rhodesiense will actively respire on proline (Srivastava and Bowman, 1971, 1972; Evans and Brown, 1972a; Brown et al., 1973). In cell-free homogenates, oxygen consumption with proline is up to two times higher than that with succinate, glycerol-3-phosphate, or α-oxoglutarate (Srivastava and Bowman, 1971, 1972; Bowman et al., 1972). Proline is present in tsetse hemolymph in concentrations as high as 150 mM under resting conditions and appears to be a major energy source during flight (Bursell, 1978, 1981). Glucose is scarce in the insect and thus proline would seem a likely physiological source of energy for the trypanosome in vivo.

Carbon-balance studies on the end products of aerobic glucose metabolism have only been reported with T. rhodesiense procyclics (Ryley, 1962). After 75 min of incubation, the major end product detected was CO_2 (55% of glucose carbon used), together with small amounts of acetate (3%) and succinate (4%). The remaining glucose carbon (38%) was presumed to have been incorporated into other unidentified metabolites. After 150 min of incubation, there was a net assimilation of succinate (-23%) with CO_2 accounting for 105% and acetate 10% of the glucose carbon used (overall balance $115 - 23 = 92\%$ recovery). The suggestion that unidentified metabolites accumulate in the earlier phase of metabolism deserves further investigation. Full carbon-balance studies have not been reported for T. brucei procyclics. However, Cross et al. (1975) have reported that 17 and 8% of the glucose carbon used by T. brucei in culture media could be recovered as succinate and alanine, respectively, similar to that reported for *T. rhodesiense* by Ryley (1962). Unfortunately, production of CO_2 was not measured.

There is virtually no information concerning the metabolism of the other two insect stages, epimastigote and metacyclic trypomastigote forms, of either *T. brucei* or *T. rhodesiense*. Njogu and Njindo (1981) reported that *T. brucei* obtained from infected tsetse salivary glands can be maintained in continuous culture for more than 700 days. Under aerobic conditions, the end products of glucose metabolism are found to be pyruvate and glycerol in a ratio of 3 : 1. However, the percentage of glucose recovered as pyruvate and glycerol can be calculated to be 49% and these authors did not analyze for other possible metabolites. Furthermore, the obligatory presence of a monolayer of bovine embryonic spleen cells raises the possibility that these organisms are in fact bloodstream rather than metacyclic trypomastigote forms. Nonetheless, the curious finding that this form produces pyruvate and glycerol from glucose under conditions where control incubations of bloodstream trypomastigotes produce solely pyruvate deserves further investigation.

4.2.2. Anaerobic Metabolism

Procyclic trypomastigotes of *T. rhodesiense* will utilize both glucose and glycerol under anaerobic conditions, provided CO_2 is present (Ryley, 1962). In the absence of CO_2, neither substrate can be utilized, underlining the vital importance of CO_2-fixation reactions in anaerobic metabolism. Under an atmosphere of 5% CO_2: 95% N_2, net assimilation of CO_2 takes place, with most of the glucose or glycerol carbon recovered as succinate (75 and 63%, respectively) together with a lesser amount of acetate (25 and 4%, respectively). Lactate was not detectable as an end product. Similar studies have not been reported for *T. brucei* procyclic forms.

5. TERMINAL RESPIRATORY SYSTEMS

The scope of this review is limited to electron transport systems in the *brucei* subgroup, and the reader is referred to other articles for information on other trypanosomatids (Bowman and Flynn, 1976; Hill, 1976a).

5.1. Mammalian Forms

Bloodstream forms of *T. brucei* and *T. rhodesiense* are highly unusual in that they lack cytochromes (Flynn and Bowman, 1973; Fairlamb and Bowman, 1977a) and respiration in whole cells is completely insensitive to inhibitors such as cyanide, azide, or antimycin A (Flynn and Bowman, 1973). Reoxidation of NADH produced during glycolysis is mediated by the *sn*-glycerol-3-phosphate oxidase system (Grant and Sargent, 1960), which is located in the mitochondrion (Opperdoes *et al.*, 1977a). Like cyanide-insensitive electron transport in certain plant and fungal mitochondria (Henry and Nyns, 1975), the glycerophosphate oxidase can be inhibited by a number of aromatic hydroxamic acids (Evans and Brown, 1973; Opperdoes *et al.*, 1976; Clarkson *et al.*, 1981), diphenylamine (Evans and Brown, 1972b), and the antifungal imidazole miconazole (Opperdoes, 1980), but not by the related ketoconazole (Opperdoes, unpublished). The glycerophosphate oxidase has a lower affinity for oxygen than cytochrome oxidase (2.1 versus 0.1 μM) (Hill, 1976b), but nonetheless apparently reduces oxygen to water as end product, since hydrogen peroxide could not be detected as a free intermediate (Fairlamb and Bowman, 1977b).

Studies using inhibitors of the glycerophosphate oxidase indicate that at least two enzyme components are present: a mitochondrial glycerophosphate dehydrogenase component that is capable of transferring electrons either to the SHAM-sensitive terminal oxidase component or to artificial electron acceptors (Fairlamb and Bowman, 1977a,b; Meshnick *et al.*, 1978). Inhibitors of the mitochondrial dehydrogenase component include thiol-reactive compounds (*p*-hydroxymercuribenzoate and melarsen oxide), heavy metal ions (Cu^{2+}, Zn^{2+}, Hg^{2+}), certain chelating compounds (*o*-phenanthroline), and the trypanocidal drug suramin (Fairlamb and Bowman, 1977b). Inhibitors acting after the site of reduction of artificial electron acceptors, which are presumed to act on the terminal oxidase component, include aromatic hydroxamic acids, 8-hydroxyquinoline, diphenylamine, and hydrogen peroxide. Other chelating compounds would appear to inhibit both components.

Difference spectra suggest that the dehydrogenase component contains FAD (Fairlamb and Bowman, 1977a), similar to its mammalian counterpart (Ringler, 1961; Dawson and Thorne, 1969). Flavin and metal analysis of mitochondrial membrane preparations yielded a flavin/Fe ratio of 1 : 8, similar to that reported for pig brain glycerophosphate dehydrogenase. The absence of acid-labile sulfide and characteristic EPR signal $g = 1.94$ indicates that nonheme iron proteins are probably not involved (Fairlamb and Bowman, 1977a). In addition to iron, 4 moles of copper per mole of flavin was found. On the basis of the differential sites of action of chelators, Fairlamb and Bowman (1977a,b) proposed that copper may be part of the terminal oxidase component. However, spectroscopic and EPR studies have failed to provide any direct evidence for the involvement of either metal in electron transport and the role of iron and copper remains controversial. Ubiquinone (CoQ_9) is present in bloodstream and culture forms of *T. rhodesiense* (Threlfall *et al.*, 1965) and has been suggested to mediate electron transfer between the glycerophosphate dehydrogenase component and SHAM-sensitive terminal oxidase (Meshnick *et al.*, 1978). A brief report by Caughey *et al.* (1979) that duroquinol, menadiol, and ubiquinol can act as alternative electron donors provides tentative support for this proposal. The nature of the terminal oxidase component remains completely unknown, paralleled only by the equally mysterious nature of the cyanide-insensitive pathway of certain plant and fungal mitochondria. The reader is referred to other articles for a survey of the many hypotheses concerning the latter (Bendall and Bonner, 1971; Bonner and Rich, 1978; Rustin *et al.*, 1983).

5.2. Insect Forms

5.2.1. Established Procyclic Trypomastigote Forms

Spectroscopic studies on established procyclic forms of *T. brucei* have demonstrated the presence of cytochromes *b*, *c*, and aa_3 (see reviews by Hill 1976a; Bowman and Flynn, 1976). Inhibitor studies further indicate that *T. brucei* procyclics possess at least two terminal oxidases (Evans and Brown, 1973; Njogu *et al.*, 1980). Cytochrome aa_3, which is inhibited by cyanide, and a second terminal oxidase, which is inhibited by SHAM, contribute 60 and 30%, respectively, of the total cell respiration (Njogu *et al.*, 1980). The nature of the latter oxidase is unknown, but probably is the same as the terminal oxidase component of the glycerophosphate oxidase system of bloodstream trypomastigotes. A third terminal oxidase, cytochrome *o*, has also been proposed to account for a carbon monoxide-binding pigment found in CO-reduced difference spectra (Hill, 1976b). However, in the absence of photochemical action spectra, the role of cytochrome *o* as a terminal oxidase remains open to question.

The cytochrome *c* of trypanosomatids is unusual in that the α-peak in the reduced form has a maximum at 555 nm compared with 550 nm for the mammalian cytochrome (Hill *et al.*, 1971a,b; Bienen *et al.*, 1983). The amino acid sequence of cytochrome *c* from a related insect trypanosomatid (*Crithidia oncopelti*) has been determined (Pettigrew, 1972). It is unusual in that the heme is linked to the apoprotein by one thio-ether linkage to cysteine (residue 27), rather than two cysteines (residues 14 and 17) found in mammalian cytochrome c_{550}. Alignment of the sequence with the "invariant" residues of other cytochromes c_{550} indicates that the second cysteine has been replaced by alanine (residue 24) in cytochrome c_{555}. In addition, there are 2 moles of ε-*N*-trimethyllysine per mole of crithidial cytochrome and the N-terminal amino acid residue is dimethylproline instead of the usual acetyl-blocking group (Pettigrew and Smith, 1977; Smith and Pettigrew, 1980). Neither of these modified amino acids is found in mammalian cytochromes. Similar studies have not been carried out for *T. brucei*.

5.2.2. Biochemical Transformation from Mammalian to Insect Forms

Mitochondrial differentiation during transformation from bloodstream to procyclic forms has been proposed to occur in two stages (Opperdoes *et al.*, 1977a). The first stage is characterized by the appearance of new mitochondrial dehydrogenases which can transfer reducing equivalents to O_2 via the alternate SHAM-sensitive oxidase. The second stage is associated with the appearance of the cytochrome system, with respiration becoming partially sensitive to inhibition by cyanide. Experimental evidence lends support to this hypothesis, although the two stages are not clearly demarcated, and tend to overlap. Furthermore, the extent and rate of acquisition of cyanide-sensitive respiration would appear to vary with the experimental conditions, although by morphological criteria transformation is complete by 48 to 72 hr in culture (Brown *et al.*, 1973; Bienen *et al.*, 1981). Overall respiration in intact *T. brucei* decreases fourfold during transformation, from 80 nmoles/min per 10^8 bloodstream forms to 22.5 nmoles/min per 10^8 procyclic forms after 14 days in culture (Bienen *et al.*, 1983).

In *T. brucei*, maximal mitochondrial dehydrogenase activities were found 2 days after initiation of transformation by Brown *et al.* (1973), whereas Bienen *et al.* (1983) found that 24–30 days in culture was necessary to obtain mitochondrial enzyme activities equivalent to established procyclic culture forms. Despite these differences, both groups found that respiration was completely insensitive to cyanide for the first 4 days, reaching a maximum inhibition by cyanide only after several weeks or months of subculture (Evans and Brown, 1971; Brown *et al.*, 1973; Bienen *et al.*, 1981, 1983). Oxygen consumption with ascorbate plus tetramethyl-*p*-phenylenediamine closely resembles the appearance of cyanide-sensitive respiration, implying the appearance of cytochrome aa_3, although this could not be observed spectroscopically until 9 days in culture (Bienen *et al.*, 1983).

Transformation of *T. rhodesiense* into procyclic forms would appear on the surface to be completely different from that of *T. brucei*. Appearance of cyanide-sensitive respiration is complete after 3 days, although maximum activity of mitochondrial dehydrogenases is not found until 7 to 14 days after initiation of transformation (Srivastava and Bowman, 1971, 1972; Bowman *et al.*, 1972). Furthermore, respiration is completely inhibited by 3 mM cyanide, unlike *T. brucei*. The reasons for these differences in findings have never been satisfactorily resolved.

Several observations suggest that cultivation conditions may play an important role. First, Brown *et al.* (1973) found that *T. brucei* fails to divide during the first 3 days in the biphasic medium of Tobie *et al.* (1950), whereas immediate growth is obtained in the all-liquid medium of Pittam (1970). Second, Bienen *et al.* (1981) noted that high levels of glucose adversely affected growth of transforming cells. Third, Simpson *et al.* (1980) and Brun and Schonenberger (1981) reported that intermediates of the citric acid cycle, notably citrate and *cis*-aconitate, can promote transformation of bloodstream *T. brucei*. These observations may be of significance in that previous studies may have used whole blood supplemented with glucose and citrate in the preparation of lysates required for

Tobie's and Pittam's culture media or used citrate as an anticoagulant during collection of bloodstream forms. Further studies could help to clarify this situation.

6. THE GLYCOSOME

6.1. Occurrence

Glycosomes are specialized microbodylike organelles found in trypanosomatids. They were initially identified in bloodstream forms of *T. brucei* (Opperdoes and Borst, 1977) and have subsequently been identified in several representatives of the order Kinetoplastida, including *T. cruzi* (Taylor *et al.*, 1980), *Crithidia* spp. (Opperdoes *et al.*, 1977a; Taylor *et al.*, 1980) and in four major representatives of *Leishmania* spp. (Hart and Coombs, 1980; Hart and Opperdoes, 1984). It is not yet known if all members of this order contain glycosomes; in particular, no studies have been reported on the free-living bodonids. Glycosomes were so named because these organelles (from all trypanosomatids examined to date) contain the first seven enzymes of the glycolytic sequence. In general, glycosomes do not contain enzymes that metabolize hydrogen peroxide, although some catalase has been found in microbodies of *Crithidia* spp. (Muse and Roberts, 1973; Opperdoes *et al.*, 1977a), as well as peroxidase in those of *T. cruzi* (Docampo *et al.*, 1976). Nonetheless, the trypanosomatid microbodies are sufficiently different from peroxisomes of other organisms to warrant the name glycosomes due to the occurrence and essential role of the glycolytic enzymes in these organelles. However, glycosomes should not be viewed solely as organelles catalyzing the initial part of glycolysis, because they contain the last two enzymes of the pyrimidine biosynthetic pathway (Hammond *et al.*, 1981), the first enzymes of the ether-lipid biosynthetic pathway (Opperdoes, 1984), as well as (in *T. brucei*) enzymes of glycerol metabolism, adenylate kinase, and enzymes of CO_2 fixation (Opperdoes and Borst, 1977; Opperdoes *et al.*, 1981; McLaughlin, 1981; Opperdoes and Cottem, 1982; Broman *et al.*, 1983).

6.2. Isolation and Purification

Trypanosomes are resistant to gentle disruption methods, due to the basketlike arrangement of subpellicular microtubules running in parallel arrays beneath the entire plasma membrane of the cells.

Efficient cell breakage can be obtained by grinding a cell paste with abrasives, such as silicon carbide (Opperdoes *et al.*, 1977b) or alumina (Oduro *et al.*, 1980a,b). While cells can also be easily disrupted in a hand-operated Dounce homogenizer after pretreatment with saponin (Fairlamb and Bowman, 1977a,b), this detergent damages the integrity of the glycosomal membrane (Fairlamb, unpublished results). The original method of purification using a combination of differential and isopycnic sucrose density gradient centrifugation (Opperdoes *et al.*, 1977a,b) has been improved by an additional centrifugation step in isotonic Percoll (Opperdoes *et al.*, 1984). Glycosomes isolated by this method are purified 12- to 13-fold, with an overall yield of 31%. Such preparations are highly homogeneous, containing less than 1% mitochondrial contamination (Opperdoes *et al.*, 1984). A convenient rapid method for isolating intact glycosomes 9-fold purified by Percoll-gradient centrifugation in a vertical rotor has also been described (Opperdoes, 1981a).

6.3. General Properties

Morphologically, glycosomes resemble microbodies found in other eukaryotes (Opperdoes *et al.*, 1984). In intact bloodstream forms of *T. brucei*, glycosomes are spherical or ellipsoid, bounded by a single membrane. Like other microbodies, these organelles have an electron-dense matrix, in which crystalloid cores can occasionally be seen (Fig. 2A,B). The glycosomes of *T. brucei* are homogeneous in size, with a mean diameter of 0.27 ± 0.03 μm, and it has been calculated from morphometric studies that a single *T. brucei* cell would contain on average 230 glycosomes, representing 4.3% of the total cell volume and about 8% of the total cell protein (Opperdoes *et al.*, 1984).

Intact glycosomes from *T. brucei* have a density of 1.087 g/cm^3 in isotonic sucrose (Opperdoes, 1981a) and equilibrate at 1.23 g/cm^3 in sucrose gradients (Opperdoes and Borst, 1977) (Fig. 3). In freshly isolated glycosomes, the glycolytic enzymes exhibit a high degree of latency (up to 95%), presumably due to the surrounding membrane presenting a permeability barrier to solutes, especially phosphorylated intermediates of glycolysis (Opperdoes and Borst, 1977). Latency can be partially abolished by freeze-thawing, sonication, osmotic shock, or with phospholipase treatment and completely abolished by treatment with detergents such as Triton X-100 (Opperdoes *et al.*, 1977b, 1981). In contrast, in culture epimastigotes of *T. cruzi*, the last two enzymes of the UMP biosynthetic pathway (orotate phosphoribosyltransferase and orotidine-5′-phosphate decarboxylase) are associated with the glycosome, but do not exhibit latency, suggesting that these enzymes may be situated on the outside of the glycosome (Hammond and Gutteridge, 1983). Similar latency studies have not been reported for these enzymes in *T. brucei*.

The phospholipid content of glycosomal membranes has been analyzed for *T. brucei* long-slender trypomastigote and procyclic trypomastigote forms. Bloodstream forms contain phosphatidylcholine and phosphatidylethanolamine in a ratio of 2 : 1 (Opperdoes *et al.*, 1984). Procyclic culture forms also contain

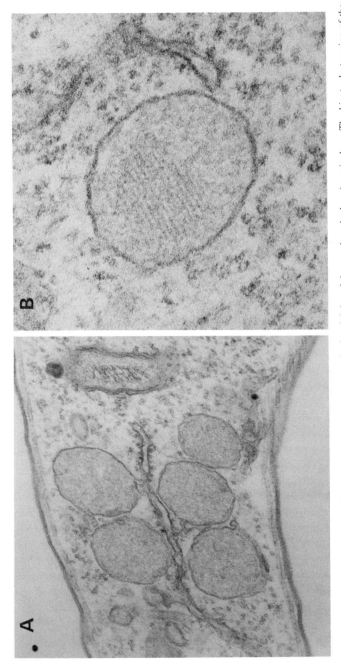

Figure 2. Glycosomes in *T. brucei.* (A) Cluster of five glycosomes in the vicinity of the rough endoplasmic reticulum. The kinetoplast region of the mitochondrion is also visible. (B) Glycosome *in situ* showing dense crystalloid core. (From Opperdoes *et al.,* 1984.)

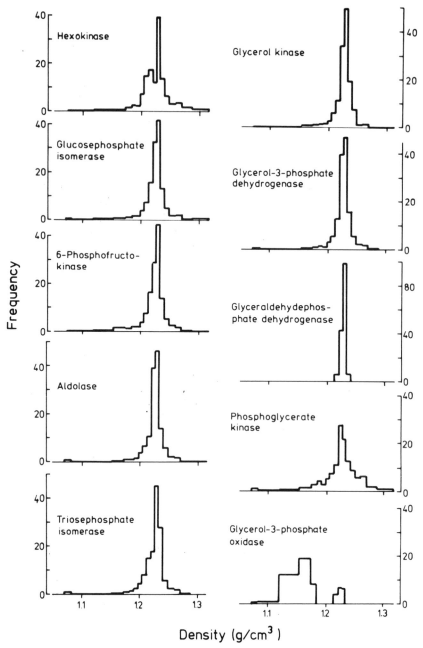

Figure 3. Distribution of enzyme activities from a bloodstream *T. brucei* large granule fraction after isopycnic centrifugation in a linear sucrose gradient. The nine enzymes involved in glycolysis equilibrate at a density of 1.23 g/cm³ and are localized in the glycosome. The mitochondrial marker enzyme activity, glycerol-3-phosphate oxidase, is included for comparison. (From Opperdoes and Borst, 1977.)

phosphatidylcholine and phosphatidylethanolamine together with lesser amounts of phosphatidylinositol and phosphatidylserine, not found in bloodstream forms (Hart *et al.*, 1984). Sphingomyelin and cardiolipin could not be detected in glycosomes from either form. The steroid content of glycosomal membranes has not been reported.

Inside the glycosomal membrane, a number of the glycolytic enzymes appear to be associated with each other (Oduro *et al.*, 1980a; Opperdoes and Nwagwu, 1980). Regardless of the preparative procedure used (grinding with abrasives, freeze-thawing, or detergent treatment), all the glycolytic enzymes, except phosphoglucose isomerase, sediment through sucrose gradients or pass through gel filtration columns as a multienzyme complex. Sodium chloride (0.15 M) dissociates the complex into its constituent components and this process can be reversed by removing added salt by dialysis (Oduro *et al.*, 1980a). These observations suggest that the isolated enzyme aggregate may represent a functional complex existing *in situ* within the glycosome.

Highly purified glycosomes have a low DNA content of heterogeneous size, which could be accounted for as partially degraded nuclear and kinetoplast DNA (Opperdoes *et al.*, 1984). The mechanism of synthesis and assembly of the glycosome is unknown but is currently under investigation.

6.4. Glycosomal Enzymes in T. brucei

The enzyme activities localized in the glycosomes of *T. brucei* bloodstream trypomastigote and procyclic trypomastigote forms are summarized in Table III. Coincident with the extremely high glycolytic rate in bloodstream forms, the nine glycolytic enzymes required for conversion of glucose to 3-phosphoglycerate and glycerol are concentrated in the glycosome to extremely high specific activities, ranging from 1 to 15 μmoles/min per mg protein. In contrast, adenylate kinase, malate dehydrogenase, phosphoenolpyruvate carboxykinase, and the enzymes of pyrimidine biosynthesis, orotate phosphoribosyltransferase and orotidine decarboxylase, show lower specific activities.

Comparison of the specific activities of glycosomal enzymes from bloodstream and procyclic forms, reveals some striking changes in enzyme levels, which correlate with both the reduced glycolytic capacity of the procyclic forms and the changes in their carbohydrate metabolism (Table III). The specific activities of hexokinase and aldolase are significantly reduced 15- and 28-fold, respectively. The remaining enzymes of the glycolytic pathway to 3-phosphoglycerate are reduced 4- to 8-fold with the exception of glycerol kinase, glycerol-3-phosphate dehydrogenase, and triose phosphate isomerase, which remain unchanged. There is also a striking increase in the specific activity of phosphoenolpyruvate carboxykinase and malate dehydrogenase in procyclics (10- and 100-fold, respectively), which is in agreement with other observations

Table III. Enzymes and Polypeptides in Glycosomes from Long-Slender
Trypomastigote (LST) and Procyclic Trypomastigote (PT) Forms of T. brucei[a]

Enzyme	Molecular weight (SDS–PAGE) × 10^{-3}	Specific activity (μmoles/min per mg)		Ratio LST/PT
		LST	PT	
Hexokinase	50	8.2	0.6	15
Phosphoglucose isomerase	64	10.3	1.4	7
Phosphofructokinase	51	6.6	1.7	4
Fructose bisphosphate aldolase	40	1.7	0.06	28
Triose phosphate isomerase	26.5	11.2	10.6	1
Glyceraldehyde-3-phosphate dehydrogenase	38	2.5	0.3	8
Phosphoglycerate kinase	46	7.0	0.9	8
Glycerol-3-phosphate dehydrogenase	37	3.5	3.5	1
Glycerol kinase	52	14.3	15.9	1
Adenylate kinase	?	0.06	0.32	0.2
Malate dehydrogenase	?	0.03	3.0	0.01
Phosphoenolpyruvate carboxykinase	?	0.23	2.4	0.1
Orotate phosphoribosyltransferase*	?	0.0025	?	?
Orotidine decarboxylase*	?	?	?	?
Unassigned polypeptides	110	−	+	
	110	−	±	
	90	+	+	
	71	+	+	
	61	±	+	
	34	−	+	
	28	+	±	

[a]Data from Hart et al. (1984) and Missel and Opperdoes (unpublished results) with additional information (*)
calculated from Hammond et al. (1981). The unassigned polypeptide bands visible on SDS–polyacrylamide gel
electrophoresis are indicated as: +, prominent band; ±, weakly visible; −, absent.

that these forms metabolize phosphoenolpyruvate by fixation of CO_2 to form
succinate via oxaloacetate, malate, and fumarate, under aerobic or anaerobic
conditions (Ryley, 1962; Klein et al., 1975; Cross et al., 1975). Apart from
malate dehydrogenase, none of the enzymes of the glyoxylate cycle could be
detected in either bloodstream or procyclic trypomastigotes and would thus appear to be absent (Opperdoes et al., 1977b; Opperdoes and Cottem, 1982).

Examination of the polypeptide composition of glycosomes by SDS–PAGE
reflects the changes in enzyme levels described above (Table III), indicating that
pathways of glucose metabolism are regulated by enzyme synthesis. All the
protein bands found in glycosomes of bloodstream forms could be identified in
procyclic forms, which contain at least three additional polypeptides not seen in
the bloodstream forms (Hart et al., 1984). Many of the protein bands have now

been assigned to particular enzyme activities, leaving only a few unassigned polypeptides, suggesting that most of the major glycosomal proteins have been identified.

Nothing is known about the physical arrangement of the genes coding for the glycosomal proteins, other than the apparent lack of DNA in the glycosome itself. The observation that the modulation of the amounts of glycosomal peptides falls into three mutually exclusive classes, in response to the different environments experienced during the life cycle, may be indicative of the occurrence of regulatory mechanisms at the level of the genome (Hart *et al.*, 1984). The factors regulating such differential gene expression are an area well worthy of further study.

7. PATHWAYS OF GLUCOSE METABOLISM

7.1. Long-Slender Trypomastigotes

7.1.1. Aerobic Glycolysis

Aerobic glycolysis in *T. brucei* proceeds rapidly with almost quantitative recovery of glucose carbon as pyruvate. All the enzymes of the classical glycolytic pathway have been identified (Opperdoes and Borst, 1977; Oduro *et al.*, 1980a,b; Hart *et al.*, 1984), as well as the concentrations of the reaction intermediates (Visser and Opperdoes, 1980; Hammond and Bowman, 1980a). The specific activities of the individual glycolytic enzymes are all in excess of the overall glycolytic rate and the relative maximal activities suggest that the rate-limiting steps in glycolysis are the reactions catalyzed by aldolase and phosphoglycerate mutase (Oduro *et al.*, 1980b). However, studies on D-glucose transport indicate that the permeation process may be the overall rate-limiting step of glucose metabolism in bloodstream forms (Gruenberg *et al.*, 1980). The overall metabolic sequence and its compartmentation within the cell is illustrated in Fig. 4. Several interesting features of this compartmentalization are apparent. First, the reoxidation of NADH generated within the glycosome is accomplished by means of an NAD-dependent glycerol-3-phosphate dehydrogenase, which catalyzes the formation of glycerol-3-phosphate from dihydroxyacetone phosphate. Glycerol-3-phosphate diffuses from the glycosome to the mitochondrion, where it is reoxidized to dihydroxyacetone phosphate by the mitochondrial glycerol-3-phosphate oxidase. Thus, by means of the glycerophosphate shuttle, NADH is reoxidized allowing the formation of 2 moles of 3-PGA from 1 mole of glucose. Second, the overall net synthesis and utilization of ATP within the glycosome is zero, because 2 moles of ATP is required to convert glucose to FDP

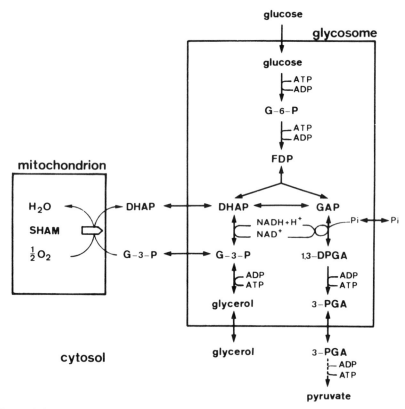

Figure 4. Compartmentation of glycolysis in *T. brucei* long-slender bloodstream trypomastigotes. (From Opperdoes, 1982.)

and 2 moles of ATP is synthesized from 2 moles of diphosphoglycerate by the action of phosphoglycerate kinase. Opperdoes and Borst (1977) have formulated the net reaction catalyzed by the glycosome as:

$$\text{glucose} + 2\,P_i + 2\,\text{DHAP} \rightarrow 2\,3\text{PGA} + 2\,\text{G3P}$$
$$\text{glycerol} + P_i + 2\,\text{DHAP} \rightarrow 3\text{PGA} + 2\,\text{G3P}$$

These equations illustrate that net consumption of ATP and NAD within the glycosome is zero and that entry of glucose, P_i, and DHAP must occur with exit of 3PGA and G3P across the glycosomal membrane. Direct evidence for the proposal (Opperdoes and Borst, 1977) that the membrane is selectively permeable to these compounds, but impermeable to other glycolytic intermediates, is still lacking. However, indirect evidence suggests that the glycosomal membrane does present a permeability barrier to glycolytic intermediates. Pulse labeling of intact cells with [^{14}C]glucose shows the presence of two compartments for glycolytic intermediates: one pool is rapidly labeled and probably represents the glycosomal compartment; a second pool is slowly labeled and probably represents the cytosol (Visser *et al.*, 1981). Furthermore, Oduro *et al.* (1980a) and Broman *et al.* (1983) found that conversion of glucose to triose phosphate *in vitro* by isolated glycosomes was stimulated by exogenous ATP and NAD. This latter approach is open to the criticism that the observed metabolism of glucose is due to a proportion of the glycosome preparation having damaged or "leaky" membranes. Further work is required before the question of the degree of permeability of the glycosome membrane can be definitively answered.

7.1.2. Anaerobic Glycolysis

Under anaerobic conditions, or when the glycerophosphate oxidase is inhibited by SHAM, long-slender *T. brucei* remain actively mobile for long periods by metabolizing glucose into equimolar amounts of glycerol and pyruvate (Ryley, 1962; Opperdoes *et al.*, 1976; Brohn and Clarkson, 1978). Originally, it was thought that glycerol was produced from G3P by the action of a phosphatase (see Bowman and Flynn, 1976), but inspection of Fig. 4 indicates that no net synthesis of ATP would occur in this scheme. Consequently, long-slender *T. brucei* would be expected to be unable to survive anaerobiosis, as they lack endogenous energy reserves. However, ATP levels were shown to be maintained in cells metabolizing glucose even though respiration was blocked by SHAM (Opperdoes *et al.*, 1976). Furthermore, thermodynamic calculations suggest that net synthesis of ATP is feasible during anaerobic glycolysis:

$$\text{glucose} \rightarrow \text{pyruvate} + \text{glycerol} \qquad -77 \text{ kJ/mole}$$
$$\text{ADP} + \text{P}_i \rightarrow \text{ATP} + \text{H}_2\text{O} \qquad +31 \text{ kJ/mole}$$
$$\text{glucose} + \text{ADP} + \text{P}_i \rightarrow \text{pyruvate} + \text{glycerol} + \text{ATP} + \text{H}_2\text{O} \qquad -46 \text{ kJ/mole}$$

Several hypothetical schemes have been proposed to account for these observations (Clarkson and Brohn, 1976; Opperdoes and Borst, 1977). However, all evidence so far supports the proposal of Opperdoes and Borst (1977) that net synthesis of ATP is achieved under anaerobic conditions by reversal of the glycerol kinase-catalyzed reaction. First, anaerobiosis or inhibition with SHAM results in a dramatic increase in the intracellular concentration of glycerol-3-phosphate, suggesting a precursor–product relationship (Hammond and Bowman, 1980a; Visser and Opperdoes, 1980; Mackenzie *et al.*, 1983). Second, no novel phosphorylated intermediates or novel enzyme activities required by other schemes have been isolated. Third, glycerol kinase is present in extremely high specific activities only in trypanosomes capable of synthesizing glycerol under anaerobic conditions (Hammond and Bowman, 1980b). Fourth, these authors identified a glycerol-3-phosphate:ADP transphosphorylase activity, which could be attributed to glycerol kinase. Fifth, in pulse-labeling experiments with [^{14}C]glycerol, Hammond and Bowman (1980a) showed that under anaerobiosis, [^{14}C]glycerol is incorporated into glycerol-3-phosphate at a much higher rate than the rate of glycerol production (Fig. 5). Since [^{14}C]glycerol-3-phosphate is not further metabolized, these results indicate that glycerol kinase catalyzes metabolite fluxes to glycerol-3-phosphate (59 nmoles/min per mg protein) and glycerol (83 nmoles/min per mg protein), resulting in a net flux of 24 nmoles glycerol/min per mg protein. Further, the total concentration of glycerol-3-phosphate in the cell is independent of the glycerol concentration. In any other scheme (Clarkson and Brohn, 1976), the presence of an active glycerol kinase would result in an increase of glycerol-3-phosphate with time, which is not found. Similar studies by Gruenberg *et al.* (1980) have confirmed the work of Hammond and Bowman (1980a) and shown that glycerol is removed from the cell by an asymmetric carrier-mediated process. The K_m for efflux of glycerol (20 μM) from the cell is about an order of magnitude lower than for entry, suggesting that the permease functions to remove glycerol from the cell, thereby preventing inhibition of the glycerol kinase-mediated reaction.

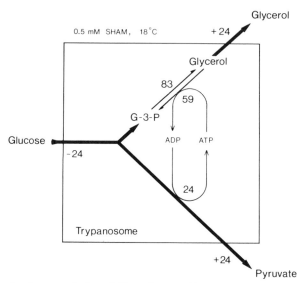

Figure 5. The involvement of glycerol kinase in anaerobic glycolysis in *T. brucei.* The figures indicate the rate of metabolic fluxes in nmoles/min per mg protein at 18°C. (Drawn from data of Hammond and Bowman, 1980a.)

7.1.3. Pentose Phosphate Pathway

Studies using isotopically labeled glucose have demonstrated that glucose is metabolized by a classical glycolytic pathway (Grant and Fulton, 1957). The pentose phosphate pathway in which the C-1 of glucose is evolved as CO_2 does not appear to be significant under the conditions used, since only 0.6% of [1-^{14}C]glucose and 0.7% of [3,4-^{14}C]glucose was found as [14]CO_2. This is in contrast to T. cruzi, where there is a marked differential release of [14]CO_2 from [1-^{14}C]glucose and [6-^{14}C]glucose of 28 and 5.2%, respectively (Bowman et al., 1963).

The first enzyme of the pentose phosphate pathway, glucose 6-phosphate dehydrogenase, has been detected in low amounts in bloodstream forms (Ryley, 1962; Opperdoes et al., 1981), but the second enzyme, 6-phosphogluconate dehydrogenase, is apparently absent (Ryley, 1962). No information has been published on the other enzyme activities of this pathway, and the significance of glucose-6-phosphate dehydrogenase is obscure. In related trypanosomatids, (C. luciliae and Leishmania spp.), this enzyme appears to be localized in the cytosol (Opperdoes, 1981b; Hart and Opperdoes, 1984).

7.2. Short-Stumpy Trypomastigotes

The metabolic activity of the short-stumpy bloodstream form is essentially the same as the long-slender forms. The major end product of glucose is pyruvate, but significant quantities of CO_2, acetate, and succinate are also produced (Flynn and Bowman, 1973). Most, if not all, of the CO_2 produced can be accounted for on the basis of decarboxylation of pyruvate to form acetate, and although all the enzymes of the citric acid cycle have been identified (Ryley, 1962; Flynn and Bowman, 1973), the citric acid cycle is thought to be inoperative, due to the low enzyme activities of succinate oxidase and citrate synthetase (Flynn and Bowman, 1973). No information is available on the anaerobic metabolism of this stage; presumably, this must be different from long-slenders, since short-stumpies are able to survive SHAM–glycerol treatment, whereas long-slenders do not (Clarkson and Brohn, 1976).

7.3. Procyclic Trypomastigotes

Glucose metabolism differs significantly between procyclic (insect) and long-slender trypomastigote (mammalian) forms, reflecting the changes in mitochondrial and glycosomal enzyme activities described above. Glucose is consumed at a much slower rate in procyclic trypomastigotes; in T. rhodesiense, for example, the rate of glucose consumption is reduced 5- to 10-fold, when both forms are assayed under identical conditions (Ryley, 1962). Several of the glycolytic enzyme activities are reduced in procyclic forms (Table III), particuarly hexokinase, which is decreased by 15-fold. As there is no evidence for the presence of important regulatory mechanisms in the glycolytic pathway (Nwagwu and Opperdoes, 1982), it has been suggested that one of the glycolytic enzymes must catalyze a rate-limiting step (Opperdoes et al., 1981). However, in long-slender bloodstream forms, the rate of glucose transport into the trypanosome is the rate-limiting step in its metabolism (Gruenberg et al., 1978) and the possibility that the decreased activity of a glucose transporter may limit the glycolytic rate in procyclics cannot be ruled out.

The changes in metabolic end products of aerobic glucose metabolism from pyruvate to CO_2, succinate, and acetate can be correlated with changes in other enzyme activities. In particular, pyruvate kinase cannot be detected in procyclics ($\leq$ 2 nmoles/min per mg protein; Opperdoes and Cottem, 1982), which contrasts sharply with the high specific activity reported in bloodstream forms (1200–1900 nmoles/min per mg protein; Oduro et al., 1980a; Flynn and Bowman, 1980). The absence of pyruvate kinase—a cytosolic enzyme—is associated with a remarkable alteration in distribution of another ATP-generating enzyme, phosphoglycerate kinase. In bloodstream forms, this enzyme is predominantly located within the glycosome (Opperdoes and Borst, 1977; Oduro et al., 1980a; Opperdoes et al., 1981), whereas in procyclics over 95% is found in the cytosol (Opperdoes and Cottem, 1982; Broman et al., 1982). This may be significant with regard to subcellular compartmentation in the glycosome as will be discussed below. In addition, the large increase in specific activity of glycosomal phosphoenolpyruvate carboxykinase suggests that the glycosome plays an important role in CO_2 fixation in this stage of the life cycle.

The presence of a functional TCA cycle has been inferred mainly from the pioneering work of Ryley (1956, 1962). However, the situation remains unclear, because of the failure to demonstrate key metabolites in radioisotope studies and the appropriate enzyme activity in sufficient amounts necessary to catalyze each step. While aconitase, α-oxoglutarate decarboxylase, succinate dehydrogenase, fumarate hydratase, and malate dehydrogenase have been identified (Ryley, 1962), reports of citrate synthetase and NAD^+-dependent isocitrate dehydrogenase activities remain conspicuously absent. Indeed, Evans and Brown (1972a) were unable to detect either of these enzyme activities. Neither could they detect labeled citrate after incubation with [^{14}C]glucose for 12 min, even though succinate, malate, and fumarate were labeled after 30 sec of incubation. The radiotracer studies of Klein et al. (1975) using either [^{14}C]glucose or $NaH^{14}CO_3$ gave similar results.

However, metabolism via CO_2 fixation to malate, fumarate, and succinate has been clearly demonstrated by the above radiolabeling studies and the enzymes necessary to catalyze this route identified (Klein et al., 1975; Opperdoes et al., 1981; Opperdoes and Cottem, 1982; Broman et al., 1983). Fumarate reductase (NADH-dependent reversal of succinate dehydrogenase) has also been detected (72 nmoles/min per mg protein) in particulate fractions from homogenates of *T. brucei* procyclics (Klein *et al.*, 1975). The subcellular localization and the properties of this enzyme have not been studied.

A tentative metabolic scheme is given in Fig. 6, which is based on a composite of available data from both *T. brucei* and *T. rhodesiense*. Examination of the pathways in Fig. 6 indicates that if no net change in ATP occurs in the glycosomal compartment during glucose metabolism, then all the diphosphoglycerate leaving the glycosome must reenter this compartment as phosphoenolpyruvate. To fulfill this requirement, pyruvate kinase should be absent— as found by Opperdoes and Cottem (1982). Similarly, if no net change of NAD$^+$ occurs, then all NADH produced in the glyceraldehyde-3-phosphate dehydrogenase reaction should be available for the reduction of oxaloacetate to malate, otherwise phosphoenolpyruvate would accumulate. Another consequence of maintaining the net internal balance of ATP and NAD as zero in the glycosome is that the glycerophosphate shuttle must be inoperative. Studies by Broman *et al.* (1983) have provided support for this proposal. Isolated glycosomes were able to catalyze conversion of glucose to 1,3-diphosphoglycerate without production of glycerol-3-phosphate, in the presence of PEP. If ATP replaced PEP in the assay mixture, then glycerol-3-phosphate was produced. Further, PEP and oxaloacetate strongly inhibited the ATP-dependent production of glycerol-3-phosphate. The mechanism by which PEP and oxaloacetate inhibit the glycerophosphate shuttle is not known. Particularly, the proposal that the glycosome regulates the flow of glycolytic intermediates by means of compartmentalization remains uncertain. Further studies on the permeability properties of the glycosomal membrane would help to elucidate this question.

The pathways by which malate, the final product of glycosomal metabolism, is converted to the final end products of glucose metabolism are far from certain. Indeed, apart from studies by Ryley (1962) of *T. rhodesiense* procyclics, no comprehensive carbon-balance studies are available. In the light of this paucity of information, some of it conflicting, the scheme in Fig. 6 is to be regarded as a tentative model to be used as the basis for further research. Evans and Brown (1972a) and Cross *et al.* (1975) both found that *T. brucei* procyclics were able to convert [^{14}C]glucose to alanine. The pathway is presumably by decarboxylation of malate, catalyzed by the cytosolic malic enzyme, followed by transamination with a suitable amino acid, catalyzed by the active alanine aminotransferase found in these organisms (Kilgour and Godfrey, 1973). Malate could also be converted to aspartate, since malate dehydrogenase and aspartate aminotransferase are present in the cytosol. Aspartate has not been reported as an end product, but could function as a shuttle between the mitochondrion and cytoplasm, since a proportion of these enzyme activities would appear to have a mitochondrial location (Opperdoes *et al.*, 1981; Opperdoes and Cottem, 1982).

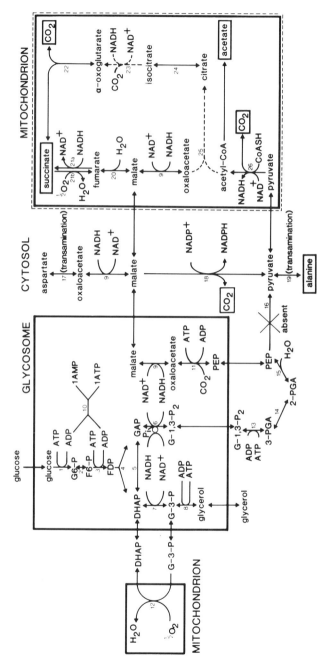

Figure 6. Pathways of glucose metabolism in procyclic trypomastigotes. The enzyme locations in the glycosome and cytosol have been established, but the location of the mitochondrial enzymes (except for glycerophosphate oxidase and malate dehydrogenase) have not yet been clearly demonstrated. End products of aerobic or anaerobic metabolism are enclosed in boxes. The dashed lines indicate enzymes whose presence remains uncertain. 1, hexokinase; 2, phosphoglucose isomerase; 3, phosphofructokinase; 4, aldolase; 5, triose phosphate isomerase; 6, glyceraldehyde-3-phosphate dehydrogenase; 7, glycerol-3-phosphate dehydrogenase; 8, glycerol kinase; 9, malate dehydrogenase; 10, adenylate kinase; 11, phosphoenolpyruvate carboxykinase; 12, glycerol-3-phosphate oxidase; 13, phosphoglycerate kinase; 14, phosphoglycerate mutase; 15, enolase; 16, pyruvate kinase; 17, aspartate aminotransferase; 18, malic enzyme; 19, alanine aminotransferase; 20, fumarate hydratase; 21a, fumarate reductase; 21b, succinate dehydrogenase; 22, α-oxoglutarate decarboxylase; 23, isocitrate dehydrogenase; 24, aconitase; 25, citrate synthetase; 26, pyruvate dehydrogenase.

The work of Klein *et al.* (1975) clearly demonstrated that malate can be converted to succinate via fumarate through the action of fumarase and fumarate reductase. The subcellular localization of these enzymes has not been determined, but the particulate nature of the latter (Klein *et al.*, 1975) suggests a mitochondrial location, by analogy with the fumarate reductase of certain helminths (Saz, 1972). This pathway would account for the production of succinate, by either *T. brucei* or *T. rhodesiense,* particularly under anaerobic conditions. Acetate is produced both aerobically (3–10% of glucose carbon used) and anaerobically (25%) in *T. rhodesiense* and the presence of pyruvate dehydrogenase has been demonstrated (Ryley, 1962), suggesting a route from malate via pyruvate and acetyl CoA. The nature of the enzyme catalyzing the last step from acetyl CoA to acetate is unknown. No information on acetate production from glucose is available for *T. brucei* procyclics. Under aerobic conditions, acetate is a minor end product of *T. rhodesiense* glucose metabolism, and CO_2 accounts for most of the glucose carbon used especially in prolonged incubations (Ryley, 1962), implying a functional TCA cycle. However, the failure to identify citrate as an intermediate or detect citrate synthetase in one strain of *T. brucei* must raise doubts as to the validity of the general assumption that all procyclic forms possess a functional TCA cycle.

8. GLYCOLYSIS AS A TARGET FOR CHEMOTHERAPY

8.1. Energy Metabolism

Long-slender bloodstream *T. brucei* are entirely dependent on substrate-level phosphorylation during glycolysis for production of ATP, because mitochondrial oxidative phosphorylation is absent and energy stores are lacking (reviewed in Fairlamb, 1981, 1982; Opperdoes, 1983a,b). Thus, the rate of ATP production is directly coupled to glycolysis and, under steady-state conditions, is directly proportional to the rate of pyruvate production under aerobic or anaerobic conditions or a mixture of the two (Fairlamb and Bowman, 1980a). This can be seen from the following equations:
Aerobically:

(a) glucose + 2 ADP + 2 P_i + O_2 → 2 pyruvate + 2 ATP + 4 H_2O
(b) glycerol + ADP + P_i + O_2 → pyruvate + ATP + 3 H_2O

Anaerobically:

(c) glycerol + ADP + P_i + O_2 → pyruvate + ATP + H_2O + glycerol

There are three enzymes catalyzing substrate-level phosphorylation: phosphoglycerate kinase, pyruvate kinase, and, under anaerobic conditions, glycerol kinase. As discussed in Sections 7.1.1 and 7.1.2, the net production of ATP within the glycosome is zero with either glucose or glycerol as substrate and thus net production of ATP occurs in the cytoplasmic compartment at the step catalyzed by pyruvate kinase.

Thus, drugs that inhibit the glycolytic sequence will shut off ATP production, leading to death of the organism within a few minutes. Two drugs (aromatic arsenicals and suramin) currently in use for the treatment of trypanosomiasis are thought to exert their chemotherapeutic action in this manner. In addition, a third experimental drug combination (SHAM–glycerol) blocks aerobic and anaerobic glycolysis *in vitro* (Fairlamb *et al.*, 1977) and *in vivo* eliminates long-slender forms from the bloodstream of an infected animal within a few minutes (Clarkson and Brohn, 1976).

8.2. Action of Trivalent Aromatic Arsenicals

Trivalent arsenicals such as melarsen oxide and phenylarsenoxide are potent inhibitors *in vitro* of a number of enzymes from bloodstream trypanosomes, including pyruvate kinase, pyruvate decarboxylase, α-ketoglutarate decarboxylase (Flynn and Bowman, 1974), glycerol kinase (Hammond and Bowman, 1980b) and the dehydrogenase component of the glycerophosphate oxidase (Fairlamb and Bowman, 1977b). However, in the intact cell, pyruvate kinase would appear to be selectively inhibited, because PEP accumulates in trypanosomes treated with concentrations of melarsen oxide that have a negligible effect on the rate of respiration or glucose consumption (Flynn and Bowman, 1974). Although nonallosteric muscle pyruvate kinase is not inhibited by melarsen oxide, the allosteric liver enzyme is as equally sensitive to inhibition *in vitro* as the allosteric trypanosome pyruvate kinase, suggesting that differential uptake of the drug must be involved in chemotherapeutic activity. Ojeda and Flynn (1982) reported that melarsen oxide is taken up by a carrier-mediated mechanism, which is absent in a drug-resistant line derived from the drug-sensitive strain, providing support for this hypothesis.

8.3. Action of Suramin

Suramin is a high-molecular-weight polysulfonated compound that has been used in the treatment of early stages (before CNS involvement) of sleeping sickness for over 60 years. Little is known of the mode of action of this drug, because, unlike the rapid trypanocidal action of arsenicals or SHAM–glycerol, suramin takes up to 24 hr to exert its effect. The reason for suramin's slow trypanocidal activity may reflect its slow rate of uptake (Fairlamb and Bowman, 1980a). Studies *in vivo* have shown that the drug is not actively concentrated within the cell, but enters possibly by endocytosis (Fairlamb and Bowman, 1980a). However, even though trypanosomes continue to show active motility and increase in number in the bloodstream of infected animals for at least 6 hr, glycolysis and respiration are progressively inhibited over this time period (Fairlamb and Bowman, 1980a).

In cell-free homogenates, suramin is a potent inhibitor of both enzymes involved in the reoxidation of NADH via the glycerophosphate shuttle (Fairlamb *et al.*, 1979). The glycosomal NAD^+-dependent glycerol-3-phosphate dehydrogenase is inhibited noncompetitively with respect to dihydroxyacetone phosphate with a K_i of 1.2 μM (Fairlamb, unpublished results). The glycerophosphate oxidase is equally sensitive with K_is of 4.1 and 8.1 μM for glycerol-3-phosphate and oxygen as variable substrate, respectively (Bowman and Fairlamb, 1976; Hill, 1976b; Fairlamb and Bowman, 1977c).

Some experimental evidence supports the notion that suramin's mode of action *in vivo* may be by inhibition of these key glycolytic enzymes. If trypanosomes are isolated from infected animals treated with suramin for varying intervals, then the amount of suramin taken up correlates with the extent of inhibition of respiration (Fairlamb and Bowman, 1980a). Only a fraction (4 to 9%) of the total taken up is required to inhibit respiration to the extent found in broken cell preparations. Further, glucose consumption and pyruvate production are inhibited 38 and 45%, respectively, after exposure to suramin (150 mg/kg for 3 hr), associated with a significant accumulation of glucose carbon used as glycerol (15%). These findings are compatible with a dual inhibitor action of suramin on the enzymes of the glycerophosphate shuttle. However, it would be premature to conclude that this is suramin's only mode of action, as these effects could possibly be secondary to some other action of the drug.

8.4. Action of SHAM–Glycerol

Based on a knowledge of the carbohydrate metabolism of bloodstream trypanosomes, the prediction that SHAM and glycerol would have a lethal effect by inhibiting glycolysis has been shown to be correct both *in vivo* (Clarkson and Brohn, 1976) and *in vitro* (Fairlamb *et al.*, 1977). Long-slender forms are completely eliminated from the bloodstream of infected rodents within 3 min of receiving an intravenous injection of SHAM–glycerol, whereas short-stumpy forms survive at least 1 hr, but are completely cleared after 24 hr (Clarkson and Brohn, 1976). A serum factor, which was noted to accelerate the lytic effect of SHAM–glycerol (Amole and Clarkson, 1981), has been identified as calcium (Clarkson and Amole, 1982). These authors have proposed that calcium activates endogenous phospholipases, which catalyze the release of lysophospholipids, thus causing lysis as a result of membrane damage.

Although experimental infections with *T. brucei, T. vivax,* and *T. evansi* have been cured by SHAM–glycerol administration (Evans *et al.*, 1977; Evans and Holland, 1978; Van der Meer *et al.*, 1979; Evans and Brightman, 1980), the treatment does not appear to be of practical use, because of the large volumes and near-lethal doses that need to be administered. Moreover, animals infected with pleomorphic strains that produce chronic infections tend to relapse about 1 week after treatment (Clarkson and Brohn, 1976; Evans and Brightman, 1980). The failure to achieve radical cure in these instances is thought to result from inadequate levels of both drugs in the tissues, particularly the brain, as had been reported for relapses following delayed treatment with Berenil (Jennings *et al.*, 1977). This has been confirmed by Abolarin *et al.* (1982), who demonstrated that parasites were able to survive SHAM–glycerol treatment in brain tissue (possibly the choroid plexus), but not other tissues. Pharmacological studies have shown that levels of SHAM in brain and blood are similar (Van der Meer *et al.*, 1979), and that SHAM is rapidly eliminated from the animal (Van der Meer and Zwart, 1980). In contrast, glycerol enters the brain slowly and the concentration in brain tissue remains considerably lower than in plasma (Van der Meer *et al.*, 1979). Systematic screening for alternatives to the SHAM–glycerol combination has not been successful, although structure–activity relationships suggest that a lipophilic aromatic iron-chelating agent would be the most likely substitute for SHAM in combination therapy (Clarkson *et al.*, 1981). So far, no alternatives have been discovered for glycerol, but in the light of glycerol's effect on anaerobic metabolism via a mass-action effect on glycerol kinase, it would seem worthwhile to screen for inhibitors of this enzyme.

9. SUMMARY AND OUTLOOK

Trypanosomes possess many unusual features in carbohydrate metabolism that could be exploited for chemotherapeutic purposes. In particular, the bloodstream forms of many salivarian trypanosomes rely entirely on glycolysis for energy production and at least one class of drugs, the arsenicals, is thought to kill these organisms by inhibiting glycolysis. Three areas deserve further attention in the search for new chemotherapeutic agents:

1. Characterization of the individual glycolytic enzymes, in particular glycerol kinase
2. Mechanistic studies on the glycerophosphate oxidase to aid development of a more effective substitute for SHAM
3. Studies on the mechanism of glucose entry and glycerol exit from the cell

Other areas that deserve attention are:

1. Studies on the properties of the glycosomal membrane, particularly to identify the proposed transport systems for glycolytic intermediates

2. The study of the enzyme–enzyme interactions within the glycosomal multienzyme complex and the tunneling of metabolites through it
3. The synthesis of glycosomal enzymes and the mechanism of assembly within the glycosome
4. The regulation of genomic expression of mitochondrial and glycosomal enzymes during different stages of the life cycle.

Chemotherapy of bacterial diseases has come a long way since the pioneering work of Ehrlich and others at the turn of the century. With the modern technologies available today and with the identification of the aforementioned new targets for chemotherapy, there is no scientific reason why the treatment of these killer diseases should go unchallenged for the want of a "magic bullet."

ACKNOWLEDGMENTS This work was supported in part by grants from the Rockefeller Foundation (RF 83067) and National Institute of Health (1RO1 AI 19428) to A.H.F. and in part by the UNDP/World Bank/WHO Special Programme for Research and Training in Tropical Diseases to F.R.O.

Addendum

The literature surveyed in this review was completed in December 1983. Since then, it has been demonstrated that the requirement for a feeder cell layer in the cultivation of bloodstream trypomastigote forms can be eliminated by the addition of thiols such as 2-mercaptoethanol or thioglycerol (Baltz *et al.*, 1985) or cysteine (Dusenko *et al.*, 1985).

REFERENCES

Abolarin, M. O., Evans, D. A., Tovey, D. G.,and Ormerod, W. E., 1982, Cryptic stage of sleeping-sickness trypanosome developing in choroid plexus epithelial cells, *Br. Med. J.* **285**:1380–1382.
Amole, B. O., and Clarkson, A. B., Jr., 1981, *Trypanosoma brucei*: Host parasite interaction in parasite destruction by salicylhydroxamic acid and glycerol in mice, *Exp. Parasitol.* **51**:133–140.
Apted, F. I. C., 1970, Treatment of human trypanosomiasis, in: *The African Trypanosomiases* (H. W. Mulligan, ed.), pp. 684–710, Allen & Unwin, London.
Balber, A. E., 1983, Primary murine bone marrow cultures support continuous growth of infectious human trypanosomes, *Science* **220**:421–423.
Barker, D. C., 1980, The ultrastructure of kinetoplast DNA with particular reference to the interpretation of dark field electron microscopy images of isolated, purified networks, *Micron* **11**:21–62.
Bendall, D. S., and Bonner, W. D., Jr., 1971, Cyanide-insensitive respiration in plant mitochondria, *Plant Physiol.* **47**:236–245.
Berens, R. L., Brun, R., and Krassner, S. M., 1976, A simple monophasic medium for axenic culture of hemoflagellates, *J. Parasitol.* **62**:360–365.

Bienen, E. J., Hammadi, E., and Hill, G. C., 1981, *Trypanosoma brucei:* Biochemical and morphological changes during *in vitro* transformation of bloodstream- to procyclic-trypomastigotes, *Exp. Parasitol.* **51**:408–417.

Bienen, E. J., Hill, G. C., and Shin, K.-O., 1983, Elaboration of mitochondrial function during *Trypanosoma brucei* differentiation, *Mol. Biochem. Parasitol.* **7**:75–86.

Bohringer, S., and Hecker, H., 1974, Quantitative ultrastructural differences between strains of the *Trypanosoma brucei* subgroup during transformation in blood, *J. Protozool.* **21**:694–698.

Bohringer, S., and Hecker, H., 1975, Quantitative ultrastructural investigations of the life cycle of *Trypanosoma brucei:* A morphometric analysis, *J. Protozool.* **22**:463–467.

Bonner, W. D., Jr., and Rich, P. R., 1978, Molecular aspects of cyanide/antimycin resistant respiration, in: *Plant Mitochondria* (G. C. Ducet and C. Lance, eds.), pp. 241–247, Elsevier/North-Holland, Amsterdam.

Borst, P., and Hoeijmakers, J. H. J., 1979, Kinetoplast DNA, *Plasmid* **2**:20–40.

Bowman, I. B. R., 1974, Intermediary metabolism of pathogenic flagellates, in: *Trypanosomiasis and Leishmaniasis with Special Reference to Chagas' Disease,* Ciba Foundation Symposium 20 (new series), pp. 255–271, Associated Scientific Publishers, Amsterdam.

Bowman, I. B. R., and Fairlamb, A. H., 1976, L-Glycerol-3-phosphate oxidase in *Trypanosoma brucei* and the effect of suramin, in: *Biochemistry of Parasites and Host–Parasite Relationships* (H. Van den Bossche, ed.), pp. 501–507, Elsevier/North-Holland, Amsterdam.

Bowman, I. B. R., and Flynn, I. W., 1976, Oxidative metabolism of trypanosomes, in: *Biology of the Kinetoplastida* (W. H. R. Lumsden and D. A. Evans, eds.), Vol. 1, pp. 435–476, Academic Press, New York.

Bowman, I. B. R., Tobie, E. J., and von Brand, T., 1963, CO_2 fixation studies with the culture form of *Trypanosoma cruzi, Comp. Biochem. Physiol.* **9**:105–114.

Bowman, I. B. R., Srivastava, H. K., and Flynn, I. W., 1972, Adaptations in oxidative metabolism during the transformation of *Trypanosoma rhodesiense* from bloodstream into culture form, *in: Comparative Biochemistry of Parasites* (H. Van den Bossche, ed.), pp. 329–342, Academic Press, New York.

Brohn, F. H., and Clarkson, A. B., Jr., 1978, Quantitative effects of salicylhydroxamic acid and glycerol on *Trypanosoma brucei* glycolysis in vitro and in vivo, *Acta Trop.* **35**:23–33.

Brohn, F. H., and Clarkson, A. B., Jr., 1980, *Trypanosoma brucei brucei:* Patterns of glycolysis at 37°C in vitro, *Mol. Biochem. Parasitol.* **1**:291–305.

Broman, K., Ropars, M., and Deshusses, J., 1982, Subcellular location of glycolytic enzymes in *Trypanosoma brucei* culture form, *Experientia* **38**:533–534.

Broman, K., Knupfer, A.-L., Ropars, M., and Deshusses, J., 1983, Occurrence and role of phosphoenolpyruvate carboxykinase in procyclic *Trypanosoma brucei brucei* glycosomes, *Mol. Biochem. Parasitol.* **8**:79–87.

Brown, R. C., Evans, D. A., and Vickerman, K., 1973, Changes in oxidative metabolism and ultrastructure accompanying differentiation of the mitochondrion in *Trypanosoma brucei, Int. J. Parasitol.* **3**:691–704.

Brun, R., and Jenni, L., 1977, A new semi-defined medium for *Trypanosoma brucei* sspp., *Acta Trop.* **34**:21–33.

Brun, R., and Schonenberger, M., 1979, Cultivation and in vitro cloning of procyclic culture forms of *Trypanosoma brucei* in a semi-defined medium, *Acta Trop.* **36**:289–292.

Brun, R., and Schonenberger, M., 1981, Stimulating effect of citrate and cis-aconitate on the transformation of *Trypanosoma brucei* bloodstream forms to procyclic forms in vitro, *Z. Parasitenkd.* **66**:17–24.

Brun, R., Jenni, L., Tanner, M., Schonenberger, M., and Schell, K.-F., 1979, Cultivation of vertebrate infective forms derived from metacyclic forms of pleomorphic *Trypanosoma brucei* stocks, *Acta Trop.* **36**:387–390.

Brun, R., Jenni, L., Schonenberger, M., and Schell, K.-F., 1981, *In vitro* cultivation of bloodstream forms of *Trypanosoma brucei, T. rhodesiense,* and *T. gambiense, J. Protozool.* **28:**470–479.

Bursell, E., 1978, Quantitative aspects of proline utilization during flight in tsetse flies, *Physiol. Entomol.* **3:**265–272.

Bursell, E., 1981, The role of proline in energy metabolism, in: *Energy Metabolism in Insects* (R. G. H. Downer, ed.), pp. 135–154, Plenum Press, New York.

Caughey, B., Hill, G. C., and Rich, P., 1979, Alternate electron donors to the α-glycerophosphate oxidase in *Trypanosoma brucei, J. Protozool.* **26:**8A.

Clarkson, A. B., Jr., and Amole, B. O., 1982, Role of calcium in trypanocidal drug action, *Science* **216:**1321–1323.

Clarkson, A. B., Jr., and Brohn, F. H., 1976, Trypanosomiasis: An approach to chemotherapy by inhibition of carbohydrate metabolism, *Science* **194:**204–206.

Clarkson, A. B., Jr., Grady, R. W., Grossman, S. A., McCallum, R. J., and Brohn, F. H., 1981, *Trypanosoma brucei brucei*: A systematic screening for alternatives to the salicylhydroxamic acid–glycerol combination, *Mol. Biochem. Parasitol.* **3:**271–291.

Cohen, S. S., 1979, Comparative biochemistry and drug design for infectious disease, *Science* **205:**964–971.

Cross, G. A. M., and Manning, J. C., 1973, Cultivation of *Trypanosoma brucei* sspp. in semi-defined and defined media, *Parasitology* **67:**315–331.

Cross, G. A. M., Klein, R. A., and Linstead, D. J., 1975, Utilization of amino acids by *Trypanosoma brucei* in culture: L-Threonine as a precursor for acetate, *Parasitology* **71:**311–326.

Cunningham, I., 1973, Quantitative studies on trypanosomes in tsetse tissue culture, *Exp. Parasitol.* **33:**34–45.

Cunningham, I., 1977, New culture medium for maintenance of tsetse tissues and growth of trypanosomatids, *J. Protozool.* **24:**325–329.

Cunningham, I., and Honigberg, B. M., 1977, Infectivity reacquisition of *Trypanosoma brucei brucei* cultivated with tsetse salivary glands, *Science* **197:**1279–1282.

Cunningham, I., Honigberg, B. M., and Taylor, A. M., 1981, Infectivity of monomorphic and pleomorphic *Trypanosoma brucei* stocks cultivated at 28°C with various tsetse fly tissues, *J. Protozool.* **67:**391–397.

Dawson, A. P., and Thorne, C. J. R., 1969, Preparation and properties of L-3-glycerophosphate dehydrogenase from pig brain mitochondria, *Biochem. J.* **111:**27–34.

De Raadt, P., 1976, African sleeping sickness today, *Trans. R. Soc. Trop. Med. Hyg.* **70:**114–116.

Dixon, H., 1966, Blood platelets as a source of enzyme activity in washed trypanosome suspensions, *Nature* **210:**428.

Docampo, R., Deboiso, J. F., Boveris, A., and Stoppani, A. O. M., 1976, Localization of peroxidase activity in *Trypanosoma cruzi* microbodies, Experientia **32:**972–975.

Englund, P. T., Hajduk, S. L., and Marini, J. C., 1982, The molecular biology of trypanosomes, *Annu. Rev. Biochem.* **51:**695–726.

Evans, D. A., and Brightman, C. A. J., 1980, Pleomorphism and the problem of recrudescent parasitaemia following treatment with salicylhydroxamic acid (SHAM) in African trypanosomiasis, *Trans. R. Soc. Trop. Med. Hyg.* **74:**601–604.

Evans, D. A., and Brown, R. C., 1971, Cyanide insensitive culture form of *Trypanosoma brucei, Nature* **230:**251–252.

Evans, D. A., and Brown, R. C., 1972a, The utilization of glucose and proline by culture forms of *Trypanosoma brucei, J. Protozool.* **19:**686–690.

Evans, D. A., and Brown, R. C., 1972b, The effect of diphenylamine on terminal respiration in bloodstream and culture forms of *Trypanosoma brucei, J. Protozool.* **19:**365–369.

Evans, D. A., and Brown, R. C., 1973, *m*-Chlorobenzhydroxamic acid—An inhibitor of cyanide-insensitive respiration in *Trypanosoma brucei, J. Protozool.* **20:**157–160.

Evans, D. A., and Holland, M. F., 1978, Effective treatment of *Trypanosoma vivax* infections with salicylhydroxamic acid (SHAM), *Trans. R. Soc. Trop. Med. Hyg.* **72:**203–204.

Evans, D. A., Brightman, C. J., and Holland, M. F., 1977, Salicylhydroxamic acid/glycerol in experimental trypanosomiasis, *Lancet* **2:**769.

Fairlamb, A. H., 1981, Alternate metabolic pathways in protozoan metabolism, *Parasitology* **82:**1–30.

Fairlamb, A. H., 1982, Biochemistry of trypanosomiasis and rational approaches to chemotherapy, *Trends Biochem. Sci.* **7:**249–253.

Fairlamb, A. H., and Bowman, I. B. R., 1977a, The isolation and characterization of particulate *sn*-glycerol-3-phosphate oxidase from *Trypanosoma brucei*, *Int. J. Biochem.* **8:**659–668.

Fairlamb, A. H., and Bowman, I. B. R., 1977b, Inhibitor studies on particulate *sn*-glycerol-3-phosphate oxidase from *Trypanosoma brucei*, *Int. J. Biochem.* **8:**669–675.

Fairlamb, A. H., and Bowman, I. B. R., 1977c, *Trypanosoma brucei:* Suramin and other trypanocidal compounds effects on *sn*-glycerol-3-phosphate oxidase, *Exp. Parasitol.* **43:**353–361.

Fairlamb, A. H., and Bowman, I. B. R., 1980a, Uptake of the trypanocidal drug suramin by bloodstream forms of *Trypanosoma brucei* and its effect on respiration and growth rate in vivo, *Mol. Biochem. Parasitol.* **1:**315–318.

Fairlamb, A. H., and Bowman, I. B. R., 1980b, *Trypanosoma brucei:* Maintenance of concentrated suspensions of bloodstream trypomastigotes in vitro using continuous dialysis for measurement of endocytosis, *Exp. Parasitol.* **49:**366–380.

Fairlamb, A. H., Opperdoes, F. R., and Borst, P., 1977, New approach to screening drugs for activity against African trypanosomes, *Nature* **265:**270–271.

Fairlamb, A. H., Oduro, K. K., and Bowman, I. B. R., 1979, Action of the trypanocidal drug suramin on the enzymes of aerobic glycolysis of *Trypanosoma brucei in vivo*, in: *FEBS Special Meeting on Enzymes,* Dubrovnik, Abstr. S4-10.

Flynn, I. W., and Bowman, I. B. R., 1973, The metabolism of carbohydrate by pleomorphic African trypanosomes, *Comp. Biochem. Physiol.* **45B:**25–42.

Flynn, I. W., and Bowman, I. B. R., 1974, The action of trypanocidal arsenical drugs on *Trypanosoma brucei* and *Trypanosoma rhodesiense, Comp. Biochem. Physiol.* **48B:**261–273.

Flynn, I. W., and Bowman, I. B. R., 1980, Purification and characterization of pyruvate kinase from *Trypanosoma brucei*, *Arch. Biochem. Biophys.* **200:**401–409.

Ford, W. C. L., and Bowman, I. B. R., 1973, Metabolism of proline by the culture midgut form of *Trypanosoma rhodesiense, Trans. R. Soc. Trop. Med. Hyg.* **67:**257.

Gardiner, P. R., Lamont, L. C., Jones, T. W., and Cunningham, I., 1980, The separation and structure of infective trypanosomes from cultures of *Trypanosoma brucei* grown in association with tsetse fly salivary glands, *J. Protozool.* **27:**182–185.

Gashumba, J., 1981, Sleeping sickness in Uganda, *New Sci.* **89:**164.

Ghiotto, V., Brun, R., Jenni, L., and Hecker, H., 1979, *Trypanosoma brucei:* Morphometric changes and loss of infectivity during transformation of bloodstream forms to procyclic culture forms *in vitro, Exp. Parasitol.* **48:**447–456.

Grant, P. T., and Fulton, J. D., 1957, The catabolism of glucose by strains of *Trypanosoma rhodesiense, Biochem. J.* **66:**242–250.

Grant, P. T., and Sargent, J. R., 1960, Properties of L-α-glycerophosphate oxidase and its role in the respiration of *Trypanosoma rhodesiense, Biochem. J.* **76:**229–237.

Gruenberg, J., Sharma, P. R., and Deshusses, J., 1978, D-Glucose transport in *Trypanosoma brucei:* D-Glucose transport is the rate limiting step of its metabolism, *Eur. J. Biochem.* **89:**461–469.

Gruenberg, J., Schwendimann, B., Sharma, P. R., and Deshusses, J., 1980, Role of glycerol permeation in the bloodstream form of *Trypanosoma brucei, J. Protozool.* **27:**484–491.

Gutteridge, W. E., and Rogerson, G. W., 1979, Biochemical aspects of the biology of *Trypanosoma cruzi,* in: *Biology of the Kinetoplastida* (W. H. R. Lumsden and D. A. Evans, eds.), Vol. 2, pp. 619–652, Academic Press, New York.

Hajduk, S. L., 1978, Influence of DNA complexing compounds on the kinetoplast of trypanosomatids, *Prog. Mol. Subcell. Biol.* **6:**158–200.

Hammond, D. J., and Bowman, I. B. R., 1980a, *Trypanosoma brucei*: The effect of glycerol on the anaerobic metabolism of glucose, *Mol. Biochem. Parasitol.* **2:**63–75.

Hammond, D. J., and Bowman, I. B. R., 1980b, Studies on glycerol kinase and its role in ATP synthesis in *Trypanosoma brucei*, *Mol. Biochem. Parasitol.* **2:**77–91.

Hammond, D. J., and Gutteridge, W. E., 1983, Studies on the glycosomal orotate phosphoribosyl transferase of *Trypanosoma cruzi*, *Mol. Biochem. Parasitol.* **7:**319–330.

Hammond, D. J., Gutteridge, W. E., and Opperdoes, F. R., 1981, A novel location for two enzymes of de novo pyrimidine biosynthesis in trypanosomes and *Leishmania*, *FEBS Lett.* **128:**22–30.

Hart, D. T., and Coombs, G. H., 1980, The importance and subcellular localization of β-oxidation, glycolysis and amino acid oxidation in *Leishmania mexicana mexicana* amastigotes and promastigotes, in: *Proceedings of the 3rd European Multicolloquium on Parasitology*, Cambridge, Abstr. 16.

Hart, D. T., and Opperdoes, F. R., 1984, The occurrence of glycosomes (microbodies) in the promastigote stage of four major *Leishmania* species of *Mol. Biochem. Parasitol.* **13:**159–172.

Hart, D. T., Misset, O., Edwards, S. W., and Opperdoes, F. R., 1984, A comparison of the glycosomes (microbodies) isolated from *Trypanosoma brucei* bloodstream form and cultured procyclic trypomastigotes, *Mol. Biochem. Parasitol.* **12:**25–35.

Hawking, F., 1973, The differentiation of *Trypanosoma rhodesiense* from *T. brucei* by means of human serum, *Trans. R. Soc. Trop. Med. Hyg.* **67:**517–527.

Henry, M.-F., and Nyns, E.-J., 1975, Cyanide insensitive respiration: An alternative mitochondrial pathway, *Sub-Cell. Biochem.* **4:**1–65.

Hill, G. C., 1976a. Electron transport systems in Kinetoplastida, *Biochim. Biophys. Acta* **456:**149–193.

Hill, G. C., 1976b, Characterization of the electron transport systems present during the life cycle of African trypanosomes, in: *Biochemistry of Parasites and Host–Parasite Relationships* (H. Van den Bossche, ed.), pp. 31–50, Elsevier/North-Holland, Amsterdam.

Hill, G. C., Gutteridge, W. E., and Matthewson, N. W., 1971a, Purification and properties of cytochromes *c* from trypanosomatids, *Biochim. Biophys. Acta* **243:**225–229.

Hill, G. C., Chan, S. K., and Smith, L., 1971b, Purification and properties of cytochrome c_{555} from a protozoan, *Crithidia fasciculata*, *Biochim. Biophys. Acta* **253:**78–87.

Hill, G. C., Shimer, S. P., Caughey, B., and Sauer, L. S., 1978a, Growth of infective forms of *Trypanosoma rhodesiense in vitro*, the causative agent of African trypanosomiasis, *Science* **202:**763–765.

Hill, G. C., Shimer, S., Caughey, B., and Sauer, L. S., 1978b, Growth of infective forms of *Trypanosoma (T.) brucei* on buffalo lung and Chinese hamster lung tissue culture cells, *Acta Trop.* **35:**201–207.

Hirumi, H., Doyle, J. J., and Hirumi, K., 1977a, African trypanosomes: Cultivation of animal-infective *Trypanosoma brucei* in vitro, *Science* **196:**992–994.

Hirumi, H., Doyle, J. J., and Hirumi, K., 1977b, Cultivation of bloodstream *Trypanosoma brucei*, *Bull. WHO* **55:**405–409.

Hoare, C. A., 1972, *The Trypanosomes of Mammals: A Zoological Monograph*, Blackwell, Oxford.

Jennings, F. W., Whitelaw, D. D., and Urquhart, G. M., 1977, The relationship between duration of infection with *Trypanosoma brucei* in mice and the efficacy of chemotherapy, *Parasitology* **75:**143–153.

Kilgour, V., and Godfrey, D. G., 1973, Species-characteristic isoenzymes of two aminotransferases in trypanosomes, *Nature New Biol.* **244:**69–70.

Klein, R. A., Linstead, D. J., and Wheeler, M. V., 1975, Carbon dioxide fixation in trypanosomatids, *Parasitology* **71:**93–107.

Lanham, S. M., 1968, Separation of trypanosomes from the blood of infected rats and mice by anion exchangers, *Nature* **218:**1273–1274.

Lanham, S. M., and Godfrey, D. G., 1970, Isolation of salivarian trypanosomes from man and other mammals using DEAE cellulose, *Exp. Parasitol.* **28**:521–534.

McGhee, R. B., and Cosgrove, W. B., 1980, Biology and physiology of the lower Trypanosomatidae, *Microbiol. Rev.* **44**:140–173.

Mackenzie, N. E., Hall, J. E., Flynn, I. W., and Scott, A. I., 1983, ^{13}C nuclear magnetic resonance studies of anaerobic glycolysis in *Trypanosoma brucei* spp., *Biosci. Rep.* **3**:141–151.

McLaughlin, J., 1981, Association of adenylate kinase with the glycosome of *Trypanosoma rhodesiense*, *Biochem. Int.* **2**:345–353.

Mendez, Y., and Honigberg, B. M., 1972, Infectivity of *Trypanosoma brucei*-subgroup flagellates maintained in culture, *J. Parasitol.* **58**:1122–1136.

Meshnick, S. R., 1984, The chemotherapy of African trypanosomiasis, in: *Parasitic Diseases* (J. M. Mansfield, ed.), Vol. 2, pp. 165–199, Dekker, New York.

Meshnick, S. R., Blobstein, S. H., Grady, R. W., and Cerami, A., 1978, An approach to the development of new drugs for African trypanosomiasis, *J. Exp. Med.* **148**:569–579.

Muse, K. E., and Roberts, J. F., 1973, Microbodies in *Crithidia fasciculata*, *Protoplasma* **78**:343–348.

Njogu, R. M., and Nyindo, M., 1981, Presence of a peculiar pathway of glucose metabolism in infective forms of *Trypanosoma brucei* cultured from salivary glands of tsetse flies, *J. Parasitol.* **67**:847–851.

Njogu, R. M., Whittaker, C. J., and Hill, G. C., 1980, Evidence for a branched electron transport chain in *Trypanosoma brucei*, *Mol. Biochem. Parasitol.* **1**:13–29.

Nwagwu, M., and Opperdoes, F. R., 1982, Regulation of glycolysis in *Trypanosoma brucei*: Hexokinase and phosphofructokinase activity, *Acta Trop.* **39**:61–72.

Oduro, K. K., Flynn, I. W., and Bowman, I. B. R., 1980a, *Trypanosoma brucei*: Activities and subcellular distribution of glycolytic enzymes from differently disrupted cells, *Exp. Parasitol.* **50**:123–135.

Oduro, K. K., Bowman, I. B. R., and Flynn, I. W., 1980b, *Trypanosoma brucei*: Preparation and some properties of a multienzyme complex catalysing part of the glycolytic pathway, *Exp. Parasitol.* **50**:240–250.

Ojeda, P. V., and Flynn, I. W., 1982, Some aspects of resistance to arsenical drugs in *Trypanosoma brucei*, in: *Abstracts of Vth International Congress of Parasitology*, Toronto, p. 724.

Opperdoes, F. R., 1980, Miconazole: An inhibitor of cyanide-insensitive respiration in *Trypanosoma brucei*, *Trans. R. Soc. Trop. Med. Hyg.* **74**:423–424.

Opperdoes, F. R., 1981a, A rapid method for the isolation of intact glycosomes from *Trypanosoma brucei* by Percoll-gradient centrifugation in a vertical rotor, *Mol. Biochem. Parasitol.* **3**:181–186.

Opperdoes, F. R., 1981b, Alternate metabolic pathways in protozoan energy metabolism, *Parasitology* **82**:1–30.

Opperdoes, F. R., 1982, The glycosome, *Ann. N.Y. Acad. Sci.* **386**:543–545.

Opperdoes, F. R., 1983a, Toward the development of new drugs for parasitic diseases, in: *Parasitology: A Global Perspective* (K. S. Warren and J. Z. Bowers, eds.), pp. 191–202, Springer-Verlag, Berlin.

Opperdoes, F. R., 1983b, Glycolysis as target for the development of new trypanocidal drugs, in: *Mechanism of Drug Action* (T. P. Singer, T. E. Mansour, and R. N. Undarza, eds.), pp. 121–131, Academic Press, New York.

Opperdoes, F. R., 1984, Localization of the initial steps in alkoxyphospholipid biosynthesis in glycosomes (microbodies) of *Trypanosoma brucei*, *FEBS Lett.* **169**:35–39.

Opperdoes, F. R., and Borst, P., 1977, Localization of nine glycolytic enzymes in a microbody-like organelle in *Trypanosoma brucei*: The glycosome, *FEBS Lett.* **80**:360–364.

Opperdoes, F. R., and Cottem, D., 1982, Involvement of the glycosome of *Trypanosoma brucei* in carbon dioxide fixation, *FEBS Lett.* **143**:60–64.

Opperdoes, F. R., and Nwagwu, M., 1980, Suborganellular localization of glycolytic enzymes in the glycosome of *Trypanosoma brucei*, in: *The Host Invader Interplay* (H. Van den Bossche, ed.), pp. 683–686, Elsevier/North-Holland, Amsterdam.

Opperdoes, F. R., Borst, P., and Fonck, K., 1976, The potential use of inhibitoes of glycerol-3-phosphate oxidase for chemotherapy of African trypanosomiasis, *FEBS Lett.* **62:**169–172.

Opperdoes, F. R., Borst, P., Bakker, S., and Leene, W., 1977a, Localization of glycerol-3-phosphate oxidase in the mitochondrion and particulate NAD^+-linked glycerol-3-phosphate dehydrogenase in the microbodies of the bloodstream form of *Trypanosoma brucei*, *Eur. J. Biochem.* **76:**29–39.

Opperdoes, F. R., Borst, P., and Spits, H., 1977b, Particle-bound enzymes in the bloodstream form of *Trypanosoma brucei*, *Eur. J. Biochem.* **76:**21–28.

Opperdoes, F. R., Markos, A., and Steiger, R. F., 1981, Localization of malate dehydrogenase, adenylate kinase and glycolytic enzymes in glycosomes and the threonine pathway in the mitochondrion of cultured procyclic trypomastigotes of *Trypanosoma brucei*, *Mol. Biochem. Parasitol.* **4:**291–309.

Opperdoes, F. R., Baudhuin, P., Coppens, I., De Roe, C., Edwards, S. W., Weijers, P. J., and Misset, O., 1984, Purification, morphometric analysis and characterization of the glycosomes (microbodies) of the protozoan hemoflagellate *Trypanosoma brucei*, *J. Cell Biol.* **98:**1178–1184.

Pettigrew, G. W., 1972, The amino acid sequence of a cytochrome *c* from a protozoan, *Crithidia oncopelti*, *FEBS Lett.* **22:**64–66.

Pettigrew, G. W., and Smith, G. M., 1977, Novel N-terminal protein blocking group identified as dimethylproline, *Nature* **265:**661–662.

Pittam, M. D., 1970, Medium for *in vitro* culture of *Trypanosoma rhodesiense* and *T. brucei*, appendix to: Dixon, H., and Williamson, J., The lipid composition of blood and culture forms of *Trypanosoma lewisi* and *Trypanosoma rhodesiense* compared to that of their environment, *Comp. Biochem. Physiol.* **33:**111–128.

Rickmann, L. R., and Robson, J., 1970, The testing of proven *Trypanosoma brucei* and *T. rhodesiense* strains by the blood incubation infectivity test, *Bull. WHO* **42:**911–916.

Rifkin, M. R., 1978, Identification of the trypanocidal factor in normal human serum: High density lipoprotein, *Proc. Natl. Acad. Sci. USA* **75:**3450–3454.

Ringler, R. L., 1961, Studies on the mitochondrial α-glycerophosphate dehydrogenase. II. Extraction and partial purification of the dehydrogenase from pig brain, *J. Biol. Chem.* **236:**1192–1198.

Robertson, D. H. H., Pickens, S., Lawson, J. H., and Lennox, B., 1980, An accidental laboratory infection with African trypanosomes of a defined stock. I. The clinical course of infection, *J. Infect.* **2:**105–112.

Rustin, P., Dupont, J., and Lance, C., 1983, A role for fatty acid peroxy radicals in the cyanide-insensitive pathway of plant mitochondria?, *Trends Biochem. Sci.* **8:**155–157.

Ryley, J. F., 1956, Studies on the metabolism of the protozoa. 7. Comparative carbohydrate metabolism of eleven species of trypanosome, *Biochem. J.* **62:**215–222.

Ryley, J. F., 1962, Studies on the metabolism of protozoa. 9. Comparative metabolism of bloodstream and culture forms of *Trypanosoma rhodesiense*, *Biochem. J.* **85:**211–223.

Saz, H. J., 1972, Comparative biochemistry of carbohydrates in nematodes and cestodes, in: *Comparative Biochemistry of Parasites* (H. Van den Bossche, ed.), pp. 33–47, Academic Press, New York.

Simpson, L., Simpson, A. M., Kidane, G., Livingston, L., and Spithill, T. W., 1980, The kinetoplast DNA of the hemoflagellate protozoa, *Am. J. Trop. Med. Hyg.* **29**(Suppl.):1053–1063.

Smith, G. M., and Pettigrew, G. W., 1980, Identification of N,N-dimethylproline as the N-terminal blocking group of *Crithidia oncopelti* cytochrome c_{557}, *Eur. J. Biochem.* **110:**123–130.

Srivastava, H. K., and Bowman, I. B. R., 1971, Adaptation in oxidative metabolism of *Trypanosoma rhodesiense* during transformation in culture, *Comp. Biochem. Physiol.* **40B**:973–981.

Srivastava, H. K., and Bowman, I. B. R., 1972, Metabolic transformation of *Trypanosoma rhodesiense* in culture, *Nature New Biol.* **57**:152–153.

Stuart, K., 1980, Cultivation of dyskinetoplastic *Trypanosoma brucei*, *J. Parasitol.* **66**:1060–1961.

Tanner, M., 1980, Studies on the mechanisms supporting the growth of *Trypanosoma (Trypanozoon) brucei* as bloodstream-like forms in vitro, *Acta Trop.* **37**:203–220.

Taylor, M. B., Berghausen, P., Heyworth, P., Messenger, N., Rees, L. J., and Gutteridge, W., 1980, Subcellular localization of some glycolytic enzymes in parasitic flagellated protozoa, *Int. J. Biochem.* **11**:117–120.

Threlfall, D. R., Williams, B. L., and Goodwin, T. W., 1965, Terpenoid quinones and sterols in parasitic and culture forms of *Trypanosoma rhodesiense*, in: *Progress in Protozoology: 2nd International Congress of Protozoology*, p. 141, Excerpta Medica, Series 91, Amsterdam.

Tobie, E. J., von Brand, T., and Mehlman, B., 1950, Cultural and physiological observations on *Trypanosoma rhodesiense* and *Trypanosoma gambiense*, *J. Parasitol.* **36**:48–54.

Trigg, P. I., 1979, Research and training in tropical diseases, *Trends Biochem. Sci.* **4**:29–30.

Turner, M. J., 1982, Biochemistry of the variant surface glycoproteins of salivarian trypanosomes, *Adv. Parasitol.* **21**:69–151.

Van der Meer, C., and Zwart, D., 1980, Pitfalls of salicylhydroxamic acid plus glycerol treatment of *T. vivax* infected goats, in: *The Host Invader Interplay* (H. Van den Bossche, ed.), pp. 687–690, Elsevier/North-Holland, Amsterdam.

Van der Meer, C., Versluijs-Broers, J. A. M., and Opperdoes, F. R., 1979, *Trypanosoma brucei*: Trypanocidal effect of salicylhydroxamic acid plus glycerol in infected rats, *Exp. Parasitol.* **48**:126–134.

Vickerman, K., 1965, Polymorphism and mitochondrial activity in sleeping sickness trypanosomes, *Nature* **208**:762–766.

Vickerman, K., 1970, Morphological and physiological considerations of extracellular blood protozoa, in: *Ecology and Physiology of Parasites* (A. M. Fallis, ed.), pp. 58–91, University of Toronto Press, Toronto.

Vickerman, K., and Preston, T. M., 1976, Comparative cell biology of the kinetoplastic flagellates, in: *Biology of the Kinetoplastida* (W. H. R. Lumsden and D. A. Evans, eds.), Vol. 1, pp. 35–130, Academic Press, New York.

Visser, N., and Opperdoes, F. R., 1980, Glycolysis in *Trypanosoma brucei*, *Eur. J. Biochem.* **103**:623–632.

Visser, N., Opperdoes, F. R., and Borst, P., 1981, Subcellular compartmentation of glycolytic intermediates in *Trypanosoma brucei*, *Eur. J. Biochem.* **118**:521–526.

Weinman, D., 1960, Cultivation of the African sleeping sickness trypanosomes from the blood and cerebrospinal fluid of patients and suspects, *Trans. R. Soc. Trop. Med. Hyg.* **54**:180–190.

WHO, 1979, The African trypanosomiases, Technical Report No. 635, World Health Organization, Geneva.

Zweygarth, E., Ahmed, J. S., and Rehbein, G., 1983, Cultivation of infective forms of *Trypanosoma (T.) brucei evansi* in a continuous culture system, *Z. Parasitenkd.* **69**:131–133.

References to Addendum

1. Baltz, T., Baltz, D., Giroud, C. and Crockett, J., 1985. Cultivation in a semi-defined medium of animal infective forms of *Trypanosoma brucei, T. equiperdum, T. evansi, T. rhodesiense* and *T. gambiense, EMBO (Eur. Mol. Biol. Org.) J.*, **4**:1273–1277.

2. Dusenko, M., Ferguson, M. A., Lamont, G. S., Rifkin, M. R., and Cross, G. A. M. Cysteine eliminated the feeder cell requirement for cultivation of *Trypanosoma brucei* bloodstream forms *in vitro, J. Exp. Med.* **162**:1256–1263.

7

Sugar Transport Systems of Baker's Yeast and Filamentous Fungi

ANTONIO H. ROMANO

1. YEAST

1.1. Monosaccharide Transport Systems

In spite of the enormous volume of work that has been done on the metabolism of sugars by *Saccharomyces cerevisiae,* the nature of the monosaccharide uptake systems has been a subject of controversy over the past three decades. The controversy has centered on the question of the role of sugar phosphorylation during transport: are sugars phosphorylated during transport via a group translocation system, or are they transported as free sugars via a carrier-mediated facilitated diffusion system and phosphorylated by the action of hexokinase or glucokinase subsequent to entry? A variation of the question is: even if a facilitated diffusion system represents the primary mode of entry of free sugars, is there some sort of intimate relationship between the sugar carriers and the phosphorylating enzymes such that the activity of the sugar carriers is affected?

The transport-associated phosphorylation hypothesis, advanced by Rothstein (1954), was based on the following observations: (1) intracellular free sugar did not accumulate during fermentation; (2) yeast cells were reportedly impermeable to nonmetabolizable sugars; (3) there was a correlation between substrate specificity for sugar uptake and sugar phosphorylation; (4) the process of sugar uptake revealed saturation kinetics; and (5) sugar metabolism was effected by certain reagents, such as uranyl ion, that did not penetrate the yeast cell mem-

ANTONIO H. ROMANO • Microbiology Section, The University of Connecticut, Storrs, Connecticut 06268.

brane. These findings were taken to mean that the uptake of sugars was dependent on phosphorylation by hexokinase at the cell surface.

It was shown subsequently that a number of nonmetabolizable sugars were in fact taken up by yeast cells (Burger *et al.*, 1959; Cirillo, 1961a,b, 1968a; Kotyk, 1961, 1967), although their affinities for transport were one to three orders of magnitude lower than metabolizable sugars. The uptake of these nonmetabolizable sugars displayed saturation kinetics, was subject to competitive inhibition, and was specifically inhibited by uranyl ion. Uptake was shown to be independent of metabolic energy, and internal concentrations reached did not exceed external concentration. Also, glucose itself was taken up (but not concentrated) by iodoacetate-poisoned yeast (Cirillo, 1962). The phenomenon of counterflow was demonstrated by the addition of D-glucose to cells preloaded with nonmetabolizable glucose analogue; moreover, free glucose could be detected in metabolizing cells by counterflow techniques (Wilkins and Cirillo, 1965). Taken together, all these findings provided strong evidence for a bidirectional carrier-mediated facilitated diffusion system, operating independently of phosphorylation. Moreover, even the earlier observations cited above, which were interpreted in terms of phosphorylation at the cell surface, appeared to be explainable on the basis of a membrane carrier system, operating prior to phosphorylation. (For a review of early studies of sugar uptake by baker's yeast, see Cirillo, 1961c.)

Ensuing studies of the specificity of hexose transport in baker's yeast have indicated that a constitutive transport system with relatively broad specificity operates for the uptake of glucose, fructose, mannose, and a number of analogues (Kotyk, 1967; Heredia *et al.*, 1968; Cirillo, 1968a). The pyranose ring is the active structure for transport; D-glucose (C1 glucopyranose chair configuration) shows the highest affinity (apparent K_m in the range of 2–6 mM). Single changes in the C1 glucopyranose structure do not result in drastic decreases in affinity; thus, transport activity is not dependent on any single ring substituent. Removal of the anomeric -OH at C-1 (1,5-anhydroglucitol) results in a decrease in affinity of one order of magnitude. Similar decreases in affinity for the D-glucose carrier occur upon substitution of an axial -OH group for the equatorial -OH of D-glucose at C-1 (D-mannose) or C-4 (D-galactose), or upon removal of the -CH$_2$OH group at C-5 (D-xylose). Changes at C-3 appear to have greater effects: the substitution of an axial -OH at C-3 (3-*O*-methyl-D-glucose) results in a 50-fold decrease in affinity. Removal of the equatorial -OH at C-2 (2-deoxy-D-glucose) has no effect on affinity; this analogue has essentially the same affinity for the carrier as D-glucose. The only single change that causes a drastic decrease in affinity is the addition of a methyl group to the anomeric -OH at C-1 (α-methyl-D-glucopyranoside). Multiple changes in the C1 D-glucopyranose structure result in marked reduction in activity. For example, a combination of an axial -OH substitution and removal of the C-5 -CH$_2$OH (L-arabinose) results in a decrease in activity of two orders of magnitude. L-Aldohexoses, that occur in solution in the 1C configuration, all have very low affinity. There is a significant correlation between activity for transport and activity for yeast hexokinase. It is of interest, however, that there is an even greater correlation between sugar transport activity of yeasts and that of human erythrocytes (Cirillo, 1968a). The only exception is D-fructose, which has little transport activity in erythrocytes, but high activity for yeast transport and hexokinase. The high activity of D-fructose remains enigmatic, and does not fit into the pattern described above for the aldohexoses. This may be due to uncertainties concerning the structure of ketose sugars in solution, as discussed by Cirillo (1968a).

While D-galactose shows significant affinity for the constitutive D-glucose transport system as described above, growth of baker's yeast on D-galactose results in induction of a D-galactose transport system that has higher specificity for galactose and its nonmetabolizable analogues. Thus, D-fucose and L-arabinose, which represent single changes in C1 D-galactopyranose structure (6-deoxy- and 5-dehydroxymethyl-, respectively) are much more rapidly transported by galactose-grown yeast than by glucose-grown yeast (Cirillo, 1968b). These latter two nonmetabolizable sugars were transported by a facilitated diffusion mechanism, as was D-galactose itself in a yeast mutant defective in galactokinase (Kuo et al., 1970).

The similarities between baker's yeast and human erythrocytes with respect to specificity for the constitutive hexose transport system were mentioned above. A further similarity is that the affinity of the yeast transport system for D-glucose is of the same order of magnitude as that found in a large number of animal cell types that carry out facilitated diffusion of D-glucose (apparent K_m in the range of 2–6 mM). It is to be noted that this range of affinity is two to three orders of magnitude below that found in many bacteria that carry out active transport or group translocation of sugars with apparent K_m in the micromolar range. This relatively low affinity of the yeast sugar transport system probably relfects evolutionary adaptation of *S. cerevisiae* to a saccharine environment.

The phosphorylation theory of sugar uptake was revived by Van Steveninck and Rothstein (1965) in the form of a hybrid carrier–phosphorylation model, whereby the sugar carrier was presumed to function in a low-affinity (high K_m, low V_{max}) facilitated diffusion system in the transport of nonmetabolizable glucose analogues (galactose in uninduced cells) or of glucose itself in iodoacetate- or Ni^{2+}-poisoned cells, and in a high-affinity (low K_m, high V_{max}) active transport system, involving phosphorylation, for glucose (galactose in induced cells) in actively metabolizing, energy-sufficient cells. This model was based on their measurements of a K_m for glucose uptake of 6 mM by unpoisoned cells, and a K_m of 75 mM for iodoacetate-poisoned cells. They also reported a differential effect of Ni^{2+} ions on uptake: Ni^{2+} did not inhibit uptake of nonmetabolized sugars or galactose in uninduced cells, but did inhibit uptake of metabolized sugars. Based on Ni^{2+} binding studies, they presumed that Ni^{2+} was bound by phosphoryl groups, probably polyphosphate in nature, on the outer surface of the cell membrane. They speculated that these phosphoryl groups might normally be involved in phosphorylation of the sugar carrier, converting it to a high-affinity carrier that functioned in the uphill transport of glucose, or galactose in induced cells. Binding of these phosphoryl groups by Ni^{2+} would then interfere with the putative phosphorylation of the sugar carrier. Iodoacetate was presumed to act by preventing the replenishment of phosphorylating capacity through its inhibition of glycolysis. However, it was subsequently shown that Ni^{2+} ions do not act only at the cell surface, but can penetrate yeast cells and block glycolysis by inhibiting alcohol dehydrogenase (Fuhrmann and Rothstein, 1968a,b).

The notion that the monosaccharide carrier can exist in more than one form or state, dependent on the metabolic state of the cell, was supported by Serrano and DelaFuente (1974). These workers reported that double reciprocal plots of initial velocity of glucose uptake by fermenting yeast versus glucose concentration showed biphasic kinetics—a high-K_m component (100 mM for glucose and fructose) at high sugar concentrations, and a low-K_m component (glucose, 3 mM; fructose, 10 mM) at sugar concentrations from 2.5 to 20 mM. Moreover, even at low sugar concentrations, stationary-phase cells incubated aerobically (conditions under which the Pasteur effect of reduced sugar utilization would be manifest) showed only the high-K_m component, whereas anaerobically incubated cells showed the low-K_m component. These workers did not ascribe a particular role to phosphorylation in influencing the state of the sugar carrier, but rather supported the view that some intermediate metabolite derived from glycolysis exerted a regulatory effect on the transport system, as had been suggested by Sols (1967).

Van Stevinck and his colleagues continued to emphasize the role of phosphorylation in transport in a series of experiments designed to establish the temporal sequence of sugar phosphorylation and sugar entry into yeast cells. In experiments using 2-deoxy-D-glucose (2-DOG), a glucose analogue that has a high affinity for the constitutive hexose transport system and can be phosphorylated by yeast hexokinase, Van Stevinck (1968) reported the following: (1) 2-DOG taken up by yeast cells was recovered inside the cells as free 2-DOG and 2-deoxy-D-glucose-6-phosphate (2-DOG-6-P). No other major product was detected. Intracellular 2-DOG and 2-DOG-6-P recovered in extracts of cells were separated by barium–zinc precipitation of the phosphorylated sugar. (2) While the intracellular level of 2-DOG-6-P was consistently higher than the free 2-DOG pool, the intracellular concentration of 2-DOG was higher than the extracellular concentration. (3) Maintenance of both the free 2-DOG and 2-DOG-6-P pools was dependent on metabolic energy; addition of iodoacetate to preloaded cells resulted in rapid decrease of both the intracellular 2-DOG-6-P and free 2-DOG when the intracellular 2-DOG concentration exceeded that of the medium. (4) In pulse-labeling experiments where [^{14}C]-2-DOG was added to suspensions of cells preloaded with nonradioactive 2-DOG and 2-DOG-6-P, the specific radioactivity of the phosphorylated sugar pool was initially higher than that of the free sugar pools. It was concluded, therefore, that transport of free 2-DOG into the cells followed by phosphorylation by the hexokinase–ATP system did not occur, but that phosphorylation of 2-DOG was associated with transport, followed by partial intracellular hydrolysis.

Similar results of pulse-labeling experiments supporting transport-associated phosphorylation were reported for α-methylglucoside in maltose-grown cells, and galactose in induced cells (Van Steveninck, 1970, 1972).

On the other hand, Kuo *et al.* (1970), using a mutant strain that lacked galactokinase but retained the inducible transport system, showed that induced cells transported galactose and the analogues L-arabinose and D-fucose by facilitated diffusion. Moreover, in pulse-labeling experiments of the type carried out by Van Steveninck, using a mutant that possessed both the galactose permease and galactokinase but lacked UDP-galactose-1-phosphate uridyltransferase (and hence would not metabolize galactose beyond galactose-1-phosphate), the intracellular free galactose pool became labeled before the galactose-1-phosphate pool (Kuo and Cirillo, 1970). Thus, a clear case for facilitated diffisuion followed by intracellular phosphorylation was demonstrated.

Kotyk and Michaljanicova (1974) criticized the experiments of Van Steveninck with 2-DOG described above on the basis that 2-DOG-6-P was not the only metabolic product derived from 6-DOG; other products, such as 2-deoxy-D-gluconate, 2-deoxy-D-glucose-1,6-bisphosphate, UDP-2-deoxy-D-glucose, and GDP-2-deoxy-D-glucose, were also formed, though in lesser amounts than 2-DOG-6-P (Biely and Bauer, 1966, 1968). Of special importance is the deoxy derivative of the nonreducing disaccharide trehalose, 2,2'-dideoxy-α,α'-trehalose, which was found to accumulate in yeast in significant amounts under aerobic conditions (Farkas *et al.*, 1969). This disaccharide would have been measured as part of the free 2-DOG pool by the analytical procedures used by Van Steveninck; thus, the sizes of free 2-DOG pools, which were used as the basis of calculations of specific radioactivity in the pulse-labeling experiments, and formed the basis of the conclusion that free 2-DOG was concentrated against a gradient, were overestimated. Kotyk and Michaljanicova (1974) also carried out pulse-labeling experiments with glucose, galactose, and α-methyl-D-glucose by Van Steveninck's method; they reported that in all cases the free sugar pools became labeled prior to the sugar phosphate pools. In response to these criticisms, Jaspers and Van Steveninck (1975) repeated their pulse-labeling experiments with 2-DOG, using *S. fragilis,* and an enzymatic method of free 2-DOG determination in which 2,2'-dideoxy-α,α'-trehalose did not interfere (though they confirmed that this disaccharide was formed in this yeast, along with several other 2-DOG derivatives). Their experiments yielded results similar to those they reported previously for *S. cerevisiae*: the sugar phosphate pool became labeled before the free sugar pools. Significantly, they noted that in some experiments the reverse order of labeling was observed; however, they presented control experiments and theoretical arguments that indicated that apparent high initial radioactivity in the free 2-DOG pool could be accounted for by extracellular radioactive 2-DOG adhering to filters.

Meredith and Romano (1977) avoided the "trehalose error" by using a respiration-deficient petite mutant of *S. cerevisiae* that did not accumulate the deoxytrehalose derivative under energy-sufficient aerobic conditions. This petite mutant concentrated 2-DOG-6-P when incubated with 2-DOG and sucrose as an energy source, but the intracellular concentration of free 2-DOG did not exceed that of the suspending medium. However, using a modified pulse-labeling procedure (intracellular 2-DOG and 2-DOG-6-P pools were preloaded by incubation of cells with [^{14}C]-2-DOG, then [^{3}H]-2-DOG at high specific activity was added), it was confirmed that 2-DOG first appeared in the cell in the phosphorylated form. These results are compatible with a transport-associated phosphorylation process, but, as pointed out by the authors, they do not rule out immediate intracellular phosphorylation by hexokinase if present in sufficient excess and in possible close association with the cell membrane, following entry by facilitated diffusion.

Further confirmation that 2-DOG first appears in baker's yeast in the sugar phosphate pool, as determined by pulse-labeling experiments, was supplied by Franzusoff and Cirillo (1982). These experiments were of special significance because they compared wild-type cells with single-kinase mutants, each possessing only one of the two hexokinase isozymes or only the glucokinase that are present in baker's yeast (Lobo and Maitra, 1977a,b,c). The wild-type and all three single-kinase strains gave similar results: when suitable corrections were made for adherence of radioactive free 2-DOG to cell surfaces and filters, intracellular 2-DOG-6-P pools became labeled before the free 2-DOG pools. Thus, they demonstrated that if transport-associated phosphorylation does occur in baker's yeast, it is not a function of a specific hexokinase or glucokinase. These authors urged caution in the interpretation of pool-labeling data, however, citing possible problems of intracellular compartmentation and multiple pools that are difficult to assess.

Romano (1982) presented evidence that phosphorylation of the transport substrate was not required for high-affinity operation of the glucose carrier: 6-DOG, which cannot be phosphorylated by kinases due to the lack of a hydroxyl group at C-6, was taken up by glucose-grown yeast cells via a facilitated diffusion system with a K_m of 2 mM. Kotyk *et al.* (1975) also reported facilitated diffusion of 6-DOG, but found a much higher affinity system in galactose-grown cells (K_m 3 mM) than in glucose-grown cells (K_m 500 mM). Reasons for this discrepancy in results for glucose-grown cells are not obvious.

The role of phosphorylating enzymes in yeast sugar uptake has been clarified considerably by Bisson and Fraenkel (1983a,b) through the use of kinaseless mutants of *S. cerevisiae*. These workers used two mutant strains, one lacking both hexokinases PI and PII (which cannot grow on fructose, but can grow on glucose since it retains glucokinase), and the other lacking glucokinase in addition (a triple-kinase mutant that cannot grow on fructose or glucose). First, it was shown that the uptake of sugars that could not be phosphorylated (6-DOG in the wild type and both mutants, fructose in the two mutants, and glucose in the triple-kinase mutant) quickly reached a plateau at or below the external sugar concentration, confirming again the facilitated diffusion nature of the uptake system. Next, they determined kinetic parameters of sugar uptake using Eadie–Hofstee plots of V versus V/S, pointing out the advantages of this method over the more widely used Lineweaver–Burk plot of $1/V$ versus $1/S$: the latter method shows bias toward data at low substrate concentration. In such V versus V/S plots, the wild-type strain showed two components, a high-affinity component (K_m of 1 mM for glucose, 6 mM for fructose) and a low-affinity component (K_m of 20 mM for glucose, 50 mM for fructose). The double-kinase mutant showed both components for glucose, but only the low-affinity component for fructose, while the triple-kinase mutant showed only the low-affinity component for either glucose or fructose. Thus, the high-affinity component for either glucose or fructose uptake was operative only in those strains possessing the appropriate kinase. These findings were confirmed by genetic analysis: the low-K_m system for fructose was absent only in strains lacking both hexokinases (hxk1/hxk2). Moreover, cloning of any single wild-type kinase gene (HXK1, HXK2, or GLK1) into the triple-kinase mutant restored the low-K_m system for glucose, while either HXK1 or HXK2 restored the low-K_m system for fructose.

While it appears clear from the above that the presence of a cognate kinase is essential for the operation or presence of a high-affinity uptake system, there is not the necessity of glycolytic metabolism of the sugar or even its initial phosphorylation. A phosphoglucose isomerase mutant that can phosphorylate glucose but cannot grow on it showed both the low-K_m and high-K_m components of glucose uptake (Bisson and Fraenkel, 1983a). More importantly, uptake of 6-DOG, which cannot be phosphorylated, also showed two components in the wild type, although the affinities were less than those found for glucose: K_m 20 mM, high affinity; K_m 250 mM, low affinity. The double-kinase mutant also showed both components, whereas the triple-kinase mutant showed only the low-affinity component. Again, cloning of any single wild-type kinase gene into the triple mutant restored the high-affinity system (Bisson and Fraenkel, 1983b).

These experiments appear to establish firmly that kinases are involved in some way in the operation of a high-affinity facilitated diffusion system for glucose that does not necessarily involve phosphorylation of the sugar during its transport. However, they do not allow a decision as to whether there are two separate systems (low K_m and high K_m), a single system that operates in the high-affinity mode in association with a kinase, or several systems, one for each of three kinases and a fourth for high-K_m uptake. Answers to these questions may come from further genetic analysis, or from studies with subcellular membrane preparations that have recently become possible. Franzusoff and Cirillo (1983a) have developed a method for the preparation of nonleaky membrane vesicles by adding exogenous phospholipid (asolectin) to isolated cytoplasmic membrane preparations from baker's yeast followed by freezing, thawing, and brief sonication. These hybrid liposome–plasma membrane vesicles showed stereospecific D -glucose uptake activity, and characteristics of zero trans influx, equilibrium exchange, and influx counterflow that are similar to those of intact cells. Further, these workers have solubilized the glucose transport system from isolated plasma membranes with octylglucoside, and reconstituted the system into proteoliposomes with removal of the detergent by dialysis, followed by addition of asolectin liposomes to the dialyzed proteins with a freeze–thaw and brief sonication step. These reconstituted proteoliposomes also showed all the characteristics of D -glucose facilitated diffusion: stereospecificity, stimulated equilibrium exchange, and influx counterflow (Franzusoff and Cirillo, 1983b). Thus, the means are now at hand for studying possible associations of membrane components with hexokinases PI, PII, or glucokinase, the regulation of sugar transport by metabolites, and the activity of isolated membrane proteins in functional reconstituted membrane systems.

1.2. Disaccharide Transport Systems

While it is now clear that *S. cerevisiae* transports monosaccharides by facilitated diffusion, the disaccharide maltose and α-methylglucoside are transported by inducible active transport systems. The importance of a permease for maltose fermentation was demonstrated in early work by Robertson and Halvorson (1957), who showed that during incubation of maltose-grown yeast cells in nitrogen-free medium in the presence of glucose (deadaptation), the ability to ferment maltose decreased more rapidly than α-glucosidase activity in cell-free extracts. They showed that this "crypticity" toward maltose metabolism that developed during deadaptation was due to the rapid loss of the ability to take up and concentrate α-methylglucoside, an analogue that was considered to be transported by the maltose uptake system on the basis of competitive inhibition of α-methylglucoside uptake by maltose. The active transport of maltose itself was subsequently demonstrated by Harris and Thompson (1961) in a strain of yeast that took up maltose, but did not metabolize it at pH 8.5; thus, they were able to show that this disaccharide was concentrated 15-fold with respect to the medium, and that this concentration was inhibited by energy uncouplers such as dinitrophenol, sodium azide, or sodium fluoride. The ability to concentrate maltose was present only in maltose-grown cells, not in α-methylglucoside-grown cells, although the latter sugar inhibited maltose uptake by maltose-grown cells when present in 3-fold excess. This indicated that the maltose and α-methylglucoside systems were distinct, though α-methylglucoside is sufficiently related structurally to maltose that the maltose uptake system shows significant activity toward it.

This was confirmed by Okada and Halvorson (1964a,b), who characterized the α-methylglucoside transport system in *S. cerevisiae* strains that fermented α-methylglucoside but were maltose negative. Using α-thioethyl-D-glucopyranoside (α-TEG) as a nonmetabolizable analogue of α-methylglucoside, they showed the following: (1) Only *S. cerevisiae* strains possessing the dominant MG_2 gene took up α-TEG. (2) Glucose-grown cells took up α-TEG, but did not concentrate it against a gradient; the process was stereospecific, with a K_m for α-TEG of 50 mM. (3) Cells that were induced for the α-methylglucoside system by growth in the presence of α-methylglucoside or α-TEG accumulated α-TEG in an unchanged state 10- to 150-fold with respect to the external medium. This active transport process was stereospecific, being competitively inhibited by α-methylglucoside, maltose, glucose, or trehalose, but not by cellobiose, β-methylglucoside, lactose, melibiose, or galactose. (4) The inducible active transport system showed a higher affinity for α-TEG (K_m 1.8 mM) than the constitutive facilitated diffusion system (K_m 50 mM). (5) The active transport of α-TEG was inhibited by dinitrophenol or sodium azide.

There is strong evidence that the active transport of maltose, α-methylglucoside, and α-TEG by *S. cerevisiae* takes place by proton symport, in response to a protonmotive force. Proton symport during maltose uptake has been demonstrated by Serrano (1977), and during maltose, α-methylglucoside, and α-TEG uptake by Brocklehurst *et al.* (1977). Similar proton symport has been demonstrated for maltose and α-methylglucoside, but not for glucose, galactose, or 2-DOG in a strain of *S. carlsbergensis* (Seaston *et al.*, 1973). Also, strains of *S. fragilis* capable of lactose fermentation absorbed protons with lactose (Seaston *et al.*, 1973) and the lactose analogue methyl-β-D-thiogalactoside (Van Den Broek and Van Steveninck, 1982).

It is not immediately obvious why *S. cerevisiae* and related species of this yeast genus appear to have evolved with facilitated diffusion systems for monosaccharides and active transport systems for disaccharides and oligosaccharides. Serrano (1977) has pointed out that the enzyme systems involved in the initial metabolic transformation of glucosides, the glucosidases, have relatively high K_m values (10^{-2} M) in comparison with the kinases that are involved in the initial step of hexose metabolism (10^{-4} M). Thus, he has speculated that the high K_m of the glucosidases requires active transport for the efficient capture of energy source, while the low K_m of the kinases obviates this need.

2. FILAMENTOUS FUNGI

Most studies of sugar transport systems of filamentous fungi have been carried out with *Aspergillus nidulans* or *Neurospora crassa*. In particular, the energetics of active transport have been investigated in *N. crassa*. For reviews of sugar transport in fungi, see Jennings (1974) and Eddy (1982).

2.1. Aspergillus nidulans

A. nidulans has individual constitutive active transport systems for glucose, fructose, and galactose (Mark and Romano, 1971; Romano, 1973). The properties of these systems were studied using 2-DOG to characterize the D-glucose uptake system in a wild-type strain, while D-fructose and D-galactose uptake were studied with mutant strains lacking fructokinase or galactokinase, respectively. The rates and extents of uptake for each of these sugars were essentially the same in cells grown in the absence or presence of the respective sugar; thus, all three uptake systems are constitutive. The uptake of all three of these sugars took place against a concentration gradient: intracellular concentrations attained were 35- to 150-fold greater than that of the external medium; intracellular accumulation was inhibited by dinitrophenol or sodium azide. The apparent K_m values for the transport of these sugars were: D-glucose, 0.04–0.06 mM; 2-DOG, 0.06 mM; D-galactose, 0.03 mM; D-fructose, 0.4 mM. Thus, they would be classified as relatively high-affinity systems, and reflect the natural habitat of this fungus (i.e., soils with low sugar concentration). Note that these K_m values reflect affinities for transport of these sugars that are two orders of magnitude higher than those found in baker's yeast (see Section 1.1).

In initial studies of 2-DOG uptake by wild-type *A. nidulans*, Romano and Kornberg (1968) interpreted an early intracellular accumulation of 2-DOG-6-P as a possible indication of transport-associated phosphorylation of glucose and its analogues. However, more critical analysis showed that this accumulation of sugar phosphate was the result of the action of intracellular kinases, and that D-glucose and its analogues entered the cell as free sugar (Brown and Romano, 1969). Evidence was: 6-DOG, which lacks the C-6 hydroxyl group and thus cannot be phosphorylated by hexokinase, competitively inhibited the uptake of 2-DOG, and was itself accumulated against a gradient in an unaltered form; pool labeling experiments showed that [^{14}C]-2-DOG appeared first in the intracellular free sugar pool and subsequently entered the sugar phosphate pool. Evidence that D-galactose and D-fructose were actively transported and concentrated as free

sugars came from experiments with mutants lacking galactokinase and fructokinase, respectively (Mark and Romano, 1971).

The pattern of specificity of the D-glucose transport system of *A. nidulans* is
broadly similar to that of baker's yeast described in Section 1.1, and to that of
mammalian red blood cells. A comparison of affinities of sugars for the glucose
transport system in *A. nidulans*, yeast, and erythrocytes is shown in Table I.
With the exception of methylation of the anomeric hydroxyl group at C-1 (α-
methyl-D-glucoside), the *A. nidulans* system tolerates single changes in the C1 D
-glucopyranose chair configuration (D-mannose, 2-DOG, D-galactose, 3-*O*-
methyl-D-glucose, D-xylose) as is the case in the yeast and erythrocyte systems,
though the specificity appears to be stricter in *A. nidulans*. This fungus does not
tolerate more than one change. Thus, the combination of an axial C-4 -OH (D-
galactose) and absence of the C-5 -CH$_2$OH (D-xylose), each of which when
occurring alone only decreased activity, resulted in complete loss of activity (D-
arabinose, D-fructose). The latter two sugars, however, which represent single
changes in the structure of D-galactose, were effective competitive inhibitors of
the active transport of D-galactose by the *A. nidulans* galactokinase mutants,
while 6-DOG and D-xylose, which represent two changes in D-galactose structure, were not (Mark and Romano, 1971). This established the distinct nature of
the D-glucose and D-galactose systems. The uptake of D-fructose was highly

Table I. *Comparative Affinities[a] of Sugars and Substrates for
Glucose Transport Systems of Aspergillus nidulans, Baker's Yeast,
and Erythrocytes*

Sugar	*A. nidulans*[b]	Yeast[c]	Erythrocytes[d]
D-Glucose	1	1	1
2-Deoxy-D-glucose	1	1	1.4
D-Fructose	<0.0012	0.2	0.0025
D-Galactose	0.005	0.13	0.2
D-Mannose	0.046	0.1	0.3
1,5-Anhydro-D-glucitol	0.04	0.1	0.2
3-*O*-Methyl-D-glucose	0.48	0.02	—
D-Xylose	0.016	0.13	0.11
D-Fucose	<0.0012	0.02	0.03
D-Arabinose	<0.0012	0.02	0.005
L-Sorbose	<0.0012	0.005	<0.0025
L-Glucose	<0.0012	0.0025	<0.0025
α-Methyl-D-glucoside	<0.0012	0.0025	<0.0025

[a]Expressed as a ratio of K_m or K_i of D-glucose to that of the individual sugars listed.
[b]From Mark and Romano (1971).
[c]From Cirillo (1968a).
[d]From LeFevre (1961).

specific in *A. nidulans*. None of the analogues listed in Table I inhibited its uptake. In this regard, *A. nidulans* differs from yeast and more closely resembles the erythrocyte system.

With the anomalous behavior of D-fructose in yeast aside, this broad pattern of similarity of specificity of the D-glucose transport system in fungal, yeast, and mammalian systems may indicate basic similarities in the monosaccharide carrier systems of eurkaryotes. A point of divergence of eukaryotes from prokaryotes is to be noted with respect to the affinity of α-methyl-D-glucoside for D-glucose transport systems. This sugar is very active in bacterial systems, and has been widely used as a nonmetabolizable D-glucose transport analogue. In contrast, it has no affinity for the D-glucose transport systems of the eukaryotes compared here.

While the monosaccharide transport systems of *A. nidulans* are constitutive, their activities appear to be subject to regulation by intracellular metabolites. Romano and Kornberg (1968) noted that when a wild-type strain of *A. nidulans* was grown with glucose and acetate as sources of carbon and energy, there was a preferential utilization of acetate, and an inhibition of the utilization of glucose. This inhibitory effect of acetate was independent of the glucose concentration in the medium; thus, the inhibition of glucose utilization by acetate was not competitive. Evidence was presented that the effect was not due to acetate itself, but to a metabolic product of acetate: the acetate effect was not manifest in a mutant that was deficient in acetyl CoA synthase. These studies of the effect of acetate on sugar utilization were extended to an *A. nidulans pdh* mutant deficient in pyruvate dehydrogenase, which could not grow on sugars but excreted pyruvate as a consequence of the metabolic block; thus, the accumulation of pyruvate was a convenient measure of sugar uptake and metabolism. It was noted that when this *pdh* mutant was incubated in the presence of acetate and any one of a number of sugars (glucose, mannose, fructose, galactose, mannitol, sorbitol, glycerol, maltose, lactose, or gluconic acid), the onset of pyruvate excretion was delayed until the acetate was utilized. An exception was sucrose; pyruvate formation from sucrose was not affected by the presence of acetate. Since the formation of pyruvate from sucrose involves the necessary action of the same glycolytic enzymes that are involved in the catabolism of other sugars, the effect of acetate might be on the transport of those sugars by *A. nidulans*. This was confirmed by studies of the effect of acetate on the transport of 2-DOG; it was shown that there was an inhibition of the uptake of this glucose analogue by wild-type cells that had been grown in the presence of acetate. On the basis of these findings, the hypothesis was formulated that acetyl CoA, as an end product of glycolysis, can regulate the utilization of a number of sugars by controlling sugar uptake.

The idea that acetyl CoA may act as a regulatory metabolite for sugar uptake was supported by parallel studies by Morgan and Kornberg (1969) with *Escherichia coli*. Working with mutants deficient in both phosphoenolpyruvate synthase and citrate synthase, which would be expected to accumulate acetyl CoA from pyruvate as a consequence of the dual metabolic block, these workers found that pyruvate strongly inhibited the uptake of a number of sugars.

A similar interpretation was made by Gill and Ratledge (1973) of their observation that n-decane and a number of alkanes inhibited glucose uptake and assimilation in *Candida* 107; the accumulation of fatty acids, fatty acyl CoA, and acetyl CoA from the metabolism of the alkanes was presented as the possible cause of inhibition of glucose uptake. However, Jennings (1974) has offered an alternative explanation of the inhibitory effect of acetate on glucose uptake in fungi: Jennings and Austin (1973) pointed out that *Dendryphiella salina* has an apparent ability to regulate hyphal osmotic pressure by maintaining a constant intracellular concentration of soluble carbohydrate, and reasoned that sugar transport may be controlled by the concentration of intracellular metabolites. Since acetate assimilation led to synthesis and accumulation of mannitol and arabitol (Holligan and Jennings, 1973), it was speculated that one of these compounds may be the metabolite regulating glucose uptake.

2.2. *Neurospora crassa*

N. crassa has two transport systems for glucose, a low-affinity facilitated diffusion system (K_m 8–25 mM) and a repressible high-affinity active transport system (K_m 0.01–0.07 mM). Scarborough (1970a) first reported that glucose-grown cells transported 3-*O*-methyl-D-glucose in an unchanged state, but did not concentrate it. Fructose-grown cells, however, concentrated 3-*O*-methyl-D-glucose and L-sorbose against considerable concentration gradients. These latter results were originally interpreted as a possible induction or derepression during growth on fructose of a system or several systems, capable of concentrating sugars, with properties that were distinct from the glucose facilitated diffusion system. However, the situation was shortly clarified by the subsequent demonstration that cells grown in high glucose concentration (50 mM) had the low-affinity (K_m 8 mM) facilitated diffusion system described previously, whereas cells grown in low glucose concentration (1 mM) possessed a high-affinity (K_m 0.01 mM) active transport system that was the same as that found in cells grown in 50 mM fructose (Scarborough, 1970b). These findings of dual systems for glucose, a low-affinity facilitated diffusion system (system I), and a high-affinity active transport system (system II) subject to repression by glucose, were quickly confirmed by Neville *et al.* (1971) and by Schneider and Wiley (1971a). Both systems were found in ungerminated conidia; germination of conidia in glucose-containing medium resulted in repression of the active transport system, followed by restoration upon depletion of glucose. Recovery from glucose repression was inhibited by cycloheximide.

While the experiments of Scarborough (1970b) were carried out using L-sorbose as the transport substrate, and Neville et al. (1971) and Schneider and Wiley (1971a) used 3-O-methyl-D-glucose, it is clear that these groups of workers were dealing with the same system. This system is also probably the same as the L-sorbose active transport system previously described by Klingmüller (1967a,b) in conidia germinated in fructose, though the argument has been made by Klingmüller and Huh (1972) that the L-sorbose transport systems in ungerminated and germinated conidia are different, based on differences in K_m and V_{max} for L-sorbose transport (ungerminated, K_m 3.5 mM, V_{max} 2.5 μg/mg dry wt per min; germinated, K_m 10 mM, V_{max} 7.6 μg/mg dry wt per min). These differences are small and may not be significant.

The glucose active transport system of N. crassa is relatively specific. It shows high affinity for 3-O-methyl-D-glucose (K_m 0.07–0.09 mM) and the uptake of this analogue was efficiently inhibited by 6-DOG and 2-DOG, indicating that single changes in D-glucose structure are well tolerated. However, D-xylose and D-mannose were less effective competitive inhibitors, and D-fructose, D-galactose, and α-methyl-D-glucose did not inhibit when present in 100-fold concentration excess (Neville et al., 1971; Schneider and Wiley, 1971a). It is to be noted that while the ketose L-sorbose was transported and concentrated by this system, the affinity (K_m 4 mM) is two orders of magnitude lower than that of D-glucose or 3-O-methyl-D-glucose (Scarborough, 1970b).

Regulation of the high-affinity system II was studied by Schneider and Wiley (1971b,c). Germination of conidia followed by growth in casein hydrolysate or tryptone media resulted in increased activity of the system. Growth on glucose or sucrose resulted in complete repression of the system; growth on fructose resulted in 80–90% repression. Thus, the system is not inducible, but is constitutive and subject to repression. Glucose concentrations as low as 2 mM brought about complete repression. [The repressive effect of fructose reported by these authors is at variance with the findings of Scarborough (1970b); there is no clear explanation of this difference, beyond possible effects of differences in germination and outgrowth times employed, or strain differences.] Starvation of glucose-grown cells in sugar-free medium resulted in derepression of the system. Addition of cycloheximide after initiation or derepression resulted in immediate cessation of formation of system II; however, addition of glucose after initiation of derepression resulted in rapid inactivation of the system ($t_{1/2}$ 40 min). Cycloheximide did not affect this glucose inactivation, indicating that de novo protein synthesis was not required for the operation of this apparently specific degradation system.

Regulation of synthesis of the high-affinity system II operates at the level of transcription (Schneider and Wiley, 1971c). Starvation of cells (derepressing conditions) in the presence of cycloheximide resulted in no synthesis of the system, but subsequent transfer of such cells to a glucose medium (repressing conditions) in the absence of cycloheximide allowed synthesis of the system. This recovery from protein synthesis inhibition under repressing conditions was not affected by actinomycin D. Thus, there was synthesis of an mRNA specific for system II under derepressing conditions in the presence of cycloheximide that was subsequently translated in the presence of glucose after removal of cycloheximide. The accumulation of mRNA during cycloheximide treatment was explained by increased half-life of the message during blockage of translation.

The energetics of the *N. crassa* high-affinity glucose active transport system have been worked out by Slayman and Slayman (1974, 1975). The importance of this work goes beyond the elucidation of the nature of active transport in this fungus, which appears to be a model for active transport systems of most if not all eukaryotic microorganisms; at the time that it was carried out, it represented perhaps the most direct experimental evidence at hand that supported the chemiosmotic hypothesis of Mitchell (1963). The experimental advantage offered by *N. crassa* that was exploited by these workers if the large hyphal cell size; this allowed insertion of microelectrodes, so that direct measurements of membrane potential could be made. The following experimental observations were made: (1) normal hyphae or carbon-starved hyphae maintained a large resting membrane potential (-180 to -250 mV); (2) addition of respiratory inhibitors resulted in a drop in both cellular ATP level and membrane potential; the rate of drop in membrane potential was in close correspondence to the drop in ATP level, indicating a dependence of membrane potential on cellular ATP; (3) there was a strong positive correlation between an outward proton flux, as measured by a pH electrode, and membrane potential; (4) addition of glucose or 3-O-methyl-D-glucose to carbon-starved cells (derepressed for the glucose active transport system) showed an immediate large drop in membrane potential; there was no significant drop in membrane potential when these sugars were added to repressed hyphae; (5) measurements of depolarization as a function of glucose or 3-O-methyl-D-glucose concentration showed saturation kinetics, with $K_{1/2}$ values that were in good agreement with the respective K_m values for transport of these two sugars; (6) the rapid depolarization brought about by addition of these two sugars were not accompanied by large drops in cellular ATP; thus, the rapid decrease in membrane potential was not due to depletion of ATP accompanying sugar phosphorylation; (7) addition of sugars to derepressed cells brought about a transient increase in extracellular pH, indicating an inward flux of protons during sugar transport. This effect was clearest with the nonmetabolizable sugars 3-O-methyl-D-glucose and 2-DOG; results with glucose were complicated, as expected, by an outward flux of protons occasioned by the metabolism of this sugar, which followed an initial small inward proton flux. All these findings are in accord with the notion that the cytoplasmic membrane of *Neurospora* contains an electrogenic pump energized by ATP via a proton-translocating ATPase, and that the active transport of glucose and its analogues take place by cotransport with H^+ ions under the influence of the membrane potential generated by the pump.

Evidence that a cytoplasmic membrane ATPase was involved in the generation of a membrane potential was supplied by Scarborough (1976), who showed that isolated cytoplasmic membrane vesicles prepared from N. crassa accumulated the permeant anion thiocyanate in the presence of Mg^{2+} and ATP. It was concluded that those vesicles in the preparation that were everted, and therefore presented the cytoplasmic side of the membrane on their external surfaces, generated a membrane potential, interior positive in this case, in response to the presence of ATP. Scarborough (1980) subsequently identified H^+ ions as the charged species in this electrogenic pump by the demonstration that addition of Mg^{2+} and ATP to the isolated membrane vesicles resulted in the concentration of imidazole, a base that readily penetrates the membrane in its uncharged form, but does not in its protonated form. Thus, hydrolysis of ATP brought about an inward flux of H^+ ions, which resulted in an accumulation of protonated imidazole to an extent corresponding to a difference of two pH units between internal and external solutions.

The Neurospora cytoplasmic membrane proton-translocating ATPase has now been studied extensively, and has been found to have properties in common with the H^+-ATPases of other fungi that have been investigated (Schizosaccharomyces pombe and Saccharomyces cerevisiae). This subject has been reviewed thoroughly by Goffeau and Slayman (1981). These fungal H^+-ATPases are quite different from the F_0F_1 H^+-ATPases of mitochondira, chloroplasts, and bacteria in terms of structure, nucleotide specificity, pH optimum, and sensitivity to inhibitors. Rather, they are more similar to the Na^+,K^+-ATPase, Ca^{2+}-ATPase, and H^+/K^+-ATPase of animal cells in that they contain a single subunit of molecular weight 100,000 (in contrast to the several polypeptides of molecular weight 7000–60,000 of the F_0F_1 ATPase), and are sensitive to vanadate inhibition.

It has been estimated that this proton-translocating ATPase accounts for 5–10% of the total protein of the N. crassa cytoplasmic membrane, and that it consumes approximately 20% of the ATP produced by the cell (Bowman et al., 1980). Thus, it plays a central role in the physiology of the cell.

The overall similarity of the H^+-ATPases of N. crassa, Schizosaccharomyces pombe, and Saccharomyces cerevisiae indicates that this enzyme is probably a common feature of all fungi, and represents the catalyst for the operation of the electrogenic proton pump that energizes the active transport of sugars and other substances in all fungi. However, as pointed out by Goffeau and Slayman (1981), all these fungi studied are ascomycetes, and it may be hazardous to extrapolate to other fungal groups before they are investigated.

REFERENCES

Biely, P., and Bauer, S., 1966, The formation of uridine diphosphate-2-deoxy-D-glucose in yeast, *Biochim. Biophys. Acta* **121**:213–214.

Biely, P., and Bauer, S., 1968, The formation of guanosine diphosphate-2-deoxy-D-glucose in yeast, *Biochim. Biophys. Acta* **156**:432–434.

Bisson, L. F., and Fraenkel, D. G., 1983a, Involvement of kinases in glucose and fructose uptake by *Saccharomyces cerevisiae*, *Proc. Natl. Acad. Sci. USA* **80**:1730–1734.

Bisson, L. F., and Fraenkel, D. G., 1983b, Transport of 6-deoxyglucose in *Saccharomyces cerevisiae*, *J. Bacteriol.* **155**:995–1000.

Bowman, B. J., Blasco, F., Allen, K. E., and Slayman, C. W., 1980, Plasma-membrane ATPase of *Neurospora*: Purification and properties, in: *Plant Membrane Transport: Current Conceptual Issues* (R. M. Spanswick, W. J. Lucas, and J. Dainty, eds.), pp. 195–209, Elsevier/North-Holland, Amsterdam.

Brocklehurst, R., Gardner, D., and Eddy, A. A., 1977, The absorption of protons with α-methyl glucoside and α-thioethyl glucoside by the yeast N.C.Y.C. 240, *Biochem. J.* **162**:591–599.

Brown, C. E., and Romano, A. H., 1969, Evidence against necessary phosphorylation during hexose transport in *Aspergillus nidulans*, *J. Bacteriol.* **100**:1198–1203.

Burger, M., Hejmova, L., and Kleinzeller, A., 1959, Transport of some monosaccharides into yeast cells, *Biochem. J.* **71**:233–242.

Cirillo, V. P., 1961a, The transport of non-fermentable sugars across the yeast cell membrane, in: *Membrane Transport and Metabolism* (A. Kleinzeller and A. Kotyk, eds.), pp. 343–351, Academic Press, New York.

Cirillo, V. P., 1961b, The mechanism of sugar transport into the yeast cell, *Trans. N.Y. Acad. Sci. Ser. II* **23**:725–734.

Cirillo, V. P., 1961c, Sugar transport in microorganisms, *Annu. Rev. Microbiol.* **15**:197–218.

Cirillo, V. P., 1962, Mechanism of glucose transport across the yeast cell membrane, *J. Bacteriol.* **84**:485–491.

Cirillo, V. P., 1968a, Relationship between sugar structure and competition for the sugar transport system in bakers' yeast, *J. Bacteriol.* **95**:603–611.

Cirillo, V. P., 1968b, Galactose transport in *Saccharomyces cerevisiae*. I. Nonmetabolized sugars as substrates and inducers of the galactose transport system, *J. Bacteriol.* **95**:1727–1731.

Eddy, A. A., 1982, Mechanisms of solute transport in selected eukaryotic microorganisms, in: *Advances in Microbial Physiology*, Vol. 23 (A. H. Rose and J. G. Morris, eds.), pp. 1–78, Academic Press, New York.

Farkas, V., Bauer, S., and Zemek, J., 1969, Metabolism of 2-deoxy-D-glucose in baker's yeast. III. Formation of 2,2′-dideoxy-α,α′-trehalose, *Biochim. Biophys. Acta* **184**:77–82.

Franzusoff, A. J., and Cirillo, V. P., 1982, Uptake and phosphorylation of 2-deoxy-D-glucose by wild-type and single-kinase strains of *Saccharomyces cerevisiae*, *Biochim. Biophys. Acta* **688**:295–304.

Franzusoff, A. J., and Cirillo, V. P., 1983a, Glucose transport activity in isolated plasma membrane vesicles from *Saccharomyces cerevisiae*, *J. Biol. Chem.* **258**:3608–3614.

Franzusoff, A. J., and Cirillo, V. P., 1983b, Solubilization and reconstitution of the glucose transport system from *Saccharomyces cerevisiae*, *Biochim. Biophys. Acta* **734**:153–159.

Fuhrmann, G. F., and Rothstein, A., 1968a, The transport of Zn^{2+}, Co^{2+}, and Ni^{2+} into yeast cells, *Biochim. Biophys. Acta* **163**:325–330.

Fuhrmann, G. F., and Rothstein, A., 1968b, The mechanism of the partial inhibition of fermentation in yeast by nickel ions, *Biochim. Biophys. Acta* **163**:331–338.

Gill, C. O., and Ratledge, C., 1973, Inhibition of glucose assimilation and transport by *n*-decane and other *n*-alkanes in *Candida* 107, *J. Gen. Microbiol.* **75**:11–22.

Goffeau, A., and Slayman, C. W., 1981, The proton-translocating ATPase of the fungal plasma membrane, *Biochim. Biophys. Acta* **639**:197–223.

Harris, G., and Thompson, C. C., 1961, The uptake of nutrients by yeasts. II. The maltose permease of a brewing yeast, *Biochim. Biophys. Acta* **52**:176–183.

Heredia, C. F., Sols, A., and De La Fuente, G., 1968, Specificity of the constitutive hexose transport in yeast, *Eur. J. Biochem.* **5**:321–329.

Holligan, P. M., and Jennings, D. H., 1973, Carbohydrate metabolism in the fungus *Dendryphiella salina*. IV. Acetate assimilation, *New Phytol.* **72**:315–319.

Jaspers, H. T. A., and Van Steveninck, J., 1975, Transport-associated phosphorylation of 2-deoxy-D -glucose in *Saccharomyces fragilis*, *Biochim. Biophys. Acta* **406**:370–385.

Jennings, D. H., 1974, Sugar transport into fungi: An essay, *Trans. Br. Mycol. Soc.* **62**:1–24.

Jennings, D. H., and Austin, S., 1973, The stimulatory effect of the non-metabolized sugar 3-*O*-methyl-glucose on the conversion of mannitol and arabitol to polysaccharide and other insoluble compounds in the fungus *Dendryphiella salina*, *J. Gen. Microbiol.* **75**:287–294.

Klingmüller, W., 1967a, Aktive Aufnahme von Zuckern durch Zellen von *Neurospora crassa* unter Beteiligung eines enzymatischen Systems mit Permeaseeigenschaften. I, *Z. Naturforsch.* **22B**:181–187.

Klingmüller, W., 1967b, Aktive Aufnahme von Zuckern von *Neurospora crassa* unter Beteiligung eines enzymatischen Systems mit Permease-Eigenschaften. II, *Z. Naturforsch.* **22B**:188–195.

Klingmüller, W., and Huh, H., 1972, Sugar transport in *Neurospora crassa*, *Eur. J. Biochem.* **25**:141–146.

Kotyk, A., 1961, The effect of oxygen on transport phenomena in a respiration-deficient mutant of baker's yeast, in: *Membrane Transport and Metabolism* (A. Kleinzeller and A. Kotyk, eds.), pp. 352–360, Academic Press, New York.

Kotyk, A., 1967, Properties of the sugar carrier in baker's yeast. II. Specificity of transport, *Folia Microbiol.* **12**:121–131.

Kotyk, A., and Michaljaničová, D., 1974, Nature of the uptake of D-galactose, D-glucose and α-methyl-D-glucoside by *Saccharomyces cerevisiae*, *Biochim. Biophys. Acta* **332**:104–113.

Kotyk, A., Michaljaničová, D., Veres, K., and Soukupova, V., 1975, Transport of 4-deoxy- and 6-deoxy-D-glucose in baker's yeast, *Folia Microbiol.* **20**:496–503.

Kuo, S. C., and Cirillo, V. P., 1970, Galactose transport in *Saccharomyces cerevisiae*. III. Characteristics of galactose uptake in transferaseless cells: Evidence against transport-associated phosphorylation, *J. Bacteriol.* **103**:679–685.

Kuo, S. C., Christensen, M. S., and Cirillo, V. P., 1970, Galactose transport in *Saccharomyces cerevisiae*. II. Characteristics of galactose uptake and exchange in galactokinaseless cells, *J. Bacteriol.* **103**:671–678.

LeFevre, P. G., 1961, Sugar transport in the red blood cell: Structure activity relationships in substrate and antagonists, *Pharmacol. Rev.* **13**:39–70.

Lobo, Z., and Maitra, P. K., 1977a, Genetics of yeast hexokinase, *Genetics* **86**:727–744.

Lobo, Z., and Maitra, P. K., 1977b, Resistance to 2-deoxyglucose in yeast: A direct selection of mutants lacking glucose-phosphorylating enzymes, *Mol. Gen. Genet.* **157**:297–300.

Lobo, Z., and Maitra, P. K., 1977c, Physiological role of glucose-phosphorylating enzymes in *Saccharomyces cerevisiae*, *Arch. Biochem. Biophys.* **182**:639–645.

Mark, C. G., and Romano, A. H., 1971, Properties of the hexose transport systems of *Aspergillus nidulans*, *Biochim. Biophys. Acta* **249**:216–226.

Meredith, S. A., and Romano, A. H., 1977, Uptake and phosphorylation of 2-deoxy-D-glucose by wild type and respiration-deficient bakers' yeast, *Biochim. Biophys. Acta* **497**:745–759.

Mitchell, P., 1963, Molecule, group and electron translocation through natural membranes, *Biochem. Soc. Symp.* **22**:142–168.

Morgan, M. J., and Kornberg, H. L., 1969, Regulation of sugar accumulation in *Escherichia coli*, *FEBS Lett.* **3**:53–56.

Neville, M. M., Suskind, S. R., and Roseman, S., 1971, A derepressible active transport system for glucose in *Neurospora crassa*, *J. Biol. Chem.* **246**:1294–1301.

Okada, H., and Halvorson, H. O., 1964a, Uptake of α-thioethyl D-glucopyranoside by *Saccharomyces cerevisiae*. I. The genetic control of facilitated diffusion and active transport, *Biochim. Biophys. Acta* **82**:538–546.

Okada, H., and Halvorson, H. O., 1964b, Uptake of α-thioethyl D-glucopyranoside by *Saccharomyces cerevisiae*. II. General characteristics of an active transport system, *Biochim. Biophys. Acta* **82**:547–555.

Robertson, J. J., and Halvorson, H. O., 1957, The components of maltozymase in yeast, and their behavior during deadaptation, *J. Bacteriol.* **73**:186–198.

Romano, A. H., 1973, Properties of the sugar transport systems in *Aspergillus nidulans* and their regulation, in: *Genetics of Industrial Microorganisms: Actinomycetes and Fungi* (Z. Vanek, Z. Hostalek, and J. Cudlin, eds.), pp. 195–212, Acdemia, Prague.

Romano, A. H., 1982, Facilitated diffusion of 6-deoxy-D-glucose in bakers' yeast: Evidence against phosphorylation-associated transport of glucose, *J. Bacteriol.* **152**:1295–1297.

Romano, A. H., and Kornberg, H. L., 1968, Regulation of sugar uptake in *Aspergillus nidulans*, *Proc. R. Soc. (London) Ser. B* **173**:475–490.

Rothstein, A., 1954, Enzyme systems of the cell surface involved in the uptake of sugars by yeast, *Symp. Soc. Exp. Biol.* **8**:165–201.

Scarborough, G. A., 1970a, Sugar transport in *Neurospora crassa*, *J. Biol. Chem.* **245**:1694–1698.

Scarborough, G. A., 1970b, Sugar transport in *Neurospora crassa*. II. A second glucose transport system, *J. Biol. Chem.* **245**:3985–3987.

Scarborough, G. A., 1976, The *Neurospora* plasma membrane ATPase is an electrogenic pump, *Proc. Natl. Acad. Sci. USA* **73**:1485–1488.

Scarborough, G. A., 1980, Proton translocation catalyzed by the electrogenic ATPase in the plasma membrane of *Neurospora*, *Biochemistry* **19**:2925–2931.

Schneider, R. P., and Wiley, W. R., 1971a, Kinetic characteristics of the two glucose transport systems in *Neurospora crassa*, *J. Bacteriol.* **106**:479–486.

Schneider, R. P., and Wiley, W. R., 1971b, Regulation of sugar transport in *Neurospora crassa*, *J. Bacteriol.* **106**:487–492.

Schneider, R. P., and Wiley, W. R., 1971c, Transcription and degradation of messenger ribonucleic acid for a glucose transport system in *Neurospora*, *J. Biol. Chem.* **246**:4784–4789.

Seaston, A., Inkson, C., and Eddy, A. A., 1973, The absorption of protons with specific amino acids and carbohydrates by yeast, *Biochem. J.* **134**:1031–1043.

Serrano, R., 1977, Energy requirements for maltose transport in yeast, *Eur. J. Biochem.* **80**:97–102.

Serrano, R., and DelaFuente, G., 1974, Regulatory properties of the constitutive hexose transport in *Saccharomyces cerevisiae*, *Mol. Cell Biochem.* **5**:161–171.

Slayman, C. L., and Slayman, C. W., 1974, Depolarization of the plasma membrane of *Neurospora* during active transport of glucose: Evidence for a proton-dependent cotransport system, *Proc. Natl. Acad. Sci. USA* **71**:1935–1939.

Slayman, C. W., and Slayman, C. L., 1975, Energy coupling in the plasma membrane of *Neurospora*: ATP-dependent proton transport and proton-dependent sugar cotransport, in: *Molecular Aspects of Membrane Phenomena* (H. R. Kaback, H. Neurath, G. K. Radda, R. Schwyzer, and W. R. Wiley, eds.), pp. 233–248, Springer-Verlag, Berlin.

Sols, A., 1967, Regulation of carbohydrate transport and metabolism in yeast, in: *Aspects of Yeast Metabolism* (A. K. Mills and H. Krebs, eds.), pp. 47–66, Blackwell, Oxford.

Van Den Broek, P. A., and Van Steveninck, J., 1982, Kinetic analysis of H$^+$/methyl-β-D-thiogalactoside symport in *Saccharomyces fragilis*, *Biochim. Biophys. Acta* **693**:213–220.

Van Steveninck, J., 1968, Transport and transport-associated phosphorylation of 2-deoxy-D-glucose in yeast, *Biochim. Biophys. Acta* **163**:386–394.

Van Steveninck, J., 1970, The transport mechanism of α-methylglucoside in yeast: Evidence for transport-associated phosphorylation, *Biochim. Biophys. Acta* **203**:376–384.

Van Steveninck, J., 1972, Transport and transport-associated phosphorylation of galactose in *Saccharomyces cerevisiae*, *Biochim. Biophys. Acta* **274**:575–583.

Van Steveninck, J., and Rothstein, A., 1965, Sugar transport and metal binding in yeast, *J. Gen. Physiol.* **49**:235–246.

Wilkins, P. O., and Cirillo, V. P., 1965, Sorbose counterflow as a measure of intracellular glucose in baker's yeast, *J. Bacteriol.* **90**:1605–1610.

8

Carbohydrate Metabolism in Yeast

JUANA M. GANCEDO

1. INTRODUCTION

The unraveling of the pathways of carbohydrate metabolism has its origins in the early observations of Buchner on fermentation of sugar by a yeast extract and on the pioneering investigations of Harden and Young (for a review see Fruton, 1972). Through the years, yeasts have remained one of the favorite organisms for studying carbohydrate metabolism and its regulation. In fact, yeasts present a number of features that make them most convenient to study. They are unicellular organisms, easily handled, usually nonpathogenic, able to grow on a variety of carbon sources, and yielding the large amounts of homogeneous material that are often required for enzymological studies. In addition, yeasts are well amenable to classical genetic techniques and can also be used in the genetic engineering field. Finally, yeasts are eukaryotic cells and as such should be useful for studying biological problems that are peculiar to eukaryotic organisms.

Although the greatest part of the work done with yeasts has employed the facultative fermentative *Saccharomyces cerevisiae,* there are many different species of yeasts with considerable variety in their metabolism. In this chapter, carbohydrate metabolism in *S. cerevisiae* will be reviewed with occasional references to the situation in other yeasts. In a first part I will give an idea of the methodological approaches that have been used or are at present being developed in the field. The second part will consist of an overview of the pathways of carbohydrate metabolism in yeasts. Aspects that have been reviewed extensively elsewhere (Sols *et al.,* 1971; Fraenkel, 1982; Gancedo and Serrano, in press)

JUANA M. GANCEDO • Instituto de Investigaciones Biomédicas, Consejo Superior de Investigaciones Científicas, Facultad de Medicina de la Universidad Autónoma, 28029 Madrid, Spain.

will not be covered in detail. In the third and last part I will discuss regulatory mechanisms that modulate carbohydrate metabolism with particular emphasis on catabolite repression and catabolite inactivation.

2. METHODOLOGICAL APPROACHES

The conclusions that can be drawn on the operation and regulation of metabolic pathways are, to a large extent, dependent on the methodology available for their study and limited by it. It is thus worthwhile to examine critically the most important experimental approaches that have been used to reach the actual state of knowledge on carbohydrate metabolism in yeasts, although many of the methods may not be specific for yeasts.

2.1. Determination of Enzymatic Activities

A first step in evaluating the role of a given pathway in metabolism is the assay of the enzymatic activities that catalyze the different reactions of the pathway. The assays are usually performed in cell-free extracts; in these conditions the enzymes are required to work in a macromolecular environment that is much less structured than that of the intact cell (Fulton, 1983). To test if this difference in macromolecular environment affected the behavior of yeast enzymes, methods have been devised for permeabilizing the yeast membrane (Bechet and Wiame, 1965; Schlenk and Zydek-Cwick, 1970; Serrano et al., 1973). These methods allow the free circulation of substrates and products between the outside and the inside of the cell without diluting the macromolecular components of the cell. Using such an in situ assay, the enzymes of the glycolytic pathway have been studied, comparing the results obtained with those found in the usual in vitro assay (Bañuelos and Gancedo, 1978). Although no major differences have been found, the in situ approach could be useful in some special cases where the conformation of an enzyme depends on its environment (Weitzman and Hewson, 1973). Also, it should always be kept in mind that some enzymatic activity, significant in vivo, can get lost in the cell-free extract. An illustrative example is that of yeast mutants that grow on glucose, but lack phosphofructokinase activity in cell-free extracts (Clifton et al., 1978). When phosphofructokinase is assayed in toluene lysates, the yeast is found to possess a particulate enzyme that catalyzes the formation of fructose-1,6-bisphosphate from fructose-6-phosphate and ATP (Lobo and Maitra, 1982a).

The situation of an enzyme that is taken outside the cell does not only change with regard to protein–protein interactions; it should also be considered that the activity of many regulatory enzymes is dependent on the presence of effectors. Therefore, it may be that the activity of an enzyme measured in vitro in some set of conditions does not account for the metabolic flux in the corresponding pathway observed *in vivo*. The addition to the assay of known effectors at a concentration similar to that found in the living cell (see below) can increase considerably the activity. That has been the case with yeast phosphofructokinase where the addition of 10 mM phosphate to an otherwise physiological mixture increased activity about 100-fold (Bañuelos *et al.*, 1977).

Another point that can be made is the following: to study the kinetic properties of an enzyme, it has been recommended that the enzyme first be purified: "First purify, then think, don't waste clean thinking on dirty things" (Racker, 1961). However, in the course of the purification, the enzyme can be modified and this danger is particularly acute in yeast where proteolytic activities are abundant (Pringle, 1975). Many an isozyme has been found after careful study to be a product of limited proteolysis. In the course of a purification where only activity in optimal conditions is monitored, it happens (and here also phosphofructokinase is a good example) that allosteric properties disappear.

As a conclusion, it can be said that although the detection of an adequate activity "*in vitro*" is strong evidence for the implication of an enzyme in a metabolic pathway, the absence of such evidence does not necessarily imply that the enzyme is not active "*in vivo*" and pains should be taken for assaying the enzyme in conditions approaching the physiological as closely as possible. In this connection the existence of yet unidentified effectors should not be overlooked. An example would be the recently discovered fructose-2,6-bisphosphate (Hers and Van Schaftingen, 1982), which is an activator of yeast phosphofructokinase (Bartrons *et al.*, 1982) and an inhibitor of yeast fructose bisphosphatase (Gancedo *et al.*, 1982) and could eventually affect the activity of other enzymes involved in carbohydrate metabolism.

2.2. Determination of Metabolite Levels

In the preceding section it was pointed out that for assessing the physiological importance of an enzymatic reaction and eventually its regulatory role, it was important to test the enzyme at the concentration of substrates and effectors found in the cell. That raises the problem of determining such concentrations.

The methodology that is most commonly used consists in arresting yeast metabolism, breaking open the cells, extracting their content, and assaying the different metabolites, usually by enzymatic methods (Bergmeyer, 1974). A compilation of the data available in 1972 has been made (Gancedo and Gancedo, 1973); however, the methods of yeast sampling used by different authors varied widely and did not always yield reliable results. A careful study has been made in an attempt to find the best conditions for yeast sampling (Saez and Lagunas, 1976). It appears that, at least for metabolites that have a high turnover, centrifugation of the yeast culture is not adequate and that rapid filtration followed by freezing in liquid nitrogen is to be preferred. Extraction of the cell contents can then be performed with perchloric or tricholoracetic acid, using two or three cycles of freezing and thawing to ensure a full breakage of the yeast cells. Using such a method the concentration of some glycolytic intermediates has been measured in yeasts, in different metabolic conditions (Saez and Lagunas, 1976; Bañuelos et al., 1977; Lagunas and Gancedo, 1983).

At this point a cautionary note should be given. Metabolites are not always distributed homogeneously in the cell, but can be found preferentially in some compartment of the cell. For example, in yeasts, vacuoles are known to accumulate polyphosphates (Indge, 1968; Shabalin et al., 1977) and amino acids (Wiemken and Durr, 1974). Mitochondria too are specialized organelles and should possess a peculiar metabolite distribution. When metabolites are predominantly cytoplasmic as occurs for glycolytic intermediates, the concentrations calculated assuming a homogeneous distribution should not be very different from the actual values. In contrast, if the concentration of a metabolite such as phosphate is measured in whole cells, for calculating the concentration of phosphate present in the cytoplasm it is very important to take into account that phosphate is accumulated in the vacuoles (Okorokov et al., 1980). Another fact that has been stressed (Sols and Marco, 1970) is the binding of metabolites to proteins. When a metabolite is present in low amounts and the possible binding sites are abundant, the free concentration of the metabolite (which is the variable to take into account for understanding regulatory phenomena) can be much lower than the total concentration.

A completely different methodology can be used for monitoring intracellular concentrations of metabolites in a suspension of whole cells. That is the technique of high-resolution nuclear magnetic resonance (NMR) using either phosphorus-31 (Salhany et al., 1975) or carbon-13 (den Hollander et al., 1979). To date, the NMR spectra have had a limited resolution; in most cases, therefore, it is not possible to calculate the actual concentration of metabolites in the cell and only qualitative changes corresponding to different metabolic situations can be recorded (Navon et al., 1979; den Hollander et al., 1981). There are, however, some exceptions to this rule; for example, trehalose concentrations have been determined using natural abundance ^{13}C NMR spectroscopy (Thevelein et al., 1982) and the yeast internal pH has been measured using the chemical shifts of phosphate and different phosphorylated metabolites (den Hollander et al., 1981). In this last case, NMR spectroscopy has an added advantage; in the chemical shift region corresponding to inorganic phosphate, three different peaks can be recorded (Navon et al., 1979). One of the peaks corresponds to external phosphate, the second was assumed to be the cytoplasmic phosphate since its position agreed within 0.2 pH unit with that calculated from the chemical shifts of the sugar phosphates, and the third was thought to be phosphate inside one of the organelles with an acidic pH. This organelle is probably the vacuole since respiring mitochondria should have a higher pH than the cytoplasm (Mitchell and Moyle, 1969). From the second and third phosphate peaks, it is then possible to calculate the pH both in the cytoplasm and in the vacuole. In most conditions the intracellular pH of the yeast is in the range 7.1–7.4 (den Hollander *et al.*, 1981). These values are higher than those obtained using other methods (Kotyk, 1963; Borst-Pauwels and Dobbelmann, 1972). It appears likely, however, that the NMR values are the most accurate since the other methods would give an average of the pH of the cytoplasm and of other organelles. Recently, a fluorescent probe has been used to study the distribution of pH inside the giant yeast *Endomyces magnusii;* the pH of the cytoplasm was 6.7–7.2, decreasing to 6.0 toward the periphery of the cell (Slavik, 1983).

2.3. Fate of Labeled Substrates: Calculation of the Proportion of Carbohydrates Used through Different Pathways

Carbohydrates can be metabolized in yeasts through alternative pathways. An approach for elucidating what proportion of a sugar is degraded through each pathway could be to compare the activities of the enzymes of each pathway. However, as we have seen before, it can often be difficult to know at what rate an enzyme would work *in vivo* and therefore better procedures had to be devised.

Some methods are limited to particular cases. For example, a pathway can have a unique end product that is excreted to the medium and can be determined. That occurs for instance with the pathways leading to ethanol, to glycerol, or to succinate. Or alternatively a pathway may be the only one to consume some compound, whose utilization can be measured. That would be the case for the products of sugar metabolism that are finally oxidized in the citric acid cycle consuming O_2 whose amount can be measured either with the classical manometric method of Warburg or by an oxygen electrode. Although limited, such simple methods have proved very useful (Pasteur, 1860) and are still being used (Lagunas and Gancedo, 1973; Oura, 1977).

A more general and powerful methodology involves the use of substrates labeled in specific carbons. Often, different pathways can give the same final product (e.g., pyruvate or CO_2); however, depending on the pathway used, different amounts of radioactivity will be present in the product. In the case of carbohydrate metabolism, this method has been mainly used for determining the amount of sugar channeled through the pentose phosphate pathway. As discussed by Gancedo and Lagunas (1973), the procedures utilized depend on some simplifying assumptions. To obtain meaningful results, it is necessary to state clearly these simplifications and to assess how far they do hold in the particular system under study. In the case of *S. cerevisiae*, the contribution of the pentose phosphate pathway to glucose metabolism can be calculated using glucose specifically labeled in C-1 or C-6. The CO_2-specific yields $G1CO_2$ and $G6CO_2$

(ratios of $^{14}CO_2$ formed to [1-^{14}C]glucose utilized and of $^{14}CO_2$ formed to [6-^{14}C]glucose utilized, respectively) are then calculated and the procedure developed by Katz and Wood (1963) is applied. Using this method, Gancedo and Lagunas (1973) found that the proportion of glucose transformed into glyceraldehyde-3-phosphate by the pentose phosphate pathway ranged from 1 to 7% depending on the nitrogen source and the level of aeration.

It is important to stress the necessity of using different approaches for evaluating metabolic pathways. For example, several methods have been applied to the study of glucose catabolism in the oxidative yeast *Rhodotorula glutinis,* and since all of them gave results consistent with the presence of an operative Embden–Meyerhof pathway, the pathway can be considered established (Mazón *et al.*, 1974a,b). On the other hand, the distribution of radioactivity in alanine isolated from *Rhodotorula* grown on labeled glucose suggests the operation of an Entner–Doudoroff pathway (M. J. Mazón and J. M. Gancedo, unpublished results). However, since the corresponding enzymes could not be detected, the operation of the Entner–Doudoroff pathway in a eukaryotic organism remains uncertain. Similarly, it has been postulated that in *S. cerevisiae* mutants lacking phosphofructokinase, glucose is metabolized by an alternative pathway (Breitenbach-Schmitt, 1981; Zimmermann, 1982). Metabolite measurements would suggest that sedoheptulose-7-phosphate is an intermediary of the pathway, but other approaches should be used before this alternative pathway is definitively established.

2.4. Continuous Cultures versus Batch Cultures

Studies on carbohydrate metabolism in yeasts have usually been carried out using batch cultures. This method offers the great advantage of its technical simplicity but is not fully adequate for investigating the behavior of yeasts in a physiological setting. Yeasts, in nature, would often have to thrive in the presence of limiting amounts of nutrients. In the laboratory, such a situation can best be mimicked using continuous cultures, where different nutrients can be made rate-limiting and where growth conditions are maintained constant for long periods (Herbert *et al.*, 1956; Melling, 1977). The setting of chemostat cultures is, however, technically demanding and it is significant that the few papers reporting work on continuous cultures of yeasts have appeared mostly in journals devoted to biotechnology. An intermediary procedure, the semicontinuous culture (Katz *et al.*, 1971), has therefore been developed. In the initial stages the culture is not diluted but allowed to grow from an inoculum as a batch culture; when a suitable density is reached, dilution is started until steady-state chemostat conditions are attained. However, even this simplified technique has not been often used.

The main conclusion that can be reached from experiments using continuous cultures is that some phenomena that had been described in batch cultures, no longer occur in chemostat cultures where the concentrations of nutrients can be severely limited. For example, at the very low glucose concentrations attainable in a chemostat, repression of respiration is not apparent (Oura, 1974; Barford and Hall, 1979). Conversely, in *S. cerevisiae* the Pasteur effect is not observed in batch cultures but is well marked in glucose-limited continuous cultures (see a later section for more details).

Regulation of invertase is also affected by the culture conditions. Derepression of invertase at low glucose concentrations is found in continuous cultures but does not occur in a batch culture even after exhaustion of glucose when the yeast is utilizing the ethanol accumulated during fermentation (Toda, 1976). It has also been reported that in a phosphate-limited continuous culture, invertase levels were increased tenfold with respect to those attained in the presence of excess phosphate while no such effect of phosphate was observed in batch cultures (Toda *et al.*, 1982).

A possible application of continuous cultures that apparently has not been put to use in yeasts is the selection of mutants able to utilize novel carbon sources. An exploration along this line would probably be worthwhile taking into account the satisfactory results obtained with bacteria (Rigby *et al.*, 1974), and the fact that in copper-limited continuous cultures of *Candida utilis* it is easy to select variants with a decreased requirement for copper (Downie and Garland, 1973).

2.5. Use of Mutants

Traditionally the operativity of a metabolic pathway in an organism has been confirmed by the use of mutants impaired at different steps of the pathway. In the case of carbohydrate metabolism in yeasts, the unraveling of the pathways involved was achieved at a very early stage and the search for mutants started much later. In this review I will not describe in detail the procedures used and the results obtained but rather will indicate what kind of questions related to carbohydrate metabolism in yeasts can best be answered by the use of genetic techniques. I shall treat this topic in two parts, one concerned with classical genetics and the other dealing with the newer techniques of genetic engineering, although the distinction is not really clear-cut in all cases.

2.5.1. Classical Genetics

If a given metabolic pathway is the only one that allows the utilization of a sugar, mutants that have lost any of the enzymes of the pathway will be unable to use this sugar as carbon source for growth. In addition, if the regulatory properties of an enzyme are crucial for the proper functioning of a pathway, the loss of these properties would impair growth even if the catalytic function of the enzyme is retained. Finally, the isolation of mutants affected in regulatory genes would help to delineate the mechanism(s) that controls the expression of the structural genes of the pathways.

The most powerful screening procedure for the isolation of mutants is that that allows growth of the mutant in conditions where the wild type will not grow. It is logical that the first glycolytic mutants described were a mutant lacking one isozyme of alcohol dehydrogenase and therefore resistant to allyl alcohol (Megnet, 1965; Lutstorf and Megnet, 1968) and a mutant lacking hexokinase activity and selected by its resistance to 2-deoxyglucose (Maitra, 1970).

The next step was to look for mutants unable to grow on glucose. This allowed the isolation of mutants impaired in most steps between glucose and ethanol (Maitra, 1971; Sprague, 1977; Lam and Marmur, 1977; Clifton *et al.*, 1978; Ciriacy and Breitenbach, 1979). The characteristics of these mutants have given the following information on the operation of the pathway. First, glucose does not support growth if the phosphoglucose isomerase step is blocked (Maitra, 1971; Herrera and Pascual, 1978). This behavior is in contrast to that observed in *E. coli* where the pentose phosphate pathway allows the slow growth on glucose of pgi⁻ mutants (Fraenkel and Vinopal, 1973). Second, the addition of glucose to an otherwise permissive medium blocks the growth of mutants lacking phosphoglucose isomerase (Maitra, 1971), triose phosphate isomerase (Ciriacy and Breitenbach, 1979), phosphoglycerate kinase, phosphoglycerate mutase (Lam and Marmur, 1977), and pyruvate kinase (Sprague, 1977; Clifton *et al.*, 1978). This growth inhibition is likely due to ATP depletion caused by the initial step(s) of glucose utilization. It would indicate that the yeast has no control mechanism for blocking glucose uptake and phosphorylation when its further metabolism is impaired.

Among the mutants unable to grow on glucose, no mutants were characterized lacking hexokinase, phosphofructokinase, aldolase, glyceraldehyde-3-phosphate dehydrogenase, or enolase. This fact could be due to the existence of several isozymes catalyzing the same step and that appears to indeed be the case for hexokinase, glyceraldehyde-3-phosphate dehydrogenase, or enolase (see Section 2.5.2). Mutants lacking phosphofructokinase could, however, be isolated using a different selection procedure. From pyruvate kinase-negative mutants unable to grow on a medium containing a gluconeogenic carbon source and glucose, glucose-resistant revertants were isolated (Clifton *et al.*, 1978). Some of these resistant strains lacked both pyruvate kinase and phosphofructokinase activity. When the phosphofructokinase mutation was segregated, mutants were obtained that lacked phosphofructokinase activity but grew normally on glucose as sole carbon source. Taken at face value, this result would imply that in yeasts there is a pathway for glucose metabolism able to bypass the phosphofructokinase reaction. However, the situation is a complicated one and will be discussed more thoroughly in Section 3.1.4 on phosphofructokinase.

As for mutants affected in regulatory properties, Maitra and Lobo (1977a) have described a mutant with a pyruvate kinase that no longer requires fructose-1,6-bisphosphate for activity. Apparently, the mutant can grow on gluconeogenic sources at a normal rate although it would have been expected that the unchecked activity of pyruvate kinase in gluconeogenic conditions would have impaired growth.

Most of the mutants isolated to date are probably affected in the structural genes; there is, however, a curious pleiotropic mutant that has decreased levels of several glycolytic enzymes (Clifton *et al.*, 1978). This mutation appears to affect a regulatory gene that controls the mRNA levels for different glycolytic enzymes (Clifton and Fraenkel, 1981).

Regulatory mutations that affect the synthesis of alcohol dehydrogenase have been well characterized; they will be briefly discussed in the next section. There also are a number of regulatory mutations that affect catabolite repression and these will be dealt with in Section 4.1.

2.5.2. Genetic Engineering

A powerful tool for studying the control of gene expression is the cloning of genes and the use of techniques of genetic engineering. The enzymes of the glycolytic pathway in yeasts appear particularly amenable to this kind of analysis since they constitute an important proportion of the protein synthesized by the cell. Therefore, it is possible to clone genes not only in the case where mutants have been isolated and complementation techniques can be applied, but also in cases where no mutants are available.

For example, a yeast glyceraldehyde-3-phosphate dehydrogenase (G3PD) gene was isolated by hybridization of randomly sheared segments of yeast DNA with a complementary DNA synthesized from partially purified G3PD mRNA (M. J. Holland and Holland, 1979). It was later found that there are three G3PD genes in yeasts (Holland and Holland, 1980). The sequences of two of the genes and of their flanking regions have been determined (J. P. Holland and Holland, 1979, 1980) and it has been established that they are very similar and do not contain introns. In a similar fashion, the primary structures of two yeast enolase genes have been determined (Holland *et al.*, 1981). These results showed that there are regions of extensive homology between the two enolase and the two G3PD genes within the 5' noncoding portions. This would suggest that the structures are involved in the regulation of gene expression but this point remains to be demonstrated. The fact that there are multiple genes for these two enzymes would explain why it has not been possible to isolate mutants affected at these steps. An immunological screening technique has been applied to the isolation of the 3-phosphoglycerokinase gene and could have wider applications (Hitzeman *et al.*, 1980).

The majority of the genes of the glycolytic pathway have been cloned by complementation of defective mutants (Williamson *et al.*, 1980; Kawasaki and Fraenkel, 1982; Burke *et al.*, 1983; Walsh *et al.*, 1983). The transformants had enzyme levels up to 20-fold higher than the parental strains. Physiological studies of such transformants have yet to be carried out but they could yield useful information on the limiting steps and in general on the regulation of the glycolytic pathway.

Some information has already been gathered on the regulation of gene expression, using cloned genes. A few examples will be given. The utilization of galactose by yeasts requires the expression of several enzymes whose synthesis is induced by galactose and repressed by glucose. Classical genetic studies had shown that the product of a regulator gene activated the synthesis of these enzymes possibly by binding to DNA sequences at or near the 5' ends of the corresponding genes. The promoter region of the gene whose product is UDP-galactose epimerase has now been identified (Guarente *et al.*, 1982). It is a region more than 130 base pairs upstream of the transcriptional start site. This region can be linked to the gene for cytochrome *c* conferring galactose-specific control to the synthesis of cytochrome *c*. On the other hand, this DNA segment does not appear to contain the sequences responsible for glucose repression of UDP-galactose epimerase synthesis.

A sequence of DNA containing a locus implicated in glucose repression has been isolated (Beier and Young, 1982). It is located between 200 and 1000 base pairs upstream of the structural gene for one of the isozymes of alcohol dehydrogenase (alcohol dehydrogenase II). If this sequence is excised and ligated in front of the gene for the alcohol dehydrogenase I (the isozyme that is abundant during fermentative growth), this gene is put under glucose control.

The transcription of cytochrome *c* mRNA is dependent on the presence of oxygen. It is possible, however, by a rearrangement of an upstream sequence to place the transcription of the cytochrome *c* gene under the control of a different modulator, in such a way that cytochrome *c* synthesis is repressed by oxygen (Lowry *et al.*, 1983).

3. OVERVIEW OF THE PATHWAYS OF CARBOHYDRATE METABOLISM

The utilization of carbohydrates by yeasts is centered on the conversion of glucose-6-phosphate into pyruvate in the Embden–Meyerhof pathway. The initial reactions that give glucose-6-phosphate as a product depend on the sugar being used, while pyruvate can be further metabolized to give a variety of different end products whose relative proportion will depend on the type of yeast and on the metabolic conditions. There are in addition some collateral pathways, like the pentose phosphate pathway or the synthesis of reserve carbohydrates that fulfill special needs of the cell. In the following sections the different pathways involved in carbohydrate metabolism by yeast will be considered in turn.

3.1. Glycolytic Pathway

3.1.1. Glucose Uptake

The plasma membrane of yeasts is not freely permeable to sugars. Therefore, the utilization of glucose requires the operation of a specific transport system. In *S. cerevisiae*, glucose transport is equilibrative and is carried out by a constitutive permease that is also involved in the uptake of fructose, mannose, and a number of nonmetabolizable analogues (Kotyk, 1967; Heredia *et al.*, 1968; Cirillo, 1968a). In other yeasts, however, like species of *Rhodotorula* and *Candida*, glucose transport is of the active type, i.e., the concentration of sugar in the cell can exceed that present in the medium (Kotyk and Höfer, 1965; Höfer and Kotyk, 1968).

In spite of considerable study, the mode of operation of the glucose transport system of *Saccharomyces* has remained controversial. The mechanisms proposed have been basically a facilitated diffusion model (Cirillo, 1961) and a transport-associated phosphorylation model (Rothstein, 1954; Van Steveninck, 1968). Although the question is still not completely settled, the actual situation can be summarized as follows.

S. cerevisiae possesses at least two functionally different hexose uptake systems, with "low affinity" and "high affinity" for sugars (Serrano and De la Fuente, 1974; Bisson and Fraenkel, 1983). Although usually sugars that can be phosphorylated are transported by the low-K_m system and sugars that cannot be phosphorylated are transported with a much higher K_m, this rule is not absolute. Recently, Romano (1982) reported that 6-deoxyglucose, an analogue of glucose that cannot be phosphorylated, is transported by *S. cerevisiae* with an affinity equivalent to that for glucose. It appears therefore that the operation of the high-affinity carrier does not require the phosphorylation of the sugar. On the other hand, Bisson and Fraenkel (1983) have studied glucose and fructose transport in mutants lacking either hexokinase I and II or both hexokinases and glucokinase. In strains lacking both hexokinases, the low-K_m system for fructose was absent and in the triple mutant both glucose and fructose showed only the high-K_m uptake. It is clear therefore that the hexose kinases are somehow involved in hexose uptake although their role is not strictly related to their ability to phosphorylate the sugar transported (uptake of 6-deoxyglucose, which is not phosphorylated, is also affected by the absence of the kinases). The mode of action of the kinases remains to be elucidated; an attractive possibility would be that any of the kinases in the presence of a substrate (or analogue) could direct the conversion of the transport system from a low-affinity into a high-affinity state.

3.1.2. Glucose Phosphorylation

In *S. cerevisiae*, glucose can be phosphorylated by three different kinases whose characteristics are summarized in Table I. In contrast to what occurs in mammals, there is no feedback inhibition of phosphorylation by glucose-6-phosphate and it is not clear whether the kinases play a regulatory role in the utilization of glucose. As discussed in the previous section, this role could be indirect through modulation of the transport system.

Any- one of the kinases appears adequate to support growth on glucose (Lobo and Maitra, 1977) but a mutant lacking all three is completely unable to grow on glucose. There are therefore probably no other enzymes for initiating the metabolism of glucose. Although the relative amounts of hexokinase I and II vary depending on the growth conditions (Kopperschläger and Hoffmann, 1969; Gancedo *et al.*, 1977), it is not clear whether the different enzymes have specific physiological functions.

In *R. glutinis* a hexokinase and a glucokinase have been identified (Mazón *et al.*, 1975). Both enzymes are constitutive; the hexokinase is similar to hexokinase II from *S. cerevisiae*, while the glucokinase has a somewhat broader specificity than the corresponding enzyme from *S. cerevisiae*.

3.1.3. Phosphoglucose Isomerase

This enzyme catalyzes the interconversion between glucose-6-phosphate and fructose-6-phosphate. It has been reported (Kempe *et al.*, 1974) that in commercial dried brewer's yeast, three isozymes were present in the proportion 40 : 45 : 15. Although precautions were taken to avoid artifacts due to proteolysis or to oxidation of sulfhydryl groups, the existence of isozymes remains doubtful. In effect, it has been shown that mutations in a structural gene eliminate over

Table I. Characteristics of Yeast Hexokinases and Glucokinase

	Hexokinase I (or A)	Hexokinase II (or B)	Glucokinase[a]
V_{max} (fructose)/V_{max} (glucose)	3.0[b]	1.2[b]	<0.1
K_m (glucose)	0.2 mM[b]	0.4 mM[b]	0.03 mM
K_m (fructose)	0.9 mM[a]	1.4 mM[a]	10 mM[c]
Inhibition by glucose-6-phosphate	No[d]	No[d]	No
Repressed during logarithmic growth on glucose	Yes[e]	No[e]	No

[a]Lobo and Maitra (1977).
[b]Barnard (1975).
[c]Determined as K_i.
[d]Serrano *et al.* (1973).
[e]Gancedo *et al.* (1977).

99% of the enzymatic activity (Maitra and Lobo, 1977b). A possible explanation for the conflicting results would be that the commercial yeast is a diploid with slightly different allelic genes for phosphoglucose isomerase. As discussed in Section 2.5.1, mutants lacking phosphoglucose isomerase are unable to grow on glucose.

3.1.4. Phosphofructokinase

Phosphofructokinase is a complex enzyme whose activity can be modulated by a variety of effectors. The enzyme from *S. cerevisiae* is known to be allosterically inhibited by ATP (Viñuela *et al.*, 1963) and activated by AMP (Ramaiah *et al.*, 1964), NH_4^+ ions (Sols and Salas, 1966), phosphate (Bañuelos *et al.*, 1977), and fructose-2,6-bisphosphate (Bartrons *et al.*, 1982). The crucial role of phosphofructokinase in the control of the glycolytic pathway is supported by another kind of evidence: an analysis of glycolytic metabolites in a yeast extract sustaining glycolytic oscillations showed a crossover point at the reaction catalyzed by phosphofructokinase (Hess *et al.*, 1969). Therefore, the isolation of mutants lacking assayable phosphofructokinase but nonetheless able to grow on glucose was very perplexing. A complete explanation is not available but the present situation can be summarized as follows.

Mutants affected in phosphofructokinase activity have been isolated and classified in three complementation groups. Strains belonging to the pfk1 class do not show phosphofructokinase activity in cell-free extracts (Clifton *et al.*, 1978; Breitenbach-Schmitt, 1981), but appear to contain a particulate phosphofructokinase (Lobo and Maitra, 1982a). Strains belonging to the pfk2 class* do not show a homogeneous behavior; in some cases the activity of phosphofructokinase is similar to that of the wild type (mutation pfk2-2 of Clifton and Fraenkel, 1982) and no longer sensitive to ATP inhibition (Lobo and Maitra, 1982b). In other cases little activity (mutation of Navon *et al.*, 1979) or some (mutations of Breitenbach-Schmitt, 1981) is found. Lastly, the only pfk3 strain reported (Nadkarni *et al.*, 1982) appears to have a normal soluble phosphofructokinase and lacks the particulate form. From these results and from the behavior of the double mutants pfk1 pfk2 and pfk1 pfk 3 that no longer grow on glucose, the following conclusions have been derived (Nadkarni *et al.*, 1982). Yeasts possess two species of phosphofructokinase: a soluble species, which corresponds to the enzyme previously purified and characterized (Kopperschläger *et al.*, 1977; Lau-

---

*It should be noted that no complementation tests have been performed between most of the mutants isolated by different groups. However, the pfk2-1 mutation of Navon *et al.* (1979) and the pfk2-2 mutation of Clifton and Fraenkel (1982) belong to the same group and show a different behavior. The pfk2 mutants of Lobo and Maitra (1982a,b) probably belong to this group but the situation of the mutants of Breitenbach-Schmitt (1981) is less clear.

rent *et al.*, 1979), and a particulate one. The first species is an octamer $\alpha_4\beta_4$ whose α and β subunits would be determined by the PFK1 and PFK2 genes, respectively. The particulate species would also be formed by two kind of subunits, β and γ, specified by the PFK2 and PFK3 genes. It has been suggested that in yeasts the soluble phosphofructokinase is composed of separate catalytic and regulatory subunits (Laurent *et al.*, 1978; Tijane *et al.*, 1980). In these latter papers, α designates the regulatory subunit and β the catalytic one. Clifton and Fraenkel (1982) postulate also that pfk2 specifies β and that this subunit has a catalytic function. However, because pfk2 mutants show a phosphofructokinase activity that is no longer inhibited by ATP and because extracts from pfk1 mutants can restore the inhibition (Lobo and Maitra, 1982b), it is more likely that α is the catalytic subunit and β the regulatory one.

The particulate phosphofructokinase can only be assayed in toluene lysates. In the test conditions it has only about 25% of the activity of the soluble form and is not sensitive to ATP inhibition. The enzyme is only found in glucose-grown cells in the early logarithmic phase (Nadkarni *et al.*, 1982). It is hypothesized that the synthesis of the γ subunit is induced by glucose. The fact that pfk1 mutants grown on glucose until stationary phase no longer show subunit β in extracts while logarithmic-phase cultures do (Clifton and Fraenkel, 1982) would suggest that the β subunit is easily degraded unless protected by either α or γ subunits.

The hypothesis of Nadkarni *et al.* (1982) would explain the growth on glucose of pfk1, pfk2, or pfk3 mutants and the lack of growth of pfk1 pfk2 or pfk1 pfk3 double mutants. It would also predict normal growth for the pfk2 pfk3 mutants. On the other hand, the absence of assayable phosphofructokinase activity in some pfk2 mutants is not easily understood unless a defective β subunit somehow makes the activity of the α subunit labile *in vitro*. Furthermore, the existence of several unlinked mutations byp1, byp2, byp3 that do not by themselves affect growth but that together with pfk2 make yeasts unable to grow on glucose (Breitenbach-Schmitt, 1981) is not easily accommodated in the framework of the two phosphofructokinase species. Further work is therefore necessary to clarify our concept of the role of phosphofructokinase in yeasts.

3.1.5. From Fructose-1,6-bisphosphate to Pyruvate

The cleavage of fructose bisphosphate into dihydroxyacetone phosphate and glyceraldehyde-3-phosphate (G3P) is catalyzed by an aldolase. It has been reported that there are aldolase isozymes in yeasts (Harris *et al.*, 1969) but there is as yet no genetic confirmation of this observation. The existence of isozymes, however, would explain the fact that no mutants lacking aldolase have been isolated up to now.

The next step is the isomerization between the triose phosphates catalyzed by a triose phosphate isomerase (Krietsch *et al.*, 1970). Mutants for this enzyme have been isolated (Ciriacy and Breitenbach, 1979).

G3PD catalyzes an important step in the pathway, since at this level the energy liberated in the oxidation is used for incorporating inorganic phosphate in the molecule to give 1,3-diphosphoglycerate. Although it could be expected that this reaction would be tightly controlled, no specific regulatory properties have been described for G3PD. However, it should be pointed out that although there are good data for the existence of different genes for G3PD (Holland and Holland, 1980), the characterization of the corresponding dehydrogenases from a regulatory point of view is still very incomplete.

Phosphoglycerate kinase and phosphoglyceromutase catalyze the next two steps in glycolysis: transfer of a phosphate group from 1,3-diphosphoglycerate to ADP and isomerization of 3-phosphoglycerate to 2-phosphoglycerate, respectively. Mutants have been isolated lacking these enzymes (Lam and Marmur, 1977; Ciriacy and Breitenbach, 1979) and the corresponding genes cloned (Hitzeman *et al.*, 1980; Kawasaki and Fraenkel, 1982). Phosphoglycerate kinase is the most abundant enzyme in *S. carlsbergensis* (Hess *et al.*, 1969).

Enolase catalyzes the isomerization of 2-phosphoglycerate to phosphoenolpyruvate. This enzyme is the only yeast glycolytic enzyme that has been shown to present different kinetic properties *in vitro*, and in an *in situ* assay (Bañuelos and Gancedo, 1978), but this question has not been further investigated. It has also been found that there are at least two isozymes for enolase, whose concentration depends on the carbon source of the medium (Fröhlich and Entian, 1982), thus suggesting that the isozymes could be used alternatively in glycolytic conditions or during gluconeogenesis (K. D. Entian, personal communication). The existence of two nontandemly repeated genes for this enzyme has also been demonstrated (Holland *et al.*, 1981) and the amino acid sequence of the protein coded by one of the genes determined (Chin *et al.*, 1981). This sequence is consistent with that deduced from the sequence of the gene (Holland *et al.*, 1981).

Pyruvate kinase is the last enzyme of the common glycolytic pathway forming pyruvate from phosphoenolpyruvate in a reaction coupled to the synthesis of ATP. The step catalyzed by this enzyme is highly regulated, but two different regulation mechanisms have evolved in different yeast genera (J. M. Gancedo *et al.*, 1967). In some cases (*Saccharomyces*), pyruvate kinase activity is dependent on fructose bisphosphate (Hess *et al.*, 1966). In other yeasts (*Candida, Rhodotorula*) the concentration of pyruvate kinase shows large variations depending on the carbon source available (Ruiz-Amil *et al.*, 1965; J. M. Gancedo *et al.*, 1967). In every case the same effect is obtained, pyruvate kinase activity decreasing in conditions of gluconeogenesis. The importance of this type of control is, however, open to doubt since a mutant has been isolated (Maitra and Lobo, 1977a) with a pyruvate kinase that does not require fructose bisphosphate for activity and this mutant grows at a normal rate on gluconeogenic sources. Many pyruvate kinase mutants have been described (Lam and Marmur, 1977; Maitra and Lobo, 1977c; Sprague, 1977; Clifton *et al.*, 1978; Ciriacy and Breitenbach, 1979), the locus has been mapped (Sprague, 1977; Sinha and Maitra, 1977), and the structural gene has been cloned (Kawasaki and Fraenkel, 1982).

3.1.6. Alcohol Production and Ancillary Pathways

The pyruvate formed in the preceding reaction can be converted into acetyl CoA and oxidized in the tricarboxylic acid cycle (see Section 3.4) or can be converted to ethanol. In the latter case, pyruvate is first decarboxylated by a pyruvate decarboxylase yielding acetaldehyde, which is in turn reduced to ethanol by an $NADH_2$-dependent alcohol dehydrogenase. Pyruvate decarboxylase, which appears to be distinct from the decarboxylase in the pyruvate dehydrogenase complex (H. D. Schmitt, personal communication), is activated by phosphate (Boiteux and Hess, 1970) and induced by growth on glucose (Ruiz-Amil *et al.*, 1966; Schmitt and Zimmermann, 1982). This enzyme is necessary for anaerobic growth on glucose. *R. glutinis*, an obligate aerobe, has low levels of pyruvate decarboxylase (J. M. Gancedo, unpublished results) and mutants of *S. cerevisiae* lacking this enzyme cannot grow in anaerobiosis (Lancashire *et al.*, 1981; Schmitt and Zimmermann, 1982). Although mutants have been isolated that lack pyruvate decarboxylase and cannot grow on glucose even in aerobic conditions (Lam and Marmur, 1977), it is likely that this particular mutation occurs in a regulatory gene with pleiotropic effects.

In *S. cerevisiae*, three isozymes of alcohol dehydrogenase have been identified by both biochemical and genetic criteria (Schimpfessel, 1968; Ciriacy, 1975, 1979; Wills and Phelps, 1975; Wills, 1976). The isozyme alcohol dehydrogenase I is constitutive and is located in the cytoplasm. It is the enzyme that is operative during fermentation of carbohydrates. The other isozymes are repressed by glucose and are connected with ethanol utilization. In *Schizosaccharomyces pombe*, mutants lacking alcohol dehydrogenase have been isolated (Megnet, 1965). It appears that in this yeast there is only one NAD-dependent enzyme (Megnet, 1967).

The balance of alcoholic fermentation can be established as follows. For each mole of glucose fermented, there is a net gain of 2 moles of ATP from ADP and phosphate. As by-product, 2 moles of ethanol and 2 moles of CO_2 are liberated into the medium. The redox balance is completely equilibrated since the NAD reduced in the G3PD step is reoxidized in the reaction catalyzed by alcohol dehydrogenase. In growing yeasts, not all the glucose metabolized can be fermented to ethanol since in addition to energy, the cells require carbon skeletons and reducing power. Other pathways should then complement the glycolytic pathway (for a more thorough discussion of this point, see Sols *et al.*, 1971; Gancedo and Serrano, in press).

From these ancillary pathways, the most important quantitatively is that leading to the production of glycerol. Dihydroxyacetone phosphate is first reduced to glycerophosphate by an NAD-dependent dehydrogenase; glycerophosphate is then hydrolyzed by a specific phosphatase to glycerol, which is excreted into the medium (Gancedo *et al.*, 1968). The proportion of glucose that is catabolized to give glycerol depends on the amount of $NADH_2$ required by the yeast for biosynthetic purposes (Lagunas and Gancedo, 1973).

Minor products of the fermentation of glucose are succinate and acetate. Two mechanisms have been proposed for succinate formation, an oxidative mechanism via citrate and α-ketoglutarate (Oura, 1977) or a reductive one from oxaloacetate, with malate and fumarate as intermediates (Chapman and Bartley, 1968; Machado *et al.*, 1975). The available evidence makes the oxidative route more probable (Heerde and Radler, 1978; Freitas-Valle *et al.*, 1981). It has been shown that substances that depolarize the plasma membrane inhibit succinate production (Duro and Serrano, 1981).

Acetate is formed from acetaldehyde in a reaction catalyzed by an aldehyde dehydrogenase. In *S. cerevisiae* an aldehyde dehydrogenase has been isolated that is activated by K^+ ions and that can utilize either NAD or NADP as cosubstrate (Steinman and Jacoby, 1968). The enzyme has an alkaline pH optimum, and this could explain the fact that acetate production increases when the culture medium has an alkaline pH. On the other hand, it has been reported that in glucose-repressed yeasts, acetaldehyde is oxidized through an NADP-linked dehydrogenase (Llorente and Nuñez de Castro, 1977). Some yeasts species such as *Zygosaccharomyces acidifaciens* or yeasts of the genus *Brettanomyces* produce large amounts of acetic acid during glucose fermentation (Nickerson and Carroll, 1945; Custers, 1940). It can be expected that in these yeasts, aldehyde dehydrogenase activity is very high.

3.1.7. Changes in the Levels of Glycolytic Enzymes

In Section 4, some general regulatory mechanisms related with carbohydrate metabolism will be described. At this point, however, some observations should be made on the regulation of the amount of glycolytic enzymes in yeasts. In *S. cerevisiae*, the levels of glycolytic enzymes do not increase much in glycolytic conditions as compared with gluconeogenic ones, usually less than 2-fold (Bañuelos and Gancedo, 1978). There have been some reports of low phosphofructokinase activity in gluconeogenic conditions (Gancedo *et al.*, 1965; Foy and Bhattacharjee, 1978) but it is not clear whether the amount of enzyme had diminished or some activator like fructose-2,6-bisphosphate was lacking. Induction of glycolytic enzymes by glucose has been reported in a hybrid strain of *S. fragilis* × *S. dobzhanski* (Maitra and Lobo, 1971), and Hommes (1966) found that in *Candida parapsilosis* the levels of some glycolytic enzymes increase between 25- and 50-fold with increasing concentrations of glucose in the growth medium. A mutation has been described, gcr1, that causes a large decrease in the levels of most glycolytic enzymes, in gluconeogenic conditions (Clifton *et al.*, 1978). In sugar-grown cultures, the decrease is much less marked, and it has been shown that the mutation affects mRNA levels (Clifton and Fraenkel, 1981).

3.2. Metabolism of Carbohydrates Other Than Glucose

Mannose and fructose are transported into the cell and phosphorylated by the same system that acts on glucose. Mannose-6-phosphate is converted into fructose-6-phosphate by a phosphomannose isomerase. Mutants unable to grow on mannose lack this isomerase (Herrera *et al.*, 1976). The fructose-6-phosphate formed from mannose or fructose, follows then the normal glycolytic pathway. It should be noted that in yeasts, fructose is a true glycolytic source in contrast to what occurs in mammalian cells where fructose enters the glycolytic pathway at the level of triose phosphates and acts therefore as a gluconeogenic substrate.

Growth on galactose requires a specific permease (Cirillo, 1968b), a galactokinase that produces galactose-1-phosphate (Schell and Wilson, 1977), and a uridyltransferase that converts galactose-1-phosphate and UDP-glucose into glucose-1-phosphate and UDP-galactose (Segawa and Fukasawa, 1979). UDP-glucose is regenerated from UDP-galactose by an epimerase (Fukasawa *et al.*, 1980), while glucose-1-phosphate enters the glycolytic pathway after isomerization to glucose-6-phosphate by phosphoglucomutase. Except for phosphoglucomutase, which does not really belong to the galactose pathway, the enzymes specific for the pathway are inducible by galactose and repressed by glucose. The galactose system has been subjected to intensive genetic analysis and has been shown to present a complex pattern of regulation (Oshima, 1982; Matsumoto *et al.*, 1983).

Some yeasts are also able to grow on *N*-acetylglucosamine. It seems that the sugar is first phosphorylated, then deacylated, and finally deaminated to give fructose-6-phosphate. In *Candida utilis* and *S. cerevisiae*, the transport for *N*-acetylglucosamine and glucosamine-6-phosphate deaminase are both inducible. In *C. utilis*, there is in addition an inducible specific kinase for *N*-acetylglucosamine (Singh and Datta, 1978).

Growth on polyols such as glucitol or mannitol would involve an oxidation by an aldolase reductase followed by the usual glycolytic pathway.

R. glutinis is able to grow on δ-gluconolactone and spontaneous mutants can be isolated able to grow on gluconate. The metabolism of gluconate involves an inducible transport and a constitutive kinase (Gancedo and Mazón, 1978). 6-

Phosphogluconate is further metabolized through the pentose phosphate pathway.

S. cerevisiae cannot grow on any pentose, but a number of yeasts of the genera Rhodotorula and Candida, among others, are able to grow on D-xylose, D-ribose, and L-arabinose, as well as on D-xylulose and D-ribulose. In fact, in most cases the pentoses must be isomerized to pentuloses before they can be phosphorylated (Barnett, 1976). The enzymes involved in the isomerization and phosphorylation are inducible; the products of the phosphorylation in turn are intermediates of the pentose phosphate pathway.

Growth of yeasts on oligosaccharides or polysaccharides depends on the presence of the corresponding hydrolases. These hydrolases can be external or intracellular; in the latter case, the oligosaccharides must first be taken up by the cells by a specific transport system. That is what occurs for maltose and isomaltose in S. cerevisiae and for lactose in Kluyveromyces fragilis (De la Fuente and Sols, 1962; Okada and Halvorson, 1964). Other disaccharides such as sucrose and melibiose are hydrolyzed by external invertase and α-galactosidase, respectively, these enzymes allowing also the utilization of raffinose. Some yeasts lacking external invertase can ferment sucrose either by making use of the maltose permease and of maltase (Khan et al., 1973) or by having a specific transport and an intracellular invertase as occurs with S. wickerhamii (Santa María, 1964). Multiple genes are known to be involved in the utilization of maltose (Federoff et al., 1983), sucrose (Hackel, 1975; Grossmann and Zimmermann, 1979), and isomaltose (Barnett, 1976) but their exact mode of action is not well understood.

Trehalase is cytosolic in S. cerevisiae but periplasmic in R. glutinis (Barnett, 1981); in the latter, therefore, trehalose is first hydrolyzed and the glucose produced taken up by the cell. On the other hand, in the strains of S. cerevisiae able to grow on trehalose, transport precedes metabolism (Kotyk and Michaljaničová, 1979).

Polysaccharides should always be hydrolyzed externally before they can be metabolized; this is the case for starch and dextrins that can be used by Candida tropicalis and S. diastaticus, for inulin used by K. fragilis, and for xylan used by Cryptococcus albidus. In the latter case, it is known that induction of extracellular β-xylanase is accompanied by induction of a transport system for xylobiose and methyl-β-D-xyloside (Krátký and Biely, 1980).

3.3. Pentose Phosphate Pathway

Some of the glucose-6-phosphate formed during carbohydrate catabolism is not metabolized through the glycolytic pathway but channeled instead into the pentose phosphate pathway. This pathway is formed by two parts: an oxidative branch from glucose-6-phosphate to ribulose-5-phosphate, which is irreversible, and a nonoxidative branch, which is reversible and allows the interconversion of pentose phosphates into fructose-6-phosphate and G3P (Fig. 1). Either of the two branches allows the synthesis of pentose phosphates from glucose-6-phosphate while only the oxidative pathway is able to generate $NADPH_2$. It should be noted, however, that the synthesis of pentose phosphates from glucose-6-phosphate through the nonoxidative branch can occur only if the glycolytic pathway from glucose-6-phosphate to G3P is operative.

It has been proposed that the pentose phosphate pathway is the principal source of the $NADPH_2$ needed for biosynthetic reactions, and this is supported by the fact that the flux through the pathway diminishes when $S.$ $cerevisiae$ grows on a rich medium that represses many biosynthetic routes (Lagunas and Gancedo, 1973). The regulation of the pathway, however, is not well understood; it is likely that the activity of glucose-6-phosphate dehydrogenase is regulated by the ratio $NADP/NADPH_2$ since $NADPH_2$ is a competitive inhibitor (Glaser and Brown, 1955), but a thorough biochemical characterization of the enzyme is lacking. The existence of mutants lacking glucose-6-phosphate dehydrogenase and able to grow normally on glucose (Lobo and Maitra, 1982c) indicates that some alternative way for supplying $NADPH_2$ must be available to yeasts. Two possibilities would be the NADP-dependent isocitrate dehydrogenase and the malic enzyme. Mutants lacking 6-phosphogluconate dehydrogenase cannot grow on glucose (Lobo and Maitra, 1982c) probably because there is an accumulation of 6-phosphogluconate, toxic for the cell. In $E.$ $coli,$ mutants lacking phosphoglucose isomerase are able to grow on glucose, using solely the pentose phosphate pathway (Fraenkel and Vinopal, 1973), while that is not the

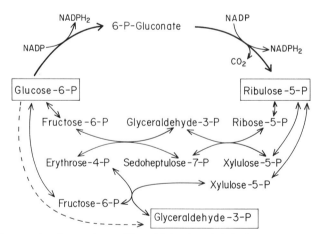

Figure 1. Pentose phosphate pathway. The oxidative branch is indicated by heavy arrows. The dashed arrow represents reactions of the glycolytic pathway.

case in *S. cerevisiae*. Since *S. cerevisiae* is able to grow slowly on δ-gluconolactone, which can be metabolized only via the pentose phosphate pathway, the complete lack of growth on glucose of the phosphoglucose isomerase-less mutants is probably not due to an intrinsic limitation of the pathway but to the toxic effects of the glucose-6-phosphate that accumulates in the mutants.

It had been proposed that *R. glutinis*, a strictly oxidative yeast, metabolizes glucose through the pentose phosphate cycle only; it was further postulated that pentose phosphate formed by the nonoxidative part of the pathway was split into a C_2 fragment and G3P (Höfer *et al.*, 1971). It was later found, however, that *R. glutinis* has a functional glycolytic pathway (Mazón *et al.*, 1974a,b) and it appears therefore that in this yeast too, only a small part of the glucose consumed is metabolized through the pentose phosphate pathway.

3.4. Citric Acid Cycle

A part of the pyruvate formed in the catabolism of carbohydrates is converted into acetyl CoA, which is in turn metabolized through the citric acid cycle. The conversion of pyruvate into acetyl CoA can occur in two ways, either directly in a reaction catalyzed by the pyruvate dehydrogenase complex (Ullrich and Wais, 1975) or indirectly (Holzer and Goedde, 1957) through decarboxylation to acetaldehyde, which is oxidized to give acetate that in turn is transformed into acetyl CoA by an acetyl CoA synthase (Klein and Jahnke, 1979).

The citric acid cycle has two major functions: the synthesis of intermediary metabolites that are precursors of biosynthetic routes and the coupling of acetyl CoA oxidation to ATP formation. In anaerobic conditions, the citric acid cycle plays only a biosynthetic role and the $NADH_2$ formed in the process should be reoxidized by a reaction that does not depend on oxygen, probably the reduction of dihydroxyacetone phosphate to glycerophosphate (see Section 3.1.6). Since intermediates are withdrawn from the cycle, there should be some reaction able to replenish the cycle. In yeasts growing on carbohydrates, this reaction is the formation of oxaloacetate from pyruvate catalyzed by pyruvate carboxylase (Ruiz-Amil *et al.*, 1965).

The enzymes of the citric acid cycle are repressed in the presence of glucose and in anaerobiosis (Polakis *et al.*, 1965; Duntze *et al.*, 1969; Wales *et al.*, 1980). The enzymes appear to be located in the mitochondria, although in some cases there are also cytoplasmic isozymes (Atzpodien *et al.*, 1968). Differential centrifugation of cell-free extracts gives different distribution patterns of the citric acid cycle enzymes under various conditions (Wales *et al.*, 1980). It is not clear if this reflects a different location in the living cell or if it is an artifact caused by the fragility of the promitochondria of anaerobically grown yeasts. Since pyruvate carboxylase is located in the cytosol (Haarasilta and Taskinen, 1977), the oxaloacetate formed by carboxylation of pyruvate must enter the mitochondria to serve as a substrate for citrate synthase. The mitochondria being impermeable to oxaloacetate itself, oxaloacetate should first be transformed into aspartate by a cytoplasmic transaminase; aspartate then crosses the mitochondrial membrane and regenerates oxaloacetate inside the mitochondria. If there is a supply of $NADH_2$ in the cytoplasm, oxaloacetate can also be reduced to malate, which crosses freely the mitochondrial membrane.

A full turn of the citric acid cycle results in the formation of a molecule of GTP from GDP, one protein-bound $FADH_2$, and three molecules of $NADH_2$. In the presence of oxygen, there is a transfer of electrons from $FADH_2$ and $NADH_2$ to O_2 and this is coupled to the synthesis of ATP. In $S.$ $carlsbergensis,$ it appears that phosphorylation takes place only at two sites in the respiratory chain al-though phosphorylation at site I can occur in some special conditions, like prolonged starvation (Ohnishi, 1970). In $C.$ $utilis$ the usual situation would be the operation of three phosphorylation sites, but iron-limited growth, for exam-ple, can cause the loss of energy coupling at site I (Light et $al.,$ 1968). The energy balance of the oxidation of a molecule of acetyl CoA through the citric acid cycle would therefore be the synthesis of either 9 or 12 energy-rich phos-phate bonds, depending on the yeast species and on its metabolic situation.

3.5. Reserve Carbohydrates

Yeasts are able to accumulate two main reserve carbohydrates: a polysac-charide, glycogen, and a disaccharide, trehalose. The pathways of synthesis and degradation of these two compounds are relatively well known, but there is still considerable uncertainty about the exact function of the two types of reserve. I will here review briefly the latest developments on this subject.

3.5.1. Glycogen

When yeast cells growing exponentially on glucose are deprived of nitro-gen, sulfur, or phosphate, glycogen accumulates (Lillie and Pringle, 1980). A low level of oxygen also increases the concentration of glycogen (Chester, 1963). In yeasts undergoing diauxic growth, glycogen is low (1% or less of the cellular dry weight) during the early exponential growth on glucose; glycogen accumulation begins when a half of the external glucose has been consumed and reaches a maximum shortly before the glucose in the medium is exhausted (Lillie and Pringle, 1980). A part of this glycogen is then degraded, apparently to provide energy for the induction of the enzymes necessary for growth on ethanol. The glycogen level then remains stable or increases slightly, reaching 6–8% of the cellular dry weight. During a prolonged stationary phase after the exhaustion of the carbon source, glycogen is slowly degraded, dropping to 3% of the cellular dry weight after 50 days (Lillie and Pringle, 1980).

The subcellular location of glycogen remains controversial, although there have been reports on two metabolically distinct pools of glycogen (Gunja-Smith *et al.*, 1977). A minor fraction of the glycogen (8–50% depending on the yeast species and on the stage of growth) is extracted by hot alkali, while the major water-insoluble fraction is solubilized by repeated extractions with hot acid. It has been suggested that the major glycogen fraction is associated with a component of the cell wall by a covalent bond, and would play a role in cell wall metabolism. If this interpretation is confirmed, the changes in glycogen concentration in different metabolic conditions should be reexamined, making a distinction between the soluble and insoluble pools.

Glycogen snythase is the enzyme that controls the rate of glycogen synthesis. It exists in two forms that are interconvertible by phosphorylation–dephosphorylation (Rothman-Denes and Cabib, 1970). Although earlier work suggested that cAMP was not involved in the phosphorylation reaction (Rothman-Denes and Cabib, 1971), some recent evidence points to cAMP protein kinase as responsible for the phosphorylation of glycogen synthase (Ortiz *et al.*, 1983). The phosphorylated enzyme is strongly inhibited by ADP and ATP and this inhibition can only be reversed by high concentrations of glucose-6-phosphate (Rothman-Denes and Cabib, 1971). Since such high concentrations are usually not attained in the cell, the phosphorylated enzyme can be considered as inactive in physiological conditions. In stationary-phase cells, the nonphosphorylated form of glycogen synthase predominates. For this form, the concentration of glucose-6-phosphate necessary for relieving ATP inhibition is much lower, so that glycogen synthase is active, at least as long as enough substrate (UDP-glucose) is present.

Glycogen degradation is catalyzed by a glycogen phosphorylase that releases glucose-6-phosphate, although in sporulating cells glycogen degradation can also be due to the action of glucosidases (Colonna and Magee, 1978). Like glycogen synthase, glycogen phosphorylase can exist in two forms interconvertible by phosphorylation–dephosphorylation (Fosset *et al.*, 1971; Lerch and Fischer, 1975). The dephosphorylated glycogen phosphorylase has only 10–20% of the activity of the phosphorylated form. The latter form is predominant during exponential growth on glucose. There is at present no information on the characteristics of the enzymes that catalyze the interconversion reactions.

Mutants have been isolated that cannot accumulate glycogen (Chester, 1968). In one such mutant, the cAMP-dependent protein kinase appears to have lost its dependence on cAMP so that glycogen synthase is permanently in the phosphorylated, nonactive, form (Ortiz *et al.*, 1983).

3.5.2. Trehalose

Like glycogen, trehalose accumulates in yeast cells deprived of nitrogen, sulfate, or phosphate. There is also a great increase in trehalose content (up to 15–20% of the cellular dry weight) in yeasts undergoing carbon and energy limitation (Lillie and Pringle, 1980). During a prolonged stationary phase, trehalose is degraded very slowly, less than 50% in 70 days. In contrast, when glucose is added to the medium, trehalose rapidly disappears (Panek, 1962; Küenzi and Fiechter, 1969).

Trehalose concentration also increases in sporulating cultures; this appears to be a sporulation-specific event since it is not observed in nonsporogenic strains transferred to a sporulation medium (Inoue and Shimoda, 1981). Conversely, trehalose breakdown is one of the first events observed upon induction of spore germination (Thevelein et al., 1982).

Although it had been postulated that trehalose was bound to the cell membranes (Souza and Panek, 1968), more recent evidence points to a cytosolic location for trehalose (Keller et al., 1982). Since trehalose can reach an intracellular concentration greater than 0.1 M, this compound could play a role in osmoregulation and function as a protective agent under conditions of stress.

Trehalose synthesis depends on the UDPG-linked trehalose-6-phosphate synthase. Although this enzyme appears to be subject to glucose repression (Panek and Mattoon, 1977), no further studies have been carried out on the regulation of trehalose-6-phosphate synthase. Mutants lacking this enzyme can still accumulate trehalose during growth on a maltose medium or in the presence of a constitutive MAL gene. This observation suggests that there is an alternative pathway for trehalose synthesis whose activity would be controlled by a MAL gene (Operti et al., 1982).

Trehalose degradation is mediated by trehalase, an enzyme that exists in two interconvertible states, a phosphorylated active form and a dephosphorylated inactive form (Van der Plaat, 1974; Van Solingen and Van der Plaat, 1975). The phosphorylation is catalyzed by a cAMP-dependent protein kinase, and the absence of trehalose accumulation in a mutant that has a protein kinase lacking dependence on cAMP, can therefore be attributed to a trehalase that is permanently in the active, phosphorylated form (Ortiz et al., 1983).

It has recently been reported that the inactive trehalase is located in the cytosol, and that when trehalase is phosphorylated it is taken into the vacuole, where it is subject to degradation by intravacuolar proteases (Wiemken and Schellenberg, 1982). Since trehalose is located in the cytosol, this scheme would imply that active trehalase and trehalose are in contact only for a short time before the trehalase enters the vacuole. It appears therefore that there are still missing elements to complete our knowledge of trehalase regulation.

4. REGULATORY MECHANISMS

In the preceding section, different factors that control the operation of the pathways of carbohydrate metabolism were considered. However, there are some regulatory mechanisms that have a special relevance in yeast and these will be studied here in more detail.

4.1. Catabolite Repression

Most yeasts utilize preferentially the fermentable hexoses: glucose, mannose, fructose. In the presence of glucose, for example, the enzymatic systems that would allow the utilization of alternative carbon sources (e.g., galactose, sucrose, ethanol) are repressed. This phenomenon has been termed catabolite repression by analogy with that occurring in bacteria (Magasanik, 1961). Repression is particularly marked in facultative anaerobes such as *Saccharomyces* (Polakis *et al.*, 1965; Duntze *et al.*, 1969; Perlman and Mahler, 1974) but it is also found in aerobic yeasts like *Rhodotorula, Hansenula* and *Pichia* (Duntze *et al.*, 1967; Gancedo and Gancedo, 1971). It has been reported, however, that the enzymes implicated in the catabolism of *N*-acetylglucosamine are repressed by glucose in *S. cerevisiae* but not in *C. albicans* (Singh and Datta, 1978). Most remarkable is the fact that in *K. lactis,* induction of β-galactosidase is not prevented by glucose (Dickson and Markin, 1980).

The molecular basis of the repression of specific enzymes by glucose has not yet been elucidated. The first point to establish is whether repression occurs at the level of transcription, translation, or both. In the few cases that have been studied in detail, glucose has been shown to affect the amount of mRNA synthesized (Zitomer *et al.*, 1979; Perlman and Halvorson, 1981; Denis *et al.*, 1981). However, there have been some claims that glucose can also repress translation of the mRNAs coding for α-galactosidase (Van Wijk and Konijn, 1971) or invertase (Elorza *et al.*, 1977). It has been thought that the signal for repression was not glucose itself but some metabolite that builds up during glucose catabolism. It has been proposed that this metabolite could be glucose-6-phosphate (C. Gancedo *et al.*, 1967) but no clear-cut evidence for this has been presented.

Since in *E. coli* the role of cAMP in relieving glucose repression has been clearly demonstrated (Botsford, 1981), it has been tempting to extend these results to yeasts. In fact, there have been several reports correlating low levels of cAMP with increased catabolite repression (Van Wijk and Konijn, 1971; Schlanderer and Dellweg, 1974; Mahler *et al.*, 1981). Recently, this theory has been challenged using mutants that depend on external cAMP for growth (Matsumoto *et al.*, 1982). Moreover, determination of cAMP in *S. cerevisiae* grown on different carbon sources has shown that the level of cAMP is indeed higher in glucose-grown yeast than in yeast grown on ethanol (Eraso and Gancedo, 1984).

An approach to the study of catabolite repression has been the search for mutants insensitive to catabolite repression or unable to overcome repression even in the absence of glucose (Zimmermann and Scheel, 1977; Ciriacy, 1977; Michels and Romanowski, 1980). Although a number of such mutants have been found, the conclusions that can be drawn from their study are still scarce. Two points, however, seem well established. First, there are several circuits for catabolite repression since mutants where maltase, invertase, and malate dehydrogenase are derepressed have still repressed levels of malate synthase, isocitrate lyase, and fructose-1,6-bisphosphatase (Entian *et al.*, 1977). Second, one of the isozymes of hexokinase, hexokinase II, appears to play a role in catabolite repression that is not due solely to its ability to phosphorylate glucose (Entian and Mecke, 1982).

Recent results with cloned genes confirmed that catabolite repression requires a functional fragment of DNA close to the structural gene (Beier and Young, 1982); however, further studies would be necessary to define the other elements of the regulatory circuit.

4.2. Catabolite Inactivation

Glucose and other fermentable hexoses not only repress the synthesis of many enzymes, but also provoke the inactivation of a small number of enzymes when derepressed yeast cells are shifted to a repressing medium. This phenomenon has been termed catabolite inactivation (Holzer, 1976) and is now being thoroughly explored.

The first observation of the effect of glucose on preexisting activities is already over 35 years old, since in 1947 Spiegelman and Reiner reported the rapid disappearance of "galactozymase" when glucose was added to yeast grown on galactose. However, this result was long ignored, due to the fact that in rapidly growing microorganisms repression of new synthesis and dilution of the preexisting enzyme appeared sufficient for the regulation of enzymes that were no longer needed. Years later, new cases of enzymes subject to rapid inactivation were found: malate dehydrogenase (Ferguson *et al.*, 1967), fructose bisphosphatase (Gancedo, 1971), phosphoenolpyruvate carboxykinase (Haarasilta and Oura, 1975; Gancedo and Schwerzmann, 1976). In these three cases, the inactivation was accompanied by a loss of specific antigenic material (Neeff *et al.*, 1978; Funayama *et al.*, 1980; Müller *et al.*, 1981). It was therefore suggested that in the presence of glucose, or more probably some metabolite derived from it, the enzymes suffered a change in conformation and were then susceptible to proteolysis. It was first thought that the enzyme responsible for the proteolysis was protease B, a vacuolar enzyme that *in vivo* inactivates readily the three enzymes; however, the finding of mutants deficient in protease B but possessing normal catabolite inactivation (Wolf and Ehmann, 1979; Zubenko and Jones, 1979) ruled out protease B as responsible for the inactivation. Although in the last few years new proteases from yeasts have been characterized (Achstetter *et al.*, 1981), the protease (or proteases) responsible for the catabolite inactivation has not been identified. It is not even known if this protease is located in the cytoplasm or in the vacuole. If the protease were vacuolar, it would be necessary to postulate that after the addition of glucose there is a conformational change in the enzymes subject to inactivation that facilitates their uptake into the vacuole. In the case of fructose-1,6-bisphosphatase, it has been found that in certain conditions the degradation triggered by glucose is preceded by a reversible partial inactivation (Lenz and Holzer, 1980). This inactivation is accompanied by a phosphorylation of the enzyme (Mazón *et al.*, 1981, 1982a; Müller and Holzer, 1981). Phosphorylation can also occur when the yeast is treated with proton ionophores. Both glucose and the ionophores increase the intracellular levels of cAMP (Mazón *et al.*, 1982b) and of fructose-2,6-bisphosphate (Lederer *et al.*, 1981; Gancedo *et al.*, 1983). These observations as well as the fact that fructose-2,6-bisphosphate activates the cAMP-dependent phosphorylation of fructose-1,6-bisphosphatase *in vitro* (Gancedo *et al.*, 1983) suggest that cata-bolite inactivation of fructose-1,6-bisphosphatase proceeds as follows. Addition of glucose increases the levels of cAMP and of fructose-2,6-bisphosphate; cAMP in turn activates a protein kinase while fructose-2,6-bisphosphate changes the conformation of the bisphosphatase, making it a better substrate for the kinase. The phosphorylated enzyme is then proteolyzed. A plausible hypothesis would be that the phosphorylation of the bisphosphatase makes it a better target for the degradation process, but as discussed above this point will remain unproven as long as the proteolytic system has not been identified.

Inactivation of fructose-1,6-bisphosphatase by glucose does not occur in other yeasts belonging to the genera *Rhodotorula*, *Pichia*, or *Torulopsis* (Gancedo, 1971). It would appear therefore that the three types of regulation that coexist in *S. cerevisiae*—repression by glucose (Gancedo *et al.*, 1965), inactivation by glucose (Gancedo, 1971), inhibition by AMP (Gancedo *et al.*, 1965, 1982)—are diminished or absent in strictly aerobic yeasts (Gancedo and Gancedo, 1971).

4.3. Pasteur Effect and Other Effects

Many yeast species are able to metabolize carbohydrates in anaerobic as well as in aerobic conditions. When oxygen is absent, all the sugar has to be fermented while in the presence of oxygen a proportion of the sugar, which varies depending on the species and on the metabolic conditions (see below), can be completely oxidized. It results from this fact that respiration decreases the rate of fermentation, a phenomenon termed the Pasteur effect (Warburg, 1926) (for an account of the significance of the early Pasteur observations, see Lagunas, 1981).

Taking in account that oxidation yields much more ATP than does fermentation, it could be expected that in aerobiosis the rate of sugar consumption would be less than in anaerobiosis. Today, the Pasteur effect is generally defined as the decrease in glucose utilization caused by aerobiosis, although occasionally the term is still applied to the decrease in the amount of glucose fermented caused by aerobiosis.

Many theories have been advanced for explaining the mechanisms that underlie the Pasteur effect (for reviews on this subject, see Dickens, 1951; Krebs, 1972; Racker, 1974; Sols, 1976). However, much less work has been devoted to the basic task of determining the extent of the Pasteur effect in different yeasts in well-defined conditions.

In *S. cerevisiae,* the situation is as follows (Lagunas, 1979). In yeast growing on glucose, the glycolytic flux is barely affected by the presence of air and the inhibition of the rate of fermentation does not exceed 10%. In yeast starved for NH_4^+, aerobiosis decreases glucose utilization to 55% of the anaerobiosis rate while the rate of fermentation is reduced to 45%. It should be noted that in chemostat cultures where a very low concentration of glucose limits the rate of growth, the effect of aerobiosis is far greater, at least a threefold decrease in glucose consumption and more than a tenfold decrease in the rate of fermentation (Oura, 1974; Rogers and Stewart, 1974).

In *C. tropicalis,* aerobiosis decreases the glycolytic flux three- to fourfold while the rate of fermentation is reduced to less than 5% of the anaerobic rate (De Deken, 1966).

From the results described above, it appears unfortunate that *S. cerevisiae* has been chosen for studying the mechanism of the Pasteur effect, since this effect is quite small in this yeast. From recent results (Lagunas and Gancedo, 1983), it would appear that in *S. cerevisiae,* aerobiosis provokes a decrease in the concentration of phosphate in the cytoplasm. The basis for this decrease does not seem to be due simply to oxidative phosphorylation since the concentrations of ADP and ATP remain stable, and would rather be related to an increase in the synthesis of polyphosphates. In any case, the decrease of phosphate, a potent activator of yeast phosphofructokinase (Bañuelos *et al.,* 1977), would diminish the activity of phosphofructokinase and consequently the glycolytic flux. Since phosphofructokinase is not the first irreversible step in the utilization of glucose, its regulation is not sufficient for inhibiting glucose utilization. In animal cells, inhibition of phosphofructokinase causes a buildup of fructose-6-phosphate and of the glucose-6-phosphate in equilibrium with it, glucose-6-phosphate inhibiting in turn the phosphorylation of glucose by hexokinase. In yeasts, none of the kinases that phosphorylate glucose are sensitive to glucose-6-phosphate inhibition and it has been postulated that glucose transport would be the regulated step (Sols, 1967; Azam and Kotyk, 1969). Inhibition of glucose transport by glucose-6-phosphate, however, does not appear to occur (Perea and Gancedo, 1978) and the control of transport is still far from understood.

As previously indicated, the decrease in glycolytic flux is generally accompanied by a more pronounced decrease in the amount of glucose fermented. This decrease in fermentation is due to a diversion of the pyruvate produced by glycolysis from the anaerobic decarboxylation to the conversion into acetyl CoA catalyzed by pyruvate dehydrogenase (Holzer, 1961). Pyruvate dehydrogenase has a much higher affinity for pyruvate than does pyruvate decarboxylase; therefore, pyruvate will be decarboxylated only insofar as pyruvate dehydrogenase cannot handle all the pyruvate formed in the glycolytic pathway and there follows a buildup of pyruvate in the cell.

In some yeast species, fermentation of glucose is inhibited by anaerobiosis. This phenomenon has been termed the Custers effect (Scheffers, 1961) and it is characteristic of the genus *Brettanomyces*. It has been postulated that the decrease in fermentation is due to a shortage of NAD, due to the action of an NAD-linked aldehyde dehydrogenase, involved in the production of acetic acid in *Brettanomyces* (Carrascosa *et al.*, 1981).

The "reverse Pasteur effect" or Crabtree effect (De Deken, 1966) is the impairment of respiratory capacity shown by many yeasts when abundant glucose is available in the medium. It seems likely that this effect is a result of the catabolite inactivation and catabolite repression brought about by glucose. These phenomena have been discussed in preceding sections.

4.4. Futile Cycles

Most major catabolite pathways have some reactions that are irreversible in physiological conditions. When the pathway is required to work in the reverse direction, the corresponding steps are reversed by a different reaction catalyzed by a separate enzyme. In the glycolytic pathway, for example, the phosphorylation of fructose-6-phosphate by ATP to give fructose-1,6-bisphosphate and ADP, catalyzed by phosphofructokinase, is reversed by the hydrolysis of fructose-1,6-bisphosphate to fructose-6-phosphate and phosphate, catalyzed by fructose-1,6-bisphosphatase. The existence of such pairs of opposite reactions can produce cycles, whose net balance is a hydrolysis of ATP. This will lead to an energy expenditure that would, in principle, be wasteful for the cell, hence the term "futile cycles." It has been suggested, however, that such cycles may serve a regulatory function (Newsholme *et al.*, 1972) and a thorough analysis of the possible roles of futile cycles in metabolism has been carried out (Stein and Blum, 1978).

To determine if futile cycles play any role in carbohydrate metabolism in yeasts, it would be necessary to test first the extent of cycling in different metabolic conditions. In yeasts, cycling could occur at two levels, the fructose-6-phosphate/fructose-1,6-bisphosphate cycle and the phosphoenolpyruvate/pyruvate/oxaloacetate cycle. In *S. cerevisiae* growing on glucose, fructose-1,6-bisphosphatase and phosphoenolpyruvate carboxykinase are strongly repressed (Gancedo *et al.*, 1965; de Torróntegui *et al.*, 1966; Gancedo and Schwerzmann,

1976) and no cycling would be expected. When *S. cerevisiae* is shifted from glucogenic to glycolytic conditions, fructose-1,6-bisphosphatase and phosphoenolpyruvate carboxykinase are inactivated (see Section 4.2) but over a 1-hr period the glycolytic and gluconeogenic enzymes would coexist and there could be some degree of cycling. Moreover, *S. carlsbergensis* mutants have been described where glucose addition did not cause fructose bisphosphatase inactivation, and since the mutants did not grow on glucose it was proposed that this was due to the deleterious effect of uninterrupted cycling (Van de Poll *et al.*, 1974). Recently, labeling experiments have been carried out to assess the degree of cycling at the step fructose-6-phosphate/fructose-1,6-bisphosphate using mutants unable to inactivate the fructose-1,6-bisphosphatase and the corresponding wild-type strains (Bañuelos and Fraenkel, 1982). Even in conditions where a high cycling would be expected, the cycling observed is marginal at best. Although no comparable experiments have been carried out for the phosphoenolpyruvate/pyruvate/oxaloacetate cycle, it is likely that no major cycling would occur there too. It can therefore be concluded that futile cycles do not play a major role in carbohydrate metabolism in yeasts and that the impairment in glucose growth of some yeast strains cannot be explained by depletion of ATP due to futile cycling (Mazón *et al.*, 1981; Bañuelos and Fraenkel, 1982).

5. CONCLUSIONS

Some years ago it would have seemed that the study of carbohydrate metabolism in yeasts was an exhausted field, where only details remained to be uncovered. Recent experimental results that have been discussed in this chapter indicate that there are several areas where further studies are needed to produce a coherent picture of yeast carbohydrate metabolism.

Even at the level of the glycolytic pathway itself, there are points that remain obscure. An example is the relationship between transport of glucose and activity of the hexokinases, which is not well understood. Another is the possible bypass to the phosphofructokinase reaction.

With regard to regulatory problems, one of the areas of current interest is the study of catabolite repression, which has been approached up to now mainly with the tools of classical genetics but where studies at the molecular level are sorely needed.

I will conclude by pointing to a field that deserves wider attention: the study of yeasts different from *S. cerevisiae*. It would be fruitful to conduct comparative studies between the diverse metabolic strategies of yeasts adapted to different ecological niches.

ACKNOWLEDGMENTS. The helpful suggestions of Dr. Carlos Gancedo are warmly acknowledged. Thanks are due to Ms. Francisca de Luchi for careful typing of the manuscript. Work from this laboratory was partially supported by grants from the Comision Asesora Científica y Técnica.

REFERENCES

Achstetter, T., Ehmann, C., and Wolf, D. H., 1981, New proteolytic enzymes in yeast, *Arch. Biochem. Biophys.* **207**:445–454.

Atzpodien, W., Gancedo, J. M., Duntze, W., and Holzer, H., 1968, Isoenzymes of malate dehydrogenase in *Saccharomyces cerevisiae*, *Eur. J. Biochem.* **7**:58–62.

Azam, F., and Kotyk, A., 1969, Glucose-6-phosphate as regulator of monosaccharide transport in baker's yeast, *FEBS Lett.* **2**:333–335.

Bañuelos, M., and Fraenkel, D. G., 1982, *Saccharomyces carlsbergensis fdp* mutant and futile cycling of fructose-6-phosphate, *Mol. Cell. Biol.* **2**:921–929.

Bañuelos, M., and Gancedo, C., 1978, In situ study of the glycolytic pathway in *Saccharomyces cerevisiae*, *Arch. Microbiol.* **117**:197–201.

Bañuelos, M., Gancedo, C., and Gancedo, J. M., 1977, Activation by phosphate of yeast phosphofructokinase, *J. Biol. Chem.* **252**:6394–6398.

Barford, J. P., and Hall, R. J., 1979, An examination of the Crabtree effect in *Saccharomyces cerevisiae:* The role of respiratory adaptation, *J. Gen. Microbiol.* **114**:267–275.

Barnard, E. A., 1975, Hexokinases from yeast, *Methods Enzymol.* **42**:6–20.

Barnett, J. A., 1976, The utilization of sugars by yeasts, *Adv. Carbohyd. Chem. Biochem.* **32**:125–234.

Barnett, J. A., 1981, The utilization of disaccharides and some other sugars by yeasts, *Adv. Carbohyd. Chem. Biochem.* **39**:347–404.

Bartrons, R., Van Schaftingen, E., Vissers, S., and Hers, H. G., 1982, The stimulation of yeast phosphofructokinase by fructose-2,6-bisphosphate, *FEBS Lett.* **143**:137–140.

Bechet, J., and Wiame, J. M., 1965, Indication of a specific regulatory binding protein for ornithinetranscarbamylase in *Saccharomyces cerevisiae*, *Biochem. Biophys. Res. Commun.* **21**:226–234.

Beier, D. R., and Young, E. T., 1982, Characterization of a regulatory region upstream of the ADR2 locus of *S. cerevisiae*, *Nature* **300**:724–728.

Bergmeyer, H. A. (ed.), 1974, *Methods of Enzymatic Analysis*, Verlag Chemie, Weinheim/ Academic Press, New York.

Bisson, L. F., and Fraenkel, D. G., 1983, Involvement of kinases in glucose and fructose uptake by *Saccharomyces cerevisiae*, *Proc. Natl. Acad. Sci. USA* **80**:1730–1734.

Boiteux, A., and Hess, B., 1970, Allosteric properties of yeast pyruvate decarboxylase, *FEBS Lett.* **9**:293–296.

Borst-Pauwels, G. W. F. H., and Dobbelmann, J., 1972, Determination of the yeast cell pH, *Acta Bot. Neerl.* **21**:149–154.

Botsford, J. L., 1981, Cyclic nucleotides in procaryotes, *Microbiol. Rev.* **45**:620–632.

Breitenbach-Schmitt, I., 1981, Genetische und physiologische Hinweise auf die Existenz eines zweiten Stoffwechselwegs neben der "klassischen" Phosphofructokinasereaktion beim Abbau von Glucose in *Saccharomyces cerevisiae*, Doctoral thesis, Technische Hochschule Darmstadt.

Burke, R. L., Tekamp-Olson, P., and Najarian, R., 1983, The isolation, characterization and sequence of the pyruvate kinase gene of *Saccharomyces cerevisiae*, *J. Biol. Chem.* **258**:2193–2201.

Carrascosa, J. M., Viguera, M. D., Nuñez de Castro, I., and Scheffers, W. A., 1981, Metabolism of acetaldehyde and Custers effect in the yeast *Brettanomyces abstinens*, *Antonie van Leeuwenhoek* **47**:209–215.

Chapman, C., and Bartley, W., 1968, The kinetics of enzyme changes in yeast under conditions that cause the loss of mitochondria, *Biochem. J.* **107**:455–465.

Chester, V. E., 1963, The dissimilation of the carbohydrate reserves of a strain of *Saccharomyces cerevisiae*, *Biochem. J.* **86**:153–160.

Chester, V. E., 1968, Heritable glycogen-storage deficiency in yeast and its induction by ultraviolet light, *J. Gen. Microbiol.* **51**:49–56.

Chin, C. C. Q., Brewer, J. M., and Wold, F., 1981, The aminoacid sequence of yeast enolase, *J. Biol. Chem.* **256**:1377–1384.

Ciriacy, M., 1975, Genetics of alcohol dehydrogenase in *Saccharomyces cerevisiae*, *Mutat. Res.* **29**:315–326.

Ciriacy, M., 1977, Isolation and characterization of yeast mutants defective in intermediary carbon metabolism and in carbon catabolite derepression. *Mol. Gen. Genet.* **154**:213–220.

Ciriacy, M., 1979, Isolation and characterization of further cis- and trans-acting regulatory elements involved in the synthesis of glucose-repressible alcohol dehydrogenase (ADHII) in *Saccharomyces cerevisiae*, *Mol. Gen. Genet.* **176**:427–431.

Ciriacy, M., and Breitenbach, I., 1979, Physiological effects of seven different blocks in glycolysis in *Saccharomyces cerevisiae*, *J. Bacteriol.* **139**:152–160.

Cirillo, V. P., 1961, Sugar transport in microorganisms, *Annu. Rev. Microbiol.* **15**:197–218.

Cirillo, V. P., 1968a, Relationship between sugar structure and competition for the sugar transport system in baker's yeast, *J. Bacteriol.* **95**:603–611.

Cirillo, V. P., 1968b, Galactose transport in *Saccharomyces cerevisiae*, *J. Bacteriol.* **95**:1727–1731.

Clifton, D., and Fraenkel, D. G., 1981, The *gcr (glycolysis regulation) mutation of Saccharomyces cerevisiae*, *J. Biol. Chem.* **256**:13074–13078.

Clifton, D., and Fraenkel, D. G., 1982, Mutant studies of yeast phosphofructokinase, *Biochemistry* **21**:1935–1942.

Clifton, D., Weinstock, S. B., and Fraenkel, D. G., 1978, Glycolysis mutants in *Saccharomyces cerevisiae*, *Genetics* **88**: 1–11.

Colonna, W. J., and Magee, P. T., 1978, Glycogenolytic enzymes in sporulating yeast, *J. Bacteriol.* **134**:844–853.

Custers, M. T. J., 1940, Onderzoekingen over het Gistgeslacht *Brettanomyces*, Ph.D. thesis, De Technische Hoogeschool, Delft.

De Deken, R. H., 1966, The Crabtree effect: A regulatory system in yeast, *J. Gen. Microbiol.* **44**:149–156.

De la Fuente, G., and Sols, A., 1962, Transport of sugars in yeasts, *Biochim. Biophys. Acta* **56**:49–62.

den Hollander, J. A., Brown, T. R., Ugurbil, K., and Shulman, R. G., 1979, ^{13}C nuclear magnetic resonance studies of anaerobic glycolysis in suspension of yeast cells, *Proc. Natl. Acad. Sci. USA* **76**:6096–6100.

den Hollander, J. A., Ugurbil, K., Brown, T. R., and Shulman, R. G., 1981, Phosphorus-31 nuclear magnetic resonance studies of the effect of oxygen upon glycolysis in yeast, *Biochemistry* **20**:5871–5880.

Denis, C., Young, E. T., and Ciriacy, M., 1981, A positive regulatory gene is required for accumulation of the functional messenger RNA for the glucose-repressible alcohol dehydrogenase from *Saccharomyces cerevisiae*, *J. Mol. Biol.* **148**:355–368.

De Torróntegui, G., Palacian, E., and Losada, M., 1966, Phosphoenolpyruvate carboxykinase in gluconeogenesis and its repression by hexoses in yeast, *Biochem. Biophys. Res. Commun.* **22**:227–231.

Dickens, F., 1951, Aerobic glycolysis, respiration, and the Pasteur effect, in: *The Enzymes* (J. B. Sumner and K. Myrback, eds.), Vol. 2, Part 1, pp. 624–683, Academic Press, New York.

Dickson, R. C., and Markin, J. S., 1980, Physiological studies of β-galactosidase induction in *Kluyveromyces lactis*, *J. Bacteriol.* **142**:777–785.

Downie, J. A., and Garland, P. B., 1973, An antimycin A- and cyanide-resistant variant of *Candida utilis* arising during copper-limited growth, *Biochem. J.* **134**:1051–1061.

Duntze, W., Atzpodien, W., and Holzer, H., 1967, Glucose dependent enzyme activities in different yeast species, *Arch. Mikrobiol.* **58**:296–301.

Duntze, W., Neumann, D., Gancedo, J. M., Atzpodien, W., and Holzer, H., 1969, Studies on the regulation and localization of the glyoxylate cycle enzymes in *Saccharomyces cerevisiae, Eur. J. Biochem.* **10**:83–89.

Duro, A. F., and Serrano, R., 1981, Inhibition of succinate production during yeast fermentation by deenergization of the plasma membrane, *Curr. Microbiol.* **6**:111–113.

Elorza, M. V., Villanueva, J. R., and Sentandreu, R., 1977, The mechanism of catabolite inhibition of invertase by glucose in *Saccharomyces cerevisiae, Biochim. Biophys. Acta* **475**:103–112.

Entian, K. D., and Mecke, D., 1982, Genetic evidence for a role of hexokinase isozyme PII in carbon catabolite repression in *Saccharomyces cerevisiae, J. Biol. Chem.* **257**:870–874.

Entian, K. D., Zimmermann, F. K., and Scheel, I., 1977, A partial defect in carbon catabolite repression in mutants of *Saccharomyces cerevisiae* with reduced hexose phosphorylation, *Mol. Gen. Genet.* **156**:99–105.

Eraso, P., and Gancedo, J. M., 1984, Catabolite repression in yeasts is not associated with low levels of cAMP, *Eur. J. Biochem.* **141**:195–198.

Federoff, H. J., Eccleshall, T. R., and Marmur, J., 1983, Regulation of maltase synthesis in *Saccharomyces carlsbergensis, J. Bacteriol.* **154**:1301–1308.

Ferguson, J., Boll, M., and Holzer, H., 1967, Yeast malate dehydrogenase and enzyme inactivation in catabolite repression, *Eur. J. Biochem.* **1**:21–25.

Fosset, M., Muir, L. W., Nielsen, L. D., and Fischer, E. H., 1971, Purification and properties of yeast glycogen phosphorylase *a* and *b, Biochemistry* **10**:4105–4113.

Foy, J. J., and Bhattacharjee, J. K., 1978, Biosynthesis and regulation of fructose-1,6-bisphosphatase and phosphofructokinase in *Saccharomyces cerevisiae* grown in the presence of glucose and gluconeogenic carbon sources, *J. Bacteriol.* **136**:647–656.

Fraenkel, D. G., 1982, Carbohydrate metabolism, in: *The Molecular Biology of the Yeast Saccharomyces: Metabolism and Gene Expression* (J. N. Strathern, E. W. Jones, and J. R. Broach, eds.), pp. 1–37, Cold Spring Harbor Laboratory, Cold Spring Harbor, N.Y.

Fraenkel, D. G., and Vinopal, R. T., 1973, Carbohydrate metabolism in bacteria, *Annu. Rev. Microbiol.* **27**:69–100.

Freitas-Valle, A. B., Menezes, R. R., Panek, A. D., and Mattoon, J. R., 1981, Relationship between succinate excretion and cytochrome levels in *Saccharomyces cerevisiae, Cell. Mol. Biol.* **27**:467–471.

Fröhlich, K. U., and Entian, K. D., 1982, Regulation of gluconeogenesis in the yeast *Saccharomyces cerevisiae, FEBS Lett.* **139**:164–166.

Fruton, J. S., 1972, *Molecules and Life,* Wiley–Interscience, New York.

Fukasawa, T., Obonai, K., Segawa, T., and Nogi, Y., 1980, The enzymes of the galactose cluster in *Saccharomyces cerevisiae:* Purification and characterization of uridine diphosphoglucose-4-epimerase, *J. Biol. Chem.* **255**:2705–2707.

Fulton, A. B., 1983, How crowded is the cytoplasm?, *Cell* **30**:345–347.

Funayama, S., Gancedo, J. M., and Gancedo, C., 1980, Turnover of yeast fructose-bisphosphatase in different metabolic conditions, *Eur. J. Biochem.* **109**:61–66.

Gancedo, C., 1971, Inactivation of fructose-1,6-diphosphatase by glucose in yeast, *J. Bacteriol.* **107**:401–405.

Gancedo, C., and Schwerzmann, K., 1976, Inactivation by glucose of phosphoenolpyruvate carboxykinase from *Saccharomyces cerevisiae, Arch. Microbiol.* **109**:221–225.

Gancedo, C., and Serrano, R., in press, Energy yielding metabolism in yeast, in: *The Yeasts* (A. H. Rose and J. S. Harrison, eds.), Vol. III, Academic Press, New York.

Gancedo, C., Salas, M. L., Giner, A., and Sols, A., 1965, Reciprocal effects of carbon sources on the levels of an AMP-sensitive fructose-1,6-diphosphate and phosphofructokinase in yeast, *Biochem. Biophys. Res. Commun.* **20**:15–20.

Gancedo, C., Gancedo, J. M., and Sols, A., 1967, Metabolite repression of fructose-1,6-diphosphatase in yeast, *Biochem. Biophys. Res. Commun.* **26**:528–531.

Gancedo, C., Gancedo, J. M., and Sols, A., 1968, Glycerol metabolism in yeasts: Pathways of utilization and production, *Eur. J. Biochem.* **5**:165–172.

Gancedo, J. M., and Gancedo, C., 1971, Fructose-1,6-diphosphatase, phosphofructokinase and glucose-6-phosphate dehydrogenase from fermenting and non-fermenting yeasts, *Arch. Mikrobiol.* **76**:132–138.

Gancedo, J. M., and Gancedo, C., 1973, Concentrations of intermediary metabolites in yeast, *Biochimie* **55**:205–211.

Gancedo, J. M., and Lagunas, R., 1973, Contribution of the pentose-phosphate pathway to glucose metabolism in *Saccharomyces cerevisiae:* A critical analysis on the use of labelled glucose, *Plant Sci. Lett.* **1**:193–200.

Gancedo, J. M., and Mazón, M. J., 1978, Transport of gluconate in *Rhodotorula glutinis:* Inactivation by glucose of the uptake system, *Arch. Biochem. Biophys.* **185**:466–472.

Gancedo, J. M., Gancedo, C., and Sols, A., 1967, Regulation of the concentration or activity of pyruvate kinase in yeasts and its relationship to gluconeogenesis, *Biochem. J.* **102**:23C–25C.

Gancedo, J. M., Clifton, D., and Fraenkel, D. G., 1977, Yeast hexokinase mutants, *J. Biol. Chem.* **252**:4443–4444.

Gancedo, J. M., Mazón, M. J., and Gancedo, C., 1982, Kinetic differences between two interconvertible forms of fructose-1,6-bisphosphatase from *Saccharomyces cerevisiae, Arch. Biochem. Biophys.* **218**:478–482.

Gancedo, J. M., Mazón, M. J., and Gancedo, C., 1983, Fructose 2,6-bisphosphate activates the cAMP-dependent phosphorylation of yeast fructose-1,6-bisphosphatase *in vitro, J. Biol. Chem.* **258**:5998–5999.

Glaser, L., and Brown, D. H., 1955, Purification and properties of D-glucose-6-phosphate dehydrogenase, *J. Biol. Chem.* **216**:67–79.

Grossmann, M. K., and Zimmermann, F. K., 1979, The structural genes of internal invertases in *Saccharomyces cerevisiae, Mol. Gen. Genet.* **175**:223–229.

Guarente, L., Yocum, R. R., and Gifford, P., 1982, A GAL-CYCl hybrid yeast promoter identifies the GAL 4 regulatory region as an upstream site, *Proc. Natl. Acad. Sci. USA* **79**:7410–7414.

Gunja-Smith, Z., Patil, N. B., and Smith, E. E., 1977, Two pools of glycogen in *Saccharomyces, J. Bacteriol.* **130**:818–825.

Haarasilta, S., and Oura, E., 1975, On the activity and regulation of anaplerotic and gluconeogenic enzymes during the growth process of baker's yeast, *Eur. J. Biochem.* **52**:1–7.

Haarasilta, S., and Taskinen, L., 1977, Location of three key enzymes of gluconeogenesis in baker's yeast, *Arch. Microbiol.* **113**:159–161.

Hackel, R. A., 1975, Genetic control of invertase formation. I. Isolation and characterization of mutants affecting sucrose utilization, *Mol. Gen. Genet.* **140**:361–370.

Harris, C. E., Kobes, R. D., Teller, D. C., and Rutter, W. J., 1969, The molecular characteristics of yeast aldolase, *Biochemistry* **8**:2442–2454.

Heerde, E., and Radler, F., 1978, Metabolism of the anaerobic formation of succinic acid by *Saccharomyces cerevisiae, Arch. Microbiol.* **117**:269–276.

Herbert, D., Elsworth, R., and Telling, R. C., 1956, The continuous culture of bacteria: A theoretical and experimental study, *J. Gen. Microbiol.* **14**:601–622.

Heredia, C. F., Sols, A., and De la Fuente, G., 1968, Specificity of the constitutive transport in yeast, *Eur. J. Biochem.* **5**:321–329.

Herrera, L. S., and Pascual, C., 1978, Genetical and biochemical studies of glucosephosphate isomerase deficient mutants in *Saccharomyces cerevisiae, J. Gen. Microbiol.* **108**:305–310.

Herrera, L. S., Pascual, C., and Alvarez, X., 1976, Genetic and biochemical studies of phosphomannose isomerase deficient mutants of *Saccharomyces cerevisiae, Mol. Gen. Genet.* **144:**223–230.

Hers, H. G., and Van Schaftingen, E., 1982, Fructose 2,6-bisphosphate 2 years after its discovery, *Biochem. J.* **206:**1–12.

Hess, B., Haeckel, R., and Brand, K., 1966, FDP-activation of yeast pyruvate kinase, *Biochem. Biophys. Res. Commun.* **24:**824–831.

Hess, B., Boiteux, A., and Krüger, J., 1969, Cooperation of glycolytic enzymes, *Adv. Enzyme Regul.* **7:**149–167.

Hitzeman, R. A., Clarke, L., and Carbon, J., 1980, Isolation and characterization of the yeast 3-phosphoglycerokinase gene (PGK) by an immunological screening technique, *J. Biol. Chem.* **255:**12073–12080.

Höfer, M., and Kotyk, A., 1968, Tight coupling of monosaccharide transport and metabolism in *Rhodotorula gracilis, Folia Microbiol.* **13:**197–204.

Höfer, M., Brand, K., Deckner, K., and Becker, J. U., 1971, Importance of the pentose phosphate pathway for D-glucose catabolism in the obligatory aerobic yeast *Rhodotorula gracilis, Biochem. J.* **123:**855–863.

Holland, J. P., and Holland, M. J., 1979, The primary structure of a glyceraldehyde-3-phosphate dehydrogenase gene from *Saccharomyces cerevisiae, J. Biol. Chem.* **254:**9839–9845.

Holland, J. P., and Holland, M. J., 1980, Structural comparison of two nontandemly repeated yeast glyceraldehyde-3-phosphate dehydrogenase genes, *J. Biol. Chem.* **255:**2596–2605.

Holland, M. J., and Holland, J. P., 1979, Isolation and characterization of a gene coding for glyceraldehyde-3-phosphate dehydrogenase from *Saccharomyces cerevisiae, J. Biol. Chem.* **254:**5466–5474.

Holland, M. J., Holland, J. P., Thill, G. P., and Jackson, K. A., 1981, The primary structures of two yeast enolase genes, *J. Biol. Chem.* **256:**1385–1395.

Holzer, H., 1961, Regulation of carbohydrate metabolism by enzyme competition, *Cold Spring Harbor Symp. Quant. Biol.* **26:**227–288.

Holzer, H., 1976, Catabolite inactivation in yeast, *Trends Biochem. Sci.* **1:**178–181.

Holzer, H., and Goedde, H. W., 1957, Zwei Wege von Pyruvat zu Acetyl-Coenzym A in Hefe, *Biochem. Z.* **329:**175–191.

Hommes, F. A., 1966, Effect of glucose on the levels of glycolytic enzymes in yeast, *Arch. Biochem. Biophys.* **114:**231–233.

Indge, K. J., 1968, Phosphates of the yeast cell vacuole, *J. Gen. Microbiol.* **51:**447–455.

Inoue, H., and Shimoda, C., 1981, Induction of trehalase activity on a nitrogen-free medium: A sporulation-specific event in the fission yeast *Schizosaccharomyces pombe, Mol. Gen. Genet.* **183:**32–36.

Katz, J., and Wood, H. G., 1963, The use of $C^{14}O_2$ yields from glucose-1- and -6-C^{14} for the evaluation of the pathways of glucose metabolism, *J. Biol. Chem.* **238:**517–523.

Katz, R., Kilpatrick, L., and Chance, B., 1971, Acquisition and loss of rotenone sensitivity in *Torulopsis utilis, Eur. J. Biochem.* **21:**301–307.

Kawasaki, G., and Fraenkel, D. G., 1982, Cloning of yeast glycolysis genes by complementation, *Biochem. Biophys. Res. Commun.* **108:**1107–1112.

Keller, F., Schellenberg, M., and Wiemken, A., 1982, Localization of trehalase in vacuoles and of trehalose in the cytosol of yeast (*Saccharomyces cerevisiae*), *Arch. Microbiol.* **131:**298–301.

Kempe, T. D., Nakagawa, Y., and Noltmann, E. A., 1974, Physical and chemical properties of yeast phosphoglucose isomerase isoenzymes, *J. Biol. Chem.* **249:**4617–4624.

Khan, N. A., Zimmermann, F. K., and Eaton, N. R., 1973, Genetic and biochemical evidence of sucrose fermentation by maltase in yeast, *Mol. Gen. Genet.* **123:**43–50.

Klein, H. P., and Jahnke, L., 1979, Effects of aeration on formation and localization of the acetyl coenzyme A synthetases of *Saccharomyces cerevisiae, J. Bacteriol.* **137**:179–184.

Kopperschläger, G., and Hofmann, E., 1969, Uber multiple Formen der Hexokinase in Hefe, *Eur. J. Biochem.* **9**:419–423.

Kopperschläger, G., Bär, J., Nissler, K., and Hofmann, E., 1977, Physicochemical parameters and subunit composition of yeast phosphofructokinase, *Eur. J. Biochem.* **81**:317–327.

Kotyk, A., 1963, Intracellular pH of baker's yeast, *Folia Microbiol.* **8**:27–30.

Kotyk, A., 1967, Properties of the sugar carrier in baker's yeast. II. Specificity of transport, *Folia Microbiol.* **12**:121–131.

Kotyk, A., and Höfer, M., 1965, Uphill transport of sugars in the yeast *Rhodotorula gracilis, Biochim. Biophys. Acta* **102**:410–422.

Kotyk, A., and Michaljaničová, D., 1979, Uptake of trehalose by *Saccharomyces cerevisiae, J. Gen. Microbiol.* **110**:323–332.

Krátký, Z., and Biely, P., 1980, Inducible β-xyloside permease as a constituent of the xylan-degrading enzyme system of the yeast *Cryptococcus albidus, Eur. J. Biochem.* **112**:367–373.

Krebs, H. A., 1972, The Pasteur effect and the relations between respiration and fermentation, *Essays Biochem.* **8**:1–35.

Krietsch, W. K. G., Pentchev, P. G., Klingenbürg, H., Hofstätter, T., and Bücher, T., 1970, The isolation and crystallization of yeast and rabbit liver triosephosphate isomerase and a comparative characterization with the rabbit muscle enzyme, *Eur. J. Biochem.* **14**:289–300.

Küenzi, M. T., and Fiechter, A., 1969, Changes in carbohydrate composition and trehalase-activity during the budding cycle of *Saccharomyces cerevisiae, Arch. Mikrobiol.* **64**:396–407.

Lagunas, R., 1979, Energetic irrelevance of aerobiosis for *S. cerevisiae* growing on sugars, *Mol. Cell. Biochem.* **27**:139–146.

Lagunas, R., 1981, Is *Saccharomyces cerevisiae* a typical facultative anaerobe?, *Trends Biochem. Sci.* **6**:201–202.

Lagunas, R., and Gancedo, J. M., 1973, Reduced pyridine nucleotide balance in glucose-growing *S. cerevisiae, Eur. J. Biochem.* **37**:90–94.

Lagunas, R., and Gancedo, C., 1983, Role of phosphate in the regulation of the Pasteur effect in *Saccharomyces cerevisiae, Eur. J. Biochem.* **137**:479–483.

Lam, K. B., and Marmur, J., 1977, Isolation and characterization of *Saccharomyces cerevisiae* glycolytic pathway mutants, *J. Bacteriol.* **130**:746–749.

Lancashire, M., Payton, A., Webber, M. J., and Hartley, B. S., 1981, Petite-negative mutants of *Saccharomyces cerevisiae, Mol. Gen. Genet.* **181**:409–410.

Laurent, M., Chaffotte, A. F., Tenu, J. P., Roucous, C., and Seydoux, F. J., 1978, Binding of nucleotides AMP and ATP to yeast phosphofructokinase: Evidence for distinct catalytic and regulatory subunits, *Biochem. Biophys. Res. Commun.* **80**:646–652.

Laurent, M., Seydoux, F. J., and Dessen, P., 1979, Allosteric regulation of yeast phosphofructokinase: Correlation between equilibrium binding, spectroscopic and kinetic data, *J. Biol. Chem.* **254**:7515–7520.

Lederer, B., Vissers, S., Van Schaftingen, E., and Hers, H. G., 1981, Fructose-2,6-bisphosphate in yeast, *Biochem. Biophys. Res. Commun.* **103**:1281–1287.

Lenz, A. G., and Holzer, H., 1980, Rapid reversible inactivation of fructose-1,6-bisphosphatase in yeast, *FEBS Lett.* **109**:271–274.

Lerch, K., and Fischer, E. H., 1975, Amino acid sequence of two functional sites in yeast glycogen phosphorylase, *Biochemistry* **14**:2009–2014.

Light, P. A., Ragan, C. I., Clegg, R. A., and Garland, P. B., 1968, Iron-limited growth of *Torulopsis utilis* and the reversible loss of mitochondrial energy conservation at site 1 and of sensitivity to rotenone and piericidin A, *FEBS Lett.* **1**:4–8.

Lillie, S. H., and Pringle, J. R., 1980, Reserve carbohydrate metabolism in *Saccharomyces cerevisiae:* Responses to nutrient limitation, *J. Bacteriol.* **143:**1384–1394.

Llorente, N., and Nuñez de Castro, I., 1977, Physiological role of yeast NAD(P)+ and NADP+-linked aldehyde dehydrogenases, *Rev. Esp. Fisiol.* **33:**135–142.

Lobo, Z., and Maitra, P. K., 1977, Physiological role of glucose-phosphorylating enzymes in *Saccharomyces cerevisiae, Arch. Biochem. Biophys.* **182:**639–645.

Lobo, Z., and Maitra, P. K., 1982a, A particulate phosphofructokinase from yeast, *FEBS Lett.* **137:**279–282.

Lobo, Z., and Maitra, P. K., 1982b, Genetic evidence for distinct catalytic and regulatory subunits in yeast phosphofructokinase, *FEBS Lett.* **139:**93–96.

Lobo, Z., and Maitra, P. K., 1982c, Pentose phosphate pathway mutants of yeast, *Mol. Gen. Genet.* **185:**367–368.

Lowry, C. W., Weiss, J. L., Wathall, D. A., and Zitomer, R. S., 1983, Modulator sequences mediate oxygen regulation of CYC1 and a neighboring gene in yeast, *Proc. Natl. Acad. Sci. USA* **80:**151–155.

Lutstorf, U., and Megnet, R., 1968, Multiple forms of alcohol dehydrogenase in *Saccharomyces cerevisiae, Arch. Biochem. Biophys.* **126:**933–944.

Machado, A., Nuñez de Castro, I., and Mayor, F., 1975, Isocitrate dehydrogenases and oxoglutarate dehydrogenase activities of baker's yeast grown in a variety of hypoxic conditions, *Mol. Cell. Biochem.* **6:**93–100.

Magasanik, B., 1961, Catabolite repression, *Cold Spring Harbor Symp. Quant. Biol.* **26:**249–262.

Mahler, H. R., Jaynes, P. K., McDonough, J. P., and Hanson, D. K., 1981, Catabolite repression in yeast: Mediation by cAMP, *Curr. Top. Cell. Regul.* **8:**455–474.

Maitra, P. K., 1970, A glucokinase from *Saccharomyces cerevisiae, J. Biol. Chem.* **245:**2423–2431.

Maitra, P. K., 1971, Glucose and fructose metabolism in a phosphoglucose-isomeraseless mutant of *Saccharomyces cerevisiae, J. Bacteriol.* **107:**759–769.

Maitra, P. K., and Lobo, Z., 1971, A kinetic study of glycolytic enzyme synthesis in yeast, *J. Biol. Chem.* **246:**475–488.

Maitra, P. K., and Lobo, Z., 1977a, Yeasts pyruvate kinase: A mutant form catalytically insensitive to fructose 1,6 bisphosphate, *Eur. J. Biochem,* **78,** 353–360.

Maitra, P. K., and Lobo, Z., 1977b, Genetic studies with a phosphoglucose isomerase mutant of *Saccharomyces cerevisiae, Mol. Gen. Genet.* **156:**55–60.

Maitra, P. K., and Lobo, Z., 1977c, Pyruvate kinase mutants of *Saccharomyces cerevisiae:* Biochemical and genetic characterization, *Mol. Gen. Genet.* **152:**193–200.

Matsumoto, K., Uno, I., Toh-E, A., Ishikawa, T., and Oshima, Y., 1982, Cyclic AMP may not be involved in catabolite repression in *Saccharomyces cerevisiae:* Evidence from mutants capable of utilizing it as an adenine source, *J. Bacteriol.* **150:**277–285.

Matsumoto, K., Yoshimatsu, T., and Oshima, Y., 1983, Recessive mutations conferring resistance to carbon catabolite repression of galactokinase synthesis in *Saccharomyces cerevisiae, J. Bacteriol.* **153:**1405–1414.

Mazón, M. J., Gancedo, J. M., and Gancedo, C., 1974a, Glucose metabolism in *Rhodotorula glutinis,* in: *Proceedings of the Fourth International Symposium on Yeasts* (H. Klaushofer and U. B. Sleytr, eds.), Part I, p. 31, Hochschülerschaft an der Hochschule für Bodenkultur, Wien.

Mazón, M. J., Gancedo, J. M., and Gancedo, C., 1974b, Identification of an unusual phosphofructokinase in the red yeast *Rhodotorula glutinis, Biochem. Biophys. Res. Commun.* **61:**1304–1309.

Mazón, M. J., Gancedo, J. M., and Gancedo, C., 1975, Hexose kinases from *Rhodotorula glutinis, Arch. Biochem. Biophys.* **167:**452–457.

Mazón, M. J., Gancedo, J. M., and Gancedo, C., 1981, Inactivation and turnover of fructose-1,6-

bisphosphatase from *Saccharomyces cerevisiae,* in: *Metabolic Interconversion of Enzymes 1980* (H. Holzer, ed.), pp. 168–173, Springer-Verlag, Berlin.

Mazón, M. J., Gancedo, J. M., and Gancedo, C., 1982a, Inactivation of yeast fructose-1,6-bisphosphatase: *In vivo* phosphorylation of the enzyme, *J. Biol. Chem.* **257:**1128–1130.

Mazón, M. J., Gancedo, J. M., and Gancedo, C., 1982b, Phosphorylation and inactivation of yeast fructose bisphosphatase *in vivo* by glucose and by protein ionophores: A possible role for cAMP, *Eur. J. Biochem.* **127:**605–608.

Megnet, R., 1965, Alkoholdehydrogenasemutanten von *Schizosaccharomyces pombe, Pathol. Microbiol.* **28:**50–57.

Megnet, R., 1967, Mutants partially deficient in alcohol dehydrogenase in *Schizosaccharomyces pombe, Arch. Biochem. Biophys.* **121:**194–201.

Melling, J., 1977, Regulation of enzyme synthesis in continuous culture, in: *Topics in Enzyme and Fermentation Biotechnology* (A. Wiseman, ed.), pp. 10–42, Horwood, Chichester.

Michels, C. A., and Romanowski, A., 1980, Pleiotropic glucose repression-resistant mutation in *Saccharomyces carlsbergensis, J. Bacteriol.* **143:**674–679.

Mitchell, P., and Moyle, J., 1969, Estimation of membrane potential and pH difference across the cristae membrane of rat liver mitochondria, *Eur. J. Biochem.* **7:**471–484.

Müller, D., and Holzer, H., 1981, Regulation of fructose-1,6-bisphosphatase in yeast by phosphorylation/dephosphorylation, *Biochem. Biophys. Res. Commun.* **103:**926–933.

Müller, M., Müller, H., and Holzer, H., 1981, Immunochemical studies on catabolite inactivation of phosphoenolpyruvate carboxykinase in *Saccharomyces cerevisiae, J. Biol. Chem.* **256:**723–727.

Nadkarni, M., Lobo, Z., and Maitra, P. K., 1982, Particulate phosphofructokinase of yeast: Physiological studies, *FEBS Lett.* **147:**251–255.

Navon, G., Shulman, R. G., Yamane, T., Eccleshall, T. R., Lam, K. B., Baronofsky, J. J., and Marmur, J., 1979, Phosphorus-31 nuclear magnetic resonance studies of wild-type and glycolytic pathway mutants of *Saccharomyces cerevisiae, Biochemistry* **18:**4487–4499.

Neeff, J., Hägele, E., Nauhaus, J., Heer, U., and Mecke, D., 1978, Evidence for catabolite degradation in the glucose dependent inactivation of yeast cytoplasmic malate dehydrogenase, *Eur. J. Biochem.* **87:**489–495.

Newsholme, E. A., Crabtree, B., Higgins, S. J., Thornton, S. D., and Start, C., 1972, The activities of fructose diphosphatase in flight muscles from the bumble-bee and the role of this enzyme in heat generation, *Biochem. J.* **128:**84–97.

Nickerson, W. J., and Carroll, W. R., 1945, On the metabolism of *Zygosaccharomyces, Arch. Biochem.* **7:**257–271.

Ohnishi, T., 1970, Induction of the site I phosphorylation *in vivo* in *Saccharomyces carlsbergensis, Biochem. Biophys. Res. Commun.* **41:**344–352.

Okada, H., and Halvorson, H. O., 1964, Uptake of alpha-thioethyl D-glucopyranoside by *Saccharomyces cerevisiae.* I. The genetic control of facilitated diffusion and active transport, *Biochim. Biophys. Acta* **82:**538–546.

Okorokov, L. A., Lichko, L. P., and Kulaev, I. S., 1980, Vacuoles: Main compartments of potassium, magnesium and phosphate ions in *Saccharomyces carlsbergensis* cells, *J. Bacteriol.* **144:**661–665.

Operti, M. S., Oliveira, D. E., Freitas-Valle, A. B., Oestreicher, E. G., Mattoon, J. R., and Panek, A. D., 1982, Relationship between trehalose metabolism and maltose utilization in *Saccharomyces cerevisiae:* Evidence for alternative pathways of trehalose synthesis, *Curr. Genet.* **5:**69–76.

Ortiz, C. H., Maia, J. C. C., Tenan, M. N., Braz-Padrao, G. R., Mattoon, J. R., and Panek, A. D., 1983, Regulation of yeast trehalase by a monocyclic, cyclic AMP-dependent phosphorylation–dephosphorylation cascade system, *J. Bacteriol.* **153:**644–651.

Oshima, Y., 1982, Regulatory circuits for gene expression: The metabolism of galactose and phosphate, in: *The Molecular Biology of the Yeast Saccharomyces: Metabolism and Gene Expression* (J. N. Strathern, E. W. Jones, and J. R. Broach, eds.), pp. 159–180, Cold Spring Harbor Laboratory, Cold Spring Harbor, N.Y.

Oura, E., 1974, Effect of aeration intensity on the biochemical composition of baker's yeast, *Biotechnol. Bioeng.* **16**:1197–1225.

Oura, E., 1977, Reaction products of yeast fermentations, *Process Biochem.* **12**(3):19–22.

Panek, A., 1962, Synthesis of trehalose by baker's yeast (*Saccharomyces cerevisiae*), *Arch. Biochem. Biophys.* **98**:349–355.

Panek, A., and Mattoon, J. R., 1977, Regulation of energy metabolism in *Saccharomyces cerevisiae:* Relationships between catabolite repression, trehalose synthesis and mitochondrial development, *Arch. Biochem. Biophys.* **183**:306–316.

Pasteur, L., 1860, Mémoire sur la fermentation alcoholique, *Ann. Chim. Phys.* Ser. 3 **58**:323–426.

Perea, J., and Gancedo, C., 1978, Glucose transport in a glucose-phosphate isomeraseless mutant of *Saccharomyces cerevisiae, Curr. Microbiol.* **1**:209–211.

Perlman, D., and Halvorson, H. A., 1981, Distinct repressible mRNAs for cytoplasmic and secreted yeast invertase are encoded by a single gene, *Cell* **25**:525–536.

Perlman, P. S., and Mahler, H. D., 1974, Derepression of mitochondria and their enzymes in yeast: Regulatory aspects, *Arch. Biochem. Biophys.* **162**:248–271.

Polakis, E. S., Bartley, W., and Meek, G. A., 1965, Changes in the activities of respiratory enzymes during aerobic growth of yeast on different carbon sources, *Biochem. J.* **97**:298–302.

Pringle, J. R., 1975, Methods for avoiding proteolytic artifacts in studies of enzymes and other proteins from yeasts, *Methods Cell Biol.* **12**:149–184.

Racker, E., 1961, IUB Congress in Moskow, personal communication.

Racker, E., 1974, History of the Pasteur effect and its pathobiology, *Mol. Cell. Biochem.* **5**:17–23.

Ramaiah, A., Hathaway, J. A., and Atkinson, D. E., 1964, Adenylate as a metabolic regulator: Effect on yeast phosphofructokinase kinetics, *J. Biol. Chem.* **239**:3619–3622.

Rigby, P. W. J., Burleigh, B. D., Jr., and Hartley, B. S., 1974, Gene duplication in experimental enzyme evolution, *Nature* **251**:200–204.

Rogers, P. J., and Stewart, P. R., 1974, Energetic efficiency and maintenance energy characteristics of *Saccharomyces cerevisiae* (wild type and petite) and *Candida parapsilosis* grown aerobically and microaerobically in continuous culture, *Arch. Mikrobiol.* **99**:25–46.

Romano, A. H., 1982, Facilitated diffusion of 6-deoxy-D-glucose in baker's yeast: Evidence against phosphorylation-associated transport of glucose, *J. Bacteriol.* **152**:1295–1297.

Rothman-Denes, L. B., and Cabib, E., 1970, Two forms of yeast glycogen synthetase and their role in glycogen accumulation, *Proc. Natl. Acad. Sci. USA* **66**:967–974.

Rothman-Denes, L. B., and Cabib, E., 1971, Glucose-6-phosphate dependent and independent forms of yeast glycogen synthetase: Their properties and interconversions, *Biochemistry* **10**:1236–1242.

Rothstein, A., 1954, Enzyme systems of the cell surface involved in the uptake of sugars by yeast, *Symp. Soc. Exp. Biol.* **8**:165–201.

Ruiz-Amil, M., de Torróntegui, G., Palacián, E., Catalina, L., and Losada, M., 1965, Properties and function of yeast pyruvate carboxylase, *J. Biol. Chem.* **240**:3485–3492.

Ruiz-Amil, M., Fernández, M. J., Medrano, L., and Losada, M., 1966, Cellular distribution of yeast pyruvate decarboxylase and its induction by glucose, *Arch. Mikrobiol.* **55**:46–53.

Saez, M. J., and Lagunas, R., 1976, Determination of intermediary metabolites in yeast: A critical examination of the effect of sampling conditions and recommendations for obtaining true levels, *Mol. Cell. Biochem.* **13**:73–78.

Salhany, J. M., Yamane, T., Shulman, R. G., and Ogawa, S., 1975, High resolution ^{31}P nuclear magnetic resonance studies of intact yeast cells, *Proc. Natl. Acad. Sci. USA* **72**:4966–4970.

Santa María, J., 1964, Utilización de sacarosa y maltosa por levaduras, *Bol. Inst. Nac. Invest. Agron. (Spain)* **50:**1–64.

Scheffers, W. A., 1961, On the inhibition of alcoholic fermentation in *Brettanomyces* yeasts under anaerobic conditions, *Experientia* **17:**10–42.

Schell, M. A., and Wilson, D. B., 1977, Purification and properties of galactose kinase from *Saccharomyces cerevisiae*, *J. Biol. Chem.* **252:**1162–1166.

Schimpfessel, L., 1968, Présence et régulation de la synthèse de deux alcool deshydrogènases chez la levure *Saccharomyces cerevisiae*, *Biochim. Biophys. Acta* **151:**317–329.

Schlanderer, G., and Dellweg, H., 1974, Cyclic AMP and catabolite repression in yeast, *Eur. J. Biochem.* **49:**305–316.

Schlenk, F., and Zyder-Cwick, C. R., 1970, Enzymatic activity of yeast cell ghosts produced by protein action on the membranes, *Arch. Biochem. Biophys.* **138:**220–225.

Schmitt, H. D., and Zimmermann, F. K., 1982, Genetic analysis of the pyruvate decarboxylase reaction in yeast glycolysis, *J. Bacteriol.* **151:**1146–1152.

Segawa, T., and Fukasawa, T., 1979, The enzymes of the galactose cluster in *Saccharomyces cerevisiae:* Purification and characterization of galactose-1 phosphate uridylyltransferase, *J. Biol. Chem.* **254:**10707–10709.

Serrano, R., and De la Fuente, G., 1974, Regulatory properties of the constitutive hexose transport in *Saccharomyces cerevisiae*, *Mol. Cell. Biochem.* **3:**161–171.

Serrano, R., Gancedo, J. M., and Gancedo, C., 1973, Assay of yeast enzymes in situ, *Eur. J. Biochem.* **34:**479–482.

Shabalin, Y. A., Vagabov, V. I., Tsiomenko, A. B., Zemlyanukhina, O. A., and Kulaev, I. S., 1977, Polyphosphate kinase activity in vacuoles of yeasts, *Biokhimiya* **42:**1642–1645.

Singh, B. R., and Datta, A., 1978, Glucose repression of the inducible catabolic pathway for N-acetylglucosamine in yeast, *Biochem. Biophys. Res. Commun.* **84:**58–64.

Sinha, P., and Maitra, P. K., 1977, Mutants of *Saccharomyces cerevisiae* having structurally altered pyruvate kinase, *Mol. Gen. Genet.* **158:**171–177.

Slavik, J., 1983, Intracellular pH topography: Determination by a fluorescent probe, *FEBS Lett.* **156:**227–230.

Sols, A., 1967, Regulation of carbohydrate transport and metabolism in yeast, in: *Aspects of Yeast Metabolism* (A. K. Mills and H. A. Krebs, eds.), pp. 47–66, Blackwell, Oxford.

Sols, A., 1976, The Pasteur effect in the allosteric era, in *Reflections on Biochemistry* (A. Kornberg, B. L. Horecker, L. Cornudella, and J. Oró, eds.), pp. 199–206, Pergamon Press, Elmsford, N.Y.

Sols, A., and Marco, R., 1970, Concentrations of metabolites and binding sites: Implications in metabolic regulation, *Curr. Top. Cell. Regul.* **2:**227–273.

Sols, A., and Salas, M. L., 1966, Phosphofructokinase. III. Yeast, *Methods Enzymol.* **9:**436–442.

Sols, A., Gancedo, C., and De la Fuente, G., 1971, Energy-yielding metabolism in yeast, in: *The Yeasts* (A. H. Rose and J. S. Harrison, eds.), Vol. 2, pp. 271–307, Academic Press, New York.

Souza, N. O., and Panek, A. D., 1968, Location of trehalase and trehalose in yeast cells, *Arch. Biochem. Biophys.* **125:**22–28.

Spiegelman, S., and Reiner, J. M., 1947, The formation and stabilization of an adaptive enzyme in the absence of its substrate, *J. Gen. Physiol.* **31:**175–193.

Sprague, G. F., Jr., 1977, Isolation and characterization of a *Saccharomyces cerevisiae* mutant deficient in pyruvate kinase activity, *J. Bacteriol.* **130:**232–241.

Stein, R. B., and Blum, J. J., 1978, On the analysis of futile cycles in metabolism, *J. Theor. Biol.* **72:**487–522.

Steinman, C. R., and Jakoby, W. B., 1968, Yeast aldehyde dehydrogenase, *J. Biol. Chem.* **243:**730–734.

Thevelein, J. M., den Hollander, J. A., and Shulman, R. G., 1982, Changes in the activity and properties of trehalase during early germination of yeast ascospores: Correlation with trehalose breakdown as studied by *in vivo* ^{13}C NMR, *Proc. Natl. Acad. Sci. USA* **79**:3503–3507.

Tijane, M. N., Chaffotte, A. F., Seydoux, F. J., Roucous, C., and Laurent, M., 1980, Sulfhydryl groups of yeast phosphofructokinase-specific localization on β subunits of fructose-6-phosphate binding sites as demonstrated by a differential chemical labeling study, *J. Biol. Chem.* **255**:10188–10193.

Toda, K., 1976, Invertase biosynthesis by *Saccharomyces carlsbergensis* in batch and continuous cultures, *Biotechnol. Bioeng.* **18**:1103–1115.

Toda, K., Yabe, I., and Yamagata, T., 1982, Invertase and phosphatase of yeast in a phosphate-limited continuous culture, *Eur. J. Appl. Microbiol. Biotechnol.* **16**:17–22.

Ullrich, J., and Wais, U., 1975, Pyruvate dehydrogenase complex from brewer's yeast: Regulation by the carbon sources, *Biochem. Soc. Trans.* **3**:920–924.

Van de Poll, K. W., Kerkenaar, A., and Schamhart, D. H. J., 1974, Isolation of a regulatory mutant of fructose-1,6-diphosphatase in *Saccharomyces carlsbergensis*, *J. Bacteriol.* **117**:965–970.

Van der Plaat, J. B., 1974, Cyclic 3′,5′-adenosine monophosphate stimulates trehalose degradation in baker's yeast, *Biochem. Biophys. Res. Commun.* **56**:580–587.

Van Solingen, P., and Van der Plaat, J. B., 1975, Partial purification of the protein system controlling the breakdown of trehalose in baker's yeast, *Biochem. Biophys. Res. Commun.* **62**:553–560.

Van Stevenink, J., 1968, Competition of sugars for the hexose transport system in yeast, *Biochim. Biophys. Acta.* **150**:424–434.

Van Wijk, R., and Konijn, T. M., 1971, Cyclic 3′,5′-AMP in *Saccharomyces carlsbergensis* under various conditions of catabolite repression, *FEBS Lett.* **13**:184–186.

Viñuela, E., Salas, M. L., and Sols, A., 1963, End-product inhibition of yeast phosphofructokinase by ATP, *Biochem. Biophys. Res. Commun.* **12**:140–145.

Wales; D. S., Cartledge, T. G., and Lloyd, D., 1980, Effects of glucose repression and anaerobiosis on the activities and subcellular distribution of tricarboxylic acid cycle and associated enzymes in *Saccharomyces carlsbergensis*, *J. Gen. Microbiol.* **116**:93–98.

Walsh, R. B., Kawasaki, G., and Fraenkel, D. G., 1983, Cloning of genes that complement yeast hexokinase and glucokinase mutants, *J. Bacteriol.* **154**:1002–1004.

Warburg, O., 1926, Uber die Wirkung von Blausäureäthylester (Athylcarbylamin) auf die Pasteursche Reaktion, *Biochem. Z.* **172**:432–441.

Weitzman, P. J. D., and Hewson, J. K., 1973, In situ regulation of yeast citrate synthase: Absence of ATP inhibition observed *in vitro*, *FEBS Lett.* **36**:227–231.

Wiemken, A., and Durr, M., 1974, Characterization of amino acid pools in the vacuolar compartment of *S. cerevisiae*, *Arch. Microbiol.* **101**:45–57.

Wiemken, A., and Schellenberg, M., 1982, Does a cyclic AMP-dependent phosphorylation initiate the transfer of trehalase from the cytosol into the vacuoles of *Saccharomyces cerevisiae?*, *FEBS Lett.* **150**:329–331.

Williamson, V. M., Bennetzen, J., Young, E. T., Nasmyth, K., and Hall, B. D., 1980, Isolation of the structural gene for alcohol dehydrogenase by genetic complementation in yeast, *Nature* **283**:214–216.

Wills, C., 1976, Production of yeast alcohol dehydrogenase isoenzymes by selection, *Nature* **261**:26–29.

Wills, C., and Phelps, J., 1975, A technique for the isolation of yeast alcohol dehydrogenase mutants with altered substrate specificity, *Arch. Biochem. Biophys.* **167**:627–637.

Wolf, D. H., and Ehmann, C., 1979, Studies on a proteinase B mutant of yeast, *Eur. J. Biochem.* **98**:375–384.

Zimmermann, F. K., 1982, Function of genetic material: Gene structure, gene function, and genetic regulation of metabolism in bacteria and fungi, in: *Fortschritte der Botanik* (H. Ellenberg, K.

Esser, K. Kubitzki, E. Schnepf, and H. Ziegler, eds.), Vol. 44, pp. 267–285, Springer-Verlag, Berlin.

Zimmermann, F. K., and Scheel, I., 1977, Mutants of *Saccharomyces cerevisiae* resistant to carbon catabolite repression, *Mol. Gen. Genet.* **154:**75–82.

Zitomer, R. S., Montgomery, D. L., Nichols, D. L., and Hall, B. D., 1979, Transcriptional regulation of the yeast cytochrome c gene, *Proc. Natl. Acad. Sci. USA* **76:**3627–3631.

Zubenko, G. S., and Jones, E. W., 1979, Catabolite inactivation of gluconeogenic enzymes in mutants of yeast deficient in proteinase B, *Proc. Natl. Acad. Sci. USA* **76:**4581–4585.

9

Regulation of Carbon Metabolism in Filamentous Fungi

WILLIAM McCULLOUGH, CLIVE F. ROBERTS,
STEPHEN A. OSMANI, and MICHAEL C. SCRUTTON

1. INTRODUCTION

Fungi are primarily soil organisms and as agents of decay produce extracellular enzymes degrading many natural polymers, especially polysaccharides (Section 2). The majority of fungi also utilize a wide range of growth substrates including poly- and oligosaccharides, hexoses, pentoses, many organic acids and alcohols, but cannot metabolize C_1 compounds such as methanol (Cooney and Levine, 1972). Although most filamentous fungi are obligate aerobes, they exhibit high affinities for oxygen and thus grow at very low oxygen concentrations (Bull and Bushell, 1976). It is therefore difficult to demonstrate true anaerobic growth, which is found only in a few fungi such as *Blastocladiella ramosa* (Held *et al.*, 1969), *Mucor* species (Bartnicki-Garcia and Nickerson, 1962), and *Fusarium oxysporum* (Gunner and Alexander, 1964). The capacity for fermentation in a limiting concentration of oxygen is, however, more common. For example, a lactic acid fermentation is found in aquatic lower fungi, an alcoholic fermentation in *F. lini*, and fermentations yielding lactic acid, ethanol, and CO_2 in *Rhizopus* species (see Cochrane, 1976). Anaerobic growth or fermentation may demand specific nutritional requirements such as thiamine and nicotinic acid in

WILLIAM McCULLOUGH • Department of Biology, University of Ulster, Newtownabbey, Co. Antrim BT37 0QB, Northern Ireland.   CLIVE F. ROBERTS • Department of Genetics, University of Leicester, Leicester LE1 7RH, United Kingdom.   STEPHEN A. OSMANI and MICHAEL C. SCRUTTON • Department of Biochemistry, King's College London, London WC2R 2LS, United Kingdom.

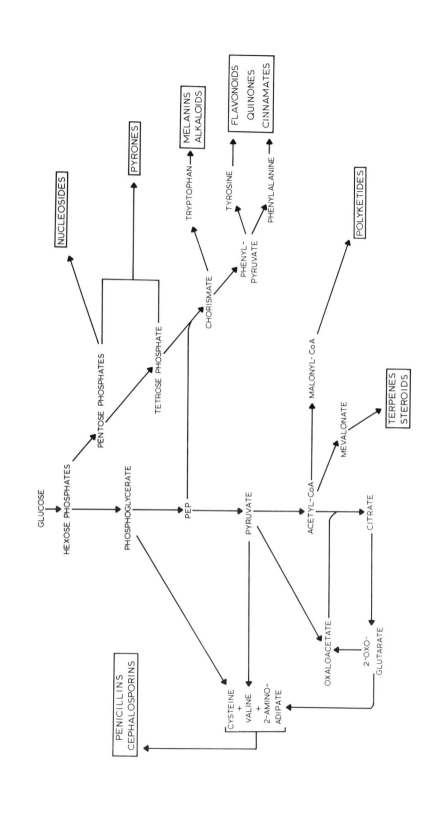

Figure 1. Relationships between primary and secondary metabolism in filamentous fungi. The origins of major classes of secondary metabolites produced by filamentous fungi are shown in relation to the central pathways of intermediary metabolism. Apart from the β-lactam antibiotics, no attempt is made to show the many fungal products that arise by further reactions between secondary metabolites of diverse origins. The biosynthesis of certain organic acids is discussed in some detail (Section 3.4) and illustrated in Figs. 5–7.

Among *nucleosides* are compounds such as cordycepin (3'-deoxyadenosine), which inhibits polyadenylation and a number of other cytotoxic and antitumor reagents (Fuska and Proksa, 1976). The *pyrones* include fermentation products such as kojic acid and the fungal toxin muscarine. The shikimate–chorismate pathway provides many derivatives of benzoic acid, the classic ergot and hallucinogenic *alkaloids*, and a vast array of derivatives of phenylpyruvic acid with the complex benzenoid ring structures of the *quinones*, *cinnamates*, and *flavonoids*. The *polyketides* comprise the largest class of fungal products, particularly among the Ascomycetes and related Deuteromycetes. Repeated C_2 condensation (mediated through decarboxylation of C_3 units) yields multiple-ring aromatic compounds of enormous variety among which are antibiotics such as patulin and griseofulvin and mycotoxins of the aflatoxin group (Moss, 1977; Smith and Hacking, 1983). Condensation of polyketides with aromatic amino acids yields the cytochalasins, which are cytotoxic reagents.

The isoprenoid pathway from mevalonic acid provides fungal hormones, such as the trisporic acids in the Mucorales, and a wide variety of fungal *steroids* and reflects the ability of fungi to effect specific steroid conversions. This pathway also provides, in addition to the carotenoids, a range of *terpenes* among which are phytotoxins such as the gibberellins. A more limited number of straight-chain derivatives of fatty acids not shown in the figure comprise the polyacetylenes and cyclized derivatives; the cyclopentanes include compounds like brefeldin A, which is structurally similar to mammalian prostaglandins.

A classic text relating fungal products to their biosynthetic origins is that of Turner (1971), and a series of reviews of the biosynthesis and variety of products in each major class is presented in Smith and Berry (1976). There is an extensive literature on fungal products frequently updated in the *Annual Review of Applied Microbiology*. It is of interest to note that a number of identical products, particularly antibiotics from the chorismate and polyketide pathways, are produced by disparate groups of microorganisms including bacteria, streptomycetes, and fungi (Lechevalier, 1975).

M. rouxii (Bartnicki-Garcia and Nickerson, 1961), or sterols and fatty acids in *Aqualinderella fermentans* (Held, 1970). Filamentous fungi are also noted for their ability to make an extraordinarily diverse array of secondary metabolic products. The relationships of the central pathways of primary metabolism (trophophase) to the biosynthesis of these products in secondary metabolism (idiophase) are illustrated in Fig. 1, although in this chapter we confine our attention to the production of certain organic acids with particular reference to the organization and regulation of these biosynthetic pathways (Section 3.4). The exuberant repertoire of biosynthetic capacities exhibited by fungi serves, however, to emphasize the heterogeneity of this group of organisms, which occupy a wide range of natural habitats and in which classification is based on morphological characteristics rather than on physiological properties. It would be hazardous to predict the extent to which the present classification schemes would survive a more searching analysis based, for example, on DNA sequence homology. However, some success in relating morphologically based classification to biochemical differences has been achieved with regard to the nature of the two major constituents of the cell wall in different orders (Bartnicki-Garcia, 1968). Accepting the limitations of the present classification, it is important to have some concept of the relationships between the various fungi discussed and this information is summarized in Table I.

Morphological differentiation is also a characteristic response of many fungi to environmental change. It should therefore be possible to utilize genetic and environmental manipulation, which is readily accessible for several filamentous fungi, to elucidate the biochemical basis for such processes. Such studies have, however, mainly consisted of observations of changes in enzyme levels occurring in the course of differentiation (see Schwalb, 1974; Hammond, 1981) and little insight has been gained thus far into the underlying biochemical relationships. Although the use of specific mutants has the potential to illuminate this area, there are few reports of such work except in *Neurospora crassa* where some association between altered (colonial) morphology and lesions in the oxidative pentose phosphate pathway has been demonstrated (Brody, 1973; Scott and Mahoney, 1976; see Section 3.1.1). The use of genetic information from organisms such as *N. crassa* and *Aspergillus nidulans,* which are well characterized in this respect, has been mainly limited to consideration of the properties resulting from known specific enzyme lesions. Certain classes of presumed regulatory gene mutations that produce striking multiple effects extending over considerable areas of nitrogen (Pateman and Kinghorn, 1977) or carbon (Kelly and Hynes, 1977) metabolism are apparently concerned with the integration of biochemical controls exerted in the preferential utilization of particular nitrogen or carbon sources for growth. However, the primary biochemical functions of these pleiotropic genes are known only in a few cases; e.g., *gdhA* in *A. nidulans,* which is the structural gene for biosynthetic NADP-glutamate dehydrogenase

Table I. Guide to the Classification of the Fungi Cited

Class	Order	Representative organisms
Chytridomycetes	Blastocladiales	*Aqualinderella fermentans*
		Blastocladiella emersonii, B. ramosa
Oomycetes		No data
Zygomycetes	Mucorales	*Mucor miehei, M. rouxii, M. rouxianus, M.*
		pusillus, M. racemosus
		Rhizopus nigricans, R. arrhizus, R. javanicus
Ascomycetes	Endomycetales	*Saccharomyces cerevisiae*
		Schizosaccharomyces pombe
Plectomycetes	Eurotiales	*Aspergillus* spp.[a]
		Chrysosporium lignorum[b]
		Penicillium spp.[a]
Pyrenomycetes	Sphaeriales	*Neurospora crassa*
	Chaetomiales	*Chaetomium*
Discomycetes	Heliotales	*Sclerotina sclerotiorum*
Basidiomycetes	Polyporales	*Coniophora cerebella*
	Agaricales	*Sporotrichum pulverulentum*[c]
		Agaricus bisporus
		Coprinus lagopus, C. cinereus, C. macrorhizus
Deuteromycetes		
Fungi Imperfecti	Monilales	*Alternaria alternata*
		Aspergillus spp.[a]
		Cephalosporium eichhorniae
		Dendryphiella salina
		Monilia
		Myrothecium verrucaria
		Penicillium spp.[a]
		Phymatotrichum omnivorum
		Piricularia oryzae
		Stachybotrys atra, S. echinata
		Trichoderma reesei (= *viride*), *Fusarium oxysporium, T. Koningii*
		Verticillium albo-atrum, F. lini, F. solani
Mycelia Sterilia		*Sclerotium*

[a]The majority of species of *Aspergillus* and *Penicillium* have no sexual stage in their life cycle. However, the close physiological similarities between the "imperfect" species and those also reproducing sexually ("perfect" stage) strongly suggest that both constitute single generic groups. This view is strengthened by the observation that mitotic recombination occurring during vegetative growth and comprising the parasexual cycle (Pontecorvo, 1956), first described in *Aspergillus nidulans* (*Emericella nidulans*) which also has a conventional sexual cycle, occurs widely among asexual species of *Aspergillus* and *Penicillium* and also other similar imperfect fungi such as *Cephalosporium, Verticillium,* and *Fusarium* (see reviews by Day, 1974; Fincham *et al.,* 1979). Detailed discussion of the relationships between the species groups in *Aspergillus* and in *Penicillium* are given by Raper and Fennell (1965) and Raper and Thom (1949), respectively.

[b]Imperfect species considered to be closely related to *Gymnoascus.*

[c]Imperfect species considered to be closely related to *Polysporus.*

and is involved in ammonium repression since this enzyme provides a major entry point for NH_4^+ into metabolism (Pateman and Kinghorn, 1977). It is not clear whether other genes of this type also have a simple biochemical explanation or whether some at least have a more general regulatory role.

Previous articles on carbohydrate metabolism in filamentous fungi (Cochrane, 1976: Watson, 1976; Blumenthal, 1976; Casselton, 1976; McCullough et al., 1977) have largely been concerned with individual pathways or with genetic approaches to metabolism. We adopt an alternate approach by considering the strategies used to enable the growth of filamentous fungi on particular groups of carbon sources with special reference to the subcellular localization of the enzymes involved and the mechanisms responsible for their regulation. This approach explicitly recognizes the unique position of the fungi as organisms that in most cases grow on simple defined media but also exhibit an intracellular organization comparable to that of higher eukaryotes. Furthermore, since the genetics of some filamentous fungi, notably *N. crassa* and *A. nidulans*, are well developed, it is feasible to apply to these eukaryotic species the powerful combination of genetic and physiological investigations, a combination that has so successfully illuminated the organization of metabolism in prokaryotes such as *Escherichia coli*.

2. EXTRACELLULAR FORMATION OF HEXOSES FROM POLYSACCHARIDES

Growth of fungi in their natural environment depends on the ability to make growth substrate available to the cells. Many microorganisms produce enzymes capable of hydrolyzing plant polysaccharides (see also Chapter 5) since such polysaccharides represent the most abundant renewable carbon source on earth. Attention has been focused in recent years on the commercial exploitation of this process using a variety of microorganisms including several filamentous fungi.

2.1. Cellulose Degradation

One of the most abundant of the natural polysaccharides is cellulose, which is a linear chain of D-glucose molecules linked by β-1,4-glucosidic bonds. The linear chains are cross-linked by hydrogen bonds to give a fibril with a diameter of approximately 3 nm, and further cross-bonding gives larger-order fibrils 5–30 nm wide that contain highly ordered or crystalline structures interspersed with amorphous regions (Sihtola and Neimo, 1975).

The most important cellulolytic fungi belong to the Ascomycetes and Deuteromycetes, and include *Trichoderma, Penicillium, Fusarium, Aspergillus,* and *Chaetomium.* Of these the most studied is *T. viride* (renamed *T. reesei:* Rifai,

1969) from which cellulase preparations are obtained commercially. All of these organisms produce extracellular endo-β-1,4-glucanase, exoglucanase (cellobiohydrolase), and β-glucosidase in culture filtrates and can therefore degrade natural cellulose to glucose (Fig. 2). Thus, for example, when the cellulolytic enzyme complex from *T. reesei* (Celluloclast, Novo Industry) is fractionated by ion-exchange chromatography, the cellulase activity can be divided into three classes, which have been identified as two endoglucanases, two exoglucanases, and a β-glucosidase (Lützen *et al.*, 1983). In this instance the exoglucanases form the major activity in the complex and only a small amount of β-glucosidase is present. The balance between the individual activities differs, however, in cellulolytic complexes obtained from other filamentous fungi. The basidiomycete *Sporotrichum pulverulentum,* which causes white rot in wood, produces a similar range of enzymes including five endoglucanases, one exoglucanase, and two β-glucosidases (Eriksson, 1981). In contrast, *Coniophora cerebella,* which causes brown rot, produces mainly endo-β-1,4-glucanase with little exoglucanase (Eidsa, 1972). The cellulase system in *S. pulverulentum* also differs in other ways from that typical of the Ascomycetes and most Deuteromycetes. The exoglucanase present in this rot fungus liberates both cellobiose and glucose from the cellulose chains and is strongly inhibited by D-glucono-(1→5)-lactone (Reese *et al.*, 1968; Wood and McCrae, 1977b; Eriksson, 1978). Moreover, the final stage of cellobiose degradation in *Sporotrichum* (Fig. 2) involves two oxidizing systems in addition to β-glucosidase activity (Eriksson, 1978). One of

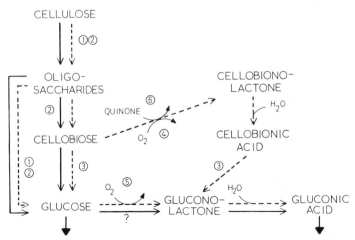

Figure 2. Pathways of cellulose degradation in *Trichoderma reesei* (*viride*) (—) and *Sporotrichum pulverulentum* (----). The enzymatic steps indicated are (1) endoglucanase, (2) exoglucanase, (3) β-glucosidase, (4) cellobiose oxidase, (5) glucose oxidase, (6) cellobiose-quinone oxidoreductase.

the enzymes is a heme protein that converts cellobiose to cellobionolactone as the final product. The other system is a cellobiose-quinone oxidoreductase that probably uses degradation products of lignin as the electron acceptor, and a lactonase that is also found in culture filtrates. Cellobiose oxidase activity has also been found in *Monilia* (Dekker, 1980) with properties similar to those of the enzyme from *Sporotrichum*. These latter oxidative enzymes may act under aerobic conditions to minimize accumulation of cellobiose and thus limit the extent to which cellulose degradation is inhibited by this disaccharide (see Section 2.1.3).

2.1.1. Endo-β-1,4-glucanase

Endoglucanase activity probably initiates cellulose degradation by random cleavage of cellulose chains within less-well-ordered regions of the structure, thereby increasing the number of free reducing ends of the polymer at which exoglucanase can act to liberate cellobiose (Wood and McCrae, 1972; Eriksson and Pettersson, 1972).

Endoglucanase purified from culture filtrates of fungi growing on cellulose shows considerable diversity of properties, even when the same commercial cellulase preparation is used (Hakansson *et al.*, 1978). Several forms of the enzyme have been identified in *T. reesei* (Berghem *et al.*, 1976), *T. koningii* (Wood and McCrae, 1977b), *S. pulverulentum* (Almin *et al.*, 1975; Eriksson and Pettersson, 1975a,b), and *A. fumigatus* (Parry *et al.*, 1983). These molecular forms divide into species having molecular weights below 20,000 and those having molecular weights approximating 50,000. Antibody raised to the high-molecular-weight endoglucanase of *T. reesei* did not cross-react with the low-molecular-weight enzyme (Hakansson *et al.*, 1979), indicating the two forms are unlikely to have a close structural relationship. The high-molecular-weight endoglucanase from *T. reesei* is itself heterogeneous since it can be resolved into several subcomponents by electrofocusing (Fagerstam and Pettersson, 1979). In this case, however, antibody raised against one subcomponent cross-reacted with the others, indicating a considerable extent of structural identity.

Marked differences in properties have also been noted when endoglucanases are isolated from culture filtrates of the same organism by different groups of workers. For example, when a freeze-dried filtrate of *T. reesei* growth medium was used (Hankansson *et al.*, 1978, 1979), the endoglucanases isolated had molecular weights of 20,000 and 51,000 and exhibited isoelectric points of 7.52 and 4.66, respectively. Neither enzyme appeared to be a glycoprotein. In contrast, Berghem *et al.* (1976), using a commercial preparation of cellulase from the same organism (Onozuka S.S., All Japan Biochemicals Co. Ltd.), purified two enzymes with molecular weights of 12,500 and 50,000 and isoelectric points of 4.6 and 3.39, respectively. Both the isolated endoglucanases were characterized as glycoproteins. The basis for such striking differences in properties is not clear although variation in culture conditions might determine the extent to which an extracellular enzyme is glycosylated. Furthermore, since proteases are present in the culture filtrates (Eriksson, 1981), modification of enzymes giving rise to artifacts may occur during the prolonged incubations that are often necessary for optimal enzyme production.

2.1.2. Exoglucanase

The exoglucanase component of the cellulase complex, which is often misleading described as "cellobiohydrolase," is probably identical with the C_1 enzyme described by Reese (1975). It can be identified by its ability to degrade disordered and hydrated cellulose derivatives, such as phospho- or carboxymethyl-cellulose, to yield the disaccharide cellobiose. The enzyme has been purified from culture filtrates of *Penicillium funiculosum* (Wood *et al.*, 1980), *T. koningii* (Wood and McCrae, 1977b), and *Fusarium solani* (Wood and McCrae, 1977a). Multiple forms of exoglucanase activity have been described in *F. solani* (Wood and McCrae, 1977a), *T. reesei* (Gwm and Brown, 1977), and *T. koningii* (Wood and McCrae, 1972) whereas only one exoglucanase has been found in *P. funiculosum* (Wood *et al.*, 1980). The molecular weights of the exoglucanases are generally in the range 45,000–48,000 and exhibit pI values in the region 3.8–4.95. All the exoglucanases are glycoproteins with a high proportion of mannose, but different forms of the enzyme from the same fungus may differ substantially in carbohydrate composition. The four different enzymes purified from *T. reesei* have similar properties and show immunological cross-reactivity, indicating structural similarity in the polypeptide component (Gwm and Brown, 1977). However, Fagerstam and Pettersson (1979) using another strain of *T. reesei* identified two immunologically distinct forms of exoglucanase. One of these forms was separated into multiple species by isoelectric focusing, although these subspecies were similar in structure since immunological cross-reactivity was demonstrated.

2.1.3. β-Glucosidase

The final enzyme activity of the cellulase complex, β-glucosidase, hydrolyzes cellobiose to glucose. This enzyme is present in low amounts in filtrates from *T. reesei* and its activity probably limits the extent of degradation since cellobiose inhibits the activity of the endo- and exoglucanases. The molecular weights reported for β-glucosidase are often considerably higher than those characteristic of the other enzymes. For example, values in the range 39,800 for *T. koningii* (Wood and McCrae, 1982) to 400,000 for *F. solani* (Wood, 1971) have been reported. Two forms of β-glucosidase purified from *T. koningii* had identical molecular weights and resembled each other closely in other properties such as pI and the extent of inhibition by gluconolactone. However, the two forms exhibited important differences in substrate specificity (Wood and Mc-Crae, 1982).

2.1.4. The Cellulase Complex

Highly ordered celluloses are typically not degraded on exposure to purified endo- or exoglucanases separately. However, when the purified enzymes were combined in the same ratio as found in culture filtrates, a quantitative recovery of cellulase activity was found when cotton was the substrate (Selby and Maitland, 1967; Wood and McCrae, 1972, 1977b; Wood et al., 1980). Such synergism between purified endo- and exoglucanases was more marked when the exoglucanase from P. funiculosum was combined with endoglucanase from those fungi that secrete both endo- and exoglucanases into the culture medium (T. koningii and F. solani) than with endoglucanase from fungi that do not secrete an exoglucanase [Myrothecium verrucaria, Stachybotrys atra, and Memnoniella (Stachybotrys) echinata] (Wood et al., 1980). This relationship has been interpreted to indicate the formation of an endo-/exoglucanase enzyme complex that acts simultaneously to cleave glucosidic bonds and to release cellobiose from crystalline regions of highly ordered cellulose (Wood et al., 1980).

2.1.5. Control of Cellulase Synthesis

Little is known about the regulation of cellulase synthesis. All the components of the complex are induced in the presence of cellulose although the actual inducer is presumably a soluble breakdown product of this polysaccharide. Cellobiose as well as other disaccharides such as lactose (Sternberg and Mandels, 1979), gentiobiose (Nisizawa et al., 1971a,b), sophorose (Mandels et al., 1962; Eriksson and Hamp, 1978) and thiocellobiose (Rho et al., 1982) have been identified as inducers in different fungi. Sophorose is the most effective inducer in Trichoderma and at a concentration as low as 1.5 μg/ml gives induction of a cellulase that degrades carboxymethylcellulose (Sternberg and Mandels, 1979). In contrast, much higher concentrations are required for induction by cellobiose. The continuous presence of free sophorose in the medium is essential for continued induction of cellulase (Loewenberg and Chapman, 1977). Sophorose and other β-disaccharides are produced in low concentration by the action of β-glucosidase (Crook and Stone, 1957). This observation prompted Sternberg and Mandels (1979) to suggest that sophorose or a related compound is the natural inducer, by analogy with the induction of the lac operon in E. coli by allolactose formed by the action of β-galactosidase (Jobe and Bourgeois, 1972). Induction of β-glucosidase activity is, however, repressed strongly by sophorose although a low level of activity that is bound to the cell wall remains (Sternberg and Mandels, 1980). Such repression occurs independently of the induction of the other enzymes of the cellulase complex. Accumulation of sophorose (or of a related disaccharide) may therefore maximize the rate of cellulase hydrolysis by its opposite effects on the levels of β-glucosidase (repression) and the endo-β-1,4-glucanases and exoglucanases (induction).

Glucose represses the formation of extracellular cellulolytic enzymes (Nisizawa *et al.*, 1972), although the underlying mechanisms are as little understood as is the case for general catabolite repression in fungi. No significant changes were found in cAMP levels in mycelia of *T. reesei* during the period of cellulase induction (Montenecourt *et al.*, 1980). However, this observation may be misleading since in some strains of *T. reesei* (RUT-C30) components of the cellulase system are secreted even in the presence of a repressing carbon source. Thus, this strain of *T. reesei* may be resistant to catabolite repression.

2.2. Hemicellulose (Xylan) Degradation

Hemicellulose or xylans are noncrystalline branched heteromorphic polymers containing a number of hexoses and pentoses linked by glycosidic bonds. The major sugar present is xylose together with significant amounts of glucose, mannose, arabinose, galactose, and uronic acids (Whistler and Richards, 1970). The sugars may also be acetylated. The straight-chain component is largely composed of xylose linked by β-1,4-glycosidic bonds with the side branches consisting of the other hexoses, pentoses, and uronic acids (Reilly, 1981). Since hemicelluloses are very heterogeneous in comparison with cellulose, numerous enzymes are involved in their degradation. This topic has been reviewed by Reilly (1981) who identified six distinct types of catalytic activity including β-xylosidases, exoxylanases, and four endoxylanases with different substrate specificities reflecting the variety of branch points within the chains.

β-Xylosidase has been identified in a number of filamentous fungi (Reese *et al.*, 1973) and has been purified from *Penicillium wortmanni* (Deleyn *et al.*, 1978) and *A. niger* (Takenishi *et al.*, 1973). Its preferred substrate is a short xylooligosaccharide and activity decreases as the chain length increases (Takenishi *et al.*, 1973). Xylose, xylonolactone, and gluconolactone are inhibitors.

The endoxylanases are difficult to classify, but Reilly (1981) identified four groups on the basis of (1) the nature of the products as xylose and xylobiose or as longer oligosaccharides, and (2) whether or not cleavage occurs at a branch point involving arabinose. All of these endoxylanases are present in filamentous fungi (Reilly, 1981) and have been investigated particularly in *A. niger* (Tsujisaka *et al.*, 1971; Takenishi and Tsujisaka, 1973; Rodionova *et al.*, 1977; Gorbacheva and Rodionova, 1977a,b).

2.3. Starch Degradation

Starch hydrolysis by fungi occurs predominantly by the action of two classes of enzyme, amyloglucosidases and α-amylases (Table II). The distribution of amylolytic enzymes in microorganisms and the enzymatic mechanisms involved have been reviewed by Fogarty and Kelly (1980) and French (1980), respectively.

Amyloglucosidases are exoenzymes that hydrolyze α-1,6 linkages in starch, releasing D-glucose. The rate of hydrolysis depends both on the molecular weight of the substrate and on the positions of the α-1,6 linkages. Dextrans containing α-1,2 or α-1,3 bonds are resistant to attack (Kobayashi and Matsuda, 1978). Different molecular forms of the enzyme are present in some fungi although their interrelationships are unclear. For example, in *A. niger,* some earlier reports (Pazur *et al.,* 1971; Lineback *et al.,* 1972) suggested the presence of two isozymes that differed in both molecular weight and carbohydrate composition whereas more recently the studies of Paszczynski *et al.* (1982) have indicated the existence of four isozymes that appear to be polymers of a common subunit having a molecular weight of 70,000. Three isozymes produced by *A. awamori* may also differ in substrate specificity (Hayashida, 1975).

The α-amylases are endoenzymes that specifically hydrolyze α-1,4 bonds in

Table II. Enzymes Produced by Filamentous Fungi Degrading Starch

Organism	Amyloglucosidases	α-Amylases
Aspergillus awamori	Hayashida *et al.* (1976), Yamasaki *et al.* (1977a)	Watanabe and Fukimbara (1967)
A. batatae	Bendetskii *et al.* (1974)	Bendetskii *et al.* (1974)
A. clavatus	Voitkova-Lepshikova and Kotskova-Kratokhvilova (1966)	
A. foetidus		Hang and Woodams (1977)
A. niger	Barton *et al.* (1972), Freedberg *et al.* (1975)	Aoki *et al.* (1971)
A. oryzae	Morita *et al.* (1966)	Bata *et al.* (1978)
Cephalosporium eichhorniae	Day (1978)	
Mucor rouxianus	Yamasaki *et al.* (1977c)	
M. pusillus		Adams and Deploey (1976)
Neurospora crassa	Murayama and Ishikawa (1973)	Gratzner (1972)
Penicillium oxalicum	Yamasaki *et al.* (1977b)	
Rhizopus javanicus	Watanabe and Fukimbara (1973)	
Trichoderma reesei (= *viride*)	Schellart *et al.* (1976)	

starch, yielding limit dextrins in addition to glucose and maltose. Such an activity has been reported in *Aspergillus*, *Mucor*, and *Neurospora* (Table II).

3. GROWTH ON GLUCOSE AND RELATED CARBON SOURCES

3.1. General Considerations

3.1.1. Conversion to Pyruvate

Filamentous fungi generally grow most effectively on hexoses, especially glucose, derived in the natural environment from extracellular breakdown of polysaccharides such as cellulose, hemicellulose, and starch (Section 2). Much of the information relevant to an understanding of carbohydrate metabolism in this group of organisms has therefore been obtained using glucose- or sucrose-grown mycelia. Although growth on other substrates can be demonstrated (see Sections 4–6), in the natural environment this would probably occur as a secondary response to accumulation of these substrates in the medium as the result of the initial metabolism of hexose. The pattern of hexose metabolism resembles that characteristic of the cells of higher eukaryotes, with added features reflecting the ability of many filamentous fungi to grow on a simple medium containing only sources of carbon, inorganic nitrogen (e.g., NH_4^+ or NO_3^-), phosphate, and some minerals.

Glucose is initially metabolized primarily via the classical glycolytic pathway in order to allow for energy production (mainly using the tricarboxylic acid cycle) and for provision of substrates for amino acid, lipid, purine, and pyrimidine biosynthesis. Metabolism also occurs via the oxidative pentose phosphate pathway to provide NADPH for reductive biosynthesis and ribose-5-phosphate for nucleotide and nucleic acid synthesis (Cochrane, 1976). There is no substantial evidence for glucose metabolism via the Entner–Doudoroff pathway in filamentous fungi although a related mechanism is used for degradation of gluconate in some species (see Section 3.4.3). Several lines of evidence support these concepts.

1. Mutant strains having lesions in one of the specific enzymes of glycolysis, such as pyruvate kinase (Payton and Roberts, 1976; Uitzetter, 1982), are unable to utilize hexoses as growth substrates but can grow on acetate, alanine, or butyrate.

2. The levels of certain glycolytic enzymes (e.g., pyruvate kinase) are elevated in hexose-grown mycelia as compared with those in mycelia grown on acetate (Stewart and Moore, 1971; Payton and Roberts, 1976).

Both points 1 and 2 only apply to the specific enzymes of glycolysis, i.e., phosphofructokinase and pyruvate kinase, with most of the data having been obtained for pyruvate kinase. Mutant strains having lesions in an enzyme required for both glycolysis and gluconeogenesis, such as enolase or aldolase, would be expected to show an entirely different phenotype. No such mutants have yet been described in filamentous fungi (however, see Chapter 7).

3. Studies measuring $^{14}CO_2$ yields from [1-^{14}C]- and [6-^{14}C] glucose have suggested that during growth, approximately 20–40% of the glucose taken up is metabolized by the oxidative pentose phosphate pathway (Mitchell and Shaw, 1966; Holligan and Jennings, 1972c; Gunasekaran, 1972). The extent of metabolism through the pentose phosphate pathway as assessed by this method is increased if NADPH is also required for the utilization of nitrogen sources such as NO_3^- (Holligan and Jennings, 1972c), and under such conditions the levels of oxidative pentose phosphate pathway enzymes are increased (Hankinson and Cove, 1974). Similar findings have been described for yeast (see Chapter 8). In a continuous culture system with A. nidulans grown under glucose limitation, an increased rate of glucose supply led to increased metabolism through the pentose phosphate pathway (Carter et al., 1971). While such observations are consistent with the proposed role for this pathway, the quantitative estimates of the extent of its contribution to glucose metabolism as provided by the [1-^{14}C]-/[6-^{14}C]glucose ratio method must be viewed with caution in light of the complications posed by recycling and diversion of carbon to other pathways (Katz and Wood, 1960; Wood et al., 1963).

The concept of the pentose phosphate pathway as an ancillary route of glucose catabolism is also supported by studies using mutant strains of N. crassa and A. nidulans although the properties of the strains used to date are far from ideal. In N. crassa, mutants having low levels of several of the enzymes of this pathway (Fuscaldo et al., 1971) or a glucose-6-phosphate dehydrogenase with atypical properties (Brody and Tatum, 1966; Scott and Tatum, 1970; Lechner et al., 1971) exhibit abnormal colonial morphology. Such mutants have low levels of NADPH and of unsaturated fatty acids (Brody and Nyc, 1970) as might be expected. However, an unequivocal link between pentose phosphate pathway flux and altered cellular structure cannot be inferred since other defects such as decreased level of cAMP occur in some of these strains (Scott, 1976). Studies using apparently selective inhibitors of cAMP phosphodiesterase indicate a relationship between cellular cAMP levels and the shape of the cells (Scott and Solomon, 1975). In A. nidulans, mutants having diminished levels of transaldolase (pppA) grow normally on glucose but fail to utilize pentoses such as xylose or ribose that are incorporated via this pathway (Hankinson, 1974). It may only be possible to recover mutant strains totally lacking one of the enzymes of the oxidative pentose phosphate pathway as heat-sensitive conditional mutations that are lethal for growth on any carbon source at the restrictive temperature, although in yeasts mutants lacking glucose-6-phosphate dehydrogenase appear to grow normally on glucose (see Chapter 8).

3.1.2. Hexose Interconversion

Growth on glucose as sole carbon source also requires the ability to synthesize other sugars needed for construction of a normal cell wall and for formation of glycoproteins and glycolipids. The importance of such interconversions is illustrated by isolation of mutants of *A. nidulans* that lack phosphoglucomutase and require addition of galactose for growth on glucose (Boschloo and Roberts, 1979), or that lack either phosphomannose isomerase or mutase and exhibit abnormal mycelial morphology (Valentine and Bainbridge, 1978). In *N. crassa,* deficiency of phosphoglucomutase is associated with colonial growth (Brody and Tatum, 1967).

Conversely, mutant strains that lack β-galactosidase, galactokinase, or galactose-1-phosphate uridyltransferase are unable to grow, or grow weakly, on galactose or lactose as sole carbon source (Roberts, 1963; Gajewski *et al.*, 1972; Fantes and Roberts, 1973) but grow effectively on glucose in the absence of galactose.

3.1.3. Pyruvate Metabolism

Several striking differences between fungi and other eukaryotes are found in the subcellular organization of pyruvate metabolism. Pyruvate dehydrogenase and pyruvate carboxylase are mitochondrial enzymes in higher eukaryotes (Bottger *et al.*, 1969; Denton and Halestrap, 1979), but in filamentous fungi pyruvate carboxylase is localized in the cytosol in all species studied thus far (Osmani and Scrutton, 1983, 1985). This situation also occurs in *Saccharomyces cerevisiae* (Haarasilta and Taskinen, 1977; see also Chapter 8) although not in certain other yeasts (Evans *et al.*, 1983). Since pyruvate dehydrogenase is localized in the mitochondrion of filamentous fungi (Harding *et al.*, 1970; Wieland *et al.*, 1972; Osmani and Scrutton, 1983, 1985), the pathway by which oxaloacetate, and hence citrate, is generated within the mitochondrion is very different from that characterizing mammalian mitochondria. A proposed scheme (Fig. 3) postulates transport of oxaloacetate carbon into the mitochondrion as L-malate. This scheme is formulated primarily on the basis of data obtained for *A. nidulans* and *Rhizopus arrhizus* and to a lesser extent for *A. niger* and *A. terreus* and is supported by a number of observations:

1. Mitochondrial and cytosolic isozymes of NAD-malate dehydrogenase and NADP-isocitrate dehydrogenase are typically present in glucose-grown mycelia (Ma *et al.*, 1981; Osmani and Scrutton, 1983).

2. Mutant strains lacking pyruvate carboxylase (*pycA*) require addition of a

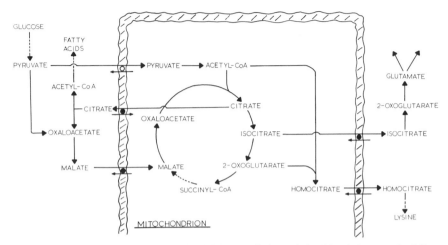

Figure 3. A proposed scheme for mitochondrial–cytosolic interrelationships during growth of filamentous fungi on glucose. The scheme is based primarily on enzyme subcellular localization data for
Aspergillus nidulans and *Rhizopus arrhizus* (Osmani and Scrutton, 1983, 1985) and on the assumption of similarity in antiports between mitochondria in fungi and mammalian liver.

dicarboxylic acid such as glutamate for growth on hexoses (Skinner and Armitt,
1972).

    3. Functional mitochondria have been obtained from several Aspergilli as
well as from other filamentous fungi with properties suggesting the presence of a
complete TCA cycle (Watson and Smith, 1967a; Hall and Greenawalt, 1967;
Watson *et al.*, 1969; Winskill, 1983). A complete TCA cycle is also suggested
by demonstration of the presence in a mitochondrial fraction from *A. nidulans* of
all of the enzymes of this cycle (Osmani and Scrutton, 1983) with the exception
of succinate thiokinase, which does not appear to have been detected in filamentous fungi. Although fungal mitochondria have often been reported to lack 2-
oxoglutarate dehydrogenase (e.g., Khouw and McCurdy, 1969; Ng *et al.*, 1973;
McCullough *et al.*, 1977), such findings are probably explained by the repression of the enzyme to low levels by increased concentrations of NH_4^+ (Rohr and
Kubicek, 1981) and by the conditions under which cell-free extracts were
prepared.

    4. *A. nidulans* when grown on glucose contains only NADP-glutamate
dehydrogenase, which is localized in the cytosol (Osmani and Scrutton, 1983)
and which is involved in glutamate biosynthesis (Kinghorn and Pateman, 1973).
Hence, carbon supplying this pathway may be exported from the mitochondrion
as isocitrate in exchange for L-malate and be further metabolized in the cytosol to
2-oxoglutarate (Fig. 3), thus providing a role for cytosolic NADP-isocitrate
dehydrogenase and generating in this compartment the NADPH required for
glutamate synthesis.

5. Export of acetyl units as citrate in exchange for L-malate (Fig. 3) is supported by detection of a cytosolic ATP-citrate lyase in *A. nidulans* (Osmani and Scrutton, 1983). The activity of this enzyme is high in glucose-grown, and low in acetate-grown, mycelia (McCullough, unpublished observations). The pyruvate/citrate cycle (Ballard and Hanson, 1967) is not required in *A. nidulans* due to the cytosolic localization of pyruvate carboxylase but instead a malate/citrate cycle may operate. Consequently, glucose-grown mycelia contain little, if any, NADP-malate dehydrogenase (EC 1.1.1.40) (McCullough and Roberts, 1974; see below).

We also propose in Fig. 3 that the initial step of lysine biosynthesis by the aminoadipate pathway (Vogel, 1964) occurs in the mitochondria and that homocitrate is then exported in exchange for L-malate using the citrate/malate antiport. This postulate is based on analogy to the situation in *S. cerevisiae* where homocitrate synthase appears to be localized in the mitochondrion (Tracy and Kohlaw, 1975). It is, however, equally likely that this step together with the remainder of lysine biosynthesis occurs in the cytosol. We have thus far been unable to detect homocitrate synthase in *A. nidulans* (Osmani and Scrutton, unpublished observations) and hence its intracellular localization is unknown.

The only major observation not accommodated by Fig. 3 is the reported presence of PEP carboxylase in *A. nidulans* (Bushell and Bull, 1981). We have, however, been unable to confirm this observation in either *A. nidulans* or *A. niger* under a variety of conditions (McCullough and Roberts, 1980; Osmani, Everard, and Scrutton, unpublished observations). The phenotype observed for *pycA* mutants in *A. nidulans* and *suc* mutants in *N. crassa,* both of which lack pyruvate carboxylase (Skinner and Armitt, 1972; Beever, 1973), is also inconsistent with the existence of an alternate route of oxaloacetate synthesis.

Less complete data suggest that Fig. 3 may provide an adequate description of the organization of pyruvate metabolism in other filamentous fungi except, however, for the role of NADP-malate dehydrogenase. In direct contrast to the situation in *A. nidulans* (see above and Section 5.2.3), this enzyme is present at significant levels in glucose-grown *N. crassa* (Zink and Shaw, 1968), *Fusarium oxysporum* (Zink and Katz, 1973), and *Rhizopus arrhizus* (Osmani and Scrutton, 1985). In *N. crassa,* growth on acetate represses NADP-malate dehydrogenase (Zink and Shaw, 1968) although this is not the case in *F. oxysporum* during growth on ethanol (Zink and Katz, 1973) or in *A. nidulans* where the enzyme is induced by growth on acetate (McCullough and Roberts, 1974). The role of NADP-malate dehydrogenase during growth on glucose is not clear since pyruvate for amino acid synthesis can be derived directly from glycolysis and at least in *R. arrhizus* a pyruvate/citrate cycle is ruled out by the cytosolic localization of pyruvate carboxylase (Osmani and Scrutton, 1985). Participation in cytosolic NADPH synthesis via a transhydrogenation cycle as illustrated in Fig. 4B appears possible in *R. arrhizus* since cytosolic NADP-isocitrate dehydrogenase is undetectable in glucose-grown mycelia from this organism although such a role

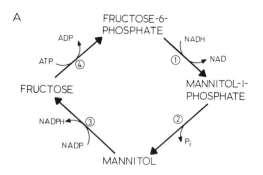

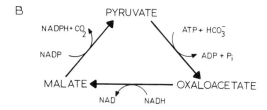

Net (A or B) :   NADH + NADP + ATP ──► NAD + NADPH + P$_i$

Figure 4. ATP-dependent transhydrogenation using (A) the proposed mannitol cycle or (B) the proposed pyruvate/malate cycle. In (A) the relevant enzymes are (1) mannitol-1-phosphate dehydrogenase, (2) mannitol-1-phosphatase, (3) mannitol dehydrogenase, (4) hexokinase.

would seem rather surprising given the high capacity of the pentose phosphate pathway in such mycelia (Osmani and Scrutton, 1985). There may therefore be an additional, and hitherto unsuspected, role for NADP-malate dehydrogenase in fungal metabolism (see also Section 5.2.3).

3.2. Enzymes of Hexose Metabolism

3.2.1. Phosphofructokinase

The catalytic and regulatory properties of phosphofructokinase have been characterized for a number of filamentous fungi including *N. crassa* (Tsao and Madley, 1972), *Mucor rouxii, Penicillium notatum* (Paveto and Passeron, 1977; Boonsaeng *et al.,* 1977a), and *A. niger* (Habison *et al.,* 1979). However, in all cases, instability of the enzyme has prevented the isolation of a pure preparation.

Hence, confirmation of results obtained in cell-free extracts has only been possible for *A. niger* from which the enzyme has been partially purified by Habison *et al.* (1983). Such studies do not reveal significant qualitative differences as compared to phosphofructokinases in higher eukaryotes. Fungal phosphofructokinases show substrate inhibition by ATP and an increasingly sigmoid relationship between initial velocity and [fructose-6-phosphate] as the ATP concentration is increased (Boonsaeng *et al.*, 1977a; Habison *et al.*, 1983). Citrate acts as a synergistic inhibitor with ATP for the enzyme from *A. niger*, and inhibition by citrate and ATP is counteracted by addition of AMP, ADP, or NH_4^+ (Habison *et al.*, 1983). Hence, the regulatory mechanisms responsible for interrelating glycolytic and TCA cycle flux appear similar to those found in yeast and other eukaryotic cells (see Krebs, 1972; see also Chapter 8) except that citrate may serve more directly as an indicator of the level of flux within the cycle itself (Paveto and Passeron, 1977). The specific relevance to citric acid production of control of phosphofructokinase by NH_4^+ and citrate is considered in Section 3.4.1.

The only major difference in regulatory properties described thus far may occur with the enzyme from *M. rouxii*, which is activated much more effectively by cAMP than by AMP (Boonsaeng *et al.*, 1977a; Paveto and Passeron, 1977). However, the significance of this observation is uncertain since the effect of cAMP is observed only at concentrations approaching the millimolar range (Paveto and Passeron, 1977; Habison *et al.*, 1983). It is not known whether fructose-2,6-bisphosphate is an activator of phosphofructokinases in filamentous fungi as it is in mammalian liver (see Hers and Van Schaftingen, 1982) and yeast (see Chapter 8).

Only certain filamentous fungi such as *M. rouxii* in the yeast phase (Paveto and Passeron, 1977) are capable of growing under anaerobic conditions and of performing classical fermentation as observed in yeast (see Chapter 8). The level of phosphofructokinase in yeast-like cells of *M. rouxii* is markedly higher than that present in the mycelia that predominate in aerobic conditions (Paveto and Passeron, 1977) in accord with the relative rates of glycolysis in the two situations. Similarly, in *N. crassa*, a decrease in the level of sucrose in the growth medium is accompanied by a fall in the level of phosphofructokinase in the mycelium although the levels of other enzymes such as glucose-6-phosphate dehydrogenase are maintained (Tsao *et al.*, 1969; see also Chapter 8). Such a pattern may relate to reutilization of acetate or ethanol initially excreted into the growth medium during sucrose utilization (Colvin *et al.*, 1973).

3.2.2. Pyruvate Kinase

In contrast to phosphofructokinase, very marked differences are observed in the properties of pyruvate kinases purified from different filamentous fungi. These differences relate primarily to the properties of activation by fructose-1,6-

bisphosphate. The enzymes obtained from *N. crassa* or *M. rouxii* are activated by fructose-1,6-bisphosphate in the range 0.1–1.0 mM and exhibit a sigmoid relationship between initial velocity and PEP concentration that becomes hyperbolic on addition of a saturating concentration of fructose-1,6-bisphosphate as described also for yeast (Passeron and Terenzi, 1970; Kapoor, 1975; Boonsaeng *et al.*, 1977a; see Chapter 8). However, in cell-free extracts of *Penicillium spp.* and *A. niger*, or in preparations purified from these fungi in the presence of fructose-1,6-bisphosphate, no activation by this hexose phosphate can be demonstrated and a hyperbolic relationship is observed between initial velocity and PEP concentration (Boonsaeng *et al.*, 1977a; Monori *et al.*, 1984). This latter response probably results from tight binding of fructose-1,6-bisphosphate since the enzyme from *A. niger* behaves like the enzymes from *N. crassa* and *M. rouxii* after prolonged incubation at 4°C. However, the extract treated in this way exhibits maximal activation by less than 1 μM fructose-1,6-bisphosphate (Monori *et al.*, 1984). Pyruvate kinase from *Coprinus cinereus* occupies an intermediate position, showing a sigmoid relationship between initial velocity and [PEP] and a K_A for fructose-1,6-bisphosphate of 1 μM (Stewart and Moore, 1971). Fungal pyruvate kinases appear insensitive to inhibition by ATP (except for the enzyme from *C. cinereus*) or by L-alanine (Stewart and Moore, 1971; Kapoor, 1975; Monori *et al.*, 1984).

Activation by fructose-1,6-bisphosphate presumably serves to coordinate flux response within the glycolytic pathway as proposed for mammalian liver and kidney (see Seubert and Schoner, 1971). The significance of the striking difference in the properties of activation by fructose-1,6-bisphosphate in different fungi is uncertain. Monori *et al.* (1984) have proposed that this difference may relate to the nature of the end products formed in "overflow" metabolism as either fermentative (ethanol, lactate) as in *M. rouxii* or *N. crassa*, or oxidative (organic acids) as in *Penicillium* or *Aspergillus*. Such a postulate does not account for the marked alteration in the K_A for fructose-1,6-bisphosphate caused by growth of *N. crassa* at 28 and 42°C (Kapoor *et al.*, 1976), or for the failure to observe differences in activation by this metabolite between the mycelial and yeastlike forms of *M. rouxii* (Passeron and Terenzi, 1970). Furthermore, the correlation of high sensitivity to activation by fructose-1,6-bisphosphate with oxidative "overflow" metabolism would not appear to be the expected relationship since fermentative metabolism should be a more direct consequence of the activation of glycolytic flux.

3.2.3. Pyruvate Carboxylase

Pyruvate carboxylases have been isolated and studied from several filamentous fungi. In *A. niger* (Bloom and Johnson, 1962; Feir and Suzuki, 1969) and *A. nidulans* (Osmani *et al.*, 1981; Osmani and Scrutton, 1981), this enzyme is activated weakly, if at all, by acetyl CoA and is inhibited by L-aspartate. Similar properties have been observed in preliminary studies of the enzyme isolated from other Ascomycetes such as *Verticillium albo-atrum* (Hartman and Keen, 1974a), *N. crassa* (Beever, 1973), and *Penicillium camemberti* (Schormuller and Stan, 1970). Detailed studies of the purified enzyme from *A. nidulans* (Osmani *et al.*, 1981) have indicated the presence of an allostéric acyl CoA activator site but have shown that occupancy of the site by an acyl CoA only causes a marked change in catalytic properties if the enzyme is assayed in the presence of a regulatory inhibitor such as L-aspartate. These studies also showed that long-chain unsaturated acyl coA's, such as oleoyl CoA, are more effective activators than acetyl CoA, and that inhibition is also observed in the presence of 2-oxoglutarate with properties that are distinct from those characterizing inhibition by L-aspartate. Distinct regulatory sites for L-aspartate and 2-oxoglutarate may therefore be present on *A. nidulans* pyruvate carboxylase. The existence of unique inhibitory sites for these two metabolites can be rationalized as a mechanism for independent regulation of the supply of oxaloacetate for synthesis of amino acids and other products derived from L-aspartate and 2-oxoglutarate (Osmani and Scrutton, 1983). In this context, we may note that neither L-aspartate nor 2-oxoglutarate causes complete inhibition of *A. nidulans* pyruvate carboxylase; that no competition is observed between the effects of these two inhibitors; and that the effects observed on parameters related to enzyme structure are consistent with the presence of two independent regulatory sites (Osmani *et al.*, 1984).

The regulatory properties of *A. nidulans* pyruvate carboxylase are similar to those described previously for the enzyme purified from *S. cerevisiae* (Cazzulo and Stoppani, 1968; Tolbert, 1970; Libor *et al.*, 1978), but differ markedly from those observed for pyruvate carboxylase isolated from *R. arrhizus* (*nigricans*) ATCC 13310, the only member of the lower fungi studied thus far. This enzyme, which was initially described as having no catalytic activity in the absence of acetyl CoA (Overman and Romano, 1969), has in fact substantial activity in the absence of this activator when examined more carefully. However, even in the presence of saturating concentrations of substrates and cofactors, addition of acetyl CoA causes a three to fourfold increase in catalytic activity when the enzyme is isolated in buffers containing 1 mM EDTA. Furthermore, the enzyme is subject to regulatory inhibition by 2-oxoadipate rather than by 2-oxoglutarate as in the Ascomycetes, as well as by L-aspartate (Osmani and Scrutton, 1985). This apparent divergence in properties between the pyruvate carboxylases isolated from Ascomycetes and the lower fungi may reflect differences in the balance and organization of amino acid and lipid synthesis in these two groups.

3.2.4. Hexokinase

Four isozymes of hexokinase have been detected in *N. crassa*. All these isozymes show similar kinetic properties, exhibit an apparent K_m for glucose of

50–100 μM, and notably are insensitive to regulatory inhibition by glucose-6-phosphate (Lagos and Ureta, 1980). Such properties might be expected since many fungi when grown in a balanced medium containing glucose or sucrose as carbon source do not restrict the metabolism of this carbohydrate, as would be the case in mammalian cells, but rather convert it to a reserve compound such as mannitol or trehalose (see Section 3.3).

3.2.5. Fructose-1,6-bisphosphate Aldolase

This enzyme in both A. niger (Jagannathan et al., 1956) and F. oxysporum (Ingram and Hochster, 1957) is activated by a divalent metal ion and hence resembles that present in S. cerevisiae.

3.2.6. Pyruvate Dehydrogenase

Lesions in the pyruvate dehydrogenase complex, which has a structure similar to that described for mammalian tissues (Reed, 1969; Harding et al., 1970), give rise to a phenotype in which growth occurs on acetate or on any substrate that can give rise to acetyl CoA such as butyrate, but not on glycolytic substrates such as glucose, glycerol or a source of pyruvate such as alanine. Mutants of this type have been described for A. nidulans (Romano and Kornberg, 1968; Payton et al., 1977) and N. crassa (Okumura et al., 1977). In A. nidulans, such mutants map at three unlinked sites and appear to correspond respectively to lesions in the α (pdhC) and β (pdhB) subunits of pyruvate decarboxylase (E1) and dihydrolipoyl transacetylase (E2, pdhA) components of the multienzyme complex (Payton et al., 1977; Bos et al., 1981).

Regulation of pyruvate dehydrogenase in N. crassa appears to result primarily from interconversion between phosphorylated (low activity) and dephosphorylated (high activity) forms of the enzyme catalyzed by a specific kinase and phosphatase that are associated with the complex (Harding et al., 1970; Wieland et al., 1972). Preliminary data suggest a similar mechanism may operate in A. nidulans (Uitzetter, 1982). This mechanism is comparable to that described for mammalian tissues (see Reed, 1981). However, the factors that control the activity of pyruvate dehydrogenase kinase and phosphatase and hence presumptively regulate flux through pyruvate dehydrogenase in filamentous fungi may differ from those characteristic of mammalian tissues (see Reed, 1981). In N. crassa, pyruvate dehydrogenase kinase is not inhibited by pyruvate or ADP, and pyruvate dehydrogenase phosphatase is insensitive to activation by Ca^{2+} (Wieland et al., 1972). In addition, in N. crassa, but not apparently in A. nidulans (Payton et al., 1977), the total level of the pyruvate dehydrogenase complex is altered by growth on different carbon sources, being elevated in glucose-grown mycelia but present at low levels in mycelia grown on acetate or glycerol (Courtright, 1977; Song *et al.*, 1978).

N. crassa pyruvate dehydrogenase has the capacity to synthesize 2-acetolactate from pyruvate, and presumptively 2-aceto-2-hydroxybutyrate from pyruvate + 2-ketobutyrate (Kuwana *et al.*, 1968; Harding *et al.*, 1970). Since these compounds are necessary intermediates in the mitochondrial biosynthesis of valine and leucine and of isoleucine, respectively (Cassady *et al.*, 1972), such a capacity could explain the presence of pyruvate dehydrogenase in acetate-grown mycelia where this enzyme has no obvious role in relation to acetyl CoA formation. However, such an explanation must be accepted with caution since mutants of *N. crassa* that are blocked in α-acetohydroxy acid synthase (*iv*-3) and hence require isoleucine and valine for growth will grow on glucose without addition of acetate. The latter phenotype would not be expected if the mutants also lacked pyruvate dehydrogenase activity (Kuwana and Wagner, 1969).

3.2.7. Citrate Synthase

This enzyme purified from *A. niger* (Kubicek *et al.*, 1980) has properties similar to those described for other eukaryotic citrate synthases and in particular fails to show regulation by NADH, AMP, and other metabolites described for the enzyme obtained from certain prokaryotes (Weitzman and Danson, 1976).

3.2.8. NAD-Isocitrate Dehydrogenase

Filamentous fungi, with the exception of certain groups of lower fungi, contain an NAD-linked isocitrate dehydrogenase that is exclusively localized within the mitochondrion (Ramakrishnan and Martin, 1955; Sanwal and Stachow, 1965; LeJohn, 1971; Osmani and Scrutton, 1983, 1985). Reports of the absence of this enzyme in *A. nidulans* (Hankinson and Cove, 1974; McCullough *et al.*, 1977) can probably be attributed to problems similar to those that caused such reports for certain vertebrate tissues (see Plaut, 1970). Fungal NAD-isocitrate dehydrogenases are typically activated by AMP, rather than ADP as in mammalian tissues, and are inhibited by ATP (Sanwal and Stachow, 1965; LeJohn, 1971). Control by the AMP/ATP ratio, which is also observed for this enzyme in *S. cerevisiae*, may serve to relate TCA cycle flux directly to the mitochondrial adenylate energy charge system.

The effect of citrate on NAD-isocitrate dehydrogenase in filamentous fungi is less well defined. Activation has been described for the enzyme from *N. crassa* (Sanwal and Stachow, 1965) and from Blastocladiales (LeJohn *et al.*, 1969). However, LeJohn (1971) has stated that most fungal NAD-isocitrate dehydrogenases are inhibited by citrate.

3.2.9. NADP-Malate Dehydrogenase (EC 1.1.1.40)

Considerable diversity exists among filamentous fungi not only in their total content of NADP-malate dehydrogenase when grown on glucose as carbon source (Section 3.1) but also in the nature of the isozymes present. For example, *F. oxysporum* contains a single enzyme (Zink and Katz, 1973) whereas *N. crassa* contains either two or three isozymes depending whether the mycelium is examined in the early or late phase of growth (Zink, 1972). The relationship between the two sets of isozymes in *N. crassa* is unclear since they have different pH optima and regulatory properties. For example, the isozymes present in the early phase of growth are inhibited by fructose-1,6-bisphosphate at a concentration similar to that that activates pyruvate kinase in *N. crassa* but are insensitive to inhibition by L-aspartate. Conversely, the isozymes present late in growth are inhibited by millimolar levels of L-aspartate but are insensitive to inhibition by fructose-1,6-bisphosphate (Zink, 1967, 1972). In the absence of insight into the role of NADP-malate dehydrogenase during growth on glucose (Section 3.1.3), we cannot suggest how these properties may contribute to flux control in glucose metabolism in *N. crassa* although they may indicate that the role of NADP-malate dehydrogenase changes during the growth cycle. Definition of the properties of mutants of *N. crassa* having lesions in NADP-malate dehydrogenase would be of considerable interest.

Three isozymes of NADP-malate dehydrogenase have also been detected in glucose-grown mycelia of *R. arrhizus*. Two of these isozymes are localized in the cytosol and the third is tightly bound to the mitochondrion but their regulatory properties have not been examined (Osmani and Scrutton, 1985).

3.3. Synthesis of Reserve Carbohydrates

Most filamentous fungi accumulate large quantities of reserve carbohydrate regardless of the carbon source on which they are grown. In addition to glycogen, which typically constitutes a minor part of the carbohydrate reserve, the major storage molecules are mannitol and trehalose, which are low-molecular-weight soluble products. This topic has been thoroughly reviewed by Blumenthal (1976) and only certain aspects that are particularly pertinent to cellular metabolism will be discussed here. For example, consideration of trehalose metabolism is omitted since although this disaccharide is widely distributed as a reserve carbohydrate in fungi, little is known about mechanisms responsible for regulating the rates of its synthesis or degradation (however, see Chapter 7).

3.3.1. Mannitol

Filamentous fungi with the exception of the lower fungi, accumulate mannitol during growth on various substrates to levels in the range 2–10% of dry weight (Yamada *et al.*, 1959; Ballio *et al.*, 1964; Lee, 1967a; Lewis and Smith, 1967; Smith *et al.*, 1969; Holligan and Jennings, 1972a). This very hydrophilic substance appears to act as an osmotic stabilizer and may be excreted into the culture medium presumably because it cannot be retained in the cells at these concentrations. During growth of *A. clavatus* on glucose, mannitol accumulates during the initial phase as glucose in the medium decreases, but subsequently itself decreases with the accumulation of ribitol and sorbitol (Corina and Munday, 1971). A similar sequence of events has been observed in other fungi such as *Dendryphiella salina* (Holligan and Jennings, 1972a,b). Mannitol utilization also occurs during conidial germination (Horikoshi *et al.*, 1965).

Mannitol synthesis is generally believed to result from reduction of fructose-6-phosphate to mannitol-1-phosphate followed by dephosphorylation (steps 1 and 2 of Fig. 4A) while mannitol is utilized by oxidation to fructose followed by phosphorylation (steps 3 and 4 of Fig. 4A). A low level of an enzyme that phosphorylates mannitol using acetylphosphate as phosphoryl donor is induced by growth of Aspergilli on mannitol (Lee, 1967b) but the significance of this latter reaction is not clear. The equilibrium of mannitol-1-phosphate dehydrogenase lies far in the direction of mannitol-1-phosphate synthesis (Kisor and Niehaus, 1981) and hence in conjunction with mannitol-1-phosphatase a virtually irreversible pathway exists for mannitol biosynthesis (Lee, 1967b). Conversely, the equilibria of both mannitol dehydrogenase (Niehaus and Dilts, 1982) and hexokinase (Robbins and Boyer, 1957) favor mannitol utilization. The operation of the biosynthetic and degradative pathways for mannitol metabolism as indicated is consistent with the findings that (1) during germination of conidia of *A. oryzae* a decrease in mannitol content is associated with a decreased level of NADP-mannitol dehydrogenase (Horikoshi *et al.*, 1965) and (2) growth of Aspergilli on mannitol causes a decrease in the level of mannitol-1-phosphatase (Lee, 1967b). However, other evidence is not so easily reconciled with this concept. For example, in *A. candidus,* growth on mannitol causes a *decrease* in the level of NADP-mannitol dehydrogenase and an *increase* in the level of mannitol-1-phosphate dehydrogenase as compared with that observed during growth on glucose (Strandberg, 1969). Furthermore, the K_m of *A. parasiticus* NADP-mannitol dehydrogenase for mannitol is 0.1 (M Niehaus and Dilts, 1982), a value that appears surprisingly unfavorable even given the high intracellular concentrations to which this metabolite is accumulated.

More recently, it has been proposed that the enzymes of mannitol metabolism operate as a cyclic system (Fig. 4A) for ATP-dependent transfer of reducing equivalents from NADH to NADPH (Hult and Gatenbeck, 1978). The evidence supporting this latter proposal can be summarized as follows:

1. Isotopic flux measurements in *Alternaria alternata* suggest that active conversion of fructose-6-phosphate to mannitol and of mannitol to fructose-6-phosphate occurs during trophophase and idiophase in this organism. The calculated fluxes, especially that from mannitol to fructose-6-phosphate, are higher in

a strain that synthesizes lipid as compared with a strain in which alternariol (a polyketide) is produced and lipid synthesis is markedly diminished thus decreasing the demand for NADPH. The calculations also indicate that NADPH production by the mannitol cycle is adequate to account for the extent of the observed lipid synthesis at all times except during the phase of active growth in a strain that does not produce alternariol (Hult and Gatenbeck, 1978).

2. All the enzymes of the mannitol cycle (Fig. 4A) are present at significant levels in a representative selection of Deuteromycetes (Fungi Imperfecti) although other classes of fungi lack either all (lower fungi) or some (Ascomycetes, Basidiomycetes) of these enzymes. Very few of the Deuteromycetes have significant levels of NAD-mannitol dehydrogenase, thus linking mannitol oxidation with NADPH production (Boonsaeng et al., 1977b; Hult et al., 1980).

3. The concentration of Zn^{2+} required to inhibit NADP-mannitol dehydrogenase ($K_i = 1$ μM) is very similar to that required to induce polyketide synthesis in A. parasiticus (Niehaus and Dilts, 1982). Although this correlation is interpreted as supporting the proposed transhydrogenation role of the mannitol cycle on the basis that polyketide synthesis is not dependent on a supply of NADPH, it is hard to reconcile with the finding that growth is activated by Zn^{2+} over the same concentration range (Turner, 1976; Niehaus and Dilts, 1982).

Although the concept of the mannitol cycle is an attractive one especially in the absence (as indicated below) of a clear definition of the source of NADPH during growth on substrates such as alanine or glycerol (Section 4), several observations are difficult to reconcile with this postulate. For example:

1. The levels of the specific enzymes of the mannitol cycle are not significantly different in strains of Alternaria alternata that do, or do not, produce alternariol despite the marked difference in demand for NADPH (Hult and Gatenbeck, 1978, 1979).

2. Growth of A. nidulans on urea as nitrogen source gives increased levels of mannitol-1-phosphate dehydrogenase as compared with those observed for growth on NO_3^- as nitrogen source. Addition of NO_3^- to the medium containing urea depresses the level of this enzyme. In contrast, the levels of glucose-6-phosphate dehydrogenase and 6-phosphogluconate dehydrogenase are severalfold higher in mycelia grown on NO_3^- as nitrogen source as compared with those grown on NH_4^+ as might be expected from the relative NADPH requirements in the two situations (Hankinson and Cove, 1974, 1975).

3. The very unfavorable K_m for NADP-mannitol dehydrogenase noted above does not appear to accord with efficient cyclic operation of the pathway shown in Fig. 4A. Moreover, the calculated flux measurements are not indicative of coordinated mannitol turnover but rather of a system in which the rates of mannitol synthesis and degradation vary independently with culture age and encompass a considerable overlap, thus giving a period when both pathways are active (Hult and Gatenbeck, 1978).

Further progress in our understanding of the relationship(s) between mannitol metabolism and growth in filamentous fungi will require isolation of mutants defective in one of the specific enzymes of the mannitol cycle or the development of selective inhibitors for one or more of these enzymes. Furthermore, we have at present no clear insight into factors responsible for regulation of mannitol biosynthesis and degradation although such regulation is implied by several of the studies cited above (Corina and Munday, 1971; Hult and Gatenbeck, 1978).

3.3.2. Polyglucans

Glycogen, starch, and other glucose polymers (e.g., nigeran and pullulan) are made by many filamentous fungi although the amounts are very variable. For example, sclerotia of *Phymatotrichum omnivorum* contain up to 37% of their dry weight as glycogen (Ergle, 1947) whereas very low incorporation of glucose into glycogen occurs in Aspergilli (Lee, 1967a). The accumulation of such polyglucans may provide another general mechanism for the storage of growth substrate when carbohydrate is present in excess in the medium although a linkage to a specific function is apparent in some cases. For example, in Aspergilli, accumulation and subsequent mobilization of an α-1,3-glucan is required for fructification (Zonneveld, 1977).

The pathway of glycogen synthesis and degradation appears similar to the well-characterized pathway in higher eukaryotes (Blumenthal, 1976). However, the nature of the mechanisms that regulate these pathways are less clearly defined. Activation of glycogen phosphorylase occurs on incubation of cell-free extracts from *N. crassa* and *Coprinus macrorhizus* with ATP + Mg^{2+} and this activation is reversed by removal of ATP and replacement of Mg^{2+}. Under appropriate conditions, the activation process is dependent on the addition of cAMP while Mg^{2+}-dependent inactivation is inhibited by fluoride. Significant activation by AMP is observed for the deactivated enzyme while the activated enzyme is inhibited by relatively high concentrations of glucose-6-phosphate (Tellez-Inon and Torres, 1970; Gold *et al.*, 1974; Uno and Ishikawa, 1976). Conversely, inactivation of glycogen synthase results from incubation of cell-free extracts of *N. crassa* and *C. macrorhizus* with ATP + Mg^{2+}; and activation occurs if the inactivated extract is incubated with Mg^{2+} in the absence of ATP. These two states of glycogen synthase in *N. crassa* differ primarily in their sensitivity to activation by glucose-6-phosphate, which is much more marked after incubation with ATP + Mg^{2+}. Furthermore, in *C. macrorhizus,* addition of cAMP accelerates the inactivation process (Tellez-Inon *et al.*, 1969; Uno and Ishikawa, 1978). Although these properties seem very similar to those described for mammalian glycogen phosphorylase and synthase for which regulation by cAMP-dependent phosphorylation is well documented (see Cohen, 1983) and also to those described for the enzymes in yeast (see Chapter 8), it should be noted that no *direct* evidence shows that phosphorylation is involved. Additionally, some questions remain unanswered. For example, a highly active form of *N. crassa* glycogen phosphorylase can be obtained that does not contain any significant amounts of phospho-amino acids and is insensitive to activation by AMP (Cuppoletti and Segel, 1979). Furthermore, in cell-free extracts from *Blastocladiella emersonii*, glycogen synthase activity is not altered by incubation with ATP + Mg^{2+} in either the presence or the absence of cAMP. This latter enzyme is, however, very sensitive to reversible inhibition by Mg ATP^{2-} and to activation by glucose-6-phosphate, the extent of this activation being considerably greater in zoospores than in the log-phase mycelium (Camargo *et al.*, 1969).

All the expected machinery for cAMP-mediated regulation exists in filamentous fungi. Membrane-bound adenylate cyclase and cAMP phosphodiesterase occur in lower fungi, Ascomycetes, and (for adenylate cyclase) Basidiomycetes. Similarly, in representative lower fungi and Ascomycetes, protein kinase A has been detected with properties similar to those described for the mammalian tissue enzyme (see Pall, 1981). The factors that regulate cAMP levels in fungi are, however, not clear, and hence the significance to these organisms of possible cAMP-dependent regulation of glycogen metabolism is also uncertain. Activation of adenylate cyclase and of glycogenolysis by glucagon and inhibition of adenylate cyclase by insulin have been demonstrated in a slime mutant of *N. crassa* (Flawia and Torres, 1972, 1973) and this is one of the very few studies in which linkage between these effects has been demonstrated. However, the physiological significance of such a response is obscure since it seems improbable that the role of glucagon or of insulin could in any way resemble that characteristic of mammals. Mutants of *N. crassa* deficient in adenylate cyclase have been described that have characteristically altered growth morphology and properties of differentiation (Terenzi *et al.*, 1974; Rosenberg and Pall, 1979) and in which a normal phenotype can be induced by growth in the presence of added cAMP (Terenzi *et al.*, 1976). However, no link can be made between the observed abnormalities in morphology and altered glycogen metabolism. Indeed, it is not clear for at least some of these mutants that a deficiency in adenylate cyclase is the primary lesion (Scott and Tatum, 1970; Scott, 1976). Possible roles for cAMP in fungal metabolism and criteria that should be applied to evaluate such postulates are discussed in more detail by Pall (1981).

3.4. Pathways of "Overflow Metabolism" and Their Control

In common with other microorganisms, filamentous fungi when cultured under conditions of nitrogen or phosphate limitation convert carbohydrate to a variety of products. This response to the exhaustion of a required nutrient is often described as "overflow metabolism" and presumably serves to maintain the organism in a viable state until it encounters conditions enabling further growth. In certain fungi, such "overflow" pathways yield organic acids some of which are, or have been, of commercial importance (e.g., production of citric acid and gluconic acid by *A. niger,* of itaconic acid by *A. terreus,* of fumaric acid by *R. arrhizus*), and strains catalyzing these conversions with very high efficiencies have been developed. The conditions required to obtain accumulation of organic acids have been reviewed for citric acid (Berry *et al.,* 1977), itaconic acid (Jakubowska, 1977), and gluconic acid (Rohr *et al.,* 1983), and a summary of some of the factors that may be of importance is presented in Fig. 5. While the concentration of Mn^{2+} in the medium appears particularly critical, the presence of higher concentrations of Zn^{2+} or Fe^{2+} ($>$ 20 μM) will also inhibit citric acid production (Berry *et al.,* 1977). However, until recently, the available information has been insufficient to enable an understanding of these observations in relation to the underlying regulatory mechanisms; in particular, how, under the conditions of accumulation, the rate of glucose metabolism is increased and in some cases the further metabolism of the organic acid or other product is depressed.

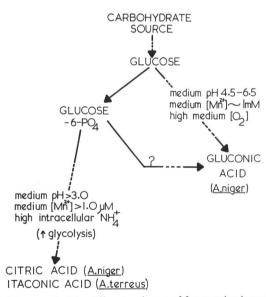

Figure 5. Conditions required for different pathways of fermentation in various Aspergilli.

3.4.1. Citric Acid Biosynthesis

Two possible mechanisms can be envisaged to explain the idiophase conversion of carbohydrate to citric acid by *A. niger* in nitrogen- and manganese-limited media. Such mechanisms must also explain why production of citric acid is often, if not always, accompanied by synthesis of oxalic acid, which accumulates as an undesirable by-product (Lockwood, 1975) when the medium is not sufficiently acidic to induce oxalate decarboxylase (Hayaishi *et al.*, 1956). These mechanisms are illustrated in Fig. 6A,B. Both versions have in common production of pyruvate from glucose by the glycolytic pathway but they differ in the pathway by which pyruvate is converted to citrate. In Fig. 6A, citrate synthesis occurs in the mitochondrion by the classical pathway after pyruvate has been converted to acetyl CoA in this organelle. Oxaloacetate is provided by carboxylation of pyruvate in the cytosol and then translocation into the mitochondrion as malate. Such translocation may occur using the malate/citrate antiport and hence be coupled to citrate export. In this mechanism, oxalate biosynthesis is regarded as an ancillary pathway that may be activated to remove excess oxaloacetate when the rate of synthesis of this intermediate exceeds the capacity to utilize it for citrate synthesis. Acetate produced as a consequence of the cleavage of oxaloacetate may be either excreted to the medium or if acetyl CoA synthase is present utilized for citrate synthesis by conversion to acetyl CoA. The proposed scheme places the enzymes of citrate biosynthesis in their expected sites within the fungal cell and suggests that since in this group of organisms pyruvate carboxylase is present in the cytosol (Osmani and Scrutton, 1983) the enzymes responsible for disposal of an excess of this metabolite are also located in the same intracellular compartment.

Alternatively, oxalate synthesis may be envisaged as an integral part of a citrate biosynthetic pathway that occurs entirely in the cytosol (Fig. 6B). The scheme shown here is a modification of that proposed by Verhoff and Spradlin (1976) and postulates that all the pyruvate produced by glycolysis is converted to oxaloacetate by pyruvate carboxylase. Half of the oxaloacetate is then cleaved to oxalate and acetate by oxaloacetase, an enzyme that is induced under conditions required for citrate production (Lenz *et al.*, 1976). The acetate produced is activated to acetyl CoA and then, together with the remaining oxaloacetate, is converted to citrate by a cytosolic citrate synthase. This latter scheme puts the relationship between citric and oxalic acid biosynthesis on a firmer basis than that provided by Fig. 6A but requires that citrate is formed by a novel pathway that is unlike that responsible for citrate biosynthesis during the phase of active growth.

A number of considerations suggest that the mechanism shown in Fig. 6A probably provides the correct explanation:

1. Citrate yields approaching 90% of carbohydrate supplied can be obtained in high-yielding strains of *A. niger* (Lockwood, 1975; Berry *et al.*, 1977). This

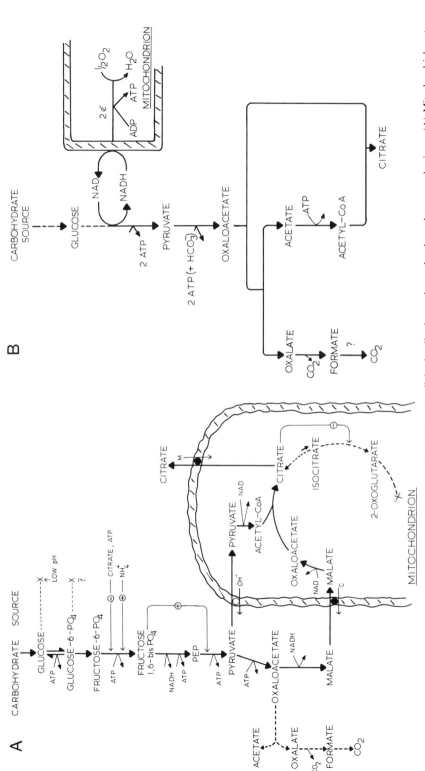

Figure 6. Proposed routes for citrate production by *Aspergillus niger* including subcellular localization and postulated regulatory mechanisms. (A) Mitochondrial route. The major pathway is shown as solid lines with ancillary pathways shown as dashed lines. The diagram is modified and updated from Rohr and Kubicek (1981). (B) Cytosolic route. The scheme shown is a modified version of that originally proposed by Verhoff and Spradlin (1976).

finding is consistent with the mitochondrial pathway (Fig. 6A) but not with the cytosolic pathway (Fig. 6B), which has a theoretical maximal yield of 66% conversion of carbohydrate to citrate.

2. $^{14}CO_2$ is incorporated selectively into C-3 and C-6 of citrate during acid production. None of the label appears in C-1 (Cleland and Johnson, 1954) as required if the pathway shown in Fig. 6B operated. Studies using glucose [3,4-^{14}C] (Cleland and Johnson, 1954) or [1-^{14}C]- or [2-^{14}C]acetate (Lewis and Weinhouse, 1951) are unable to distinguish between the two proposed pathways although they are often cited as favoring the mitochondrial route (Fig. 6A). However, these studies did not use the genetically modified strains that have been developed more recently and the results could be complicated by this factor, although this seems unlikely.

3. Figure 6A provides a role for the cytosolic and mitochondrial isozymes of NAD-malate dehydrogenase that are present in citrate-producing mycelia (Ma et al., 1981).

4. Sensitivity of citric acid synthesis to inhibitors of the alternative (cyanide-insensitive) respiratory pathway, for example salicyl-hydroxamate (Zehentgruber et al., 1980), is explained by the pathway shown in Fig. 6A since this pathway has a positive energy balance and generates excess cytosolic NADH that must be reoxidized if citric acid synthesis is to proceed efficiently. In contrast, the cytosolic pathway (Fig. 6B) has a negative energy balance and will depend on mitochondrial ATP production driven directly by NADH generated in the cytosol as is characteristic of fungal mitochondria (Watson and Smith, 1967b; Weiss et al., 1970; Ahmed et al., 1972). The alternative respiratory pathway is typically associated with "overflow" metabolism in plants but the evidence cited in favor of its association with citric acid synthesis in A. niger should be viewed with caution since the specificity of inhibitors of this pathway has been questioned (Lambers, 1980).

5. The level of oxaloacetase, the key enzyme of the cytosolic pathway (Fig. 6B), is elevated prior to active citrate synthesis in a citrate-producing strain of A. niger and declines to an undetectable level during the phase of rapid citric acid accumulation (Joshi and Ramakrishnan, 1959). This result is totally inconsistent with a role for the enzyme in citrate synthesis and suggests instead that it plays an important role in the transition from active growth to the stationary phase associated with "overflow metabolism." Since this transition might well be accompanied by overproduction of oxaloacetate resulting from decreased biosynthetic demand, the role of oxaloacetase can be envisaged as a mechanism counteracting such an accumulation (Fig. 6A).

The regulatory mechanisms responsible for the increased rate of citric acid biosynthesis during idiophase, and in particular the basis for differences in citrate yields between strains of A. niger, is far from clear although recent studies especially by Rohr and his co-workers provide some insight (see Rohr and

Kubicek, 1981), primarily by examination of the effects of changes in growth medium [Mn^{2+}] on various cellular parameters. A decrease in medium [Mn^{2+}] to a level compatible with active citric acid synthesis (see Fig. 5) causes widespread changes in mycelial morphology and composition including decreases in nucleic acid, protein, and lipid content (Kubicek et al., 1979a; Orthofer et al., 1979; Kisser et al., 1980). Marked and selective changes in enzyme levels also occur in idiophase as a consequence of deficiency of Mn^{2+} in the growth medium. Among these changes, decreased medium [Mn^{2+}], and hence enhanced citric acid production, is associated with some increase in the specific activities of hexokinase, phosphofructokinase, fructose-1,6-bisphosphate aldolase, pyruvate kinase, and citrate synthase and with marked decreases in the specific activities of glucose-6-phosphate dehydrogenase, 6-phosphogluconate dehydrogenase, transketolase, and NAD-isocitrate dehydrogenase together with less marked decreases for aconitase and NADP-isocitrate dehydrogenase (Kubicek and Rohr, 1977). These changes in enzyme levels are consistent with enhanced conversion of glucose to citrate at the expense of carbon flow into the pentose phosphate pathway or the TCA cycle. In addition, the increase in protein degradation, or decrease in protein synthesis, caused by Mn^{2+} deficiency leads to elevation of amino acid levels, and hence of NH_4^+, in the mycelium. This effect might also enhance citric acid synthesis by the mechanism illustrated in Fig. 7. This latter proposal (Rohr and Kubicek, 1981) is supported by several observations:

1. Inhibition of protein synthesis by cycloheximide also causes enhanced citric acid production (Rohr and Kubicek, 1981). Although this effect was initially interpreted as indicating that protein synthesis was required to permit inhibition of citric acid synthesis by Mn^{2+}, it is equally consistent with the proposal that when greater than 10^{-6} M Mn^{2+} is present in the growth medium, protein synthesis occurs at a rate sufficient to prevent a marked increase in amino acid levels. Alternatively, as shown in Fig. 7, protein degradation may be inhibited by higher levels of Mn^{2+} in the medium.

2. Phosphofructokinase catalyzes a ''nonequilibrium'' reaction during citric acid synthesis (Habison et al., 1979).

3. NH_4^+ activates A. niger phosphofructokinase (Habison et al., 1983) as has been observed for this enzyme in many other organisms (Sugden and Newsholme, 1975). Addition of NH_4^+ both makes the $[S]_{0.5}$ for fructose-6-phosphate more favorable and also diminishes inhibition by citrate, an effect that may be of crucial importance in the face of the massive accumulation of this metabolite (Rohr and Kubicek, 1981; Habison et al., 1983).

4. The increase in mycelial [NH_4^+] correlates during early idiophase with the increased rate of citric acid synthesis and is in the range required to activate phosphofructokinase although the correlation becomes less good at later times (Habison et al., 1979).

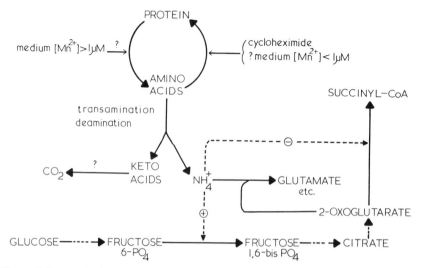

Figure 7. Proposed relationships between an increased rate of protein turnover and citric acid accumulation in *Aspergillus niger*. The diagram is based on the concepts suggested by Rohr and Kubicek (1981).

However, despite these considerations, phosphofructokinase may not be the major site for flux generation during citric acid synthesis since the mycelial concentration of fructose-6-phosphate closely approximates the concentration of this substrate required to achieve half-maximal velocity in the presence of NH_4^+ (Habison *et al.*, 1979, 1983; Newsholme and Crabtree, 1981).

Other possible sites of regulation in the pathway of citric acid biosynthesis based on analogy with other organisms would include pyruvate kinase, citrate synthase, and pyruvate carboxylase. However, evidence for the involvement of any of these enzymes is far from convincing. Citrate synthase does not appear to be a regulatory enzyme in *A. niger* (Kubicek and Rohr, 1980) and although *A. niger* pyruvate kinase is activated by fructose-1,6-bisphosphate the high affinity for this activator suggests that pyruvate kinase is unlikely to be responsive to changes in its cellular concentration (Monori *et al.*, 1984). Less information is available for *A. niger* pyruvate carboxylase although the importance of this reaction is indicated by a stoichiometric relationship between the rates of CO_2 fixation and citric acid synthesis (Kubicek *et al.*, 1979b). The enzyme is fully active in the absence of acetyl CoA and is inhibited by L-aspartate, an amino acid that increases only slightly during idiophase in a Mn^{2+}-deficient medium (Feir and Suzuki, 1969; Kubicek *et al.*, 1979a). Additionally, pyruvate carboxylase in the related organism *A. nidulans* shows little sensitivity to inhibition by citrate (Osmani *et al.*, 1981), in contrast to the enzyme from vertebrate liver (Scrutton and White, 1974). If this situation also applies in *A. niger*, it eliminates, as would be desirable for "overflow" metabolism, a further potential site in the pathway for feedback inhibition by citrate.

A decreased rate of citrate utilization via the TCA cycle may also contribute to the enhanced rate of citric acid accumulation in idiophase. Despite claims to the contrary (Ramakrishnan *et al.*, 1955), aconitase and both NAD- and NADP-isocitrate dehydrogenases are present in *A. niger* during citric acid production (La Nauze, 1966; Ahmed *et al.*, 1972). However, 2-oxoglutarate dehydrogenase, although active under other conditions, is repressed by conditions similar to those encountered during citric acid production (Rohr and Kubicek, 1981). Furthermore, a marked rise in mycelial [2-oxoglutarate] occurs during idiophase in Mn^{2+}-deficient medium without as marked a change in the levels of other TCA cycle intermediates (Kubicek and Rohr, 1978). These effects seem likely to prevent an increase in mycelial $[NH_4^+]$ to toxic levels since enhanced levels of the glutamate family of amino acids are also observed under these conditions (Kubicek *et al.*, 1979a). However, they also have the effect of depressing citrate utilization (Fig. 7). Other mechanisms proposed for manipulation of citrate utilization have largely focused on effects on the activity of NADP-isocitrate dehydrogenase although this enzyme is unlikely to be of major importance to citrate utilization via the TCA cycle (Smith and Plaut, 1979). The suggestions arose initially since citrate at concentrations achieved during idiophase inhibits NADP-isocitrate dehydrogenase from *A. niger* (Mattey, 1977) and since the extent of this inhibition in different strains is correlated with the ability of these strains to accumulate citrate (Bowes and Mattey, 1980). Furthermore, the activation of mitochondrial NADP-isocitrate dehydrogenase by Mn^{2+} has been proposed as being important (Bowes and Mattey, 1979) although the concentration required is three orders of magnitude greater than that that inhibits citrate production. There is furthermore no evidence to suggest that *A. niger* mycelia concentrate Mn^{2+} (Rohr and Kubicek, 1981). It therefore seems unlikely that citrate accumulation is influenced in any major way by changes in the activity of NADP-isocitrate dehydrogenase. The marked decrease in the level of NAD-isocitrate dehydrogenase observed during idiophase in Mn^{2+}-deficient medium (Kubicek and Rohr, 1977) may therefore be the effect that is important to depression of citrate utilization. It is, however, puzzling that only a small increase in total mycelial [isocitrate] occurs under conditions of citric acid synthesis although this observation is difficult to interpret since compensatory changes could be occurring in the mitochondrial and cytosolic isocitrate concentrations.

Changes in the levels of the enzymes of glycolysis, the pentose phosphate pathway, and the TCA cycle (Kubicek and Rohr, 1977) together with the effects related to increased protein turnover and raised mycelial $[NH_4^+]$ shown in Fig. 7 (Rohr and Kubicek, 1981) appear therefore to provide a reasonable basis for an understanding of the citric acid biosynthesis by *A. niger* during idiophase. Although other factors have been suggested as being important such as increased cAMP (Wold and Suzuki, 1973), no mechanism is apparent that relates these factors to the increased rate of citric acid synthesis unless it is considered that an increased rate of glycogenolysis might be important (see Section 3.3). In addition, accumulation of citrate must also require depression of the activity of citrate lyase, and hence of utilization of citrate for lipid synthesis although this aspect has received no attention in *A. niger*. Such a situation contrasts with lipid-accumulating yeast in which nitrogen limitation is associated with an enhanced rate of lipid synthesis.

3.4.2. Itaconic Acid Biosynthesis

The evidence now available suggests that a scheme similar to that shown in Fig. 6A for citrate biosynthesis can also explain the conversion of carbohydrate to itaconate by *A. terreus*. The evidence supporting this postulate is as follows:

1. Mitochondria that oxidize 2-oxoglutarate, succinate, and pyruvate + malate as well as NADH and that exhibit reasonable P : O ratios during such oxidations can be prepared from the mycelium of *A. terreus* harvested either in the growth phase or in the itaconate production phase (Winskill, 1983). Earlier reports of failure to obtain fully functional mitochondria from such strains (Nowakowska-Waszczuck, 1973) may be due to damage during the preparative procedures.

2. The pattern of [14C] incorporation into itaconate from specifically labeled [14C]glucose, -acetate, and -succinate (Bentley and Thiessen, 1957a; Winskill, 1983) as well as synthesis of [14C]itaconate from [14C]citrate (Corzo and Tatum, 1953) are consistent with a pathway in which itaconate is derived from citrate formed as shown in Fig. 6A. The pattern of [14C] incorporation into itaconate from [1-14C]- and [2-14C]acetate is specifically not consistent with a pathway in which condensation of pyruvate and acetyl CoA yields citramalate that in turn is dehydrated to form itaconate (Jakubowska, 1977), despite evidence presented in favor of this latter pathway (Lal and Bhargava, 1962; Shimi and Nour El Dein, 1962).

3. Some of the key enzymes of the pathway shown in Fig. 6A, for example citrate synthase as well as *cis*-aconitate hydrolase and *cis*-aconitate decarboxylase, are present in strains of *A. terreus* that produce itaconate (Bentley and Thiessen, 1957b; Neilson, 1955; Winskill, 1983).

4. A potential mechanism exists for diversion of citrate carbon to itaconate since in mycelium harvested during itaconate production the specific activity of citrate synthase increases approximately twofold whereas that of NADP-isocitrate dehydrogenase falls to 10% of the level observed in trophophase mycelium (Winskill, 1983). However, as indicated in Section 3.4.1, changes in the level of

NAD-isocitrate dehydrogenase are more likely to be relevant but the enzyme has not been demonstrated in *A. terreus* (Winskill, 1983).

Although this evidence supports a pathway for itaconate synthesis involving formation of citrate within the mitochondrion, many questions remain to be answered. For example, the subcellular localization of the enzymes responsible for conversion of citrate to itaconate is not known and this information will have an important bearing on the nature of the mechanism(s) responsible for diversion of the flow of citrate metabolism toward itaconate. If *A. terreus* contains both mitochondrial and cytosolic isozymes of NADP-isocitrate dehydrogenase as has been found for *A. nidulans* (Osmani and Scrutton, 1983), it will be important to know whether the decrease in specific activity detected by Winskill (1983) affects both, or only one, of these isozymes. Furthermore, we do not know whether the flow of glucose carbon through the glycolytic pathway increases as the rate of synthesis of itaconate increases and if so what regulatory mechanisms are involved. The only clue available is the similarity in the effect of the levels of certain metal ions, particularly Mn^{2+}, in the culture medium on citrate and itaconate biosynthesis (Miall, 1978), which suggests that the conclusions summarized in Section 3.4.1 for citrate synthesis may also apply to synthesis of itaconate.

3.4.3. Gluconic Acid Biosynthesis

Certain of the conditions required for gluconate production, for example medium pH > 4.5, medium $[Mn^{2+}]$ > 1 mM, are the converse of those required for citrate production by *A. niger* (Fig. 5). These conditions will tend to suppress the rate of glycolysis (see Section 3.4.1) and will also stabilize glucose oxidase, which is inactivated at pH's below 3. Induction of glucose oxidase, which is responsible for conversion of glucose to δ-gluconolactone, appears to be due to the presence of high glucose and oxygen concentrations in a nitrogen-limited medium (Rohr *et al.*, 1983).

Gluconate utilization may occur either (1) by phosphorylation and subsequent metabolism via the pentose phosphate pathway, or (2) by a nonphosphorylated pathway in which gluconate is first oxidized to 2-oxo-3-deoxygluconate (KDG) and then cleaved to pyruvate and glyceraldehyde (Elzainy *et al.*, 1973a). The required enzymes for the former pathway have been detected in gluconate-adapted *A. niger* (Lakshminarayana *et al.*, 1969) and in yeast (see Chapter 8). The key enzyme of the nonphosphorylated pathway, KDG-aldolase, is induced in *A. niger* by growth on gluconate as sole carbon source (Allam *et al.*, 1975) and is present in some, but not all, other fungi when grown on gluconate (Elzainy *et al.*, 1973b). Both pathways may therefore be used for growth on gluconate in different fungi. A decrease in medium pH accelerates the activity of the nonphosphorylated pathway (Elzainy et al., 1973a) and it is possible that this factor may be critical to the requirement for near-neutral pH in gluconate accumulation (Fig. 5).

3.4.4. Fumaric Acid Biosynthesis

The culture medium conditions required for production of fumaric acid by R. arrhizus described by Miall (1978) are similar to those necessary for synthesis of citric or itaconic acids by the appropriate Aspergillus strains shown in Fig. 5, thus suggesting that acceleration of glycolysis may also occur in this organism in the nitrogen-limited state. Since in R. arrhizus pyruvate carboxylase together with one of the two isozymes of NAD-malate dehydrogenase and of fumarase are located in the cytosol (Osmani and Scrutton, 1985), the entire pathway for conversion of glucose to fumaric acid probably occurs in this compartment. Hence, no further pathways of metabolism can complicate the issue in this instance. The failure of malate to enter the mitochondrion under these conditions may be explicable on the basis of equilibrium between the mitochondrial and cytosolic malate pools or of limitation of transport by restricted availability of the antiport counterion, for example citrate as in Fig. 6A.

4. GROWTH ON THREE-CARBON SUBSTRATES

These growth substrates can be divided into two categories: (1) glycerol and other substrates that give rise directly to the triose phosphates, and (2) pyruvate and substrates such as lactate or alanine giving rise directly to pyruvate.

4.1. Glycerol

Although a considerable number of studies have been reported in which glycerol was a carbon source for growth of filamentous fungi, few addressed the major areas in which the utilization of glycerol imposes constraints that may be different from those characterizing growth on glucose. In addition to the pathways by which glycerol is converted to an intermediate of glycolysis, such differences are likely to involve:

1. The provision of hexose and pentose phosphates for carbohydrate, nucleotide, and nucleic acid synthesis
2. The provision of NADPH for general reductive biosynthesis

In all other respects, the properties of growth on glycerol are expected to be the same as those observed for growth on glucose. Such a similarity is indicated in several studies (Skinner and Armitt, 1972; Payton and Roberts, 1976; Kelly and Hynes, 1981, 1982).

4.1.1. Conversion of Glycerol to Glycolytic Intermediates

The pathways that may be responsible for the conversion have been studied in some detail in *N. crassa*. Two possible pathways for which the necessary enzymes are present are shown in Fig. 8. Analogy with higher eukaryotes would suggest incorporation via glycerol-3-phosphate as an intermediate and both the relevant enzymes (glycerokinase and glycerol-3-phosphate dehydrogenase) in-

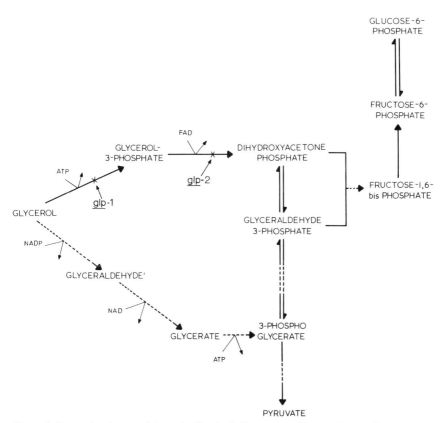

Figure 8. Proposed pathways of glycerol utilization in *Neurospora crassa* and *Aspergillus japonicus*. The pathways for which the data appear fairly secure (see text) are shown as solid arrows. Neither NAD-glycerol dehydrogenase nor glyceraldehyde kinase has been detected in *N. crassa* (Viswanath-Reddy *et al.*, 1977; Tom *et al.*, 1978).

crease markedly in activity during growth of the fungus on glycerol (North, 1973; Courtright, 1975). Glycerokinase is a cytosolic enzyme in *N. crassa* whereas glycerol-3-phosphate dehydrogenase is a flavoprotein that is tightly bound to the mitochondrial membrane (Courtright, 1975). Non-glycerol-utilizing mutants have been isolated that lack glycerokinase (*glp*-1) or glycerol-3-phosphate dehydrogenase (*glp*-2).

Tom *et al.* (1978) have, however, suggested that glycerol may also be utilized by oxidation to glycerate followed by phosphorylation as shown in Fig. 8. This latter pathway seems unlikely to be a major route for glycerol assimilation since in contrast to glycerokinase and FAD-dependent glycerol-3-phosphate dehydrogenase, the levels of NADP-glycerol dehydrogenase and of glycerate kinase are not increased by growth on glycerol (Tom *et al.*, 1978), and a non-glycerol-utilizing mutant (*glp*-2) has levels of NADP-glycerol dehydrogenase similar to those observed in the wild type (Denor and Courtright, 1978). The role of NADP-glycerol dehydrogenase during growth on glycerol is therefore not clear, although a pathway of this type has been described in *Schizosaccharomyces pombe* (May and Sloan, 1981).

A third route may exist since the enzyme glycerol oxidase, which uses molecular oxygen to oxidize glycerol to glyceraldehyde and H_2O_2, is induced by growth of certain Aspergilli, notably *A. japonicus,* on glycerol as carbon source (Uwajima *et al.,* 1980). If glyceraldehyde kinase is also induced under these conditions, then the synthesis of triose phosphate would be possible.

4.1.2. Hexose Phosphate Synthesis

Synthesis of hexose phosphates from triose phosphates requires the presence of fructose-1,6-bisphosphate aldolase and fructose-1, 6-bisphosphatase. Both enzymes are present in glycerol-grown *N. crassa* at levels similar to or greater than those found in glucose-grown mycelia (Mattoo and Parikh, 1975; Tom *et al.,* 1978). Levels of the pentose phosphate pathway enzymes have apparently not been measured in mycelia after growth on glycerol although the requirement for ribose-5-phosphate formation in order to permit nucleotide synthesis, etc., indicates the need for their presence.

4.1.3. NADPH Formation

The mechanism(s) responsible for provision of NADPH for reductive biosynthesis during growth on glycerol is uncertain. In *A. nidulans,* it is improbable that NADP-isocitrate dehydrogenase makes a significant contribution since growth on glycerol fails to increase the level of this enzyme above that observed during growth on glucose or sucrose (Kelly and Hynes, 1981, 1982). The evidence regarding NADP-malate dehydrogenase in glycerol-grown *A. nidulans* is conflicting since Kelly and Hynes (1981) have reported a very low level not significantly different from that obtained during growth on glucose, whereas McCullough *et al.* (1977) found levels of this enzyme that were approximately 25% of those obtained in acetate-grown mycelia. Furthermore, mutants (*acuK; acuM*) that lack NADP-malate dehydrogenase activity grow poorly on glycerol as carbon source (Armitt *et al.*, 1976; McCullough *et al.*, 1977). If a significant level of NADP-malate dehydrogenase is present, then the potential exists for ATP-dependent NADPH formation by the malate/pyruvate cycle (Fig. 4B) provided that the organism also contains pyruvate carboxylase. Alternatively, NADPH may be generated primarily by operation of the oxidative pentose phosphate pathway after formation of fructose-6-phosphate. This latter route is a perfectly feasible mechanism for NADPH generation during growth on carbon sources other than hexoses if net carbon utilization is minimized by recycling through the non-oxidative section of this pathway.

4.1.4. Other Enzymes

Certain filamentous fungi, for example *N. crassa,* grow poorly on glycerol unless acetate is also added, suggesting that under these conditions growth is limited by the availability of acetyl CoA (Courtright, 1975). This effect is due to the presence of a very low level of the pyruvate dehydrogenase complex in glycerol-grown mycelia and its increase by growth on glycerol plus acetate (Courtright, 1977).

4.2. Pyruvate and Related Substrates

Very few studies have been reported in which the enzyme content of mycelia has been analyzed after growth on pyruvate or related carbon sources such as lactate or alanine. This situation has probably arisen since many filamentous fungi do not grow well on these substrates and it is often considered that the requirements for growth are strictly analogous to those for growth on glucose. The latter point is valid for ATP generation via the TCA cycle, for provision of acetyl CoA in the cytosol for lipid and steroid synthesis, and for synthesis of amino acids and other compounds derived from oxaloacetate, aspartate, and 2-oxoglutarate as intermediates. However, it neglects the requirement for the provision of PEP for the synthesis of hexoses and also of certain amino acids. This provision, which cannot be met by reversal of pyruvate kinase, is satisfied in certain bacteria, such as *E. coli,* by the presence of PEP synthase (Kornberg, 1966a). However, this latter enzyme has not been reported in fungi and it seems reasonable to assume that elevation of the levels of PEP carboxykinase and fructose-1,6-bisphosphatase may be observed in mycelia grown on pyruvate or related carbon sources as is the case in acetate- or ethanol-grown mycelia (Kelly and Hynes, 1981; see Section 5). Such an increase in the level of fructose-1,6-bisphosphatase has been reported for *N. crassa* grown on pyruvate as carbon source (Reinert and Marzluf, 1974). Furthermore, mutants of *A. nidulans* lacking either of these enzymes (*acuF*, PEP carboxykinase; *acuG*, fructose-1,6-bisphosphatase) fail to grow effectively on lactate or alanine as carbon source (Payton *et al.*, 1976).

It is less easy to predict how mycelia grown on pyruvate or related carbon sources generate NADPH required for reductive biosynthesis although a number of possible routes may be considered:

1. The oxidative pentose phosphate pathway in conjunction with ribose synthesis for formation of nucleotides and nucleic acids. The importance of this pathway is suggested by the observation that the level of glucose-6-phosphate dehydrogenase in *Penicillium* is similar whether glucose or lactate is used as carbon source (Boonsaeng *et al.*, 1974).
2. The mannitol cycle (Fig. 4A).
3. Cytosolic NADP-isocitrate dehydrogenase in conjunction with provision of 2-oxoglutarate for amino acid biosynthesis.

Although it seems unlikely that NADP-malate dehydrogenase would be present in pyruvate-grown mycelia, mutants of *A. nidulans* lacking this enzyme (*acuK; acuM*) grow poorly, if at all, on lactate or alanine (Payton *et al.*, 1976). This observation could reflect the operation of a malate/pyruvate transhydrogenation cycle (Fig. 4B) as suggested above.

Studies on the pathways and regulatory mechanisms involved in growth on pyruvate and related substrates in Aspergilli may be facilitated by the finding that growth of *A. nidulans* on alanine is enhanced by a very low concentration of glucose. Such mycelia contain pyruvate carboxylase at levels similar to those present in mycelia grown on glucose as sole carbon source (Selmes and Scrutton, unpublished observations) but their content of other relevant enzymes has not been examined.

5. GROWTH ON ACETATE OR ETHANOL

5.1. General Considerations

The utilization by an organism of two-carbon growth substrates, such as acetate or ethanol, presents a very different situation than that encountered in growth on hexoses. Acetyl CoA derived from acetate can provide substrate directly for ATP production by the TCA cycle and for fatty acid and steroid biosynthesis, but it is necessary to consider how the organism provides:

1. NADPH for reductive biosynthesis and also for production of NH_4^+ if a more oxidized nitrogen source such as NO_3^+ is used, since the levels of

glucose-6-phosphate dehydrogenase and 6-phosphogluconate dehydrogenase with NH_4^+ as nitrogen source are typically decreased to 25–30% of those obtained in glucose-grown mycelia with NO_3^+ (Hankinson, 1972; Dunn-Coleman and Pateman, 1979).

2. Malate—and hence aspartate, PEP, and pyruvate—for the synthesis of amino acids, pyrimidines, and hexoses.
3. 2-Oxoglutarate for the synthesis of amino acids.

Therefore, as in bacteria (Kornberg, 1966b), growth of fungi on acetate or ethanol is associated with marked increases in the levels of a number of enzymes that are typically not present, or present at lower levels, during growth on glucose and in most cases on other carbon sources such as glycerol that feed into glycolysis (Casselton, 1976; O'Connell and Paznokas, 1980). In filamentous fungi, these enzymes include acetyl CoA synthase, isocitrate lyase, malate synthase, PEP carboxykinase, fructose-1,6-bisphosphatase, and, in some but not all fungi, NADP-malate dehydrogenase. The essential role of these enzymes for growth on acetate has been indicated by the isolation in A. nidulans (Armitt et al., 1976), N. crassa (Flavell and Fincham, 1968; Beever and Fincham, 1973), and A. terreus (Das and Sen, 1983) of non-acetate-utilizing (acu) mutants lacking these individual enzyme activities. In addition, fluoroacetate-resistant (fac) strains which lack acetyl CoA synthase and so fail to grow on acetate have been obtained in A. nidulans (Apirion, 1965; Romano and Kornberg, 1969) and Coprinus lagopus (Casselton and Casselton, 1974). Other enzymes, such as NAD- and NADP-isocitrate dehydrogenases, citrate synthase, aconitase, NAD-malate dehydrogenase, and glutamate-oxaloacetate transaminase, also increase in concentration during growth on acetate although they are present in glucose-grown mycelia (Flavell and Woodward, 1970; Benveniste and Munkres, 1970; McCullough et al., 1977; Schwitzguebel et al., 1981b; Osmani and Scrutton, 1983).

The importance of the glyoxalate bypass for growth of filamentous fungi on acetate is therefore not questioned, but the subcellular organization of acetate metabolism in these eukaryotes is less clearly defined. Structures resembling the glyoxosomes present during the germination of lipid-rich seeds (Beevers, 1969) have been detected in various fungi (Maxwell et al., 1970, 1977; Wergin, 1972; McLaughlin, 1973) but attempts to isolate these intracellular organelles in an intact state from fungi have been less successful. Particulate isocitrate lyase and malate synthase activities have been reported in cell-free extracts prepared from N. crassa (Kobr et al., 1969; Flavell and Woodward, 1970; Wanner and Theirmer, 1982), C. lagopus (O'Sullivan and Casselton, 1973) and several aspergilli (Graves et al., 1976; Osmani and Scrutton, unpublished observations). However, it is often not emphasized that the percentage of the total activity present in the particulate fraction is very low especially for isocitrate lyase. It is not uncommon that this enzyme is reported entirely present in the soluble fraction (e.g., Cotter et al., 1970) and we have not found any study in which more than 50% of the total isocitrate lyase content of the mycelium is present in the particulate fraction after the preparation of the cell-free extracts. Higher percent recovery of total activity in the particulate fraction is usually observed for malate synthase (e.g., Wanner and Theirmer, 1982) since this enzyme appears tightly bound to the glyoxosomal membrane (Koller and Kindl, 1977). The inability to isolate intact glyoxosomes is not surprising given the problems encountered in obtaining intact mitochondria from filamentous fungi (see Watson, 1976). Difficulties of achieving a clear separation between mitochondria and peroxisomelike particles (e.g., glyoxosomes) compound the situation as does the recent demonstration that multiple peroxisomal particles exist in at least some fungi (Wanner and Theirmer, 1982). If fungal glyoxosomes are defined as particulate structures that contain isocitrate lyase and malate synthase, it is far from clear which other enzymes may be present in these organelles. The evidence available from different studies is consistent with at least two possibilities.

1. Fungal glyoxosomes contain a complete glyoxalate cycle similar to that found in these structures in higher plants (Beevers, 1969), with citrate synthase, aconitase, and NAD-malate dehydrogenase being present in addition to isocitrate lyase and malate synthase. This situation has been described for citrate synthase and NAD-malate dehydrogenase in glyoxosomes isolated from a slime mutant of N. crassa by Theirmer (1982). It was initially reported by Kobr et al. (1969) for NAD-malate dehydrogenase in wild-type N. crassa although subsequent investigations have cast doubt on the initial conclusions (Kobr and Vanderhaeghe, 1973).

2. Fungal glyoxosomes contain only the enzymes of the glyoxalate bypass and hence take up isocitrate and release succinate and malate as suggested by Casselton (1976) and shown in Fig. 9. This situation appears to occur in Aspergilli (Graves et al., 1976; see also subsequent discussion) and in C. lagopus (O'Sullivan and Casselton, 1973) and is suggested for N. crassa by studies showing that its glyoxsomes do not contain citrate synthase (Schwitzguebel et al., 1981a). Furthermore, mitochondria prepared from N. crassa grown on acetate oxidize isocitrate and 2-oxoglutarate less well, but succinate more readily, than mitochondria obtained from glucose-grown mycelia (Schwitzguebel and Palmer, 1981; Schwitzguebel et al., 1981b).

Unequivocal definition of the enzyme content of the glyoxosomes in filamentous fungi is therefore unlikely to be achieved until methods are devised that permit the isolation of these organelles in an intact state. In the meantime, less direct approaches taken together with the earlier data have been used to formulate the model for the subcellular organization of acetate metabolism shown in Fig. 9. This model is supported for A. nidulans by the following observations:

1. Analysis of acu mutants shows that mutations at single gene loci cause total loss of isocitrate lyase (acuD) and malate synthase (acuE) activity (Armitt et

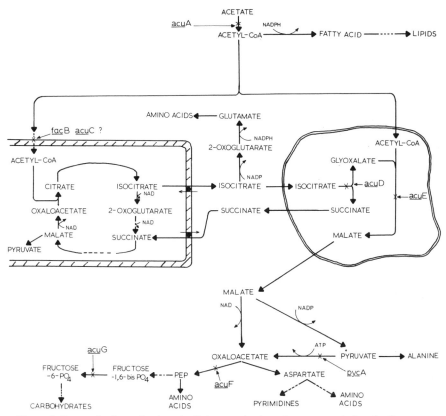

Figure 9. Proposed scheme for the subcellular organization of acetate metabolism in filamentous fungi. The diagram represents an attempt to formulate the data presently available on the subcellular organization of acetate metabolism in *Aspergillus nidulans* and *Neurospora crassa*. Enzymes localized in the mitochondria are shown enclosed on the left of the figure and those enzymes thought to be localized in the glyoxosomes enclosed on the right. A considerable degree of uncertainty is inherent in this exercise since a reasonable yield of intact glyoxosomes has not yet been isolated from any filamentous fungus. In this situation, quantitative subcellular fractionation data are difficult, if not impossible, to obtain. Ethanol is oxidized to acetate by alcohol dehydrogenase and acetaldehyde dehydrogenase and thereafter follows the pathways of acetate metabolism.

al., 1976). Furthermore, no evidence could be obtained for multiple enzymes in either case when a cell-free extract was analyzed by cellulose acetate strip electrophoresis at pH's in the range 5.6 to 8.4 and then stained for the appropriate enzyme activity even though only 10–20% of total isocitrate lyase and 80–85% of total malate synthase are recovered in a particulate fraction prepared from these extracts (Campa, Osmani, and Scrutton, unpublished observations).

2. Electrophoretic analysis has failed to demonstrate the presence of more

than one citrate synthase in acetate-grown mycelia although only 80–85% of the total activity is obtained in the particulate fraction (Campa, Osmani, and Scrutton, unpublished observations).

3. A similar number of isozymes of NAD-malate dehydrogenase (three at acid pH and two at alkaline pH) are detected on electrophoretic analysis of cell-free extracts prepared from mycelia grown on glucose or acetate as carbon source. There is no evidence from these studies of an additional isozyme being present in the mycelia grown on acetate (Campa, Osmani, and Scrutton, unpublished observations).

4. Mutants (*pycA*) lacking pyruvate carboxylase (Skinner and Armitt, 1972) grow somewhat less well on acetate than do wild-type strains. Furthermore, pyruvate carboxylase is present in mycelia grown on acetate as sole carbon source at approximately half the level found for glucose-grown mycelia and is a cytosolic enzyme (Osmani and Scrutton, 1983). These findings are accommodated in Fig. 9 by the proposal that oxaloacetate can be synthesized in the cytosol either directly using the cytosolic isozyme of NAD-malate dehydrogenase or indirectly using a cytosolic isozyme of NADP-malate dehydrogenase (see below) and pyruvate carboxylase. Although this latter pathway is more costly in energetic terms, it provides a cytosolic source of NADPH for lipid and steroid biosyntheses, which are presumably located in this subcellular compartment. Furthermore, unlike NAD-malate dehydrogenase, the overall equilibrium is favorable for oxaloacetate synthesis in this latter pathway.

5. Mutations at a single gene locus in *A. nidulans* give rise to acetate nonutilization and fluoroacetate resistance (*acuA* = *facA*) resulting from the loss of acetyl CoA synthase (Apirion, 1965; Romano and Kornberg, 1969; Armitt et al., 1976). Similar mutants have been obtained in *N. crassa* (Flavell and Fincham, 1968) and *C. lagopus* (Casselton and Casselton, 1974). The subcellular localization of acetyl CoA synthase has not been determined but the genetic evidence suggests a single enzyme and we propose that this is localized in the cytosol and that specific translocating systems make the acetyl-CoA formed available to the mitochondrial and glyoxosomal pathways.

The essential role of PEP carboxykinase and fructose-1,6-bisphosphatase for growth on acetate has been demonstrated by isolation of *acu* mutants at single gene loci that lack these enzyme activities (Beever and Fincham, 1973; Armitt et al., 1976). No subcellular localization studies have been reported for either enzyme in filamentous fungi, but in *S. cerevisiae* fructose-1,6-bisphosphatase is a cytosolic enzyme (Haarasilta and Taskinen, 1977). The essential role for NADP-malate dehydrogenase for growth of *A. nidulans* on acetate, which is indicated by the isolation of two *acu* mutant strains lacking this activity (Armitt et al., 1976), is accommodated in Fig. 9 on the basis that the mitochondrial isozyme is required to provide pyruvate for the synthesis of amino acids such as alanine, valine, leucine, and isoleucine (McCullough and Roberts, 1974). Although pyruvate could potentially be formed from malate via PEP, this latter route is improbable since levels of pyruvate kinase in *A. nidulans* are markedly decreased during growth on acetate as has been observed in yeast (see Chapter 8) and mutants lacking the enzyme (*pkiA*) grow effectively on this substrate (Payton and Roberts, 1976). Hence, Fig. 9 accommodates most of the available evidence on the pathways and organization of acetate metabolism in certain Ascomycetes and Basidiomycetes.

5.2. Enzymes of Acetate or Ethanol Metabolism

5.2.1. Isocitrate Lyase

The levels of isocitrate lyase observed in mutant strains of *A. nidulans* with known biochemical lesions upon transfer from hexose- to acetate-containing growth media suggest that the synthesis of this enzyme may be regulated by a different mechanism than that previously described for *E. coli* (Kornberg, 1966b). Levels of isocitrate lyase approximating those present in the wild type are found for mutant strains lacking malate synthase and PEP carboxykinase in which marked changes would be expected in the levels of PEP and dicarboxylic acids such as malate. In contrast, the level of isocitrate lyase is low in a mutant strain lacking acetyl CoA synthase (Armitt *et al.*, 1976). Similar observations have been reported for an *acu* mutant of *C. cinereus* (King and Casselton, 1977) and might suggest that isocitrate lyase is induced by acetyl CoA or some closely related metabolite (Armitt *et al.*, 1976). In contrast, Beever (1975) has proposed that in *N. crassa*, isocitrate lyase synthesis is repressed by a dicarboxylic acid; while in *A. terreus*, Das and Sen (1983) have suggested that acetate itself acts as the inducer for isocitrate lyase since levels of the enzyme approximating those observed in the wild type are found in a mutant strain lacking acetyl CoA synthase. It seems unlikely that such different regulatory mechanisms would occur in closely related organisms and further studies are required to clarify this situation.

Mutants that form isocitrate lyase constitutively have been isolated in *A. nidulans* (McCullough and Roberts, 1980). The mutations identify two unlinked genes (icl^CA and icl^CB), which when present separately result in low-level constitutive enzyme formation. However, when combined in a double mutant strain, synergism is observed resulting in a level of enzyme activity in growth on sucrose that is 30% of that observed in acetate-grown wild type and is comparable to that found in the wild type grown on a mixture of sucrose and acetate. Hence, catabolite repression may still operate in the double mutant strain. Alternatively, it was suggested that the level of the presumed intracellular inducer is elevated in these constitutive strains due to lesions in other enzymes of intermediary metabolism (McCullough and Roberts, 1980). The situation is further complicated by the presence of glyoxosomes in fungi (see Section 5.1), and it is important to know whether the constitutive isocitrate lyase activity is particulate in sucrose-grown mycelia. It is of interest in this context that isocitrate lyase is present at low, but significant, levels in *N. crassa* grown on glucose (Sjogren and Romano, 1967).

The catalytic activity of isocitrate lyase in filamentous fungi appears to be regulated by the same mechanism as described for *E. coli* (Ashworth and Kornberg, 1963). Thus, PEP is a noncompetitive inhibitor of isocitrate lyase obtained from *N. crassa* (Sjogren and Romano, 1967; Johanson *et al.*, 1974), *R. arrhizus* (Romano *et al.*, 1967), and *A. nidulans* (McCullough, unpublished observations) although in the context of Fig. 9 it is not clear how this effect might operate since the formation of PEP occurs in the cytosol. Other inhibitors have been described for the *N. crassa* and *A. nidulans* enzymes, for example malate and fructose-1,6-bisphosphate (Johanson *et al.*, 1974; McCullough, unpublished observations) but their *in vivo* significance is not clear. Isozymes of isocitrate lyase have been described in *N. crassa* (Sjogren and Romano, 1967) and could account for the inability to isolate isocitrate lyase-deficient mutant strains (Flavell and Fincham, 1968), but a subsequent report suggests that this observation is not correct (Rougemont and Kobr, 1973) and, as noted above, no such isozymes can be detected in *A. nidulans*.

Although isocitrate lyase is generally considered part of the glyoxalate cycle in fungi, it may provide a mechanism for glycine synthesis in *B. emersonii* (McCurdy and Cantino, 1960).

5.2.2. Malate Synthase

Considerations similar to those described in Section 5.2.1 for isocitrate lyase apply to the mechanisms responsible for regulation of the levels of malate synthase in *A. nidulans* (Armitt *et al.*, 1976; Das and Sen, 1983). No mechanisms that might regulate the catalytic activity of malate synthetase in fungi have been described.

5.2.3. NADP-Malate Dehydrogenase

This enzyme can only be regarded specific to acetate metabolism in certain filamentous fungi such as *A. nidulans* since as noted in Section 3.2.9, it is present in many other fungi during growth on glucose. Because no clear conclusions can be drawn from the data presently available for *A. nidulans*, a brief summary of the position may be useful.

1. Two cytosolic isozymes of NADP-malate dehydrogenase can be detected by electrophoretic analysis of cell-free extracts (Campa, Osmani, and Scrutton, unpublished observations).

2. Preliminary subcellular fractionation studies suggest that NADP-malate dehydrogenase is present in both particulate and soluble fractions as is the case for *R. arrhizus* grown on glucose (see Section 3.2.9). It is not possible on the basis of the available data to determine whether the particulate enzyme is localized in the mitochondria or in the glyoxosome (Osmani and Scrutton, unpublished observations), although a mitochondrial localization seems more probable and is shown in Fig. 9 because the enzymes involved in leucine and valine synthesis are also localized in this organelle in *N. crassa* (Cassady *et al.*, 1972).

3. Among the *acu* mutants of *A. nidulans*, two, *acuK* and *acuM*, are totally devoid of NADP-malate dehydrogenase after transfer to growth media containing acetate while several other mutant strains such as *acuA*, *acuH*, and *acuJ* contain levels of enzyme that are much lower than those present in the wild type treated in the same way. Elevated levels of NADP-malate dehydrogenase are present after transfer only in strains lacking PEP carboxykinase (*acuF*) and fructose-1,6-bisphosphatase (*acuG*) (Armitt *et al.*, 1976). The precise functions of the genes *acuK* and *acuM* remain ill-defined because it might be expected that the isozymes of NADP-malate dehydrogenase would be coded by different genes and that a single mutation would not lead to a total loss of enzyme activity. The elevated levels of the enzyme observed in the *acuF* and *acuG* mutants suggest that the NADP-malate dehydrogenases either are induced by a metabolite prior to PEP or are repressed by a hexose phosphate. Since total NADP-malate dehydrogenase levels are increased under conditions that minimize the likelihood of repression by hexose phosphates or related metabolites, an inductive mechanism seems more likely (Kelly and Hynes, 1981).

Further studies are required to resolve these problems and in addition to define mechanisms that may regulate the catalytic activity of NADP-malate dehydrogenase during growth on acetate (see Section 3.2.9).

5.2.4. PEP Carboxykinase

Figure 9 suggests that PEP carboxykinase lies at a branch-point in acetate metabolism. However, little information is available on mechanisms that may regulate its catalytic activity although the enzyme from *Verticillium albo-atrum* was not affected by acetyl CoA or L-aspartate (Hartman and Keen, 1974b).

The level of PEP carboxykinase in *A. nidulans* increases during growth on acetate, ethanol, or other substrates that give rise to acetyl CoA (Kelly and Hynes, 1981), and upon transfer from sucrose to acetate as growth substrate is markedly elevated in a mutant strain (*acuG*) that lacks fructose-1,6-bisphosphatase (Armitt *et al.*, 1976). Although repression by a hexose phosphate, or another glycolytic intermediate, was originally suggested as the control mechanism (Beever, 1975; Armitt *et al.*, 1976), Kelly and Hynes (1981) have shown that under certain conditions addition of sucrose has little effect on PEP carboxykinase levels unless NH_4^+ is present. Hence, it seems possible that synthesis of this enzyme is also primarily under inductive control and that repression by hexose phosphates does not play a major role.

5.2.5. Fructose-1,6-bisphosphatase

No information is available on the mechanisms responsible for regulation of the level of this enzyme although in *N. crassa* this level is much increased by growth on acetate or ethanol as carbon source (Reinert and Marzluf, 1974). Fructose-1,6-bisphosphatase catalyzes a reaction that does not lie at a branchpoint in acetate metabolism, but its activity is inhibited by AMP, an effect that can probably be related to concomitant control with phosphofructokinase (Reinert and Marzluf, 1974). It is not known whether fructose-1,6-bisphosphatase in filamentous fungi is regulated by fructose-2,6-bisphosphate although such an effect would be expected (see Chapter 8).

5.2.6. NADP-Isocitrate Dehydrogenase

Although no *acu* mutants have been found that lack NADP-isocitrate dehydrogenase (Flavell and Fincham, 1968; Armitt *et al.*, 1976) the level of this enzyme increases very markedly during growth on acetate in both *A. nidulans* (McCullough *et al.*, 1977; Kelly and Hynes, 1982) and *N. crassa* (Kobr *et al.*, 1965; Flavell and Fincham, 1968). In *A. nidulans*, this increase is almost entirely due to the cytosolic isozyme (Osmani and Scrutton, 1983), and may be presumed, together with increased NADP-malate dehydrogenase, to provide a mechanism for cytosolic synthesis of NADPH during growth on a carbon source that can less readily supply substrate to the pentose phosphate pathway. The rate of NADPH generation is not necessarily linked to the rate of production of 2-oxoglutarate for synthesis of amino acids and other products since Turian (1963) has shown that in *N. crassa* growth on acetate is associated with accumulation of 2-oxoglutarate.

The regulatory mechanism(s) involved in the control of synthesis of NADP-isocitrate dehydrogenase appears similar to those described above for NADP-malate dehydrogenase (Section 5.2.3) and for PEP carboxykinase (Section 5.2.4). The metabolite involved appears to precede succinate or glyoxalate in the pathway of assimilation of acetate since induction of NADP-isocitrate dehydrogenase was not markedly reduced in *acu* strains lacking isocitrate lyase or malate synthase (Kelly and Hynes, 1982). However, the mechanism appears to differ from that involved in induction of isocitrate lyase and malate synthase in *A. nidulans*, but possibly not in *A. terreus* (see Sections 5.2.1 and 5.2.2). The extent of induction is markedly reduced by a mutation that confers fluoroacetate resistance (*facB*) and that appears to identify an important regulatory gene for acetate utilization (Kelly and Hynes, 1982), but not by a mutation (*facA* = *acuA*) that confers fluoracetate resistance by loss of acetyl CoA synthase.

6. GROWTH ON SUBSTRATES UTILIZED VIA THE TRICARBOXYLIC ACID CYCLE

Many filamentous fungi grow on carbon sources that are TCA cycle intermediates, such as succinate, or are converted to an intermediate of the cycle as an obligatory part of their metabolism, for example proline, propionate, and glutamate. Not much is known about the intracellular organization of the metabolism of these substrates but their utilization requires that certain provisions are met as discussed above for growth on other carbon sources (Sections 3 and 5). Synthesis of the glutamate and aspartate families of amino acids is readily achieved in growth on proline or glutamate but mechanisms are also required to provide:

1. Pyruvate for the synthesis of amino acids such as alanine and valine
2. Acetyl CoA both in the mitochondria for continued operation of the TCA cycle and in the cytosol for fatty acid and steroid biosynthesis
3. NADPH for many biosynthetic reactions and also for NH_4^+ production when NO_3^- is the nitrogen source
4. PEP for both amino acid and hexose synthesis

The speculative scheme shown in Fig. 10 illustrates how these provisions might be met on the basis of the subcellular localization of enzymes in glucose-grown *A. nidulans* (Fig. 3) and the minimal data available for mycelia grown on carbon sources such as succinate and glutamate. In this scheme, malate is a key intermediate that, after formation from 2-oxoglutarate or succinate in the mitochondrion, either can be metabolized in that organelle to produce oxaloacetate or pyruvate, or may be exported to the cytosol to give rise in that compartment to oxaloacetate with consequent production of PEP and aspartate. In accord with this part of the scheme, growth of *A. nidulans* on proline, glutamate, or 4-aminobutyrate gives elevated levels of PEP carboxykinase and decreased levels of pyruvate kinase (Payton and Roberts, 1976; Kelly and Hynes, 1981), while growth of *N. crassa* on succinate gives levels of fructose-1,6-bisphosphatase similar to those found in acetate-grown mycelia (Reinert and Marzluf, 1974). Furthermore, mutants of *A. nidulans* that lack PEP carboxykinase (*acuF*) or fructose-1,6-bisphosphatase (*acuG*) grow poorly if at all on succinate or glutamate whereas mutants lacking pyruvate kinase grow well on glutamate (Payton *et al.*, 1976). Growth of *A. nidulans* on proline or glutamate increases the levels of NADP-malate dehydrogenase (McCullough and Roberts, 1974; Kelly and Hynes, 1981) and mutants lacking this enzyme (*acuK; acuM*) fail to grow on glutamate or succinate as sole carbon source (Payton *et al.*, 1976). The roles of

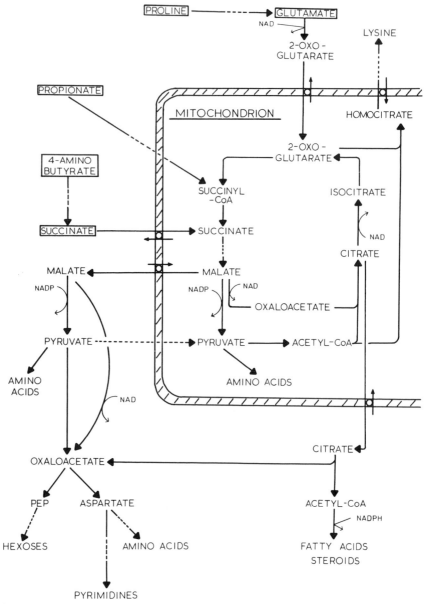

Figure 10. Proposed scheme for the subcellular organization of fungal metabolism in growth on substrates utilized via the tricarboxylic acid cycle. The growth substrates considered are shown enclosed in boxes. The scheme is largely theoretical since little data are available on the level of enzymes present during growth on these substrates and no studies of the subcellular localization of enzymes have been reported.

PEP carboxykinase and fructose-1,6-bisphosphatase are almost certainly production of PEP for amino acid and hexose synthesis, and fructose-6-phosphate for hexose and pentose synthesis, respectively. However, the role of NADP-malate dehydrogenase may be more complex since the pyruvate produced by this enzyme may be used within the mitochondrion for synthesis of amino acids such as valine, leucine, or isoleucine or for conversion to acetyl CoA and thence citrate for: (1) continued operation of the TCA cycle; (2) net synthesis of 2-oxoglutarate (for substrates entering at the level of succinate or succinyl CoA); (3) export to the cytosol in order to provide acetyl CoA for fatty acid and steroid synthesis. Hence, a mitochondrial isozyme of NADP-malate dehydrogenase should be present as is the case in *R. arrhizus* grown on glucose and *A. nidulans* grown on acetate (see Sections 3 and 5). The failure of *A. nidulans* mutants lacking pyruvate dehydrogenase (*pdhA, pdhB,* or *pdhC*) to grow on succinate is consistent with the scheme proposed (Fig. 10), but their ability to grow on glutamate or aspartate (Payton *et al.,* 1976; Uitzetter, 1982) remains a puzzle.

The organization proposed in Fig. 10 does not provide a mechanism to make NADPH available for reductive biosynthesis in the cytosol. This consideration may be of importance since growth on proline does not cause an increase in the level of NADP-isocitrate dehydrogenase (Kelly and Hynes, 1982), although some evidence suggests that the pentose phosphate pathway may be surprisingly active. In *Penicillium,* growth on succinate yields a level of glucose-6-phosphate dehydrogenase similar to that observed during growth on glucose (Boonsaeng *et al.,* 1974). However, if cytosolic isozyme(s) of NADP-malate dehydrogenase are present, as in *R. arrhizus* grown on glucose or *A. nidulans* grown on acetate, oxaloacetate may be formed via pyruvate from malate with the generation of NADPH as well as directly using NAD-malate dehydrogenase (Fig. 10; see also Section 3). The scheme implies that pyruvate carboxylase should be present in mycelia grown on substrates such as proline or succinate and that mutants lacking pyruvate carboxylase should still grow on these substrates. This latter point has been established for *pycA* mutants of *A. nidulans,* which grow effectively on succinate and L-glutamate. The presence of pyruvate carboxylase has not been demonstrated in mycelia grown on such carbon sources but it has been suggested that the enzyme is produced constitutively in this organism (Skinner and Armitt, 1972).

The proposed scheme (Fig. 10) implies that ATP-citrate lyase should be present in mycelia grown on succinate or glutamate although this has not been reported. It also suggests in accordance with the scheme for growth of *A. nidulans* on glucose (Fig. 3), that the condensation of 2-oxoglutarate with acetyl CoA to form homocitrate occurs in the mitochondrion as the initiating step for lysine biosynthesis by the 2-aminoadipic acid pathway. The proposed localization of homocitrate synthase in the mitochondrion is not essential for growth substrates such as proline and glutamate that yield 2-oxoglutarate in the cytosol, but appears much more reasonable for substrates such as propionate that enter at the level of succinyl CoA.

7. CONCLUSIONS

This survey of the subcellular organization and the regulation of carbohydrate metabolism in filamentous fungi reveals the extent to which our ignorance persists on many fundamental points for this group of eukaryotic organisms. The situation has in many respects not changed greatly since Cochrane (1976) pointed out the need both for examination of a wider range of fungal species and for indepth studies of single species. It is perhaps unfortunate that the most complete knowledge exists for those filamentous fungi (*N. crassa, A. nidulans,* and *A. niger*) that belong to the same class (Ascomycetes) as the yeasts, themselves subject to very intensive analysis (see Chapter 8). By contrast, as is clear from this chapter, other classes of both higher and lower fungi are relatively neglected. Although this situation has arisen for reasons of both investigative and commercial interest, it has done little to contribute to the level of general biological understanding for a major group of eukaryotes, many of which are uniquely available for combined genetic and physiological manipulation and analysis.

ACKNOWLEDGMENTS. Unpublished data cited in this review were obtained in studies supported by grants from the Science and Engineering Research Council to S.A.O. and M.C.S. (King's College London). We are most grateful to Mrs. Sheila Mackley (Leicester University) for her care and attention in taking the manuscript through many revisions with enthusiasm and good-humor.

REFERENCES

Adams, P. R., and Deploey, J. J., 1976, Amylase production by *Mucor miehei, Mycologia* **68**:934–938.

Ahmed, S. A., Smith, J. E., and Anderson, J. G., 1972, Mitochondrial activity during citric acid production by *Aspergillus niger, Trans. Br. Mycol. Soc.* **59**:51–60.

Allam, A. M., Hassan, M. M., and Elzainy, T. A., 1975, Formation and cleavage of 2-keto-3-deoxygluconate by 2-keto-3-deoxygluconate aldolase of *Aspergillus niger, J. Bacteriol.* **124**:1128–1131.

Almin, K. E., Eriksson, K.-E., and Pettersson, B., 1975, Extracellular enzyme system utilized by the fungus *Sporotrichum pulverulentum* for the breakdown of cellulose. (2) Activities of the five endo-1,4-β-glucanases towards carboxymethyl cellulose, *Eur. J. Biochem.* **51**:207–211.

Aoki, K., Arai, M., Minoda, Y., and Yamada, K., 1971, Acid-soluble α-amylase of black *Aspergilli*. VII. Sulfhydryl group, *Agric. Biol. Chem.* **35**:1913–1920.

Apirion, D., 1965, The two-way selection of mutants and revertants in respect to acetate utilization and resistance to fluoro-acetate in *Aspergillus nidulans, Genet. Res.* **6**:317–329.

Armitt, S., McCullough, W., and Roberts, C. F., 1976, Analysis of acetate non-utilizing (*acu*) mutants in *Aspergillus nidulans, J. Gen. Microbiol.* **92**:263–282.

Ashworth, J. M., and Kornberg, H. L., 1963, Fine control of the glyoxylate cycle by allosteric inhibition of isocitrate lyase, *Biochim. Biophys. Acta* **73**:519–522.

Ballard, F. J., and Hanson, R. W., 1967, The citrate cleavage pathway and lipogenesis in rat adipose tissue: Replenishment of oxaloacetate, *J. Lipid Res.* **8**:73–79.

Ballio, A., di Vittorio, V., and Russi, S., 1964, The isolation of trehalose and polyols from the conidia of *Penicillium chrysogenum* Thom, *Arch. Biochem. Biophys.* **107**:177–183.

Bartnicki-Garcia, S., 1968, Cell wall chemistry, morphogenesis and taxonomy in fungi, *Annu. Rev. Microbiol.* **22**:87–108.

Bartnicki-Garcia, S., and Nickerson, W. J., 1961, Thiamine and nicotinic acids: Anaerobic growth factors for *Mucor rouxii, J. Bacteriol.* **82**:142–148.

Bartnicki-Garcia, S., and Nickerson, W. J., 1962, Induction of yeast-like development in *Mucor* by carbon dioxide, *J. Bacteriol.* **84**:829–840.

Barton, L. L., Georgi, C. E., and Lineback, D. R., 1972, Effect of maltose on glucoamylase formation by *Aspergillus niger, J. Bacteriol.* **111**:771–777.

Bata, J., Vallier, P., and Colobert, L., 1978, α-Amylase activity in lysosomes of *Aspergillus oryzae, Experientia* **34**:572–573.

Beever, R. E., 1973, Pyruvate carboxylase and *N. crassa suc* mutants, *Neurospora Newsl.* **20**:15–16.

Beever, R. E., 1975, Regulation of 2-phosphoenolpyruvate carboxykinase and isocitrate lyase synthesis in *Neurospora crassa, J. Gen. Microbiol.* **86**:197–200.

Beever, R. E., and Fincham, J. R. S., 1973, Acetate non-utilizing mutants of *Neurospora crassa: acu-6,* the structural gene for PEP carboxylase and inter-allelic complementation at the *acu-6* locus, *Mol. Gen. Genet.* **126**:217–226.

Beevers, H., 1969, Glyoxysomes of castor bear endosperm and their relation to gluconeogenesis, *Ann. N.Y. Acad. Sci.* **168**:313–324.

Bendetskii, K. M., Tarovenko, V. L., Korchagina, G. T., Senatovora, T. P., and Khakhanova, T. S., 1974, The action of transglucosylase from *Aspergillus awamori* on maltose, *Biokhimiya* **39**:557–564.

Bentley, R., and Thiessen, C. P., 1957a, Biosynthesis of itaconic acid in *Aspergillus terreus.* I. Tracer studies with C¹⁴-labeled substrates, *J. Biol. Chem.* **226**:673–687.

Bentley, R., and Thiessen, C. P., 1957b, Biosynthesis of itaconic acid in *Aspergillus terreus.* III. The properties and reaction mechanism of *cis*-aconitic acid decarboxylase, *J. Biol. Chem.* **226**:703–720.

Benveniste, K., and Munkres, K. D., 1970, Cytoplasmic and mitochondrial malate dehydrogenases of *Neurospora:* Regulatory and enzymic properties, *Biochim. Biophys. Acta* **220**:161–177.

Berghem, L. E. R., Pettersson, L. G., and Axio-Fredriksson, W.-B., 1976, The mechanism of enzymatic cellulose degradation: Purification and some properties of two different 1,4-β-glucan glucanohydrolases from *Trichoderma viride, Eur. J. Biochem.* **61**:621–663.

Berry, D. R., Chmiel, A., and Al Obaidi, Z., 1977, Citric acid production by *Aspergillus niger,* in: *Genetics and Physiology of Aspergillus* (J. E. Smith and J. A. Pateman, eds.), pp. 405–426, Academic Press, New York.

Bloom, S. J., and Johnson, M. J., 1962, The pyruvate carboxylase of *Aspergillus niger, J. Biol. Chem.* **237**:2718–2720.

Blumenthal, H. J., 1976, Reserve carbohydrates in fungi, in: *The Filamentous Fungi,* Vol. 2 (J. E. Smith and D. R. Berry, eds.), pp. 292–307, Arnold, London.

Boonsaeng, V., Sullivan, P. A., and Shepherd, M. G., 1974, Succinate dehydrogenase of *Mucor rouxii* and *Penicillium duponti, Can. J. Biochem.* **52**:751–761.

Boonsaeng, V., Sullivan, P. A., and Shepherd, M. G., 1977a, Phosphofructokinase and glucose catabolism of *Mucor* and *Penicillium* species, *Can. J. Microbiol.* **23:**1214–1224.

Boonsaeng, V., Sullivan, P. A., and Shepherd, M. G., 1977b, Mannitol production in fungi during glucose catabolism, *Can. J. Microbiol.* **22:**808–816.

Bos, C. J., Slakhorst, M., Visser, J., and Roberts, C. F., 1981, A third unlinked gene controlling the pyruvate dehydrogenase complex in *Aspergillus nidulans, J. Bacteriol.* **148:**594–599.

Boschloo, J. G., and Roberts, C. F., 1979, D-Galactose requiring mutants in *Aspergillus nidulans* lacking phosphoglucomutase, *FEBS Lett.* **104:**17–20.

Bottger, I., Wieland, O., Brdiczka, D., and Pette, D., 1969, Intracellular localization of pyruvate carboxylase and phosphoenol pyruvate carboxykinase in rat liver, *Eur. J. Biochem.* **8:**113–119.

Bowes, I., and Mattey, M., 1979, The effect of manganese and magnesium ions on mitochondrial NADP⁺-dependent isocitrate dehydrogenase from *Aspergillus niger, FEMS Microbiol. Lett.* **6:**219–222.

Bowes, I., and Mattey, M., 1980, A study of mitochondrial NADP⁺-specific isocitrate dehydrogenase from selected strains of *Aspergillus niger, FEMS Microbiol. Lett.* **7:**323–325.

Brody, S., 1973, Metabolism, cell walls, and morphogenesis, in: *Developmental Regulation: Aspects of Cell Differentiation* (S. J. Coward, ed.), pp. 107–154, Academic Press, New York.

Brody, S., and Nyc, J. F., 1970, Altered fatty acid distribution in mutants of *Neurospora crassa, J. Bacteriol.* **104:**780–786.

Brody, S., and Tatum, E. L., 1966, The primary effect of a morphological mutation in *Neurospora crassa, Proc. Natl. Acad. Sci. USA* **56:**1290–1297.

Brody, S., and Tatum, E. L., 1967, Phosphoglucomutase mutants and morphological changes in *Neurospora crassa, Proc. Natl. Acad. Sci. USA* **58:**923–930.

Bull, A. T., and Bushell, M. E., 1976, Environmental control of fungal growth, in: *The Filamentous Fungi,* Vol. 2 (J. E. Smith and D. R. Berry, eds.), pp. 1–31, Arnold, London.

Bushell, M. E., and Bull, A. T., 1981, Anaplerotic metabolism of *Aspergillus nidulans* and its effect on biomass synthesis in carbon limited chemostats, *Arch. Microbiol.* **128:**282–287.

Camargo, E. P., Meuser, R., and Sonneborn, D., 1969, Kinetic analyses of the regulation of glycogen synthetase activity in zoospores and growing cells of the water mold *Blastocladiella emersonii, J. Biol. Chem.* **244:**5910–5919.

Carter, B. L. A., Bull, A. T., Pirt, S. J., and Rowley, B. I., 1971, Relationship between energy substrate utilisation and specific growth rate in *Aspergillus nidulans, J. Bacteriol.* **108:**309–313.

Cassady, W. E., Leiter, E. H., Bergquist, A., and Wagner, R. P., 1972, Separation of mitochondrial membranes of *Neurospora crassa.* 11. Submitochondrial localization of the isoleucine–valine biosynthetic pathway, *J. Cell Biol.* **53:**66–72.

Casselton, L. A., and Casselton, P. J., 1974, Functional aspects of fluoracetate resistance in *Coprinus* with special reference to acetyl-CoA synthetase deficiency, *Mol. Gen. Genet.* **132:**255–264.

Casselton, P. J., 1976, Anaplerotic pathways, in: *The Filamentous Fungi,* Vol. 2 (J. E. Smith and D. R. Berry, eds.), pp. 121–136, Arnold, London.

Cazzulo, J. J., and Stoppani, A. O. M., 1968, The regulation of yeast pyruvate carboxylase by acetyl-coenzyme A and L-aspartate, *Arch. Biochem. Biophys.* **127:**563–567.

Cleland, W. W., and Johnson, M. J., 1954, Tracer experiments on the mechanism of citric acid formation by *Aspergillus niger, J. Biol. Chem.* **208:**679–689.

Cochrane, V. W., 1976, Glycolysis, in: *The Filamentous Fungi,* Vol. 2 (J. E. Smith and D. R. Berry, eds.), pp. 65–91, Arnold, London.

Cohen, P., 1983, Control of enzyme activity, in: *Outlines in Biology,* 2nd ed. (W. J. Brammar and M. Edidin, eds.), Chapman & Hall, London.

Colvin, H. J., Sauer, B. L., and Munkres, K. D., 1973, Glucose utilization and ethanolic fermentation by wild-type and extrachromosomal mutants of *Neurospora crassa, J. Bacteriol.* **116**:1322–1328.

Cooney, C. L., and Levine, D. W., 1972, Microbial utilization of methanol, *Adv. Appl. Microbiol.* **15**:337–365.

Corina, D. L., and Munday, K. A., 1971, Studies on polyol function in *Aspergillus clavatus:* A role for mannitol and ribitol, *J. Gen. Microbiol.* **69**:221–227.

Corzo, R., and Tatum, E. L., 1953, Biosynthesis of itaconic acid, *Fed. Proc.* **12**:470.

Cotter, D. A., La Clave, A. J., Wegener, W. S., and Niederpruem, D. J., 1970, CO_2 control of fruiting in *Schizophyllum commune:* Non-involvement of sustained isocitrate lyase derepression, *Can. J. Microbiol.* **16**:605–608.

Courtright, J. B., 1975, Intracellular localization and properties of glycerokinase and glycerophosphate dehydrogenase in *Neurospora crassa,* Arch. Biochem. Biophys. **167**:21–33.

Courtright, J. B., 1977, Characteristics of a glycerol utilization mutant of *Neurospora crassa, J. Bacteriol.* **124**:497–502.

Crook, E. H., and Stone, B. A., 1957, The enzymatic hydrolysis of β-glucosides, *Biochemical Journal* **65**:1–12.

Cuppoletti, J., and Segel, I. H., 1979, Glycogen phosphorylase from *Neurospora crassa:* Purification of a high-speed-activity, nonphosphorylated form, *J. Bacteriol.* **139**:411–417.

Das, T. K., and Sen, K., 1983, Studies on the control of enzymes for the glyoxylate cycle in *Aspergillus terreus* IRRL 16043, *Curr. Microbiol.* **9**:55–58.

Day, D. F., 1978, A thermophilic glucoamylase from *Cephalosporium eichhorniae, Curr. Microbiol.* **1**:181–184.

Day, P. R., 1974, *Genetics of Host–Parasite Interaction,* Freeman, San Francisco.

Dekker, R. F. H., 1980, Interaction and characterization of a cellobiose dehydrogenase produced by a species of *Monilia, J. Gen. Microbiol.* **120**:309–316.

Deleyn, F., Claeyssens, M., Van Beeumen, J., and De Bruyne, C. K., 1978, Purification and properties of β-xylosidase from *Penicillium wortmanni, Can. J. Biochem.* **56**:43–50.

Denor, P. F., and Courtright, J. B., 1978, Isolation and characterization of glycerol-3-phosphate dehydrogenase-defective mutants of *Neurospora crassa, J. Bacteriol.* **136**:960–968.

Denton, R. M., and Halestrap, A. P., 1979, Regulation of pyruvate metabolism in mammalian tissues, *Essays Biochem.* **15**:37–77.

Dunn-Coleman, N. S., and Pateman, J. A., 1979, The regulation of hexokinase and phosphoglucomutase activity in *Aspergillus nidulans, Mol. Gen. Genet.* **171**:69–73.

Eidsa, G., 1972, Dissertation in microbiology, University of Bergen, Norway.

Elzainy, T. A., Hassan, M. M., and Allam, A. M., 1973a, A new pathway for nonphosphorylated degradation of gluconate by *Aspergillus niger, J. Bacteriol.* **114**:457–459.

Elzainy, T. A., Hassan, M. M., and Allam, A. M., 1973b, Occurrence of the nonphosphorylative pathway for gluconate degradation in different fungi, *Biochem. Syst.* **1**:127–128.

Ergle, D. R., 1947, The glycogen content of *Phymatotrichum schleroti, J. Am. Chem. Soc.* **69**:2061–2062.

Eriksson, K.-E, 1978, Enzymatic mechanisms involved in cellulose hydrolysis by the rot fungus *Sporotrichum pulverulentum, Biotechnol. Bioeng.* **20**:317–332.

Eriksson, K.-E., 1981, Cellulases of fungi, in: *Trends in the Biology of Fermentations for Fuels and Chemicals* (A. Hollaender, ed.), pp. 19–32, Plenum Press, New York.

Eriksson, K.-E., and Hamp, S. G., 1978, Regulation of endo-1,4-β-glucanase production in *Sporotrichum pulverulentum, Eur. J. Biochem.* **90**:183–190.

Eriksson, K.-E., and Pettersson, B., 1972, Extracellular enzyme system utilised by the fungus *Chrysosporium lignorum* for the breakdown of cellulose, in: *Biodeterioration of Materials* (A. H. Walters and E. H. Hueck-Van Der Plas), Vol. 2, pp. 116–120, Applied Sciences Publishers, London.

Eriksson, K.-E., and Pettersson, B., 1975a, Extracellular enzyme system utilized by the fungus *Sporotrichum pulverulentum*. 1. Separation, purification and physico-chemical characterization of five endo-1,4-β-glucanases, *Eur. J. Biochem.* **51**:193–206.

Eriksson, K.-E., and Pettersson, B., 1975b, Extracellular enzyme system utilized by the fungus *Sporotrichum pulverulentum* for the breakdown of cellulose. 3. Purification and physico-chemical characterisation of an exo-1,4-β-glucanase, *Eur. J. Biochem.* **51**:213–218.

Evans, C. T., Scragg, A. H., and Ratledge, C., 1983, A comparative study of citrate efflux from mitochondria of oleaginous and nonoleaginous yeasts, *Eur. J. Biochem.* **130**:195–204.

Fagerstam, L. G., and Pettersson, L. G., 1979, The cellulolytic complex of *Trichoderma reesei* QM9414, *FEBS Lett.* **98**:363–367.

Fantes, P. A., and Roberts, C. F., 1973, β-Galactosidase activity and lactose utilization in *Aspergillus nidulans*, *J. Gen. Microbiol.* **77**:471–486.

Feir, H. A., and Suzuki, I., 1969, Pyruvate carboxylase of *Aspergillus niger:* Kinetic study of a biotin-containing carboxylase, *Can. J. Biochem.* **47**:697–710.

Fincham, J. R. S., Day, P. R., and Radford, A., 1979, *Fungal Genetics*, Blackwell, Oxford.

Flavell, R. B., and Fincham, J. R. S., 1968, Acetate non-utilizing mutants in *Neurospora crassa*. II. Biochemical deficiencies and the role of certain enzymes, *J. Bacteriol.* **95**:1063–1068.

Flavell, R. B., and Woodward, D. O., 1970, The concurrent regulation of metabolically related enzymes: The Krebs cycle and glyoxylate shunt enzymes in *Neurospora, Eur. J. Biochem.* **17**:284–291.

Flawia, M. M., and Torres, N. H., 1972, Activation of membrane-bound adenylate cyclase by glucagon in *Neurospora crassa, Proc. Natl. Acad. Sci USA* **69**:2870–2873.

Flawia, M. M., and Torres, N. H., 1973, Adenylate cyclase activity in *Neurospora crassa*. II. Modulation by glucagon and insulin, *J. Biol. Chem.* **248**:4517–4520.

Fogarty, W. M., and Kelly, C. T., 1980, Amylases, amyloglucosidases and related glucanases, in: *Microbial Enzymes and Bioconversions* (A. H. Rose, ed.), pp. 115–158, Academic Press, New York.

Freedberg, I. M., Levin, Y., Kay, C. M., McCubbin, W. D., and Katchalski-Katzir, E., 1975, Purification and characterisation of *Aspergillus niger* exo-1,4-glucosidase, *Biochim. Biophys. Acta* **391**:361–381.

French, D., 1980, Amylases: Enzymatic mechanisms, in: *Trends in the Biology of Fermentations for Fuels and Chemicals* (A. Hollaender, ed.), pp. 151–182, Plenum Press, New York.

Fuscaldo, K. E., Lechner, J. F., and Bazinet, G., 1971, Genetic and biochemical studies of the hexose monophosphate shunt in *Neurospora crassa*. I. The influence of genetic defects in the pathway on colonial morphology, *Can. J. Microbiol.* **17**:783–788.

Fuska, J., and Proksa, B., 1976, Cytotoxic and antitumour antibiotics produced by microorganisms, *Adv. Appl. Microbiol.* **20**:259–370.

Gajewski, W., Litwinska, J., Paszewski, A., and Chojnacki, T., 1972, Isolation and characterization of lactose non-utilizing mutants of *Aspergillus nidulans, Mol. Gen. Genet.* **116**:99–106.

Gold, M. H., Farrand, R. J., Levoni, J. P., and Segel, I. H., 1974, *Neurospora crassa* glycogen phosphorylase: Interconversion and kinetic properties of the "active" form, *Arch. Biochem. Biophys.* **161**:515–527.

Gorbacheva, I. V., and Rodionova, N. A., 1977a, Studies on xylan degrading enzymes. I. Purification and characterisation of endo-1,4-β-xylanase from *Aspergillus niger* str14, *Biochim. Biophys. Acta* **484**:79–93.

Gorbacheva, I. V., and Rodionova, N. A., 1977b, Studies on xylan-degrading enzymes. II. Action pattern on endo-1,4-β-xylanase from *Aspergillus* niger str.14 on xylan and xylooligosaccharides, *Biochim. Biophys. Acta* **484**:94–102.

Gratzner, H. G., 1972, Cell wall alterations associated with the hyperproduction of extracellular enzymes in *Neurospora crassa, J. Bacteriol.* **111**:443–446.

Graves, L. B., Armentrout, V. N., and Maxwell, D. P., 1976, Distribution of glyoxylate cycle enzymes between microbodies and mitochondria in *Aspergillus tamarii, Planta* **132**:143–148.

Gunasekaran, M., 1972, Physiological studies on *Phymatotrichum omnivorum*. I. Pathways of glucose catabolism, *Arch. Mikrobiol.* **83**:328–331.

Gunner, H. B., and Alexander, M., 1964, Anaerobic growth of *Fusarium oxysporum, J. Bacteriol.* **87**:1309–1316.

Gwm, E. K., and Brown, R. D., 1977, Comparison of four purified extracellular 1,4-β-D-glucan cellobiohydrolase enzymes from *Trichoderma viride, Biochim. Biophys. Acta* **492**:225–231.

Haarasilta, S., and Taskinen, L., 1977, Location of three key enzymes of gluconeogenesis in bakers yeast, *Arch. Microbiol.* **113**:159–161.

Habison, A., Kubicek, C. P., and Rohr, M. 1979, Phosphofructokinase as a regulatory enzyme in citric acid producing *Aspergillus niger, FEMS Microbiol. Lett.* **5**:39–42.

Habison, A., Kubicek, C. P., and Rohr, M., 1983, Partial purification and regulatory properties of phosphofructokinase from *Aspergillus niger, Biochem. J.* **209**:669–676.

Hakansson, U., Fagerstam, L. G., Pettersson, L. G., and Andersson, L., 1978, Purification and characterisation of a low molecular weight 1,4-β-glucan glucanohydrolase from the cellulolytic fungus *Trichoderma viride* QM9414, *Biochim. Biophys. Acta* **524**:385–392.

Hakansson, U., Fagerstam, L. G., Pettersson, L. G., and Andersson, L., 1979, A 1,4-β-glucan glucanohydrolase from the cellulolytic fungus *Trichoderma viride* QM9414, *Biochem. J.* **179**:141–149.

Hall, D. O., and Greenawalt, J. W., 1967, The preparation and biochemical properties of mitochondria from *Neurospora crassa, J. Gen. Microbiol.* **48**:419–430.

Hammond, J. B. W., 1981, Variations in enzyme activity during periodic fruiting of *Agaricus bisporus, New Phytol.* **89**:419–428.

Hang, Y. D., and Woodams, E. E., 1977, Baked-bean waste: A potential substrate for producing fungal amylases, *Appl. Environ. Microbiol.* **33**:1293–1294.

Hankinson, O., 1972, Regulation of the pentose phosphate pathway and of mannitol-1-phosphate dehydrogenase in *Aspergillus nidulans*, Ph.D. thesis, University of Cambridge.

Hankinson, O., 1974, Mutants of the pentose phosphate pathway in *Aspergillus nidulans, J. Bacteriol.* **117**:1121–1130.

Hankinson, O., and Cove, D. J., 1974, Regulation of the pentose phosphate pathway in the fungus *Aspergillus nidulans, J. Biol. Chem.* **249**:2344–2353.

Hankinson, O., and Cove, D. J., 1975, Regulation of mannitol-1-phosphate dehydrogenase in *Aspergillus nidulans, Can. J. Microbiol.* **21**:99–101.

Harding, R. W., Caroline, D. F., and Wagner, R. P., 1970, The pyruvate dehydrogenase complex from the mitochondrial fraction of *Neurospora crassa, Arch. Biochem. Biophys.* **138**:653–661.

Hartman, R. E., and Keen, N. T., 1974a, The pyruvate carboxylase of *Verticillium albo-atrum, J. Gen. Microbiol.* **81**:15–19.

Hartman, R. E., and Keen, N. T., 1974b, The phosphoenol pyruvate carboxykinase of *Verticillium albo-atrum, J. Gen. Microbiol.* **81**:21–26.

Hayaishi, O., Shimazono, H., Katagri, M., and Saito, Y., 1956, Enzymatic formation of oxalate and acetate from oxaloacetate, *J. Am. Chem. Soc.* **78**:5126–5127.

Hayashida, S., 1975, Selective submerged productions of three types of glucoamylases by a Black-koji mould, *Agric. Biol. Chem.* **39**:2093–2099.

Hayashida, S., Nomura, T., Yoshino, E., and Hongo, M., 1976, The formation and properties of subtilisin-modified glucoamylase, *Agric. Biol. Chem.* **40**:141–146.

Held, A. A., 1970, Nutrition and fermentative energy metabolism of the water mould *Aqualinderella fermentans, Mycologia* **62**:339–358.

Held, A. A., Emerson, R., Fuller, M. S., and Gleason, F. H., 1969, *Blastocladiella* and *Aqualinderella:* Fermentative water moulds with high carbon dioxide optima, *Science* **165**:706–709.

Hers, H. G., and Van Schaftingen, E., 1982, Fructose, 2,6-bisphosphate 2 years after its discovery, *Biochem. J.* **206**:1–12.

Holligan, P. M., and Jennings, D. H., 1972a, Carbohydrate metabolism in the fungus *Dendryphiella salina*. II. The influence of different carbon and nitrogen sources on the accumulation of mannitol and arabitol, *New Phytol.* **71**:583–594.

Holligan, P. M., and Jennings, D. H., 1972b, Carbohydrate metabolism in the fungus *Dendryphiella salina*. I. Changes in the levels of soluble carbohydrates during growth, *New Phytol.* **71**:569–582.

Holligan, P. M., and Jennings, D. H., 1972c, Carbohydrate metabolism in the fungus *Dendryphiella salina*. III. The effect of the nitrogen source on the metabolism of [1-^{14}C]- and [6-^{14}C]-glucose, *New Phytol.* **71**:1119–1133.

Horikoshi, K., Iida, S., and Ikeda, Y., 1965, Mannitol and mannitol dehydrogenase in conidia of *Aspergillus oryzae*, *J. Bacteriol.* **89**:326–330.

Hult, K., and Gatenbeck, S., 1978, Production of NADPH in the mannitol cycle and its relation to polyketide formation in *Alternaria alternata*, *Eur. J. Biochem.* **88**:607–612.

Hult, K., and Gatenbeck, S., 1979, Enzyme activities of the mannitol cycle and some connected pathways in *Alternaria alternata*, with comments on the regulation of the cycle, *Acta Chem. Scand. Ser. B* **33**:239–243.

Hult, K., Veide, A., and Gatenbeck, S., 1980, The distribution of NADPH regenerating mannitol cycle among fungi, *Arch. Microbiol.* **128**:253–255.

Ingram, J. M., and Hochster, R. M., 1957, Purification and properties of fructose diphosphate aldolase from *Fusarium oxysporum* F. *lycopersici, Can. J. Biochem.* **45**:929–936.

Jagannathan, V., Singh, K., and Damodaran, M., 1956, Carbohydrate metabolism in citric acid fermentation. 4. Purification and properties of aldolase from *Aspergillus niger, Biochem. J.* **63**:94–101.

Jakubowska, J., 1977, Itaconic and itatartaric acid biosynthesis, in: *Genetics and Physiology of Aspergillus* (J. E. Smith and J. A. Pateman, eds.), pp. 427–451, Academic Press, New York.

Jobe, A., and Bourgeois, S., 1972, *Lac* repressor–operator interaction. VI. The natural inducer of the *lac* operon, *J. Mol. Biol.* **69**:397–408.

Johanson, R. A., Hill, J. M., and McFadden, B. A., 1974, Isocitrate lyase from *Neurospora crassa*. I. Purification, kinetic mechanism, and interaction with inhibitors, *Biochim. Biophys. Acta* **364**:327–340.

Joshi, A. P., and Ramakrishnan, C. V., 1959, Mechanism of formation and accumulation of citric acid in *Aspergillus niger*. I. Citric acid formation and oxaloacetic hydrolase in the citric acid producing strain of *Aspergillus niger, Enzymologia* **21**:43–51.

Kapoor, M., 1975, Subunit structure and some properties of pyruvate kinase of *Neurospora, Can. J. Biochem.* **53**:109–119.

Kapoor, M., O'Brien, M., and Braun, A., 1976, Modification of the regulatory properties of pyruvate kinase of *Neurospora* by growth at elevated temperatures, *Can. J. Biochem.* **54**:398–407.

Katz, J., and Wood, H. G., 1960, The use of glucose-C^{14} for the evaluation of the pathways of glucose metabolism, *J. Biol. Chem.* **235**:2165–2177.

Kelly, J. M., and Hynes, M. J., 1977, Increased and decreased sensitivity to carbon catabolite repression of enzymes of acetate metabolism in mutants of *Aspergillus nidulans, Mol. Gen. Genet.* **156**:87–92.

Kelly, J. M., and Hynes, M. J., 1981, The regulation of phosphoenolpyruvate carboxykinase and the NADP-linked malic enzyme in *Aspergillus nidulans, J. Gen. Microbiol.* **123**:371–375.

Kelly, J. M., and Hynes, M. J., 1982, The regulation of NADP-linked isocitrate dehydrogenase in *Aspergillus nidulans, J. Gen. Microbiol.* **128**:23–28.

Khouw, B. T., and McCurdy, H. D., 1969, Tricarboxylic acid cycle enzymes and morphogenesis in *Blastocladiella emersonii, J. Bacteriol.* **99**:197–205.

King, H. B., and Casselton, L. A., 1977, Genetics and function of isocitrate lyase in *Coprinus, Mol. Gen. Genet.* **157**:319–325.

Kinghorn, J. R., and Pateman, J. A., 1973, NAD and NADP L-glutamate dehydrogenase activity in ammonium regulation in *Aspergillus nidulans, J. Gen. Microbiol.* **78**:39–46.

Kisor, R. C., and Niehaus, W. G., 1981, Purification and kinetic characterization of mannitol-1-phosphate dehydrogenase from *Aspergillus niger, Arch. Biochem. Biophys.* **211**:613–621.

Kiser, M., Kubicek, C. P., and Rohr, M., 1980, Influence of manganese on morphology and cell wall composition of *Aspergillus niger* during citric acid fermentation, *Arch. Microbiol.* **128**:26–33.

Kobayashi, M., and Matsuda, M., 1978, Action of the glucoamylase on dextrans as an exo-dextranase, *Agric. Biol. Chem.* **42**:181–183.

Kobr, M. J., and Vanderhaeghe, F., 1973, Changes in density of organelles from *Neurospora, Experientia* **29**:1221–1223.

Kobr, M. J., Turian, G., and Zimmerman, E. J., 1965, Changes in enzymes regulating isocitrate breakdown in *Neurospora crassa, Arch. Mikrobiol.* **52**:169–177.

Kobr, M. J., Vanderhaeghe, F., and Combepine, G., 1969, Particulate enzymes of the glyoxylate cycle in *Neurospora crassa, Biochem. Biophys. Res. Commun.* **37**:640–645.

Koller, W., and Kindl, H., 1977, Glyoxylate cycle enzymes of the glyoxysomal membrane from cucumber cotyledons, *Arch. Biochem. Biophys.* **181**:236–248.

Kornberg, H. L., 1966a, Anaplerotic sequences and their role in metabolism, *Essays Biochem.* **2**:1–31.

Kornberg, H. L., 1966b, The role and control of the glyoxylate cycle in *Escherichia coli, Biochem. J.* **99**:1–11.

Krebs, H. A., 1972, The Pasteur effect and the relations between respiration and fermentation, *Essays Biochem.* **8**:1–34.

Kubicek, C. P., and Rohr, M., 1977, Influence of manganese on enzyme synthesis and citric acid accumulation in *Aspergillus niger, Eur. J. Appl. Microbiol.* **4**:167–175.

Kubicek, C. P., and Rohr, M., 1978, The role of the tricarboxylic acid cycle in citric acid accumulation by *Aspergillus niger, Eur. J. Appl. Microbiol. Biotechnol.* **5**:263–271.

Kubicek, C. P., and Rohr, M., 1980, Regulation of citrate synthase from the citric acid-accumulating fungus *Aspergillus niger, Biochem. Biophys. Acta* **615**:449–457.

Kubicek, C. P., Hampel, W., and Rohr, M., 1979a, Manganese deficiency leads to elevated amino acid pools in citric acid accumulating *Aspergillus niger, Arch. Microbiol.* **123**:73–79.

Kubicek, C. P., Zehentgruber, O., and Rohr, M., 1979b, An indirect method for studying the fine control of citric acid formation by *Aspergillus niger, Biotechnol. Lett.* **1**:47–52.

Kubicek, C. P., Zehentgruber, O., El-Kalak, H., and Rohr, M., 1980, Regulation of citric acid production by oxygen: Effect of dissolved oxygen tension on adenylate levels and respiration in *Aspergillus niger, Appl. Microbiol. Biotechnol.* **9**:101–115.

Kuwana, H., and Wagner, R. P., 1969, The *iv-3* mutants of *Neurospora crassa.* I. Genetic and biochemical characteristics, *Genetics* **62**:479–485.

Kuwana, H., Caroline, D. F., Harding, R. W., and Wagner, R. P., 1968, An acetohydroxy acid synthetase from *Neurospora crassa, Arch. Biochem. Biophys.* **128**:184–193.

Lagos, R., and Ureta, T., 1980, The hexokinases from wild-type and morphological mutant strains of *Neurospora crassa, Eur. J. Biochem.* **104**:357–365.

Lakshminarayana, K., Modi, V. V., and Shah, V. K., 1969, Studies on gluconate metabolism in *Aspergillus niger.* II. Comparative studies on the enzyme make-up of the adapted and parent strains of *Aspergillus niger, Arch. Mikrobiol.* **66**:396–405.

Lal, M., and Bhargava, P. M., 1962, Reversal by pyruvate of fluoride inhibition of *Aspergillus terreus, Biochim. Biophys. Acta* **58**:628–630.

Lambers, H., 1980, The physiological significance of cyanide-resistant respiration in higher plants, *Plant Cell Environ.* **3**:293–302.

La Nauze, I. M., 1966, Aconitase and isocitric dehydrogenases of *Aspergillus niger* in relation to citric acid production, *J. Gen. Microbiol.* **44**:73–81.

Lechevalier, H. A., 1975, Production of the same antibiotics by members of different genera of microorganisms, *Adv. Appl. Microbiol.* **19**:25–45.

Lechner, J. F., Fuscaldo, K. E., and Bazinet, G., 1971, Genetic and biochemical studies on the hexose monophosphate shunt in *Neurospora crassa*. II. Characterization of biochemical defects of the morphological mutants colonial 2 and colonial 3, *Can. J. Microbiol.* **17**:789–794.

Lee, W. H., 1967a, Carbon balance of mannitol fermentation and the biosynthetic pathway, *Appl. Microbiol.* **15**:1206–1210.

Lee, W. H., 1967b, Mannitol acetyl phosphate phosphotransferase of *Aspergillus, Biochem. Biophys. Res. Commun.* **29**:337–342.

LeJohn, H. B., 1971, Enzyme regulation, lysine pathways and cell wall structures as indicators of major lines of evolution in fungi, *Nature* **231**:164–169.

LeJohn, H. B., McCrea, B. E., Suzuki, I., and Jackson, S., 1969, Association–dissociation reactions of mitochondrial isocitric dehydrogenase induced by protons and various ligands, *J. Biol. Chem.* **244**:2484–2493.

Lenz, H., Wunderwald, P., and Eggerer, H., 1976, Partial purification and some properties of oxalacetase from *Aspergillus niger, Eur. J. Biochem.* **65**:225–236.

Lewis, D. H., and Smith, D. C., 1967, Sugar alcohols (polyols) in fungi and green plants. I. Distribution, physiology and metabolism, *New Phytol.* **66**:143–184.

Lewis, K. F., and Weinhouse, S., 1951, Studies on the mechanism of citric acid production in *Aspergillus niger, J. Am. Chem. Soc.* **73**:2500–2503.

Libor, S. M., Sundaram, T. K., and Scrutton, M. C., 1978, Pyruvate carboxylase from a thermophilic *Bacillus, Biochem. J.* **169**:543–558.

Lineback, D. R., Aira, L. A., and Horner, R. L., 1972, Structural characterisation of the two forms of glucoamylase from *Aspergillus niger, Cereal Chem.* **49**:283–298.

Lockwood, L. B., 1975, Organic acid production, in: *The Filamentous Fungi*, Vol. 1 (J. E. Smith and D. R. Berry, eds.), pp. 140–157, Arnold, London.

Loewenberg, J. R., and Chapman, C. M., 1977, Sophorose metabolism and cellulase induction in *Trichoderma, Arch. Microbiol.* **113**:61–64.

Lützen, N. W., Nielsen, M. H., Oxenboell, K. M., Schulein, M., and Stentebjerg-Olesen, B., 1983, Cellulases and their application in the conversion of lignocellulose to fermentable sugars, *Philos. Trans. R. Soc. London Ser. B* **300**:283–291.

Ma, H., Kubicek, C. P., and Rohr, M., 1981, Malate dehydrogenase isoenzymes in *Aspergillus niger, FEMS Microbiol. Lett.* **12**:147–151.

McCullough, W., and Roberts, C. F., 1974, The role of malic enzyme in *Aspergillus nidulans, FEBS Lett.* **41**:238–242.

McCullough, W., and Roberts, C. F., 1980, Genetic regulation of isocitrate lyase activity in *Aspergillus nidulans, J. Gen. Microbiol.* **120**:67–84.

McCullough, W., Payton, M. A., and Roberts, C. F., 1977, Carbon metabolism in *Aspergillus nidulans*, in *Genetics and Physiology of Aspergillus* (J. E. Smith and J. A. Pateman, eds.), pp. 97–129, Academic Press, New York

McCurdy, H. D., and Cantino, E. C., 1960, Isocitritase, glycine–alanine transaminase, and development in *Blastocladiella emersonii, Plant Physiol.* **35**:463–476.

McLaughlin, D. J., 1973, Ultrastructure of sterigma growth and basidiospore formation in *Coprinus* and *Boletus, Can. J. Bot.* **51**:145–150.

Mandels, M., Parrish, F. W., and Reese, E. T., 1962, Sophorose as an inducer of cellulase in *Trichoderma viride, J. Bacteriol.* **83**:400–408.

Mattey, M., 1977, Citrate regulation of citric acid production in *Aspergillus niger, FEMS Microbiol. Lett.* **2**:71–74.

Mattoo, A. K., and Parikh, N. R., 1975, Influence of sodium pyruvate on *Neurospora* fructose diphosphatase, *Neurospora Newsl.* **22:**9–10.

Maxwell, D. P., Williams, P. H., and Maxwell, M. D., 1970, Microbodies and lipid bodies in the hyphal tips of *Sclerotina sclerotiorum, Can. J. Bot.* **48:**1689–1691.

Maxwell, D. P., Armentrout, V. N., and Graves, L. B., 1977, Microbodies in plant pathogenic fungi, *Annu. Rev. Phytopathol.* **15:**119–134.

May, J. W., and Sloan, J., 1981, Glycerol utilization by *Schizosaccharomyces pombe:* Dehydrogenation as the initial step, *J. Gen. Microbiol.* **123:**183–185.

Miall, L. M., 1978, Organic acids, in: *Primary Products of Metabolism* (A. H. Rose, ed.), pp. 47–119, Academic Press, New York.

Mitchell, D., and Shaw, M., 1966, Metabolism of glucose-^{14}C, pyruvate-^{14}C, and mannitol-^{14}C by *Melampsora lini.* II. Conversion to soluble products, *Can. J. Bot.* **46:**453–460.

Monori, B. M., Kubicek, C. P., and Rohr, M., 1984, Pyruvate kinase from *Aspergillus niger:* A regulatory enzyme in glycolysis?, *Can. J. Microbiol.* **30:**16–22.

Montenecort, B. S., Nhlapo, S. D., Trimino-Vazques, H., Cuskey, S., Schamhart, D. H. J., and Eveleigh, D. E., 1980, Regulatory controls in relation to overproduction of fungal cellulases, in: *Trends in the Biology of Fermentations for Fuels and Chemicals* (A. Hollaender, ed.), pp. 33–53, Plenum Press, New York.

Morita, T., Shimizu, K., Ohga, M., and Korenaga, T., 1966, Studies on amylases of *Aspergillus oryzae* cultured on rice. I. Isolation and purification of glucoamylase, *Agric. Biol. Chem.* **30:**114–121.

Moss, M. O., 1977, *Aspergillus* mycotoxins, in: *Genetics and Physiology of Aspergillus* (J. E. Smith and J. A. Pateman, eds.), pp. 499–524, Academic Press, New York.

Murayama, T., and Ishikawa, T., 1973, Mutation in *Neurospora crassa* affecting some of the extracellular enzymes and several growth characteristics, *J. Bacteriol.* **115:**796–804.

Neilson, N. E., 1955, The aconitase of *Aspergillus niger, Biochim. Biophys. Acta* **17:**139–140.

Newsholme, E. A., and Crabtree, B., 1981, Theoretical considerations into the regulation of the flux through metabolic pathways, in: *Short-term Control of Liver Metabolism* (L. Hue and G. Vander Werve, eds.), pp. 3–17, Elsevier/North-Holland, Amsterdam.

Ng, A. M. L., Smith, J. E., and McIntosh, A. F., 1973, Changes in activity of tricarboxylic acid cycle and glyoxylate cycle enzymes during synchronous development of *Aspergillus niger, Trans. Br. Mycol. Soc.* **61:**13–20.

Niehaus, W. G., and Dilts, R. P., 1982, Purification and characterization of mannitol dehydrogenase from *Aspergillus parasiticus, J. Bacteriol.* **151:**243–250.

Nisizawa, T., Suzuki, H., Nakayama, M., and Nisizawa, K., 1971a, Inductive formation of cellulase by sophorose in *Trichoderma viride, J. Biochem.* **70:**375–385.

Nisizawa, T., Suzuki, H., and Nisizawa, K., 1971b, De novo synthesis of cellulase induced by sophorose in *Trichoderma viride* cells, *J. Biochem.* **70:**387–393.

Nisizawa, T., Suzuki, H., and Nisizawa, K., 1972, Catabolite repression of cellulase formation in *Trichoderma viride, J. Biochem.* **71:**999–1007.

North, M. J., 1973, Cold-induced increase of glycerol kinase in *Neurospora crassa, FEBS Lett.* **35:**67–70.

Nowakowska-Waszczuck, A., 1973, Utilization of some tricarboxylic-acid-cycle intermediates by mitochondria and growing mycelium of *Aspergillus terreus, J. Gen. Microbiol.* **79:**19–29.

O'Connell, B. T., and Paznokas, J. L., 1980, Glyoxylate cycle in *Mucor racemosus, J. Bacteriol.* **143:**416–421.

Okumura, R., Tokuda, K., Uchii, Y., and Kuwana, H., 1977, Pyruvate dehydrogenase complex mutants of *Neurospora crassa, Jpn. J. Genet.* **52:**469–470.

Orthofer, R., Kubicek, C. P., and Rohr, M., 1979, Lipid levels and manganese deficiency in citric acid producing strains of *Aspergillus niger, FEMS Microbiol. Lett.* **5:**403–406.

Osmani, S. A., and Scrutton, M. C., 1981, Activation of pyruvate carboxylase from *Aspergillus nidulans* by acetyl coenzyme A, *FEBS Lett.* **135:**253–256.

Osmani, S. A., and Scrutton, M. C., 1983, The sub-cellular localisation of pyruvate carboxylase and some other enzymes in *Aspergillus nidulans, Eur. J. Biochem.* **133:**551–560.

Osmani, S. A., and Scrutton, M. C., 1985, The subcellular localisation and regulatory properties of pyruvate carboxylase from *Rhizopus arrhizus, Eur. J. Biochem.* **147:**119–128.

Osmani, S. A., Marston, F. A. O., Selmes, I. P., Chapman, A. G., and Scrutton, M. C., 1981, Pyruvate carboxylase from *Aspergillus nidulans:* Regulatory properties, *Eur. J. Biochem.* **118:**271–278.

Osmani, S. A., Mayer, F., Marston, F. A. O., Selmes, I. P., and Scrutton, M. C., 1984, Pyruvate carboxylase from *Aspergillus nidulans:* Effects of regulatory modifiers on the structure of the enzyme, *Eur. J. Biochem.* **139:**509–518.

O'Sullivan, J., and Casselton, P. J., 1973, The subcellular localization of glyoxylate cycle enzymes in *Coprinus lagopus, J. Gen. Microbiol.* **75:**333–337.

Overman, S. A., and Romano, A. H., 1969, Pyruvate carboxylase of *Rhizopus nigricans* and its role in fumaric acid production, *Biochem. Biophys. Res. Commun.* **37:**457–463.

Pall, M. L., 1981, Adenosine 3′,5′-phosphate in fungi, *Microbiol. Rev.* **45:**462–480.

Parry, J. B., Stewart, J. C., and Heptinstall, J., 1983, Purification of the major endoglucanase from *Aspergillus fumigatus* Fresenius, *Biochem. J.* **213:**437–444.

Passeron, S., and Terenzi, H., 1970, Activation of pyruvate kinase of *Mucor rouxii* by manganese ions, *FEBS Lett.* **6:**213–216.

Paszczynski, A., Miedziak, I., Lobarzewski, J., Kochmanska, J., and Trojanowski, J., 1982, A simple method of affinity chromatography for the purification of glucoamylase obtained from *Aspergillus niger* C, *FEBS Lett.* **149:**63–66.

Pateman, J. A., and Kinghorn, J. R., 1977, Genetic regulation of nitrogen metabolism, in: *Genetics and Physiology of Aspergillus* (J. E. Smith and J. A. Pateman, eds.), pp. 203–241, Academic Press, New York.

Paveto, E., and Passeron, S., 1977, Some kinetic properties of *Mucor* rouxii phosphofructokinase: Effect of cyclic adenosine 3′,5′-monophosphate, *Arch. Biochem. Biophys.* **178:**1–7.

Payton, M. A., and Roberts, C. F., 1976, Mutants of *Aspergillus nidulans* lacking pyruvate kinase, *FEBS Lett.* **66:**73–76.

Payton, M. A., McCullough, W., and Roberts, C. F., 1976, Agar as a carbon source and its effect on the utilization of other carbon sources by acetate non-utilizing (*acu*) mutants of *Aspergillus nidulans, J. Gen. Microbiol.* **94:**228–233.

Payton, M. A., McCullough, W., Roberts, C. F., and Guest, J. R., 1977, Two unlinked genes for the pyruvate dehydrogenase complex in *Aspergillus nidulans, J. Bacteriol.* **129:**1222–1226.

Pazur, J. H., Knull, H. R., and Cepure, A., 1971, Glycoenzymes: Structure and properties of the two forms of glucoamylase from *Aspergillus niger, Carbohydr. Res.* **20:**83–96.

Plaut, G. W. E., 1970, DPN-linked isocitrate dehydrogenase of animal tissues, *Curr. Top. Cell. Regul.* **2:**1–27.

Pontecorvo, G., 1956, The parasexual cycle in fungi, *Annu. Rev. Microbiol.* **10:**393–400.

Ramakrishnan, C. V., and Martin, S. M., 1955, Isocitric dehydrogenase in *Aspergillus niger, Arch. Biochem. Biophys.* **55:**403–407.

Ramakrishnan, C. V., Steel, R., and Lentz, C. P., 1955, Mechanism of citric acid formation and accumulation in *Aspergillus niger, Arch. Biochem. Biophys.* **55:**270–273.

Raper, K. B., and Fennell, D. I., 1965, *The Genus Aspergillus*, Williams & Wilkins, Baltimore.

Raper, K. B., and Thom, C., 1949, *A Manual of the Penicillia*, Williams & Wilkins, Baltimore.

Reed, L. J., 1969, Pyruvate dehydrogenase complex, *Curr. Top. Cell. Regul.* **1:**233–251.

Reed, L. J., 1981, Regulation of mammalian pyruvate dehydrogenase complex by a phosphorylation–dephosphorylation cycle, *Curr. Top. Cell. Regul.* **18:**95–106.

Reese, E. T., 1975, Polysaccharases and the hydrolysis of insoluble substrates, in: *Biological Transformation of Wood by Microorganisms* (W. Leise, ed.), pp. 165–181, Springer-Verlag, Berlin.

Reese, E. T., Maguire, A. H., and Parrish, F. W., 1968, Glucosidases and exo-glucanases, *Can. J. Biochem.* **46**:25–34.

Reese, E. T., Maguire, A. H., and Parrish, F. W., 1973, Production of β-D-xylopyranosidases by fungi, *Can. J. Microbiol.* **19**:1065–1074.

Reilly, P. J., 1981, Xylanases: Structure and function, in: *Trends in the Biology of Fermentations for Fuels and Chemicals* (A. Hollaender, ed.), pp. 111–129, Plenum Press, New York.

Reinert, W. R., and Marzluf, G. A., 1974, Fructosediphosphatase of *Neurospora crassa*, *Neurospora Newsl.* **21**:16.

Rho, D., Desrochers, M., Jubasek, L., Driguez, H., and Defaye, J., 1982, Induction of cellulase in *Schizophyllum commune:* Thiocellobiose as a new inducer, *J. Bacteriol.* **149**:47–53.

Rifai, M. A., 1969, A Revision of the Genus *Trichoderma* Mycological Papers, Commonwealth Mycological Institute, Kew, Surrey, England, No. 116, pp. 37–42.

Robbins, E. A., and Boyer, P. D., 1957, Determination of the equilibrium of the hexokinase reaction and the free energy of hydrolysis of adenosine triphosphate, *J. Biol. Chem.* **224**:121–135.

Roberts, C. F., 1963, The adaptive metabolism of D-galactose in *Aspergillus nidulans*, *J. Gen. Microbiol.* **31**:285–295.

Rodionova, N. A., Gorbacheva, I. V., and Buivid, V. A., 1977, Fractionation and purification of endo-1,4-β-xylosidases of *Aspergillus niger, Biochemistry (USSR)* **42**:505–519.

Rohr, M., and Kubicek, C. P., 1981, Regulatory aspects of citric acid fermentation by *Aspergillus niger, Process Biochem.* **16**:34–37.

Rohr, M., Kubicek, C. P., and Kominek, J., 1983, Gluconic acid, in: *Biotechnology* (H. J. Rehm and G. Reed, eds.), Vol. 3, pp. 456–465, Verlag Chemie, Weinheim.

Romano, A. H., and Kornberg, H. L., 1968, Regulation of sugar utilization by *Aspergillus nidulans*, *Biochim. Biophys. Acta* **158**:491–493.

Romano, A. H., and Kornberg, H. L., 1969, Regulation of sugar uptake in *Aspergillus nidulans*, *Proc. R. Soc. London Ser. B* **173**:475–490.

Romano, A. H., Bright, M. M., and Scott, W. E., 1967, Mechanism of fumaric acid accumulation by *Rhizopus nigricans*, *J. Bacteriol.* **93**:600–604.

Rosenberg, G., and Pall, M. L., 1979, Properties of two cyclic nucleotide-deficient mutants of *Neurospora crassa*, *J. Bacteriol.* **137**:1140–1144.

Rougemont, A., and Kobr, M. J., 1973, Isocitrate lyase-2 from *Neurospora crassa*, *Neurospora Newsl.* **20**:28–29.

Sanwal, B. D., and Stachow, C. S., 1965, Allosteric activation of nicotinamide adenine dinucleotide specific isocitrate dehydrogenase in *Neurospora*, *Biochim. Biophys. Acta* **96**:28–44.

Schellart, J. A., van Arem, E. J. F., van Boekel, M. A. J. S., and Middlehoren, W. J., 1976, Starch degradation by the mould *Trichoderma viride*. II. Regulation of enzyme synthesis, *Antonie van Leeuwenhoek* **42**:239–244.

Schormuller, J., and Stan, H. J., 1970, Stoffwechsel-untersuchungen an lebenismitteltechnologische wiehtigen mikroorganismen pyruvatcarboxylase aus *Penicillium camemberti* var candidum 3, *Z. Lebensm. Unters. Forsch.* **142**:321–330.

Schwalb, M. N., 1974, Changes in activity of enzymes metabolizing glucose 6-phosphate during development of the basidiomycete *Schizophyllum commune*, *Dev. Biol.* **40**:84–89.

Schwitzguebel, J. P., and Palmer, J. M., 1981, Properties of mitochondria isolated from *Neurospora crassa* grown with acetate, *FEMS Microbiol. Lett.* **11**:273–277.

Schwitzguebel, J. P., Moller, I. M., and Palmer, J. M., 1981a, Changes in density of mitochondria and glyoxysomes from *Neurospora crassa:* A re-evaluation using silica sol gradient centrifugation, *J. Gen. Microbiol.* **126**:289–295.

Schwitzguebel, J. P., Moller, I. M., and Palmer, J. M., 1981b, The oxidation of tricarboxylate anions by mitochondria isolated from *Neurospora crassa*, *J. Gen. Microbiol.* **126**:297–303.

Scott, W. A., 1976, Adenosine 3′ : 5′-cyclic monophosphate deficiency in *Neurospora crassa*, *Proc. Natl. Acad. Sci. USA* **73**:2995–2999.

Scott, W. A., and Mahoney, E., 1976, Defects of glucose-6-phosphate and 6-phosphogluconate and their pleiotropic effects, *Curr. Top. Cell. Regul.* **10**:205–236.

Scott, W. A., and Solomon, B., 1975, Adenosine 3′,5′-cyclic monophosphate and morphology in *Neurospora crassa:* Drug-induced alterations, *J. Bacteriol.* **122**:454–463.

Scott, W. A., and Tatum, E. L., 1970, Glucose-6-phosphate dehydrogenase and *Neurospora* morphology, *Proc. Natl Acad. Sci. USA* **66**:515–522.

Scrutton, M. C., and White, M. D., 1974, Pyruvate carboxylase: Inhibition of mammalian and avian liver enzymes by α-ketoglutarate and L-glutamate, *J. Biol. Chem.* **249**:5405–5414.

Selby, K., and Maitland, C. C., 1967, The cellulase of *Trichoderma viride:* Separation of the components involved in the solubilization of cotton, *Biochem. J.* **104**:716–724.

Seubert, W., and Schoner, W., 1971, Regulation of pyruvate kinase, *Curr. Top. Cell. Regul.* **3**:237–267.

Shimi, I. R., and Nour El Dein, M. S., 1962, Biosynthesis of itaconic acid by *Aspergillus terreus*, *Arch. Mikrobiol.* **44**:181–188.

Sihtola, H., and Neimo, L., 1975, The structure and properties of cellulose, in: *Proceedings, Symposium on Enzymatic Hydrolysos of Cellulose* Anlanko, Finland (M. Bailey, T.-M. Enari, and M. Rinko, eds.), pp. 9–21, Sitra.

Sjogren, R. E., and Romano, A. H., 1967, Evidence of multiple forms of isocitrate lyase in *Neurospora crassa*, *J. Bacteriol.* **93**:1638–1643.

Skinner, V. A., and Armitt, S., 1972, Mutants of *Aspergillus nidulans* lacking pyruvate carboxylase, *FEBS Lett.* **20**:16–18.

Smith, C. M., and Plaut, G. W. E., 1979, Activities of NAD-specific and NADP-specific isocitrate dehydrogenases in rat-liver mitochondria: Studies with D-*threo*-α-methylisocitrate, *Eur. J. Biochem.* **97**:283–295.

Smith, D., Muscatine, L., and Lewis, D,, 1969, Carbohydrate movement from autotrophs to heterotrophs in parasitic and mutualistic symbiosis, *Biol. Rev.* **44**:17–90.

Smith, J. E., and Berry, D. R. (eds.), 1976, *The Filamentous Fungi*, Vol. 2, Arnold, London.

Smith, J. E., and Hacking, A., 1983, Fungal toxicity, in: *The Filamentous Fungi*, Vol. 4 (J. E. Smith, D. R. Berry, and A. Kristiansen, eds.), pp. 238–265, Arnold, London.

Song, E., Briggs, J., and Courtright, J. B., 1978, Alterations in the pyruvate dehydrogenase complex during adaptation to glucose by *Neurospora*, *Biochim. Biophys. Acta* **544**:453–461.

Sternberg, D., and Mandels, G. R., 1979, Induction of cellulolytic enzymes in *Trichoderma reesei* by sophorose, *J. Bacteriol.* **139**:761–769.

Sternberg, D., and Mandels, G. R., 1980, Regulation of the cellulolytic system in *Trichoderma reesei* by sophorose: Induction of cellulase and repression of β-glucosidase, *J. Bacteriol.* **144**:1197–1199.

Stewart, G. R., and Moore, D., 1971, Factors affecting the level and activity of pyruvate kinase from *Coprinus lagopus sensu* **Buller,** *J. Gen. Microbiol.* **66**:361–370.

Strandberg, G. W., 1969, D-Mannitol metabolism by *Aspergillus candidus*, *J. Bacteriol.* **97**:1305–1309.

Sugden, P. H., and Newsholme, E. A., 1975, The effects of ammonium, inorganic phosphate and potassium ions on the activity of phosphofructokinases from muscle and nervous tissues of vertebrates and invertebrates, *Biochem. J.* **150**:113–122.

Takenishi, S., and Tsujisaka, Y., 1973, On the modes of three xylanases produced by a strain of *Aspergillus niger* van Tiegham, *Agric. Biol. Chem.* **39**:2315–2323.

Takenishi, S., Tsujisaka, Y., and Fukumoto, J., 1973, Studies on hemicelluloses. IV. Purification and properties of the β-xylosidase produced by *Aspergillus niger* van Tiegham, *J. Biochem.* **73**:335–343.

Tellez-Inon, M. T., and Torres, H. N., 1970, Interconvertible forms of glycogen phosphorylase in *Neurospora crassa*, *Proc. Natl. Acad. Sci. USA* **66**:459–463.

Tellez-Inon, M. T., Terenzi, H., and Torres, H. N., 1969, Interconvertible forms of glycogen synthetase in *Neurospora crassa*, *Biochim. Biophys. Acta* **191**:765–768.

Terenzi, H. F., Flawia, M. M., and Torres, H. N., 1974, A *Neurospora crassa* morphological mutant showing reduced adenylate cyclase activity, *Biochem. Biophys. Res. Commun.* **58**:990–996.

Terenzi, H. F., Flawia, M. M., Tellez-Inon, M. T., and Torres, H. N., 1976, Control of *Neurospora crassa* morphology by cyclic adenosine3,5′-monophosphate and dibutyryl cyclic adenosine 3′,5′-monophosphate, *J. Bacteriol.* **126**:91–99.

Theirmer, R. R., 1982, Discussion, *Ann. N.Y. Acad. Sci.* **386**:283.

Tolbert, B., 1970, Purification and regulatory properties of yeast pyruvate carboxylase, *Ph.D. thesis, Case Western Reserve University.*

Tom, G. D., Viswanath-Reddy, M., and Howe, H. B., 1978, Effect of carbon source on enzymes involved in glycerol metabolism in *Neurospora crassa*, *Arch. Microbiol.* **117**:259–264.

Tracy, J. W., and Kohlaw, G. B., 1975, Reversible co-enzyme-A-mediated inactivation of biosynthetic condensing enzymes in yeast: A possible regulatory mechanism, *Proc. Natl. Acad. Sci. USA* **72**:1802–1806.

Tsao, M. U., and Madley, T. J., 1972, Kinetic properties of phosphofructokinase of *Neurospora crassa, Biochim. Biophys. Acta* **258**:99–105.

Tsao, M. U., Smith, M. W., and Borondy, P. E., 1969, Metabolic response of *Neurospora crassa* to environmental change, *Microbios* **1**:37–43.

Tsujisaka, Y., Takenishi, S., and Fukumoto, J., 1971, Studies on the hemicellulases. II. The mode of action of three hemicellulases produced from *Aspergillus niger*, *Nippon Nogei Kagaku Kaishi* **45**:253–260.

Turian, G., 1963, Sur le mecanisme de l'induction isocitratasique chez *Allomyces* et *Neurospora, Pathol. Microbiol.* **26**:553–563.

Turner, W. B., 1971, *Fungal Metabolites,* Academic Press, New York.

Turner, W. B., 1976, Polyketides and related metabolites, in: *The Filamentous Fungi,* Vol. 2 (J. E. Smith and D. R. Berry, eds.), pp. 445–459, Arnold, London.

Uitzetter, J. H. A. A., 1982, Studies on carbon metabolism in wild type and mutants of *Aspergillus nidulans,* Ph.D. thesis, University of Wageningen.

Uno, I., and Ishikawa, T., 1976, Effect of cyclic AMP on glycogen phosphorylase in *Coprinus macrorhizus, Biochim. Biophys. Acta* **452**:112–120.

Uno, I., and Ishikawa, T., 1978, Effect of cyclic AMP on glycogen synthetase in *Coprinus macrorhizus, J. Gen. Appl. Microbiol.* **24**:193–197.

Uwajima, T., Akita, H., Ito, K., Mihara, A., Aisaka, K., and Terada, O., 1980, Formation and purification of a new enzyme, glycerol oxidase and stoichiometry of the enzyme reaction, *Agric. Biol. Chem.* **44**:399–406.

Valentine, B. P., and Bainbridge, B. W., 1978, The relevance of a study of a temperature-sensitive ballooning mutant of *Aspergillus nidulans* defective in mannose metabolism to our understanding of mannose as a wall component and carbon/energy source, *J. Gen. Microbiol.* **109**:155–168.

Verhoff, F. H., and Spradlin, J. F., 1976, Mass and energy balance analysis of metabolic pathways applied to citric acid production by *Aspergillus niger, Biotechnol. Bioeng.* **18**:425–433.

Viswanath-Reddy, M., Bennett, S. N., and Howe, H. B., 1977, Characterization of glycerol non-utilizing protoperithecial mutants of *Neurospora, Mol. Gen. Genet.* **153**:29–38.

Vogel, H. J., 1964, Distribution of lysine pathways among fungi: Evolutionary implications, *Am. Nat.* **98**:435–446.

Voitkova-Lepshikova, A., and Kotskova-Kratokhvilova, A., 1966, Glucamylases of the *Aspergilli*, *Microbiology (USSR)* **35**:653–659.

Wanner, G., and Theirmer, R. R., 1982, Two types of mitochondria in *Neurospora crassa*, *Ann. N.Y. Acad. Sci.* **386**:269–284.

Watanabe, K., and Fukimbara, T., 1967, Saccharogenic amylase produced by *Aspergillus awamori.* V3. Inhibition by acid-stable and less acid-stable saccharogenic amylase, *J. Ferment. Technol.* **45**:226–232.

Watanabe, K., and Fukimbara, T., 1973, The composition of saccharogenic amylase from *Rhizopus javanicus* and the isolation of glycopeptides, *Agric. Biol. Chem.* **37**:2755–2761.

Watson, K., 1976, The biochemistry and biogenesis of mitochondria, in: *The Filamentous Fungi*, Vol. 2 (J. E. Smith and D. R. Berry, eds.), pp. 92–120, Arnold, London.

Watson, K., and Smith, J. E., 1967a, Oxidative phosphorylation and respiratory control in mitochondria from *Aspergillus niger*, *Biochem. J.* **104**:332–339.

Watson, K., and Smith, J. E., 1967b, Rotenone and amytal insensitive coupled oxidation of NADH by mitochondria from *Aspergillus niger*, *J. Biochem.* **61**:527–530.

Watson, K., Paton, W. P., and Smith, J. E., 1969, Oxidative phosphorylation and respiratory control in mitochondria from *Aspergillus oryzae*, *Can. J. Microbiol.* **15**:975–981.

Weiss, H., von Jagow, G., Klingenberg, M., and Bucher, T., 1970, Characterization of *Neurospora crassa* mitochondria prepared with a grind-mill, *Eur. J. Biochem.* **14**:75–82.

Weitzman, P. D. J., and Danson, M. J., 1976, Citrate synthase, *Curr. Top. Cell. Regul.* **10**:161–204.

Wergin, W. P., 1972, Ultrastructural comparison of microbodies in pathogenic and saprophytic hyphae of *Fusarium oxysporum* sp. *lycopersici*, *Phytopathology* **62**:1045–1051.

Whistler, R. L., and Richards, E. L., 1970, Hemicelluloses, in: *The Carbohydrates IIA* (A. J. Pigman and D. Horton, eds.), pp. 447–462, Academic Press, New York.

Wieland, O. H., Hartmann, U., and Siess, E. A., 1972, *Neurospora crassa* pyruvate dehydrogenase: Interconversion by phosphorylation and dephosphorylation, *FEBS Lett.* **27**:240–244.

Winskill, N., 1983, Tricarboxylic acid cycle activity in relation to itaconic acid biosynthesis by *Aspergillus terreus*, *J. Gen. Microbiol.* **129**:2877–2883.

Wold, W. S. M., and Suzuki, I., 1973, Cyclic AMP and citric acid accumulation by *Aspergillus niger*, *Biochem. Biophys. Res. Commun.* **50**:237–244.

Wood, H. G., Katz, J., and Landau, B. R., 1963, Estimation of the pathways of carbohydrate metabolism, *Biochem. Z.* **338**:809–847.

Wood, T. M., 1971, The cellulase of *Fusarium solani:* Purification and specificity of the β-(1→4)-glucanase and the β-D-glucosidase components, *Biochem. J.* **121**:353–362.

Wood, T. M., and McCrae, S. I., 1972, The purification and properties of the C1 component of *Trichoderma koningii* cellulase, *Biochem. J.* **128**:1183–1192.

Wood, T. M., and McCrae, S. I., 1977a, Cellulase from *Fusarium solani:* Purification and properties of the C1 component, *Carbohydr. Res.* **57**:117–133.

Wood, T. M., and McCrae, S. I., 1977b, The mechanism of cellulase action with particular reference to the C1 component, in: *Proceedings of Bioconversion Symposium IIT, Delhi* (T. K. Ghose, ed.), pp. 111–141, Thompson Press, India.

Wood, T. M., and McCrae, S. I., 1982, Purification and some properties of the extracellular β-D-glucosidase of the cellulolytic fungus *Trichoderma koningii*, *J. Gen. Microbiol.* **128**:2973–2982.

Wood, T. M., McCrae, S. I., and MacFarlane, C. C., 1980, The isolation, purification and properties of the cellobiohydrolase component of *Penicillium funiculosum* cellulase, Biochem. J. **189**:51–65.

Yamada, H., Okamoto, K., Kodama, K., and Tanaka, S., 1959, Mannitol formation by *Piricularia oryzae, Biochim. Biophys. Acta* **33**:271–273.
Yamasaki, Y., Suzuki, Y., and Ozawa, J., 1977a, Three forms of α-glucosidase and a glucoamylase from *Aspergillus awamori, Agric. Biol. Chem.* **41**:2149–2161.
Yamasaki, Y., Suzuki, Y., and Ozawa, J., 1977b, Properties of two forms of glucoamylase from *Penicillium oxalicum, Agric. Biol. Chem.* **41**:1443–1449.
Yamasaki, Y., Tsuboi, A., and Suzuki, Y., 1977c, Two forms of glucoamylase from *Mucor rouxianus*. II. Properties of the two flucoamylases, *Agric. Biol. Chem.* **41**:2139–2148.
Zehentgruber, O., Kubicek, C. P., and Rohr, M., 1980, Alternative respiration of *Aspergillus niger, FEMS Microbiol. Lett.* **8**:71–74.
Zink, M. W., 1967, Regulation of the ''malic'' enzyme in *Neurospora crassa, Can. J. Microbiol.* **13**:1211–1221.
Zink, M. W., 1972, Regulation of the two ''malic'' enzymes in *Neurospora crassa, Can. J. Microbiol.* **18**:611–617.
Zink, M. W., and Katz, J. S., 1973, Malic enzyme of *Fusarium oxysporum, Can. J. Microbiol.* **19**:1187–1196.
Zink, M. W., and Shaw, D. A., 1968, Regulation of ''malic'' isozymes and malic dehydrogenase in *Neurospora crassa, Can. J. Microbiol.* **14**:907–912.
Zonneveld, B. J. M., 1977, Biochemistry and ultrastructure of sexual development in *Aspergillus*, in: *Genetics and Physiology of Aspergillus* (J. E. Smith and J. A. Pateman, eds.), pp. 59–80, Academic Press, New York.

10

The Bacterial Phosphoenolpyruvate:Sugar Phosphotransferase System of Escherichia coli and Salmonella typhimurium

PIETER W. POSTMA

1. INTRODUCTION

Bacteria such as *Escherichia coli* and *Salmonella typhimurium* can grow in simple defined media consisting of inorganic salts and any one of a wide variety of carbon sources. But the cells must be able to respond to internal and external variations, including the depletion of nutrients, since they face continuous changes in the external environment and changing internal needs. One of the most challenging problems for a cell is to regulate and integrate its metabolism.

In this chapter I will review a bacterial phosphotransferase system (PTS) that catalyzes the transport and concomitant phosphorylation of carbohydrates, resulting in the intracellular accumulation of the corresponding sugar phosphates. This sugar PTS consists of a number of cytoplasmic and membrane-bound proteins. Each of these proteins can exist in the phosphorylated or un-phosphorylated state. A hypothesis previously formulated assigns an important role in the regulation of cellular metabolism to the phosphorylation state of one of these PTS proteins.

It is the purpose of this review to discuss this bacterial PTS as both a

---

PIETER W. POSTMA • Laboratory of Biochemistry, B.C.P. Jansen Institute, University of Amsterdam, 1018TV Amsterdam, The Netherlands.

transport system and a regulatory system and to point out the possible differences and similarities with other regulatory systems.

2. THE PEP:SUGAR PTS: AN OVERVIEW

Bacterial cells can accumulate solutes in many different ways. Specific transport systems, localized primarily in the cytoplasmic membrane, can each recognize one or at most a few compounds and are able to catalyze the translocation of these solutes into the cell. In most cases the solute is accumulated against its electrochemical potential, which implies that energy has to be supplied. As discussed elsewhere in this volume (Chapter 11), a number of mechanisms have been identified in bacteria that deal with the problem of supplying energy for solute transport. These transport systems, whether they are energized by an electrochemical proton or sodium gradient or directly by ATP, have one thing in common: the solute is accumulated in an unchanged form. In this respect the PTS differs fundamentally from these transport systems because it catalyzes the transport and concomitant phosphorylation of sugars at the expense of phosphoenolpyruvate (PEP), resulting in the intracellular accumulation of sugar phosphates.

This unique transport system was discovered approximately 20 years ago by Roseman and co-workers (Kundig *et al.,* 1964) in *E. coli* and has since been studied most extensively in the Enterobacteriaceae, *Staphylococcus aureus,* and *Streptococcus.* As reviewed more extensively in Chapter 7, the PTS has been identified in many obligate and facultative anaerobes and a few obligate aerobes. The basic observation that led to the discovery of the PTS was the phosphorylation of a number of carbohydrates in crude cell extracts at the expense of PEP. Very soon it became clear that this phosphorylating activity was associated with a system that catalyzes the uptake of many carbohydrates concomitant with their phosphorylation. One essential feature of this system is that no free sugar is found as an intermediate in the cell during this process. In that respect the PTS differs from a transport system coupled to a subsequent kinase (such as the galactose and glycerol transport systems; see Chapter 11) although formally the same product, a sugar phosphate, results from both processes.

The subsequent biochemical and genetic analysis of the PTS revealed it to consist of a number of components, some of which are common for all PTS sugars. A schematic picture is given in Fig. 1. In general, two proteins, enzyme I and HPr, are required for the first step in transport and phosphorylation of all PTS sugars, resulting in the phosphorylation of HPr at the expense of PEP. Sugar specificity resides in the membrane-bound enzymes II, each of which can recognize one or a few PTS sugars. During the translocation step, the phosphoryl group is transferred from phosphorylated HPr to the sugar via the enzymes II. Whether the enzymes II are phosphorylated during this process will be discussed

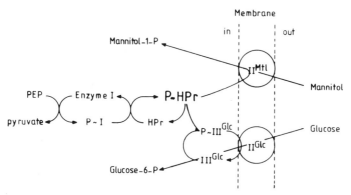

Figure 1. The PEP:sugar phosphotransferase system. Only two enzymes II are shown, specific for mannitol (II^Mtl) and glucose (II^Glc). The membrane contains many other enzymes II, each recognizing one or a few sugars. P~I, P~HPr, and P~III^Glc are the phosphorylated forms of enzyme I, HPr, and III^Glc, respectively.

later. The following general remarks can be made on the PTS in various organisms.

    1. A particular carbohydrate might be a PTS sugar in one organism but a non-PTS sugar in another organism. For instance, lactose is transported in *Staphylococcus aureus* by the PTS but in *E. coli* by an H^+–lactose symport system.

    2. The sugar-specific enzymes II do not necessarily consist of a single membrane-bound component. In some organisms an enzyme II might consist of more than one membrane-bound polypeptide. In other organisms an additional (cytoplasmic) protein, positioned between HPr and enzyme II, participates in sugar phosphorylation and transport. I will discuss some examples in more detail later.

    3. In some organisms (*Rhodospirillum rubrum, Rhodopseudomonas sphaeroides,* and *Pseudomonas aeruginosa*) no distinct enzyme I and HPr molecules are present but rather a ''composite'' protein fulfills a similar role (Saier *et al.,* 1971b; Brouwer *et al.,* 1982; Durham and Phibbs, 1982).

    It is apparent that many variations on a theme exist. In this review I will concentrate on the properties of the PTS in the Enterobacteriaceae *E. coli, S. typhimurium,* and *Klebsiella aerogenes,* which have been studied most extensively. Only occasionally will I digress to the PTS in other organisms such as *Staphylococcus aureus* and Streptococci. There is an additional reason for this rather limited view. Apart from its function as a transport system, the PTS is involved in the regulation of cellular metabolism. Knowledge about this regulatory role of the PTS in the expression and synthesis of many non-PTS systems has been almost exclusively restricted to *E. coli* and *S. typhimurium.* Various aspects of the PTS have been reviewed (Postma and Roseman, 1976; Saier, 1977; Dills *et al.,* 1980; Robillard, 1982; Postma and Lengeler, 1985).

3. COMPONENTS OF THE PTS

3.1. Soluble Proteins

As a first step in the transport and phosphorylation of all PTS sugars, the phosphoryl group of PEP is transferred to enzyme I. A histidine is phosphorylated at the N-3 position in the imidazole ring. Enzyme I has been purified from *E. coli* (Robillard *et al.*, 1979; Waygood and Steeves, 1980) and *S. typhimurum* (Weigel *et al.*, 1982a). Under nondenaturing conditions, enzyme I is a dimer consisting of identical subunits (Weigel *et al.*, 1982a). The molecular weight of the monomer as estimated from SDS–polyacrylamide gels ranges between 70,000 (Waygood and Steeves, 1980; Misset *et al.*, 1980) and 60,000 (Kukuruzinska *et al.*, 1982). Sedimentation equilibrium studies provided a molecular weight for the monomer of approximately 58,000 (Kukuruzinska *et al.*, 1982). Under certain conditions the polypeptide chains associated to dimers, which is most likely the catalytically active form (Waygood and Steeves, 1980; Misset *et al.*, 1980). This association is favored by PEP, Mg^{2+}, and higher temperature, for example. Conflicting results have been reported on the number of phosphoryl groups incorporated in enzyme I. Whereas Misset and Robillard (1982) found only one phosphoryl group incorporated per dimer in *E. coli*, two have been reported for enzyme I from *S. typhimurium* (Weigel *et al.*, 1982b). This discrepancy remains to be resolved.

Phosphorylated enzyme I has been isolated and used in several kinetic studies. It can donate its phosphoryl group to both HPr (see below) and pyruvate. In the latter case, PEP is formed and the apparent equilibrium constant of this reaction is 1.5 ± 1 (Weigel *et al.*, 1982b).

The next step in the set of reactions catalyzed by the PTS is the transfer of the phosphoryl group from phosphorylated enzyme I to the low-molecular-weight protein HPr. In contrast to enzyme I, the phosphoryl group is attached to the N-1 of an imidazole ring of histidine in HPr. *S. typhimurium* HPr, M_r 9017, contains two histidines at positions 15 and 75 (Weigel *et al.*, 1982c). The phosphoryl group is attached to His-15. Phosphorylation of HPr by phospho-enzyme I requires no Mg^{2+}. The apparent equilibrium constant for the overall reaction

$$PEP + HPr \rightleftharpoons \text{phospho-HPr} + \text{pyruvate}$$

is 11 ± 7.7 (Weigel *et al.*, 1982b). Thus, both phospho-enzyme I and phospho-HPr have very high apparent standard free energies of hydrolysis, almost equal to that of PEP.

HPr from *S. typhimurium* has been sequenced (Weigel *et al.*, 1982c) and the gene coding for HPr from *E. coli* has been cloned (Lee *et al.*, 1982; Bitoun *et al.*,

1983). The protein from *E. coli* is identical to that from *S. typhimurium* but different from the *Staphylococcus aureus* HPr. *E. coli* and *S. typhimurium* HPr on the one hand and *Staphylococcus aureus* HPr on the other hand catalyze similar reactions but cannot substitute for each other. The residues around the active site of *S. typhimurium* HPr are identical to those in the *Staphylococcus aureus* HPr as determined by Beyreuther and co-workers (Beyreuther *et al.*, 1977). ^{1}H NMR studies have also revealed the similarities of the HPr molecules from various bacteria (Kalbitzer *et al.*, 1982).

At least one additional cytoplasmic protein has been described that is required in addition to enzyme I and HPr and the enzymes II. Transport and phosphorylation of glucose and its nonmetabolizable analogue methyl α-glucoside requires III^{Glc}, which acts as a phosphoryl carrier between phospho-HPr and the membrane-bound enzyme II^{Glc} (Fig. 1). Surprisingly, III^{Glc} is also required for uptake and phosphorylation of sucrose by *E. coli*. A plasmid containing genes coding for an enzyme II^{Sucrose} and a hydrolase, allows *E. coli* K12 to grow on sucrose only when III^{Glc} is present (Lengeler *et al.*, 1982). It would be interesting to know how closely related II^{Glc} and II^{Sucrose} are.

III^{Glc} has been isolated from *S. typhimurium* (Scholte *et al.*, 1981; Meadow and Roseman, 1982). The monomer has an apparent M_r of 20,000 under denaturing conditions and can aggregate to form dimers, trimers, and hexamers (Scholte *et al.*, 1982). III^{Glc} accepts 1 mole of phosphate from phosphorylated HPr and the phosphoryl group is linked at the N-3 position of the imidazole ring in a histidine residue (Meadow and Roseman, 1982). Mutants have been described that produce an altered III^{Glc} molecule and are unable to form these aggregates at low temperature (Scholte *et al.*, 1982). Other forms of III^{Glc} exist (Scholte *et al.*, 1981, 1982; Meadow and Roseman, 1982; Nelson *et al.*, 1984a). A small percentage of purified III^{Glc} has a slightly higher apparent M_r (21,000) on SDS–polyacrylamide gels but a somewhat higher mobility in nondenaturing gels. Meadow and Roseman (1982) have shown that a fraction of III^{Glc} moves faster on nondenaturing gels and has lost its first seven N-terminal amino acids. Concomitantly, it has lost its activity in *in vitro* phosphorylation but not its ability to be phosphorylated by phospho-HPr. III^{Glc} derivatized with fluorescein-5-isothiocyanate at the N-terminal amino acid also accepts one phosphoryl group from phospho-HPr but is only 20% active in catalyzing the transfer of the phosphoryl group to methyl α-glucoside via III^{Glc} (Jablonski *et al.*, 1983). These results suggest that the N-terminus of III^{Glc} is important for the interaction with II^{Glc}. Mutants defective in the *crr* gene, coding for III^{Glc} (see below), lack both forms of III^{Glc} (Scholte *et al.*, 1982). Unexpectedly, membranes from mutants with a deletion of or a Tn10 insertion in the *crr* gene, exhibit III^{Glc}-like activity. Methyl α-glucoside is phosphorylated by these membranes in the presence of enzyme I, HPr, and PEP and this activity is abolished by antibodies against soluble III^{Glc} (Nelson *et al.*, 1982). The activity is dependent on II^{Glc} as is also shown by

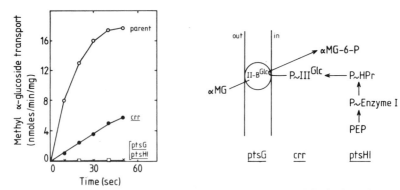

Figure 2. Transport of methyl α-glucoside in *S. typhimurium* mutants defective in various components of the PTS. *ptsHI, ptsG,* and *crr* mutants lack HPr/enzyme I, II^Glc, and III^Glc, respectively. αMG, methyl α-glucoside.

transport studies (Fig. 2; P. W. Postma, unpublished results). Mutants lacking II^Glc (*ptsG*) or enzyme I/HPr (*ptsI,H*) do not transport methyl α-glucoside at all, but strains devoid of the soluble III^Glc (*crr*) still exhibit approximately 20% of the wild-type uptake rate. I will discuss the properties of III^Glc in more detail later since it plays a central role in the regulation of expression of some genes and activity of some enzymes.

Using antibodies, III^Glc has been detected in the Enterobacteriaceae but also in *Vibrio parahaemolyticus* (P. W. Postma, unpublished results). Tanaka and co-workers (Kubota *et al.*, 1979) characterized the glucose PTS in this organism and found evidence for four components including III^Glc-like activity. Some *Vibrio* mutants, unable to grow on glucose (Kubota *et al.*, 1979), lack a protein that cross-reacts with antibody against III^Glc from *S. typhimurium.* No cross-reacting material has been detected in more distantly related organisms but this might be due to incorrect growth conditions.

No solid evidence for other enzymes III like III^Glc is available in the Enterobacteriaceae although an inducible fructose factor has been suggested in *K. aerogenes* (Walter and Anderson, 1973) and in *S. typhimurium* (Saier *et al.*, 1976; Waygood, 1980). III^Fru might be required for phosphorylation of fructose via the inducible enzyme II^Fru. Some genetic evidence (Sarno *et al.*, 1984) suggests that uptake and phosphorylation of glucitol in *S. typhimurium* may involve an enzyme III^Gut as well as the membrane-bound II^Gut. The closely related mannitol system has been shown to require only a membrane-bound II^Mtl (Jacobson *et al.*, 1983a). Perhaps we should consider the three enzymes IIA from *E. coli,* and reported to be specific for mannose, fructose, and glucose, respectively, as having a similar function (Kundig and Roseman, 1971). But after their initial isolation, no subsequent work has been published on these proteins.

Table I. The Enzymes II of the PEP:Sugar Phosphotransferase System

Enzyme II	Genetic symbol	Map position (min)	Substrates[a]	Refs.[b]
II^{Glc}	ptsG	24	Glucose, methyl α-glucoside, thioglucose, mannose, glucosamine, sorbose, 2-deoxyglucose	1–3
II^{Man}	ptsM	40	Mannose, glucose, 2-deoxyglucose, fructose, N-acetylglucosamine, glucosamine, methyl α-glucoside	1, 2, 4
II^{Fru}	ptsF	46	Fructose, glucose, sorbose	3, 5
II^{Mtl}	mtlA	81	Mannitol, glucitol	6
II^{Gat}	gatA	47	Galactitol, glucitol	6
II^{Gut}	srlA	58	Glucitol, galactitol, mannitol, fructose	6
II^{Bgl}	bglC	83	β-Glucosides, glucose, methyl α-glucoside	7
II^{Nag}	nagE	16	N-Acetylglucosamine, glucosamine, glucose	8
II^{Scr}	—	—	Sucrose	9
II^{Dha}	—	—	Dihydroxyacetone	10

[a]The underlined compounds are recognized by the enzyme II with high affinity and are most likely the "natural" substrates. The other compounds listed are recognized with a much lower affinity. The genetic symbols and map positions on the *E. coli* chromosome are taken from Bachmann (1983).
[b]References: 1, Curtis and Epstein (1975); 2, Stock *et al.* (1982); 3, Slater *et al.* (1981); 4, Jones-Mortimer and Kornberg (1980); 5, Ferenci and Kornberg (1974); 6, Lengeler (1975a,b); 7, Schaefler (1967); 8, White (1970); 9, Lengeler *et al.* (1982); 10, Jin and Lin (1984).

3.2. Membrane-Bound PTS Proteins

Membranes from *E. coli* and *S. typhimurium* contain many different sugar-specific enzymes II. Some of these are present under all growth conditions, whereas others are induced only after growth on a particular sugar. Based on genetic and biochemical studies, at least ten different enzymes II have been identified in both organisms. In one case (enzyme II^{Sucrose}), the gene coding for the enzyme II is located on a plasmid and might not originate from *E. coli* K12 (Lengeler *et al.*, 1982). The recently described PTS for dihydroxyacetone (Jin and Lin, 1984) is quite unexpected. In Table I a summary is given of the different enzymes II and some of the substrates that can be recognized by the enzymes II. It is clear that some enzymes II have rather broad specificity although the apparent affinity for some analogues is quite low. For example, the mannose-specific enzyme II, II^{Man}, is able to recognize mannose, glucose, N-acetylglucosamine, fructose, galactose, 2-deoxygalactose, D-fucose, glucosamine, 2-deoxyglucose, and methyl α-glucoside (Curtis and Epstein, 1975; Postma, 1976; Postma and van Thienen, 1978; Stock *et al.*, 1982). Table I also shows that almost any substrate can be recognized by at least two different enzymes II. This complicates the biochemical analysis considerably. Mutations in most enzymes II have been isolated. These mutations abolish the overall activity of these transport systems but have in most cases not solved the problem of the subunit composition of the enzymes II, i.e., whether they are composed of one or more different polypeptides. The recent purification of two enzymes II, II^{Mtl} from *E. coli* (Jacobson *et al.*, 1979, 1983a) and II^{Glc} from *S. typhimurium* (Erni *et al.*, 1982), now allows a more detailed study. In both cases the enzyme II consists of a single polypeptide chain that upon incubation with the various soluble PTS proteins (enzyme I, HPr, and, in the case of II^{Glc}, III^{Glc}) and PEP catalyzes the phosphorylation of its specific substrate. II^{Mtl} has been studied in some detail. The polypeptide chain, M_r approximately 68,000 (Jacobson *et al.*, 1983a), seems to be a transmembrane protein. This conclusion is based partly on inactivation studies (Jacobson *et al.*, 1983b). The bulk of the protein seems to be oriented toward the cytoplasmic side. To date, II^{Mtl} is the only enzyme II that has been cloned and sequenced (Lee *et al.*, 1981; Lee and Saier, 1983). From the sequence, one can also deduce a more hydrophobic part and a larger hydrophilic sequence that might be facing the cytoplasm. II^{Mtl} has been reconstituted in liposomes and shows both the PEP-dependent phosphorylation of mannitol and a mannitol-1-phosphate-dependent transphosphorylation of mannitol (Leonard and Saier, 1983). This transphosphorylation reaction will be discussed more extensively later.

As mentioned earlier, phosphorylated forms of the soluble PTS proteins have been demonstrated and isolated. There is less certainty about the membrane-bound enzymes II. A number of observations, all indirect, suggest that most enzymes II are probably phosphorylated. From kinetic measurements a phosphorylated enzyme II has been inferred as an intermediate. Most enzymes II conform to a Ping-Pong mechanism, i.e., phospho-HPr binds first to enzyme II, phosphorylates it, and is released (Rose and Fox, 1971; Perret and Gay, 1979; Hudig and Hengstenberg, 1980; Misset *et al.*, 1983). In contrast, from studies with II^{Man}, Rephaeli and Saier (1980a) concluded that the sugar phosphorylation reaction exhibited sequential Bi-Bi kinetics and does not involve a phosphorylated intermediate. A second indication that at least II^{Glc} is transiently phosphorylated comes from elegant studies by Knowles and co-workers (Begley *et al.*, 1982). They determined the overall stereochemical course of the reactions leading to phosphorylation of methyl α-glucoside by PEP (PEP → enzyme I → HPr → III^{Glc} → II^{Glc} → methyl α-glucoside). Starting with PEP chiral at the phosphorus and ending with methyl α-glucoside phosphate, they observed an overall inversion at the phosphorus. The authors concluded that phosphorylated II^{Glc} is an intermediate. Robillard and co-workers (Misset *et al.*, 1983) showed that incubation of membranes with enzyme I, HPr, and PEP leads to a burst of pyruvate, proportional to the amount of membranes. Subsequent addition of glucose to these membranes resulted in the formation of glucose-6-phosphate. Waygood and co-workers (Mattoo *et al.*, 1984) have recently introduced methods to keep the rather labile phospho-histidines intact during isoelectric focusing and SDS–polyacrylamide gel electrophoresis. Using these methods they demonstrated that II^{Glc} from *E. coli* is also phosphorylated (Peri *et al.*, 1984).

3.3. Levels of PTS Proteins

Whether a cell can transport and phosphorylate a carbohydrate via the PTS depends on two factors: (1) Is there an enzyme II specific for this sugar? (2) Are the PTS proteins present in the cell under the particular growth conditions? I have mentioned earlier that some systems are inducible in one type of strain, but seemingly constitutive in others. For example, in most strains of *E. coli* and *S. typhimurium* the enzymes II for mannitol, glucitol, and galactitol are inducible 10-fold or more (Lengeler and Steinberger, 1978b), depending on growth conditions. Results with respect to the enzyme II catalyzing glucose or mannose uptake are variable. Whereas Rephaeli and Saier (1980b) find that, for example, II^{Man} can be induced 8-fold in *S. typhimurium* comparing cells grown on glycerol and glucose, respectively, Stock *et al.* (1982) report only a 2-fold stimulation. II^{Glc} can be induced approximately 10-fold but a 50-fold induction was reported by Waygood *et al.* (1979). These values were obtained by measuring *in vitro* phosphorylation of PTS sugars. Transport rates in these cells show only a 3-fold stimulation at most. In *E. coli* the glucose phosphotransferase is inducible in some strains and almost ''constitutive'' in others whereas the fructose phosphotransferase is inducible in all strains tested (Kornberg and Reeves, 1972). A low level of induction is seen for the general proteins of the PTS, enzyme I and HPr, 3-fold at most (Rephaeli and Saier, 1980b; Stock *et al.*, 1982). Lastly, the III^{Glc} level is virtually independent of the growth conditions (Scholte *et al.*, 1981). One should interpret these data with some caution, however, since most results are based upon activity measurements. As pointed out earlier, the PTS is a very complex system and the overall *in vitro* phosphorylation activity depends on a number of proteins, some of which are at the same time substrates for a subsequent step. A better way to measure the amount of the various PTS proteins is the use of specific antibodies. Antibodies against enzyme I, HPr, and III^{Glc} have been raised and used to determine the number of molecules in the cell (Scholte *et al.*, 1981; Mattoo and Waygood, 1983). At least in the case of the cytoplasmic proteins, the levels as determined with antibodies vary in agreement with those determined by *in vitro* phosphorylation. No quantitative measurements of an enzyme II with specific antibodies are yet available.

In two publications, the level of the PTS was determined in cells grown under anaerobic conditions. Roehl and Vinopal (1980) found that a few *E. coli* strains cannot grow anaerobically on mannose, due to the absence of II^{Man}. A suppressor mutation, *dgs,* was found near the *man* gene (coding for phosphomannose isomerase) that allowed anaerobic growth on mannose. Curiously enough, the *E. coli* strains do grow anaerobically on mannose in the presence of nitrate as an electron acceptor. Lengeler and Steinberger (1978a) found that anaerobically grown *E. coli* contained quite high levels of Hpr, using glycerol as carbon source and fumarate as electron acceptor.

4. *IN VITRO* PHOSPHORYLATION OF SUGARS BY THE PTS

Transport systems are vectorial systems, catalyzing the movement of solutes from one compartment to another. The activity of purified components can only be measured after reconstitution in (for example) liposomes. The PTS is unique among bacterial solute transport systems in that at least part of its activity can be measured in a cell-free extract. Incubation of a PTS sugar with PEP, the soluble and membrane-bound proteins, and a radioactively labeled sugar results in the formation of a sugar phosphate that can be separated easily from the sugar and determined quantitatively. This has helped greatly in the purification of the various components. It also allows the determination of one of the PTS components in the presence of an excess of the others. The study of the partial reactions of the PTS is also possible since the various phosphoproteins, phosphorylated enzyme I, HPr, and III^{Glc} can be isolated and used as substrate for the subsequent steps. From these studies a rather detailed kinetic picture has emerged.

These *in vitro* studies have also allowed the testing of various proteins in heterologous systems, i.e., can HPr of one organism be phosphorylated by an enzyme I of another and can phospho-HPr serve as phosphoryl donor for a heterologous enzyme II? In general, these proteins cannot be exchanged unless the strains are closely related as in the case of the Enterobacteriaceae. For example, Simoni *et al.* (1973) showed that HPr from *S. typhimurium* and *Staphylococcus aureus* could not be exchanged. The main reason is the inability of the enzymes II to accept the phosphoryl group of the heterologous phospho-HPr. Phosphorylation of HPr by the heterologous enzyme I was much less affected. This might be important if one wants to clone the genes of certain PTS proteins in a heterologous organism.

Convenient as the *in vitro* phosphorylation assay using cell-free extract is, one should be aware that the conditions under which those proteins function are quite different from those in the cell. In general, the concentration of the proteins will be at least a factor of 50–100 lower in a cell-free extract and diluted even more during the actual assay. Second, after sonication or the use of a French pressure cell to break the cells, the orientation of the membrane inverts. Although in the membrane preparations the added PTS proteins are located on the same side of the membrane as in the intact cell, the sugar clearly is on the wrong side. One wonders whether it has to cross the membrane first to bind to the sugar-binding site of the enzyme II at the inside of the vesicles (which was originally at the outside of the intact cell). This would undoubtedly have an effect on the apparent K_m as measured in *in vitro* phosphorylation. Although it is difficult to compare directly the apparent K_m for sugar uptake in intact cells and for phosphorylation by membrane preparations, in a number of cases the values are close (Lengeler, 1975b; Stock *et al.*, 1982). One could argue, however, that under the *in vitro* conditions, only that part of the membrane population is active that represents leaky or not closed vesicles. The problem is obvious but no systematic study has been made.

An alternative assay system has been developed to measure phosphorylation of PTS sugars as a function of added PEP. Gachelin (1970) described how cells could be made permeable to PEP, sugars, and sugar phosphates by treatment with low concentrations of toluene. These conditions allow entrance of PEP, normally not a permeant solute, but prevent the leakage of enzymes, preserving the high cellular concentration of the various PTS components and the right orientation of the sugar substrate with respect to enzyme II. Under these conditions the phosphorylating activities approach the overall activities as measured by transport in intact cells (see Section 5).

5. TRANSPORT VIA THE PTS

As stated in the Introduction, many sugars can be taken up via the PTS. The sugar phosphates formed during this process are subsequently metabolized. This metabolism interferes with transport studies since many products are formed intracellularly that can leave the cell. Examples are pyruvate, lactate, acetate, CO_2, and others. The rate of uptake in cells as measured with radiolabeled substrates can be severely underestimated in this way. Measuring the disappearance of the label from the medium is no solution because labeled products appear outside the cell. A chemical determination of the solute is not always easy. The use of nonmetabolizable analogues or mutants that are inhibited in subsequent metabolism is only a partial solution since in this case only a limited amount of the solute can be accumulated, which hampers the determination of the initial rates of uptake. For example, the initial rate of 2-deoxyglucose transport is estimated to be about 400 nmoles/min per mg (dry weight) at room temperature (Stock *et al.*, 1982). Since the final level of the accumulated solute is approximately 20 nmoles/mg (dry weight), one can calculate that the reaction is finished within a few seconds. Finally, Kaback (1968) has shown that vesicles prepared from lysed spheroplasts and devoid of most of the cytoplasmic proteins, can take up PTS sugars provided PEP is supplied intravesicularly. Clearly, sufficient enzyme I and HPr remain inside the vesicles.

Using these various assay methods, the kinetic parameters of a number of PTS systems have been estimated, including those for glucose, mannose, fructose, and hexitols (Lengeler, 1975b; Rephaeli and Saier, 1978; Stock *et al.*,

1982). In general, one finds that the various transport systems have a high apparent affinity for their "natural" substrates, ranging between 0.5 μM and 20 μM. Sometimes the affinities are much lower, however. This can be the case with nonmetabolizable analogues or when an enzyme II, specific for a particular sugar, can recognize a few other sugars. II^{Man} of *S. typhimurium* recognizes mannose and glucose with an apparent K_m of 20 μM, but has a K_m in excess of 10 mM for fructose (Stock *et al.*, 1982). Similarly, II^{Glc}, specific for glucose, recognizes mannose with a K_m of 5 mM.

An important feature of the PTS is the concomitant phosphorylation of sugar during transport. No free sugar can be detected initially in the cell. In this respect, the PTS differs from an active transport system followed by a kinase although the final product is the same. This observation is related to the question as to whether the PTS can phosphorylate sugars from the inside. Quite early, Kaback (1968) showed that exogenous [^{14}C]glucose was phosphorylated in preference to internal [^{3}H]glucose. Recently, a more extensive study has been published using vesicles prepared from *S. typhimurium* that have the same orientation as intact cells. Beneski *et al.* (1982) showed that both 2-deoxyglucose and methyl α-glucoside are taken up by vesicles via II^{Man} and II^{Glc}, respectively. Enzyme I, HPr, and III^{Glc} have to be trapped inside the vesicles since the strain from which the vesicles are prepared lacks enzyme I, HPr, and III^{Glc} due to a *ptsHI–crr* deletion. Curiously, whereas the vesicles are able to phosphorylate 2-deoxyglucose via II^{Man} when enzyme I and HPr are added to the outside, methyl α-glucoside is not phosphorylated in the presence of enzyme I, HPr, and III^{Glc}. These results suggest that II^{Glc} is asymmetrically oriented in the membrane but that II^{Man} is able to phosphorylate its substrate from both sides of the cytoplasmic membrane. It is not clear whether this reaction occurs in intact cells (in the case of glucose generated from disaccharides). Certainly, it is not necessary since the cells contain high levels of glucokinase.

After some time the free sugar is found in the cell, possibly due to a phosphatase (Haguenauer and Kepes, 1971). Free sugar is also found outside the cell if, after uptake of a nonmetabolizable analogue such as methyl α-glucoside, the cell is chased with the unlabeled analogue or glucose. A problem that still has not been solved is whether the methyl α-glucoside phosphate is dephosphorylated before leaving the cell or whether the intracellular methyl α-glucoside phosphate donates its phosphoryl group to an incoming sugar via its enzyme II. It has been shown that *E. coli* cells take up methyl α-glucoside to a certain level and thereafter continuously exchange the analogue with a rate similar to the initial rate of uptake (Gachelin, 1970). Consequently, the absolute amount of sugar (phosphate) in the cell is constant. This process (an apparent transphosphorylation) has been analyzed in *in vitro* systems (Saier *et al.*, 1977a). A sugar can be phosphorylated via its specific enzyme II by its corresponding sugar phosphate. Enzyme I and HPr are not required for this reaction. The question is whether this *in vitro* transphosphorylation reaction plays any role in the intact cell. Recent measurements using purified II^{Mtl} show that mannitol-1-phosphate-dependent transphosphorylation of mannitol is only 0.1% of the rate of the PEP-dependent phosphorylation (Jacobson *et al.*, 1983a). Similarly, the trans-phosphorylation in membrane vesicles is only á very small fraction of the uptake rate.

Transport via the PTS involves a number of proteins, both cytoplasmic and membrane-bound. Since all PTS sugars use the common enzyme I and HPr, can a cell take up several PTS sugars at the same time with the maximal velocity of each of the sugars separately? In practice, one finds that uptake of certain PTS sugars can be inhibited by other PTS sugars. For example, *E. coli* cells growing on a mixture of glucose and fructose incorporate preferentially glucose (Amaral and Kornberg, 1975). Transport of mannitol and glucitol is inhibited strongly by methyl α-glucoside (Lengeler and Steinberger, 1978a). Similarly, methyl α-glucoside inhibits in *S. typhimurium* the uptake of mannose but mannose has almost no effect on the rate of methyl α-glucoside uptake (Scholte and Postma, 1981). This has been interpreted as follows: (1) under the experimental conditions the flow of phosphoryl groups through enzyme I/HPr is the rate-limiting step and (2) some enzymes II (or enzyme II/enzyme III complexes) have a higher affinity for phospho-HPr than others (Scholte and Postma, 1981). Interestingly, mutants have been isolated in which glucose is no longer able to prevent fructose utilization (Amaral and Kornberg, 1975). The mutation is highly cotransducible with the gene coding for II^{Fru} and may have resulted in an altered II^{Fru} with increased affinity for phospho-HPr. A similar mutation was reported for the enzyme II involved in β-glucoside uptake (Elvin and Kornberg, 1982).

The rate of uptake via the PTS is also influenced by the electrochemical H^+ gradient, arising from electron transport via the respiratory chain or ATP hydrolysis. Quite early it was noted that oxidizable substrates inhibit uptake of PTS substrates such as methyl α-glucoside (Hagihara *et al.*, 1963; Hoffee *et al.*, 1964) and it was suggested that the exit of sugars was accelerated by metabolic energy. Uncouplers stimulated uptake as well as inhibitors of the respiratory chain. In the older experiments, mostly the steady-state uptake values were determined but similar effects were found in *S. typhimurium*, measuring initial rates of methyl α-glucoside uptake (Stock *et al.*, 1982). It has been proposed that this effect of the electrochemical H^+ gradient (brought about by oxidizable substrates or lipophilic oxidizing agents) occurs via a dithiol–disulfide interchange in the enzyme II. The model has been extended to a number of non-PTS transport systems, including the lactose carrier (Konings and Robillard, 1982; Robillard and Konings, 1982). Central in the proposal is a change in affinity of these membrane proteins dependent on the oxidation states of the redox centers. For example, the apparent affinity of II^{Glc} for its substrate becomes 100- to 1000-fold lower in the oxidized form (Robillard and Konings,

1981). A 100-fold lowering in affinity was reported for the lactose carrier. In both cases the maximal velocity was not much affected. Different results were obtained by Neuhaus and Wright (1983) for the lactose permease. Using a more sensitive assay (increased levels of lactose permease in the cytoplasmic membrane) and employing the same modifying agents, they found that during partial inhibition the loss of transport paralleled the loss in binding sites. In other words, the maximal velocity changed but not the affinity. It is possible that a similar explanation is applicable to the PTS enzymes II. Although the energy state of the membrane clearly affects the PTS, the mechanism is by no means clear. One wonders how the entry of glucose in the cell escapes inhibition by its own oxidation during metabolism.

Inhibition of the enzymes II by sugar phosphates has sometimes been suggested as another regulatory mechanism. Although some enzymes II are quite sensitive to sugar phosphates, others are inhibited only at very high concentrations. Lengeler and Steinberger (1978a) found that whereas II^{Mtl} is not sensitive to glucose-1-phosphate, glucose-6-phosphate, or fructose-6-phosphate, II^{Gut} is inhibited strongly by these sugar phosphates. Glucose phosphorylation in *E. coli* is not very sensitive to inhibition by glucose-6-phosphate or fructose-6-phosphate and fructose phosphorylation is not affected at all (Clark and Holms, 1976). Similar results have been reported for mannose phosphorylation in *S. typhimurium* (Scholte and Postma, 1981). One can conclude that in most cases inhibition of PTS-mediated sugar phosphorylation by sugar phosphates is not an important regulatory mechanism. One should also keep in mind, as pointed out by Saier *et al.* (1977b), that inhibition by these sugar phosphates might be due to the hydrolytic activities of sugar phosphate phosphatases.

Are the enzymes II capable of transporting sugars in the absence of phosphorylation, i.e., when enzyme I or HPr or both are absent due to mutation? Using deletion mutants of *S. typhimurium* devoid of these proteins (see below), one can show that PTS sugars such as glucose, mannose, or fructose cannot enter the cell under these conditions, even when the outside concentration is as high as 50 mM (Postma and Stock, 1980). Thus, the enzymes II are unable to catalyze facilitated diffusion of PTS sugars. This is especially clear in the case of glucose since *S. typhimurium* and *E. coli* cells contain an active glucokinase that can trap any glucose that enters the cell (Saier *et al.*, 1973; Curtis and Epstein, 1975). Although these experiments are clear-cut, the picture is complicated by the following observations.

1. It appears that at least galactose can be transported by an enzyme II in the absence of phosphorylation. Although galactose is under normal conditions not a PTS sugar, elimination of all known galactose transport systems in *S. typhimurium* results in a cell that can take up galactose via II^{Man} (Postma, 1976). *In vitro* phosphorylation studies show that galactose can be phosphorylated by the PTS to galactose-6-phosphate, for which there is no further metabolic path-

way, however. Introduction of a *ptsI* mutation in this strain does not impair growth on galactose (Postma, 1976) but introduction of a mutation that eliminates II^{Man} inhibits growth (P. W. Postma, unpublished results). Presumably, the galactose-6-phosphate that is formed in the wild-type strain is hydrolyzed by a phosphatase before it can be metabolized further. This shows that galactose can enter by facilitated diffusion via II^{Man} and can serve as a carbon source by subsequent phosphorylation via galactokinase to galactose-1-phosphate. A similar observation has been made in *E. coli* although the authors in that case concluded that II^{Glc} was involved rather than II^{Man} (Kornberg and Riordan, 1976).

2. Facilitated diffusion of PTS sugars via the appropriate enzyme II can be induced by mutation. A mutation in II^{Glc} has been isolated that allows entry of glucose in the absence of PTS-mediated phosphorylation (Postma, 1981). The mutation maps in the gene that codes for II^{Glc} and results in an altered II^{Glc} that has lost its capacity to phosphorylate glucose even in the presence of enzyme I, HPr, and III^{Glc}. The apparent affinity for glucose has changed from approximately 5 μM in the parent strain to at least 10 mM in the mutant. However, the maximal rate of uptake is the same as in the parent strain. It has been suggested that II^{Glc} is a pore that in the absence of phosphorylation is closed. Upon phosphorylation the pore is opened. II^{Glc} recognizes glucose with high affinity and allows entry and phosphorylation (Postma, 1981). The mutation would result in a pore that is always open. In this connection it is interesting to mention a mutant of *E. coli* isolated by Bourd *et al.* (1975). They describe a strain that is unable to transport methyl α-glucoside via II^{Glc} but can still phosphorylate this sugar after the cells have been broken. Similar to the case described above, the mutation results in a partially defective II^{Glc}. Although from these results the conclusion has been drawn that two proteins might be involved in glucose and methyl α-glucoside transport and phosphorylation, one recognizing the sugar at the outside of the cell and the other involved in phosphorylation, one can as well visualize a single polypeptide with these two functions.

3. Most transport studies deal with the uptake of solutes in the cell. A pertinent question is whether sugars can leave the cell via the enzymes II. If the enzyme II is a symmetrical protein, the results discussed above would argue against an outward-directed movement of solutes via enzyme II when it is not phosphorylated. Lin and co-workers (Solomon *et al.*, 1973) have presented some data that mannitol might be able to leave the cell in strains lacking enzyme I. They measured the uptake of [^{3}H]galactosyl-mannitol via the lactose carrier and the subsequent splitting of this compound into galactose and mannitol. From the lower rate of galactosyl-mannitol in cells induced for II^{Mtl}, the authors concluded that mannitol could leave the cell via II^{Mtl}. Unfortunately, no efflux rate can be calculated to compare with the rate of uptake via the PTS. Reizer and Saier (1983) have studied methyl-β-D-thiogalactoside phosphate (TMG phosphate)

expulsion from *Streptococcus pyogenes*. Glucose is able to elicit the expulsion of TMG. This is not due to a vectorial exchange transphosphorylation but involves the dephosphorylation of intracellular TMG phosphate and subsequent efflux of TMG, presumably via II^{Lac} (Reizer *et al.*, 1983). In both cases discussed above, nonmetabolizable analogues have been used. A more physiologically significant condition is the following. According to Huber *et al.* (1980), *E. coli* cells metabolizing lactose secrete glucose and galactose. These authors suggest that neither the specific galactose transport systems (galactose permease and methyl β-galactoside permease) nor the glucose PTS (II^{Glc} and II^{Man}) are involved since a mutant lacking these four systems secretes glucose and galactose with the same rate. Another case is the following. *pts–crr* deletion mutants of *S. typhimurium* completely lacking the PTS proteins enzyme I, HPr, and III^{Glc} (see below) are able to grow on maltose and melibiose. The glucose produced internally from maltose or melibiose is phosphorylated via glucokinase. Elimination of glucokinase by mutation does not hurt the cell. In fact, they produce glucose in the medium with a rate equal to the rate of melibiose uptake (P. W. Postma, unpublished experiments). This result is difficult to reconcile with the experiments in which external glucose was unable to enter the cell. It suggests that perhaps enzyme II is functionally an asymmetrical enzyme that allows the efflux of sugars in the absence of phosphorylation via the PTS but not the influx.

It should be clear that our knowledge about the PTS-mediated transport process, in particular the role of the enzymes II, is still sketchy. Two enzymes II, those for mannitol and glucose, have now been purified. Reconstitution studies in liposomes may yield some answers to the questions raised above, i.e., whether enzyme II forms a pore that can be opened and closed by (de)phosphorylation and whether it is functionally a symmetrical enzyme.

6. ROLE OF THE PTS IN CHEMOTAXIS

Bacteria contain many receptors that are involved in the chemotactic response toward solutes. The receptors are often part of the transport systems for these solutes. For example, the periplasmic binding proteins are primary receptors and bind their substrates tightly. Subsequently, these and other proteins or solutes interact with methyl-accepting proteins (MCPs) in the cytoplasmic membrane. This alters the methylation pattern of these MCPs, which results via a series of partly unknown reactions in regulation of flagellar rotation. PTS sugars act also as chemoattractants. The enzymes II are the primary receptors (Adler and Epstein, 1974; Lengeler *et al.*, 1981) but in contrast to other solutes, the MCPs are not involved (Niwano and Taylor, 1982). It has been suggested that the phosphorylation and dephosphorylation of the enzymes II might be the trigger (Lengeler *et al.*, 1981).

7. GENETICS OF THE PTS

Dissection of the PTS into its various components and the study of its role in both bacterial transport and regulation has benefited very much from the isolation of mutants defective in one or more components of the PTS. Among others, these mutations served to test the prediction that mutants defective in enzyme 1 or HPr or both would be unable to grow on all PTS sugars, whereas mutants defective in an enzyme II are only unable to grow on the sugar for which that enzyme II is specific. These predictions have been fulfilled (for an earlier review, see Postma and Roseman, 1976).

The genes coding for the proteins involved in transport and phosphorylation of PTS compounds are located at many separate positions on the *E. coli* and *S. typhimurium* chromosome [Table I; for the most recent genetic maps, see Bachmann (1983) and Sanderson and Roth (1983)]. The genes for the general PTS proteins enzyme I, *ptsI*, and HPr, *ptsH*, are linked in an operon at approximately 52 min in *E. coli* and 50 min in *S. typhimurium* and are closely linked with the *crr* gene, coding for III^{Glc}. In *S. typhimurium* a promoter, *ptsP*, has also been identified (Cordaro *et al.*, 1974). This promoter controls the transcription of *ptsH* and *ptsI* but not of *crr*. In *E. coli*, transcription of *ptsI* is from the *ptsH* promoter but part is from a second site within or after *ptsH* (Britton *et al.*, 1984). This might explain why *ptsH* deletion strains of *E. coli* still contain enzyme I (Gershanovitch *et al.*, 1977). A detailed fine structure map of the *ptsH,I* region has been constructed by Cordaro *et al.* (1976) for *S. typhimurium*, and by Gershanovitch and co-workers (Rusina *et al.*, 1981; Rusina and Gershanovitch, 1983) for *E. coli*. The gene order *cysA cysK ptsP ptsH ptsI crr* has been proposed for *S. typhimurium* based upon deletion mapping (Cordaro and Roseman, 1972). In *E. coli* the order is *cysA crr ptsI ptsH cysK* (Epstein *et al.*, 1970; Britton *et al.*, 1983). Recent results with *S. typhimurium* DNA cloned from a lambda bank into a high-copy plasmid have thrown doubt on the gene order in *Salmonella* (Nelson *et al.*, 1984a). A 9.6-kb *Salmonella* chromosomal fragment overproduced III^{Glc}, the product of the *crr* gene, and complemented *cysA* mutations in both *E. coli* and *S. typhimurium*. No complementation of *ptsH,I* mutations was found, nor was enzyme I produced from the plasmid. Together with mapping data using phage P22 that place *crr* in between *cysA* and *ptsI*, these results suggest strongly that the gene order in *Salmonella* is *cysA crr ptsI,H cysK*, as in *E. coli*.

The genes coding for the various enzymes II are scattered across the chromosome. Table I gives the map positions of the known genes in *E. coli*. A complicating factor is the nomenclature. As far as known, the genes coding for the various enzymes II are regulated completely independent from the *ptsH,I* genes and Lin (1970) has suggested the enzymes II be designated according to their respective metabolic pathways. Nevertheless, it has been proposed that "*pts*" be used for all these genes (Cordaro, 1976). Since some were originally given other designations, we are faced with two different nomenclatures, i.e., *ptsG* and *ptsM* for II^{Glc} and II^{Man}, respectively, and *mtlA, srlA (gutA)*, and *gatA* for II^{Mtl}, II^{Gut}, and II^{Gat}, respectively. The symbols used in this chapter are given in Table I.

Some enzyme II genes such as those for the hexitols just mentioned are organized in operons that also contain the genes for the respective dehydrogenases required for the oxidation of the hexitol phosphates. Not all enzyme II genes are linked to genes coding for closely related metabolic enzymes, i.e., the genes for II^{Glc} and II^{Man}. The gene for II^{Fru}, *ptsF,* might be linked to the gene coding for fructose-1-phosphate kinase (Jones-Mortimer and Kornberg, 1974).

Some of the enzyme II genes exhibit an unexpected behavior. (1) The β-glucoside, *bgl,* operon is normally cryptic in *E. coli.* Reynolds *et al.* (1981) observed that the *bgl* operon is activated at a high frequency due to insertion of an IS1 or IS5 element. The operon is inducible. (2) *E. coli* K12 is unable to grow on sucrose. But sucrose genes (*sac* genes) can be introduced into *E. coli* K12 from a wild strain (Alaeddinoglu and Charles, 1979), and insert in the chromosome at the *dsd* site close to the *ptsH,I* genes. Upon insertion, the *dsd*$^{+}$ genes are no longer expressed. Sucrose uptake and metabolism via the *sac* genes requires enzyme I and HPr, similar to sucrose uptake dependent on plasmid-encoded II^{Sucrose} (Lengeler *et al.,* 1982). (3) The galactitol operon (*gat*) present in *E. coli* K12 can be replaced by a block of genes coding for ribitol and arabitol uptake and metabolism, which are not present in *E. coli* K12 and originate from *E. coli* C. Link and Reiner (1982, 1983) found that these two clusters had identical locations in the chromosome and are surrounded by the same sequence. They proposed an insertion/deletion model in which the arabitol–ribitol genes might act as a transposable element. In some natural *E. coli* strains, both blocks of genes are present on adjacent genes. Similar observations have been reported by Woodward and Charles (1983).

Only a few genes coding for these PTS proteins have been cloned and even less have been studied in any detail. The *ptsH* gene, coding for HPr, has been cloned from *E. coli* (Lee *et al.,* 1982; Bitoun *et al.,* 1983) as well as the *ptsI* gene, coding for enzyme I (Bitoun *et al.,* 1983). The *crr* gene, coding for III^{Glc}, has been cloned from *E. coli* (Meadow *et al.,* 1982a) and *S. typhimurium* (Nelson and Postma, 1984; Nelson *et al.,* 1984a) and the *Salmonella* gene has been sequenced (Nelson *et al.,* 1984a). The gene coding for II^{Mtl} has been cloned from *E. coli* and sequenced (Lee and Saier, 1983).

8. PHENOTYPE OF *pts* MUTANTS

What is the phenotype of the various mutants? Mutants defective in a specific enzyme II do not grow on that particular sugar unless another enzyme II can recognize the sugar and is present in the cell under the particular growth conditions. For example, glucose is transported by both II^{Glc} and II^{Man}. Mutants defective in either of these proteins are still able to grow on glucose. A mutation in II^{Mtl} on the other hand prevents growth on mannitol [unless II^{Gut} is present due to a mutation that results in its constitutive synthesis (Lengeler, 1975a,b)].

ptsH,I mutants behave differently. These mutants are pleiotropic, i.e., they cannot grow on any PTS sugar. They easily acquire suppressor mutations, however, that allow growth on some but not all PTS sugars. The most common suppressor mutation is one that results in the constitutive synthesis of the galactose permease due to either a *galR* (Saier *et al.*, 1973) or a *galC* (Postma, 1977; Postma and Stock, 1980) mutation. Glucose is taken up via the constitutive galactose permease, and glucokinase, present in the cell at high levels, catalyzes the formation of glucose-6-phosphate. The suggestion (Wagner *et al.*, 1979) that there exists in *E. coli* still another glucose uptake system, independent of the PTS, was shown to be incorrect (Postma *et al.*, 1982). The same *galC* or *galR* mutation together with a second, *mak*, mutation that results in high constitutive levels of manno(fructo)kinase (Saier *et al.*, 1971a) allows *ptsI* mutants and *ptsHI* deletion mutants to grow on mannose or fructose (Saier *et al.*, 1971a; Postma and Stock, 1980). No such suppressor mutations are known for the hexitols, however, in *E. coli* or *S. typhimurium*. In *K. aerogenes* on the other hand, *pts* mutants are able to grow on mannitol when the cells synthesize high constitutive levels of arabitol dehydrogenase and presumably the arabitol transport system (Tanaka *et al.*, 1967; Lengeler, 1975b). Mannitol is oxidized to fructose by arabitol dehydrogenase.

An unexpected pathway for glucose metabolism was recently described in *K. aerogenes*. From studies with chemostat cultures it was concluded that the rate of glucose uptake could not be accounted for by the PTS activity alone (O'Brien *et al.*, 1980). A subsequent search for another glucose-metabolizing pathway resulted in the discovery of a quinoprotein glucose dehydrogenase that uses 2,7,9-tricarboxy-1*H*-pyrrolo(2,3-*f*) quinoline-4,5-dione (PQQ) as its prosthetic group (Neyssel *et al.*, 1983). The glucose dehydrogenase converts glucose to gluconate, which serves subsequently as a carbon source. This pathway may also explain the old observation (Tanaka and Lin, 1967) that *ptsI* mutants of *K. aerogenes* can still grow on glucose in contrast to *E. coli* or *S. typhimurium ptsI* mutants. The maximal growth rate of a *K. aerogenes ptsI* mutant on glucose was about one-third of that of the parent strain (Neyssel *et al.*, 1983). Interestingly, the glucose dehydrogenase apoenzyme but not PQQ is synthesized in several *E. coli* strains. That this pathway can function in growth is shown in an experiment in which addition of PQQ to *ptsI* mutants, derived from these *E. coli* strains, allows growth on glucose (Hommes *et al.*, 1984). There are thus many ways in which *pts* mutants can grow on PTS compounds but for each of these another pathway has to be created. Although this complicates the study of these mutants, it allows one to discover new pathways, especially when looking for suppressor mutations in *pts* deletion strains.

The phenotype of *pts* mutants is more comprehensive than we have discussed up till now. It has been observed from the beginning that strains containing a tight *ptsI* mutation are unable to grow on a number of non-PTS compounds
including lactose, maltose, and glycerol. Each of these compounds has its own
transport system and I will discuss in the remainder of this chapter how this
phenotype arises and what it can tell us about regulation of gene expression and
enzyme activity in the Enterobacteriaceae.

9. REGULATION BY THE PTS

As mentioned in the previous section, *pts* mutants of *E. coli* and *S. typhimurium* are unable to grow on many non-PTS compounds [e.g., lactose,
melibiose, maltose, glycerol, xylose, rhamnose, and citric acid cycle intermediates such as succinate, malate and citrate (Table II)]. Further examples will be
discussed later. What is common between all these solutes and how is growth on
these compounds regulated by the PTS? One of the first to address this question
was Gershanovitch and co-workers (Gershanovitch *et al.*, 1967) who observed
that in *pts* mutants of *E. coli* the synthesis of β-galactosidase, one of the products
of the *lac* operon, is repressed. Expression of the *lac* operon, like many other
operons, requires the presence of both an inducer and cAMP. The inducer reacts
with a repressor or an activator whereas cAMP in a complex with the cAMP-

Table II. Phenotype of pts and crr Mutants

	Growth on[a]			
Relevant genotype	PTS sugars[b]	Class I compounds[c]	Class II compounds[d]	Galactose
Wild type	+	+	+	+
ptsI (leaky)	−	+	+	+
ptsI (leaky) + PTS sugar	−	−	−	+
pts(HI)	−	−	−	+
pts(HI) crr	−	+	−	+
pts(HI) crp*	−	+	+	+
pts(HI) + cAMP	−	+	+	+
crr	+	+	−	+
crr crp*	+	+	+	+
crr + cAMP	+	+	+	+

[a] +, growth after 48 hr at 37°C; −, no growth.
[b] PTS sugars include glucose, fructose, mannose, and hexitols.
[c] Class I compounds include maltose, melibiose, glycerol, and lactose.
[d] Class II compounds include xylose, rhamnose, and citric acid cycle intermediates.

binding protein is required for initiation of transcription (for a review, see Ullmann and Danchin, 1983). Other workers also noted that induction of various catabolic enzymes was impaired in *pts* mutants (Fox and Wilson, 1968; Pastan and Perlman, 1969; Tyler and Magasanik, 1970; Berman and Lin, 1971). Subsequent work showed that the PTS is most likely involved in the regulation of both intracellular cAMP and inducer levels.

9.1. The crr Gene

A clue to the role of the PTS in the regulation of cellular metabolism was found in the isolation of suppressor mutations in *S. typhimurium* that allowed growth of *pts* mutants on non-PTS compounds such as maltose, melibiose, or glycerol (Saier *et al.*, 1971a; Saier and Roseman, 1976a). Most mutations restore growth on all three carbon sources at the same time (and lactose in *E. coli*). Mapping of the suppressor mutation showed it to be closely linked to the *pts* operon (Saier and Roseman, 1972). The gene was designated *crr* for carbohydrate *repression resistant*, a somewhat unfortunate name since it resulted in some confusion about a possible identity between PTS-mediated repression and the general phenomenon of catabolite repression. These phenomena will be discussed below. The only biochemical defect connected with the *crr* mutation was shown to be a lower level of III^{Glc}, the PTS protein involved in glucose transport via II^{Glc} (Fig. 1; Saier and Roseman, 1976a). Later studies showed that *crr* is indeed the structural gene for III^{Glc}. This conclusion is based on the study of *crr* mutants producing "altered" III^{Glc} molecules (Scholte *et al.*, 1982; Meadow et al., 1982b) and analysis of the *crr* gene products (Meadow and Roseman, 1982; Nelson *et al.*, 1984a).

A second clue in elucidating the regulatory role of the PTS was found in the behavior of leaky *ptsI* mutants that contained less than 1% enzyme I activity. These leaky mutants are different from tight *ptsI* mutants in that they still can grow on maltose, melibiose, or glycerol. However, addition of a PTS sugar to the growth medium inhibits growth and results in the tight *ptsI* phenotype (Saier and Roseman, 1972, 1976a,b). Both metabolizable and nonmetabolizable PTS sugars elicit this phenotype. This observation allowed a simple experiment since the leaky *ptsI* strain could be grown on the particular non-PTS carbon source and induced for the transport system and subsequent catabolic enzymes. Transport studies showed that uptake of, for example, maltose in maltose-grown mutant cells or glycerol in glycerol-grown mutant cells is normal but addition of a PTS sugar, even nonmetabolizable analogues such as methyl α-glucoside or 2-deoxyglucose, inhibited transport severely (Saier and Roseman, 1972, 1976b). Mutants containing both the leaky *ptsI* mutation and a *crr* mutation are resistant to this inhibition by all PTS sugars. Strictly speaking, in most of the studies discussed here solute uptake and not transport was measured since the solute is metabolized further and the accumulated label in the cell represents the solute and all compounds derived from it. Thus, this does not allow one to eliminate the possibility that inhibition by PTS compounds is at the level of metabolism instead of transport. In the study of the melibiose transport system, a non-metabolizable analogue, thiomethylgalactoside, was used, however, and it was shown that the influx of this compound was inhibited by PTS substrates in a leaky *ptsI* mutant (Saier and Roseman, 1972). From these experiments it was concluded that PTS sugars can inhibit the uptake of non-PTS compounds via an indirect interaction since none of these non-PTS transport systems has specificity for the PTS sugars. In addition, a PTS sugar inhibits only when its enzyme II is present. Mutants lacking a particular enzyme II are resistant to inhibition by this PTS sugar, excluding direct inhibition of the various transport systems by the PTS sugars (Saier and Roseman, 1972, 1976b). The term *inducer exclusion* was introduced. This has also led to some confusion since this general term has been used mainly for a rather restricted process, i.e., the inhibition of uptake of certain non-PTS solutes by PTS sugars. Inhibition of uptake of one carbon source by another can be accomplished by different mechanisms: (1) direct competition for the same binding site of the transport system, i.e., glucose and galactose in the case of the galactose permease (Adhya and Echols, 1966; Rotman *et al.*, 1968); (2) competition of solutes for a common intermediate, e.g., phospho-HPr in the case of PTS sugars (Scholte and Postma, 1981); (3) indirect interaction via an intracellular component, either a protein or a small metabolite (e.g., a sugar phosphate). In the remainder the term *PTS-mediated inducer exclusion* will be used to describe the inhibition of maltose, melibiose, and glycerol uptake by PTS sugars. Exclusion of one PTS sugar by another (described in Section 5) is not included in this term.

The observations described above, i.e., the behavior of tight and leaky *ptsI* mutants, the inhibition of non-PTS transport systems by PTS sugars, and the isolation of *crr* mutants, have resulted in a model in which the phosphorylation state of III^{Glc} plays an important role (Saier and Feucht, 1975; Saier and Stiles, 1975; Saier and Roseman, 1976b; Postma and Roseman, 1976). Figure 3 pictures this model. As explained in a previous section, each of the cytoplasmic PTS proteins can be phosphorylated and dephosphorylated. The ratio between phosphorylated and unphosphorylated protein depends on the influx of phosphoryl groups from PEP and the efflux via the sugar/enzyme II complexes. It is easy to visualize that tight *ptsI* mutants and *ptsHI* deletion mutants will contain only unphosphorylated III^{Glc}. Leaky *ptsI* mutants, with a residual 1% enzyme I level but a normal dephosphorylation capacity via the enzymes II, will also contain most likely unphosphorylated II^{Glc} in the presence of any PTS sugar. This is immediately obvious for glucose and methyl α-glucoside, which can dephosphorylate P~III^{Glc}. All other PTS sugars can dephosphorylate P~HPr via their specific enzymes II. Since the reaction between the (un)phosphorylated

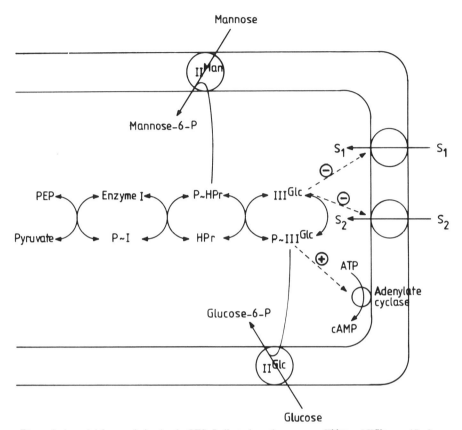

Figure 3. A model for regulation by the PTS. Indicated are the enzymes II^{Man} and II^{Glc}, specific for mannose and glucose, respectively. Activation (+) of adenylate cyclase by phosphorylated III^{Glc} (P~III^{Glc}) and inhibition (−) of two different non-PTS uptake systems by III^{Glc} are indicated. S$_1$ and S$_2$ represent maltose, melibiose, lactose, or glycerol.

forms of HPr and III^{Glc} is reversible (Meadow and Roseman, 1982), these PTS sugars will be able to dephosphorylate P~III^{Glc}. The following hypothesis has been put forward to explain PTS-mediated regulation. Unphosphorylated III^{Glc} is an inhibitor of the various non-PTS uptake systems (Fig. 3). *crr* mutants lack III^{Glc} and are thus resistant to inhibition by PTS sugars, irrespective of the other phosphorylating enzymes.

Inhibition of inducer uptake can explain the *pts* phenotype only partially, however. First, the actual uptake of some carbon sources that are unable to support growth of tight *pts* mutants is not inhibited by PTS sugars in a leaky *ptsI* mutant. This applies for example to the uptake of galactose and methyl β-

galactoside via the methyl β-galactoside permease and of citric acid cycle intermediates (Postma *et al.*, 1981). Second, it has been observed that at least under
some conditions cAMP is able to reverse partially the defect caused by a *pts*
mutation. The connection between the PTS and cAMP metabolism is discussed
next.

9.2. Adenylate Cyclase

cAMP is necessary for the expression of many genes in *E. coli* and *S.
typhimurium*. Together with the cAMP-binding protein, it forms a complex that
binds to the regulatory region of many operons, and is necessary for RNA
polymerase binding (for a review, see Ullmann and Danchin 1983). Mutants
defective in adenylate cyclase (*cya*) or the cAMP-binding protein (*crp*) are unable to grow on many carbon sources. Both mutations result in the same phenotype but whereas addition of cAMP to the growth medium restores growth of
cya mutants on these carbon sources, cAMP cannot restore growth of the *crp*
mutants. Upon comparing the phenotype of *cya* and *crp* mutants on the one hand
and that of *pts* mutants on the other, one is struck by the many similarities.
Indeed, it was reported by Pastan and Perlman (1969) that cAMP was able to
restore growth of *pts* mutants of *E. coli* on at least some of the carbon sources
unable to support growth of these mutants. These authors also found that the
nonmetabolizable analogue methyl α-glucoside elicited transient and permanent
repression of β-galactosidase synthesis in a *pts* mutant of *E. coli* using isopropyl
thio-β-galactoside (IPTG) as an inducer. Both types of repression were abolished
by cAMP.

An important discovery was made by Peterkofsky and co-workers (Peterkofsky and Gazdar, 1973, 1975; Harwood and Peterkofsky, 1975). These
authors found that adenylate cyclase activity, measured either in intact cells or in
cells made permeable by toluene treatment, was inhibited by PTS sugars. Basically, any PTS sugar for which the appropriate enzyme II is present, can inhibit
adenylate cyclase. In mutants lacking a particular enzyme II, adenylate cyclase is
not inhibited by the sugar specific for the enzyme II (Harwood *et al.*, 1976).
Whether a particular PTS sugar can actually inhibit, might be dependent on the
particular strain used. For example, whereas Peterkofsky and co-workers (Harwood *et al.*, 1976) reported that mannose inhibited adenylate cyclase in *E. coli*,
Bourd and colleagues (Voloshin *et al.*, 1981) found no inhibition by mannose at
all.

At first, inhibition of adenylate cyclase was correlated with the absence of
phosphorylated enzyme I. It was postulated that phosphorylated enzyme I and
adenylate cyclase formed an active complex or alternatively that adenylate
cyclase was phosphorylated by enzyme I (Peterkofsky and Gazdar, 1975). This
conclusion was based on results obtained with some *E. coli ptsI* and *ptsH*
mutants.

Later studies using the *crr* mutants described in a previous section, made the direct involvement of enzyme I less likely. Mutants that lack only III^{Glc} but possess normal enzyme I levels also have low adenylate cyclase activity (Feucht and Saier, 1980; Nelson *et al.*, 1982). The more likely interpretation is that III^{Glc} interacts with adenylate cyclase. In fact, the hypothesis on the regulation of non-PTS transport systems by unphosphorylated III^{Glc} has its counterpart in the activation of adenylate cyclase by phosphorylated III^{Glc} (Fig. 3). Recent experiments by Roy *et al.* (1983b) have shown that adenylate cyclase probably contains two domains. The N-terminal domain is involved in the catalytic activity, whereas the C-terminus contains a regulatory domain. Glucose inhibits the intact protein but not a truncated protein, lacking the C-terminus. Unfortunately, these truncated proteins also have much lower activity. The interaction of adenylate cyclase with III^{Glc} and/or the cAMP-binding protein might thus reside in the C-terminal part. Koop *et al.* (1984) on the other hand found that synthesis of truncated adenylate cyclase molecules (consisting of the N-terminal end) was still very sensitive to glucose.

It should be stressed that the modes of regulation by the PTS discussed above, i.e., the inhibition of uptake systems by III^{Glc} and activation of adenylate cyclase by phosphorylated III^{Glc}, are not sufficient to explain the whole of catabolite and transient repression (for earlier reviews, see Magasanik, 1970; Paigen and Williams, 1970). Although the PTS is clearly involved in regulation, catabolite and transient repression can be elicited by non-PTS compounds and can be observed in cells lacking enzyme I (Yang *et al.*, 1979). One would predict no changes in the phosphorylation state of III^{Glc} in these mutants. On the other hand, studies on the expression of several operons in *E. coli* strains lacking the cAMP-binding protein suggest that catabolite repression can occur in the absence of the cAMP/cAMP-binding protein complex (Guidi-Rontani *et al.*, 1980; Guidi-Rontani and Gicquel-Sanzey, 1981).

Finally, it should be pointed out that the intracellular cAMP levels are not only dependent on the adenylate cyclase activity but also on the breakdown of intracellular cAMP by phosphodiesterase and efflux from the cell. Whereas the former is probably not very important since the affinity of phosphodiesterase for cAMP is very low (Botsford, 1981), an energy-requiring export of cAMP from the cell has been reported. Metabolizable substrates stimulated cAMP efflux and uncouplers of oxidative phosphorylation inhibited (Saier *et al.*, 1975).

9.3. In Vitro Reconstitution of the Regulatory System

Thus far, I have described a hypothesis of which the main feature, III^{Glc} as the central regulatory molecule, is based mainly on genetic and some biochemical data obtained with intact cells. Indeed, using similar approaches, quite different proposals have been put forward. In particular, I will discuss those by Kornberg and co-workers (Kornberg and Watts, 1978, 1979; Kornberg *et al.*,

1980; Parra et al., 1983) and Gershanovitch, Bourd, and co-workers (Kalachev et al., 1980, 1981) more extensively below. In brief, Kornberg and co-workers propose that activation of adenylate cyclase and inhibition of the non-PTS uptake systems are brought about by different molecules. Gershanovitch, Bourd, and co-workers have suggested that the interaction between II^{Glc} and the non-PTS transport systems determines the activity of the latter. Maximal inhibition would be dependent on the ratio of the proteins involved.

Specific predictions can be made based on the various proposals. Due to the complex nature of the regulatory systems and the many proteins involved, the intact cell system is not suitable for resolution of these problems. Instead, the complete system should be reconstituted from the purified components. The purification of III^{Glc} (Scholte et al., 1981) and of the lactose carrier (Newman et al., 1981; Wright et al., 1983) have made this approach possible. The following results have been obtained (Nelson et al., 1983).

1. III^{Glc} binds to E. coli membranes that contain the lactose carrier at elevated levels (due to a plasmid that overproduces the carrier). Binding is stimulated only when a substrate of the lactose carrier is present. Osumi and Saier (1982) have obtained the same results.

2. Phosphorylated III^{Glc} does not bind to the lactose carrier.

3. III^{Glc} binds to the lactose carrier with an apparent K_d of 5–15 μM and a stoichiometry of 1–1.3 molecules of III^{Glc} (M_r 20,000) per molecule of lactose carrier (M_r 45,000).

4. III^{Glc}, but not phosphorylated III^{Glc}, increases the apparent affinity of the lactose carrier for β-galactosides in both membrane particles and liposomes containing the purified lactose carrier.

5. III^{Glc} inhibits β-galactoside transport via the lactose carrier in inside-out membrane particles of E. coli. It also inhibits active transport of β-galactosides in liposomes reconstituted with the purified lactose carrier.

6. III^{Glc} does not bind to a mutant lactose carrier that contains three mutations in the N-terminal amino acids.

These results provide clear evidence that at least in the case of the lactose uptake system, regulation of transport and metabolism is at the level of III^{Glc}– carrier interaction. The inhibition of lactose transport in liposomes containing the purified lactose carrier shows that III^{Glc}, the lactose carrier and its substrate are sufficient for the observed effect. No other proteins are required. Figure 4 summarizes these findings schematically. The lactose carrier in its uninhibited form binds galactosides tightly and catalyzes the translocation across the cytoplasmic membrane. Unphosphorylated III^{Glc} can bind to the carrier when a galactoside is present and prevents translocation.

From transport studies with intact cells it has been inferred that in each case the transport of the inducer is regulated by the PTS. As pointed out earlier, this

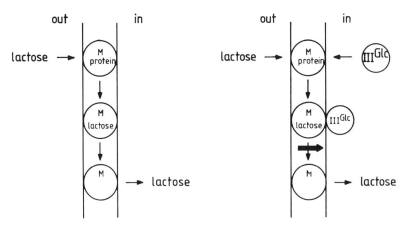

Figure 4. Interaction between III^Glc of the PTS and the lactose carrier.

conclusion is difficult to draw in the case of metabolizable compounds such as maltose and glycerol. Only studies with the melibiose carrier using the non-metabolizable analogue thiomethylgalactoside support the idea that indeed the translocation step is sensitive to regulation by unphosphorylated III^Glc (Saier and Roseman, 1976b). Glycerol uptake and metabolism is a case in point. Glycerol is taken up via the glycerol facilitator, which displays a rather broad specificity (Heller et al., 1980). It has been suggested that it acts as a pore. Subsequent metabolism of glycerol occurs via glycerol kinase. Tight pts mutants are unable to grow on glycerol whereas uptake of glycerol-3-phosphate, the product of glycerol kinase, and the true inducer of the glycerol (glp) regulon is not affected. Regulation of glycerol uptake and metabolism by the PTS is in principle possible on the level of the facilitator or glycerol kinase or both. It was recently shown that purified III^Glc inhibits glycerol kinase but not the facilitator (Postma et al., 1984). Phosphorylated III^Glc did not inhibit glycerol kinase. These results provide an even more complex picture of PTS-mediated regulation. III^Glc not only interacts with membrane-bound transport proteins such as the lactose carrier but also with cytoplasmic proteins like glycerol kinase. In addition, III^Glc recognizes of course both the cytoplasmic PTS protein HPr and the membrane-bound II^Glc.

The second hypothesized mode of interaction of III^Glc, i.e., with adenylate cyclase, has been much more refractory to reconstitution. One complication is the observation that adenylate cyclase activity is very low in cell-free extracts compared to the activity in intact or toluene-treated cells (Harwood and Peterkofsky, 1975). Obviously, one can argue that preparation of the cell-free extract results in considerable dilution of the putative activator of adenylate cyclase, phosphorylated III^Glc. However, addition of a large excess of purified

III^{Glc} in the presence of PEP and the phosphorylating proteins enzyme I and HPr had no stimulatory effect (S. O. Nelson, unpublished results). These experiments have not yet been tried with the recently purified adenylate cyclase from *E. coli* (Yang and Epstein, 1983). Although these negative results may mean that optimal conditions for activation have not yet been defined, it is equally possible that phosphorylated III^{Glc} is not the actual activator or that another, additional protein is involved. The latter suggestion is not without support. Mainly based on studies with intact cells, it has been suggested that the cAMP-binding protein is involved in the regulation of adenylate cyclase activity. It has long been known that *crp* mutants, lacking the cAMP-binding protein, have elevated levels of adenylate cyclase activity and secrete large amounts of cAMP (for a review, see Ullmann and Danchin, 1983). This increased activity could be due to either an increased synthesis of adenylate cyclase molecules or an activation of the same number of molecules. Since no antibodies against adenylate cyclase are yet available, the absolute number of molecules cannot be determined directly. Several lines of evidence, including the construction of fusions between adenylate cyclase and β-galactosidase (Bankaitis and Bassford, 1982), suggest that regulation is not on the level of transcription/translation but rather on the level of enzyme activation. Studies using the cloned *cya* gene also suggest that neither the cAMP-binding protein nor cAMP has a negative effect on *cya* expression (Roy *et al.*, 1983a). Dobrogosz and co-workers in particular have proposed that the cAMP-binding protein, in one conformation, is an inhibitor of adenylate cyclase (Dobrogosz *et al.*, 1983). The role of phosphorylated III^{Glc} might be the removal of this inhibitor. If this is true, one has to assume a rather tight complex between the cAMP-binding protein and adenylate cyclase since this low activity persists in cell-free extracts. Since the number of cAMP-binding protein molecules [estimated to be 3000 molecules per cell (Guiso and Blazy, 1980)] is larger than the number of adenylate cyclase molecules [estimated to be 15 molecules/cell by Yang and Epstein (1983) but 400 molecules/cell by Roy *et al.* (1983a)], a stoichiometric inactive complex between these two proteins seems possible. In fact, Yang and Epstein (1983) reported that during purification of adenylate cyclase from *E. coli* an inhibitor was seemingly removed. The observation that phosphorylated III^{Glc}, added in excess, could not stimulate adenylate cyclase activity, remains unexplained.

The problem is viewed from a slightly different angle by Gershanovitch and co-workers (Bolshakova *et al.*, 1978; Glesyna *et al.*, 1983). These authors have studied the expression of the *lac* operon in various *pts* mutants for quite some time. Using *E. coli ptsHI* deletion mutants and *ptsH* mutants, they observed the same decrease in the rate of β-galactosidase synthesis in both mutants, although the *ptsH* strain has the same intracellular cAMP level as the *pts*$^+$ parent, in contrast to the *ptsHI* deletion strain, which has only 20% of that level left. Using a *cya ptsI* double mutant, they found that β-galactosidase was repressed even

when excess cAMP was added. Finally, they had shown earlier that *pts* mutations repress β-galactosidase synthesis in *lac*⁺, *lacI*, and *lacO^c* strains (the latter
two mutants express the *lac* operon constitutiyely and do not require an inducer).
The same *pts* mutations have no effect in *lacP* and *crp* mutants (*lacP* mutants
have an altered promoter that no longer requires the cAMP/cAMP-binding protein complex). From these results the authors concluded that *pts* mutations interfere with the cAMP-binding protein and regulate transcription. cAMP is not
directly involved since changes in the level of this nucleotide are not correlated
directly with the level of expression. No mechanism is offered but it is suggested
that perhaps P~III^Glc interacts with the cAMP-binding protein. In that case the
P~III^Glc–cAMP-binding protein complex, present in wild-type cells, might
have different properties as effector of transcription than the III^Glc–cAMP-binding protein complex present in *pts* mutants. These problems can only be solved
by reconstitution of the complete system, both on the level of adenylate cyclase
activity as well as transcription.

It was mentioned in the beginning of this section that alternative mechanisms have been proposed to explain the role of the PTS in regulation. In the
view of Kornberg and co-workers, it is not III^Glc but an unknown protein, the
product of the *iex* gene, that is involved in the interaction with non-PTS transport
systems, including the lactose carrier (Fig. 5). They have isolated two types of
mutations in *E. coli*, one of which, *tgs* [Kornberg and Watts, 1979; later renamed
gsr (Parra *et al.*, 1983)], results in the absence of methyl α-glucoside transport

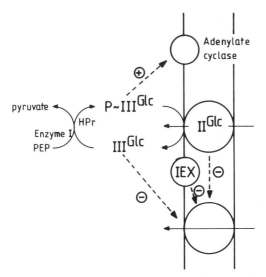

Figure 5. Schematic representation of various proposals to explain the regulatory properties of the
PTS. For details see text.

and low adenylate cyclase activity (Kornberg *et al.*, 1980). Uptake of either PTS sugars like fructose or non-PTS compounds like maltose or glycerol is resistant to inhibition by methyl α-glucoside. The other mutation, *crr* [Kornberg *et al.*, 1980; later renamed *iex* (Parra *et al.*, 1983)], is different from the *crr* mutation described in *S. typhimurium* and the *tgs* mutation in *E. coli* in that *iex* mutants have normal methyl α-glucoside transport and normal or almost normal adenylate cyclase activity (Kornberg *et al.*, 1980). However, uptake of non-PTS compounds is not inhibited by methyl α-glucoside in *iex* strains despite the presence of III^{Glc}. Fructose uptake is normally inhibited by this analogue in *iex* strains but not in the *tgs* strains that lack III^{Glc}. The authors conclude from these and other data that III^{Glc} might be involved in the activation of adenylate cyclase but that another, *i*nducer *ex*clusion, protein, the product of the *iex* gene, is involved in the regulation of non-PTS uptake systems. The *tgs* and *iex* genes of *E. coli* are reported to be localized at opposite sides of the *pts* genes (Britton *et al.*, 1983). No such mutations have been isolated in *S. typhimurium*.

A detailed analysis of the properties of III^{Glc} in *iex* mutants has produced the following results (Nelson *et al.*, 1984b). *iex* mutants contain the same level of III^{Glc} as the *iex*$^+$ parent, as measured with specific antibodies. The III^{Glc} was also active in methyl α-glucoside transport (as reported by Kornberg and Watts, 1979) and in *in vitro* phosphorylation, although in the latter assay it was only 60% as active. Further studies showed that III^{Glc} in the *iex* mutant is temperature-sensitive. Heating for 6 min at 54°C completely destroys the mutant III^{Glc} but leaves the parent III^{Glc} intact. Finally and most important, the mutant III^{Glc} is unable to bind to the lactose carrier. Introduction of a plasmid containing only the *crr* gene, coding for III^{Glc}, restored the mutant phenotype to that of the parent. The III^{Glc} synthesized from the plasmid is not temperature-sensitive. These results suggest that *iex* mutants contain an altered III^{Glc} that is still active in transport and phosphorylation but not in the inhibition of non-PTS transport systems (Nelson *et al.*, 1984b). Most likely the *iex* mutation is an allele of *crr*, suggesting that there is no difference between *E. coli* and *S. typhimurium* with respect to the gene(s) involved in PTS-mediated regulation. The mutation isolated in *E. coli* is very interesting in that it allows a separation of the various functions of III^{Glc}. It is still active in phosphorylation and regulation of adenylate cyclase (*iex* mutants still grow on for example succinate, in contrast to *crr* mutants that lack III^{Glc} completely) but the mutant III^{Glc} is unable to bind to and inhibit the lactose carrier (and presumably the other non-PTS systems). If different domains are present in III^{Glc}, it should be possible to select for the reverse, i.e., a mutant protein active in inhibition of non-PTS systems but unable to stimulate adenylate cyclase.

Still another model has been proposed by Gershanovitch, Bourd, and co-workers (Kalachev *et al.*, 1980, 1981). It is less well defined but includes the interaction between enzyme II and the various non-PTS transport systems, although the lactose carrier has been the system most studied (Fig. 5). It was concluded that interaction between II^{Glc} (the system specific for methyl α-glucoside) and the lactose carrier resulted in an inactive complex. The ratio between II^{Glc} and the lactose carrier was found to be important. From genetic studies this proposal seems unlikely since mutants lacking II^{Glc} are not generally resistant to PTS-mediated inducer exclusion as are *crr* mutants, but only with methyl α-glucoside as a substrate. Other PTS sugars still inhibit and *pts* deletion strains lacking in addition II^{Glc} do not regain growth on non-PTS compounds. The important concept of stoichiometric interaction between the putative inhibitor and the various systems will be discussed in a later section.

In conclusion, reconstitution experiments have provided experimental evidence for the hypothesis of Saier and Roseman that III^{Glc} is a regulatory molecule that interacts with many different metabolic systems. In the case of the lactose carrier, it has been shown conclusively that the purified components III^{Glc} and the lactose carrier were sufficient for the proposed interaction. III^{Glc} is also required for the inhibition of glycerol kinase. The case is less clear with respect to regulation of adenylate cyclase. Although genetic evidence suggests that III^{Glc} is involved, no biochemical evidence has been provided for the direct interaction between these two proteins.

9.4. Regulation of Carbohydrate Uptake in Vivo by the PTS

In the previous section, I have discussed some of the basic observations that have led to the model described above. A closer examination of the regulatory model and the various other mutations that have been isolated, reveals that the system is more complicated. I will first describe other suppressor mutations that have been isolated and have helped to define the system. Next I will discuss some of the mutations in detail and explain their phenotype. Finally, I will try to integrate all these facts in the model already described.

9.5. Other Mutations That Suppress the pts Phenotype

In this section I will discuss some of the mutations that have been isolated as suppressors of the *pts* phenotype. Some of these suppress the *pts* phenotype completely, others only partly (Table II).

1. The *crr* mutation. This mutation has already been discussed extensively. In *S. typhimurium* and *E. coli* it allows growth of *pts* mutants on the non-PTS compounds maltose, melibiose, and glycerol (and lactose in *E. coli*). I will call these compounds class I compounds (Postma, 1982). Whereas the *crr* mutation restores growth of mutants on certain carbon sources, it does not restore growth of *pts* mutants on citric acid cycle intermediates (malate, succinate, and citrate), xylose, or rhamnose (class II compounds). In fact, a strain containing only a *crr*

mutation is also unable to grow on these class II compounds (Scholte and Postma, 1980), most likely because these mutants also have a lowered adenylate cyclase activity. The differences between class I and class II compounds will be discussed below.

2. The *crp** mutation. cAMP was shown to restore growth of *pts* mutants of *E. coli* on non-PTS compounds (Pastan and Perlman, 1969) but no such effect was found in *S. typhimurium* (Saier *et al.*, 1970; Saier and Roseman, 1976a,b). Later, suppressor mutations were found in *S. typhimurium* that restored growth of several *pts* mutants on non-PTS compounds (both class I and II substrates). cAMP was shown to have the same effect (Scholte and Postma, 1980). The *crp** mutation also restores the growth of *crr* mutants on class II compounds (Scholte and Postma, 1980; Postma *et al.*, 1981). The mutation, *crp**, was mapped closely linked to *cysG*. Most likely it is a mutation in the gene for the cAMP-binding protein. It results in an altered cAMP-binding protein that has become independent of cAMP. The same *crp** mutations also suppress *cya* mutations that result in a defective adenylate cyclase (Postma and Scholte, 1979). However, several classes of suppressors were found. Some allowed growth on only a few compounds, others restored growth on almost all. Alper and Ames (1978) have suggested a hierarchy of operons, some requiring less cAMP and others more for complete expression. In the example described above, class I compounds require less cAMP than class II compounds since *crr* mutants grow on the first but not on the second class. Kolb and co-workers (Kolb *et al.*, 1983) have indeed found that different promoter regions have different affinities for the cAMP/cAMP-binding protein complex. At a constant cAMP-binding protein concentration, the cAMP concentration required for half-maximal binding to the DNA site increased inversely with the affinity for the cAMP-binding protein. It should be realized that the changes in the intracellular cAMP concentration are rather small. Joseph *et al.* (1982) reported a threefold change when cells were shifted from glucose to a glycerol medium. The amount of the cAMP-binding protein is constant (Guiso and Blazy, 1980). This means that the threefold change is sufficient for turning on a number of operons.

A similar *crp** mutation was reported in *E. coli* (Alexander and Tyler, 1975; Alexander, 1980). Earlier, *crp** mutations were isolated as suppressors of a *cya* mutation (for a review, see Botsford, 1981). One can also isolate mutations that restore growth of *cya* mutants on one carbon source only. They allow transcription of that particular operon which is still inducible, in the absence of cAMP. It was tested whether such a mutation is able to suppress the *pts* phenotype (Nelson *et al.*, 1982). *ptsI* deletion strains of *S. typhimurium* containing a *melP* mutation (isolated as restoring growth of a *cya* mutant on melibiose only) could still not grow on melibiose. When the strain contained the leaky *ptsI17* mutation, transport of the melibiose analogue thiomethylgalactoside was sensitive to inducer exclusion by PTS substrates. Clearly, this type of suppressor is not as strong as the *crp** mutation just described.

3. Specific suppressors. Mutations have been described that restore growth of *pts* mutants on only one carbon source. Some of these mutations have been mapped closely linked to or in a particular operon. Lin and co-workers (Berman *et al.*, 1970; Berman and Lin, 1971) have characterized some mutations that allowed *pts* mutants of *E. coli* to grow on glycerol but not on other class I compounds. Three classes of mutations were found, two of which resulted in elevated levels of the enzymes coded for by the *glp* regulon. One class resulted in a faulty repressor (*glpR*), the other was a promoter up mutation. As will be explained below, mutants with elevated levels of enzymes sensitive to PTS-mediated repression can escape from this regulation. Mutations resulting in the constitutive expression of the *lac* operon have the same effect (Wang and Morse, 1968; Pastan and Perlman, 1969; Wang *et al.*, 1970). The third class of mutations was quite unexpected. Lin and co-workers found that this suppressor mutation mapped in the *glpK* gene, coding for glycerol kinase, and resulted in an altered glycerol kinase, insensitive to feedback inhibition by fructose-1,6-bisphosphate. At that time, PTS-mediated repression was not well understood. The authors suggested that in normal cells glycerol kinase is inhibited to some extent by fructose-1,6-bisphosphate but sufficient glycerol kinase activity is left to allow growth. A *pts* mutant on the other hand already has lower glycerol kinase activity and the additional inhibition by fructose-1,6-bisphosphate would not leave sufficient glycerol kinase activity for growth. Making the glycerol kinase insensitive to feedback inhibition by mutation would increase the glycerol kinase activity above a certain threshold and allow growth. Now that the interaction between the PTS and some metabolic systems is better understood, another interpretation is suggested by the recent observation that glycerol kinase is inhibited by III^{Glc} (Postma *et al.*, 1984). It might well be that the *glpK* mutation has made the kinase less sensitive to inhibition by III^{Glc}. The lower sensitivity to fructose-1,6-bisphosphate might be a secondary consequence of the mutation.

Some other specific suppressors have been mapped close to or in the maltose, melibiose, and lactose operons (Saier and Roseman, 1976a,b; Saier *et al.*, 1978). They might represent alterations in the uptake systems such that they become insensitive to inhibition by III^{Glc}. For example, uptake of maltose in a Maltose$^+$ *pts* mutant has become insensitive to inhibition by PTS sugars. Glycerol transport is still inhibited by PTS sugars in this strain. One related mutation has been analyzed in terms of III^{Glc} binding. A mutant lactose carrier was unable to bind III^{Glc} (Nelson *et al.*, 1983). It should be added, however, that this particular lactose carrier was also unable to catalyze the translocation of β-galactosides across the membrane although the binding of β-galactosides was not impaired.

4. The *cya** mutation. Based on the proposed interaction between III^{Glc} and adenylate cyclase, one would predict that mutations exist in the *cya* gene that result in an altered adenylate cyclase independent of phosphorylated III^{Glc} or other proteins involved in its regulation. By localized mutagenesis, mutations

were isolated close to or in the *cya* gene that allowed the growth of *pts* mutants on class I compounds (Postma, 1982). These mutations have not as yet been characterized further.

9.6. A Model Involving Stoichiometric Interactions between PTS and Non-PTS Components

I have discussed in detail the regulation of solute uptake and metabolism by the PTS in various mutants. A pertinent question is whether it has any relevance to the situation in a normal cell. The answer is affirmative. PTS-mediated inducer exclusion is observed in wild-type (pts^+) cells under certain conditions that are important for the growth of the cell. From the model discussed, it follows that at least two conditions have to be fulfilled before inducer exclusion can occur: (1) III^{Glc} should be dephosphorylated either directly via II^{Glc} or indirectly via the other enzymes II and HPr and (2) the amount of III^{Glc} in the cell should be sufficient to inactivate all molecules of the system to be regulated.

The first requirement follows directly from the model and fulfillment depends on the rate of phosphorylation via enzyme I/HPr and the rate of dephosphorylation via enzyme II/sugar. Obviously, the rate of dephosphorylation is favored in *ptsH,I* mutants but this might also be the case in wild-type (pts^+) strains. Most likely the phosphorylation of III^{Glc} via enzyme I/HPr is the rate-limiting step in the presence of a PTS sugar (Scholte and Postma, 1981). As stated before, the presence of PTS sugars results only in inducer exclusion when the appropriate enzyme II is present. Since some are inducible, pregrowth on the particular PTS sugar is required for the effect. In fact, even before the discovery of the PTS, it had been observed (Cohn and Horibata, 1959; Kepes, 1960; Koch, 1964) that glucose inhibited expression of the *lac* operon and lactose uptake in *E. coli* only after pregrowth of the cells on glucose.

The second condition requires some explanation. The amount of III^{Glc} in *S. typhimurium* is relatively constant under all growth conditions as measured with antibodies (Scholte *et al.*, 1981). Since most systems regulated by the PTS are inducible, one can imagine a situation in which the number of transport molecules (or number of polypeptides with which III^{Glc} interacts) exceeds the number of III^{Glc} molecules. Under these conditions a certain percentage of the uptake system will escape inhibition, provided a stoichiometric complex is required for inactivation. It has been shown that PTS-mediated inducer exclusion occurs in wild-type cells when the uptake system, for example that for maltose or glycerol, is partially induced. Conversely, mutants that have lowered levels of III^{Glc} escape more quickly from this inhibition (Nelson *et al.*, 1982). Increased levels of III^{Glc}, on the other hand, can result in cells hypersensitive to repression and inducer exclusion. Umyarov *et al.* (1978) observed that *E. coli* containing an F′ with the *pts* genes was hypersensitive to repression by glucose. The most likely interpretation of these results is that the F' also contains the *crr* gene and consequently the cell has higher levels of III^{Glc}. Introduction of a plasmid that contains the *crr* gene but not the *ptsH,I* genes and that results in overproduction of III^{Glc}, has the same result. Inhibition of the uptake of non-PTS compounds is much stronger in these plasmid-containing cells, due to the increased III^{Glc} level (Nelson and Postma, 1984).

   Full induction can result in cells that are resistant to inducer exclusion. For example, when the leaky *ptsI* mutant of *S. typhimurium* also contains the *crp** mutation, non-PTS uptake systems are no longer sensitive to inhibition by PTS sugars (Scholte and Postma, 1980). Similar observations were reported by Saier and co-workers (Saier *et al.*, 1982) and Mitchell *et al.* (1982). Growth of cells in the presence of an inducer and cAMP resulted in non-PTS uptake systems insensitive to inhibition by PTS sugars. Although it has been suggested that this desensitization involves a novel regulatory mechanism (Saier *et al.*, 1982), in my opinion the same explanation given above can be applied, i.e., under certain growth conditions the number of PTS-sensitive uptake systems exceeds the number of regulatory molecules.

   It has been mentioned earlier that membranes of *crr* deletion mutants seem to contain a III^{Glc}-like activity that is inhibited by antibodies against soluble III^{Glc}. Is this otherwise uncharacterized protein involved in inducer exclusion? Nelson *et al.* (1982) have found that in *crr* deletion strains, maltose uptake is still sensitive to the PTS analogue 2-deoxyglucose but much less so than the *crr*$^+$ parent. In other words, *crr* deletion strains escape much earlier from inducer exclusion when the transport systems are being induced. Transport studies (Fig. 2) show that the activity of this unknown protein in methyl α-glucoside uptake is 20% or less but nothing is known about the actual number of molecules. Still, a *crr* strain escaped from inducer exclusion with about 20% of the number of maltose transport systems of that of the parent.

   Another complication arises when more than one of these systems is present in the cell at the same time. In this case, III^{Glc} must be shared between the various systems. Indeed, one finds experimentally that induction of one uptake system can render a second one insensitive to PTS-mediated inhibition, provided the substrate for the first system is present (Nelson *et al.*, 1984a). For instance, when both the glycerol and the maltose system are induced at the same time, the presence of glycerol renders the maltose system less sensitive to inhibition by PTS sugars. From these experiments one can conclude that these systems interact only with III^{Glc} when a substrate of the uptake system is present, as was found with the lactose carrier in *in vitro* studies. Saier and co-workers (Saier *et al.*, 1983) reported similar findings although they only observed release of inhibition of the glycerol and maltose uptake systems by PTS sugars when the lactose carrier was present in the cell at elevated levels, due to a plasmid carrying the *lacY* gene.

9.7. Growth in Batch Cultures versus Chemostat Cultures

The concept of a limited number of regulatory molecules from which an uptake system might escape under certain conditions might have relevance to the behavior of cells under more defined conditions than those of the much used batch culture system, i.e., in chemostat culture in which low but constant nutrient concentrations can be maintained. In classical diauxic growth experiments in batch culture, two solutes are provided to the cell at the same time as carbon source, for example glucose and lactose. In general, the cell has a preference (in the example mentioned, for glucose) and does not use lactose until all glucose has been consumed. After a certain lag time, lactose consumption starts. One finds that during glucose consumption, no lactose is used nor are the enzymes involved in lactose metabolism induced. Similar findings have been reported for the glucose/melibiose diauxie (Okada *et al.*, 1981). If cAMP is present during the experiment, no diauxie is observed. Melibiose consumption starts immediately after the glucose is exhausted. Under these conditions α-galactosidase is synthesized. One can calculate that when all glucose is consumed, the cells contain 15–30% of the α-galactosidase of the fully induced cell. Apparently, it does not function in melibiose metabolism because the melibiose is excluded from the cell. Still sufficient inducer must have entered the cell. What happens when the cell can synthesize the proteins coded for by the *lac* operon constitutively? Again, lactose is not used as long as glucose is present and no lag time is observed in lactose consumption after the glucose is exhausted (Fraser and Yamazaki, 1982). The enzymes required for metabolism are present but uptake of the solute is inhibited. Measurements of the actual levels of the synthesized enzymes show that they are much lower than in the fully induced cell. This resembles the partially induced cells discussed earlier.

Most experiments described in the previous sections involve cells grown in batch culture, i.e., cells growing under continuously changing conditions. In general, one starts with a high substrate condition, which becomes depleted in the stationary phase. Under these conditions the surroundings of the cell change continuously. It is difficult to compare the results obtained with isolated cells from batch culture with those obtained with cells growing under steady-state conditions with a low but constant concentration of the different solutes. How well does the PTS-mediated regulatory model, based upon results with cells from batch culture, apply to conditions in which cells are competing for limiting carbon source? Only a few studies have been published that deal with this particular problem. Some chemostat experiments have been described in which *E. coli* or *K. aerogenes* was grown on mixed substrates, for example glucose and lactose or glucose and maltose (Harte and Webb, 1967). One can ask whether under carbon limitation, bacteria like *E. coli* still use preferentially glucose and are unable to utilize maltose or lactose at the same time as is found in batch (diauxic growth). Harte and Webb (1967) found that maltose utilization in *K. aerogenes* was prevented by the presence of glucose at concentrations as low as a few micromolar. Only after a long lag maltose utilization started. Increasing this glucose concentration increased the lag time linearly (Edwards, 1969). It would be interesting to test whether this external concentration of glucose that could no longer prevent maltose utilization was too low to keep all III^{Glc} dephosphorylated in the cell. Unfortunately, such quantitative measurements are not yet possible. In fact, not much is known about the levels of the various PTS proteins in chemostat cultures. Interestingly, there is a large difference between adenylate cyclase activity in succinate- and glucose-grown chemostat cultures and cells grown in batch culture on the same carbon sources (Botsford and Drexler, 1978; Harman and Botsford, 1979). The latter type of cells have much higher activity but the activity of the chemostat cultures can be stimulated by adding the carbon source on which the cells were grown.

9.8. Final Remarks on PTS-Mediated Regulation

Regulation of solute uptake by the PTS has been described in some detail. Mostly methyl α-glucoside but also other PTS sugars are used to elicit inhibition. Surprisingly, quite low concentrations of for example methyl α-glucoside are effective. Whereas the apparent K_m for methyl α-glucoside transport in *S. typhimurium* is around 170 μM (Stock *et al.*, 1982), Saier (Saier and Roseman, 1976b) reported that concentrations as low as 1 μM inhibited glycerol uptake in a *ptsI* mutant of *S. typhimurium*. In some other strains, 70 nM methyl α-glucoside inhibited half-maximally (Feucht and Saier, 1980). It is difficult to determine whether the low flux of phosphoryl groups via the mutated enzyme I is actually larger or smaller than the low rate of dephosphorylation due to the very low methyl α-glucoside concentration. An intriguing possibility is the type of regulation described by Koshland and co-workers (Goldbeter and Koshland, 1981; Koshland *et al.*, 1982). They introduced the concept of zero-order ultrasensitivity to explain that small changes in the rates of phosphorylation and dephosphorylation of a phosphoprotein by a kinase and a phosphatase can result in large changes in the ratio of phosphorylated to unphosphorylated enzyme. Formally, one can describe the phosphorylation and dephosphorylation of III^{Glc} in a similar manner. HPr is the kinase and III^{Glc} is the phosphatase. Koshland and co-workers showed that under conditions in which the protein to be regulated is in the same concentration range as the kinase/phosphatase, a small change in the phosphatase activity for example can result in a large change in the phosphorylation state of this protein. That is exactly the condition I have described above. III^{Glc} is most likely in the same concentration range as enzyme I/HPr (Scholte *et al.*, 1981; Mattoo and Waygood, 1983) and II^{Glc} and low concentrations of methyl α-glucosidase can elicit small changes in the "phosphatase" activity. It would be interesting to analyze the PTS in these terms and it might explain how regulation via the PTS can occur by suboptimal effector concentrations. Why has such a complex regulatory system evolved (at least in the Enterobacteriaceae) in which the uptake and metabolism of some solutes are controlled by intracellular cAMP levels whereas the metabolism of others involves regulation of both the intracellular cAMP levels and the inducer levels? One explanation is suggested by the observation that many carbohydrates, both PTS and non-PTS, can lower intracellular cAMP levels and that different operons require different concentrations of cAMP for full expression. Consequently, some carbohydrates can lower the cAMP level sufficiently to prevent utilization of other solutes without doing too much harm to their own operons that in general also require cAMP but less. How to prevent utilization of these carbon sources by even more preferred solutes? In this case, lowering of the cAMP level even more might be dangerous for the cell because most PTS sugars for example also require cAMP for full expression. The solution is inhibition of inducer entry, rendering maximal expression of these operons much more difficult. Finally, within the group of PTS sugars, preference for a particular sugar derives from the fact that all PTS sugars have to compete for P~HPr and that the various enzyme II/enzyme III complexes may have different affinities for the common pool of phosphorylated HPr. By using these various controls on different levels, the cell is able to regulate its solute uptake and metabolism.

10. COMPARISON WITH OTHER REGULATORY MECHANISMS

Do similar regulatory systems as described for *E. coli* and *S. typhimurium* exist in other organisms, or are these uniquely limited to the Enterobacteriaceae? With respect to the prokaryotes, the PTS has been found in most obligate and facultative anaerobes and some obligate aerobes (see Chapter 7). Although the various PTS systems have their own modifications, the mechanism is basically the same. Regulation via the PTS is another matter. Not much evidence exists for a regulatory role of the PTS as described in this chapter in bacteria other than the enterobacteriaceae. This might indicate a basic distinction between different bacteria or might be due to the fact that this type of regulatory mechanism has been studied in only a few other bacteria. Recently, a similar type of regulation has been found in *Bacillus subtilis*. It has long been known that mutants of *B. subtilis* lacking enzyme I are unable to grow on glycerol in addition to PTS sugars (Gay *et al.*, 1973). Suppressor mutations were isolated that restored growth of these mutants on glycerol but not further characterized. It has now been shown that in *B. subtilis* mutants containing a temperature-sensitive enzyme I (Niaudet *et al.*, 1975), the uptake of glycerol is inhibited by methyl α-glucoside upon inactivation of enzyme I (Reizer *et al.*, 1984). Since suppressor mutations have been isolated (Gay *et al.*, 1973), it is possible that a molecule with a role similar to that of III^Glc in the Enterobacteriaceae exists in *B. subtilis*.

Evidence was obtained that in Streptococci the PTS might regulate uptake and efflux of solutes via another mechanism, involving a protein kinase and protein phosphatase. It has been shown that HPr can be phosphorylated by an ATP-dependent kinase on a serine residue (Deutscher and Saier, 1983); recall that PEP-dependent phosphorylation is on a histidine. This P(ser)–HPr would be involved in the expulsion of certain sugars from Streptococci. Deutscher and co-workers (Deutscher *et al.*, 1984) have studied in more detail the reactivity of *Streptococcus lactis* HPr when phosphorylated on a serine and found that this P(ser)–HPr is only very slowly phosphorylated in a PEP/enzyme I-dependent manner. Thus, the normal phosphoryl transfer via HPr seems blocked. However, in the presence of an enzyme III (e.g., the III^Gluc of *Streptococcus faecalis*, specific for gluconate, or III^Lac from *Staphylococcus aureus*), this block is relieved partly or completely, suggesting an interaction between P(ser)–HPr and enzymes III. It is not clear how these *in vitro* data can be fitted into the physiology of the cell. Although glucose-induced phosphorylation of HPr might inactivate HPr and thus explain the inhibition of other PTS sugars, one should remember that glucose itself is also taken up by the PTS and should be able to enter the cell. It is not clear how it can inactivate HPr for other sugars but not affect its own uptake. No evidence exists at the moment to indicate whether a similar system involving phosphorylation of HPr or another PTS protein at a second site different from the histide occurs in the Enterobacteriaceae. Several ATP-dependent kinases as well as phosphatases have been described in *S. typhimurium* (Wang and Koshland, 1978, 1981). In all cases serine phosphates have been found but with the exception of isocitrate dehydrogenase (Garnak and Reeves, 1979), none of the proteins phosphorylated has been identified.

Are similar regulatory mechanisms as described for the Enterobacteriaceae present in cells other than bacteria? Phosphorylation and dephosphorylation of proteins is of course widespread in eukaryotic cells as an important mechanism to regulate many processes (for reviews, see Ingebritsen and Cohen, 1983; Nestler and Greengard, 1983). Part of these phosphorylation/dephosphorylation cycles are dependent on cAMP. The kinases and phosphatases described in bacteria, on the other hand, are independent of cAMP (Wang and Koshland, 1981).

Here, I will compare PTS-mediated regulation in the Enterobacteriaceae in some more detail with the process of catabolite repression in yeasts since there are some striking similarities. A more detailed description of this process can be found in Chapter 8. The synthesis of certain enzymes in yeasts is repressed by sugars like glucose, a process called catabolite repression, as in prokaryotes. Although the same term is used, it is by no means clear that similar mechanisms underlie the processes in both types of organisms. In fact, a direct involvement of cAMP in this process is not very likely in yeasts. Yeast mutants have been isolated that are resistant to catabolite repression (Zimmerman and Scheel, 1977; Entian, 1980, 1981). These mutations also affect in one way or another hexokinase PII, one of the three kinases that are able to phosphorylate glucose. The other two are hexokinase PI and glucokinase. The mutations have different effects on hexokinase PII. In some cases the enzyme is absent (Entian and Mecke, 1982) whereas in other cases it is present at elevated levels in the cell or has altered properties (Entian and Fröhlich, 1984). In the latter case, the hexokinase PII has lost its regulatory function but is still active catalytically. Similar mutations have been found in *E. coli* III^{Glc} (Nelson *et al.*, 1984b). It has been suggested by Entian and co-workers (Entian and Fröhlich, 1984) that hexokinase PII is a bifunctional enzyme, involved both in phosphorylation of glucose and in regulation. A similar role has been attributed to III^{Glc} but in contrast to the latter case no reconstitution of the system in yeasts has been possible. This similar role may be completely accidental, however.

Finally, it might be interesting to compare briefly regulation of adenylate cyclase in prokaryotes with that in eukaryotes. Recent progress in the study of hormone-sensitive adenylate cyclase has revealed a complex interaction between a number of proteins (for a review, see Gilman, 1984). Two guanine nucleotide-binding regulatory components have been identified, a stimulatory one (G_s) and an inhibitory one (G_i). Each of these consists of at least two polypeptides, α and β, one of which (β) is most likely the same in both complexes G_s and G_i.

Guanine nucleotides regulate adenylate cyclase by "activating" G_s and G_i, which results in dissociation of the subunit. The resulting $G_{s(\alpha)}$ can activate adenylate cyclase. The β subunit inhibits by forming an inactive complex with $G_{s(\alpha)}$. The inhibitory effect of G_i is brought about by the release from G_i of extra β subunits, which can bind $G_{s(\alpha)}$ in an inactive complex (Gilman, 1984). One can speculate as to whether the bacterial adenylate cyclase is regulated in an analogous manner. Formally, one can consider the cAMP-binding protein (CRP) as the inhibitory component that can be "complexed" by phosphorylated III^{Glc} in a noninhibiting complex. CRP and III^{Glc}, on the other hand, might form only a weak complex, thus releasing the inhibitory component. The ratio between the two forms is determined by the phosphorylation of III^{Glc} via the PTS and dephosphorylation via the enzymes II. This would mean that adenylate cyclase is by itself active but that CRP inhibits its activity, unless the phosphorylated III^{Glc} binds CRP (see also Dobrogosz *et al.*, 1983).

REFERENCES

Adhya, S., and Echols, H., 1966, Glucose effect and the galactose enzymes of *Escherichia coli:* Correlation between glucose inhibition of induction and inducer exclusion, *J. Bacteriol.* **92:**601–608.

Adler, J., and Epstein, W., 1974, Phosphotransferase-system enzymes as chemoreceptors for certain sugars in *Escherichia coli* chemotaxis, *Proc. Natl. Acad. Sci. USA* **71**:2895–2899.

Alaeddinoglu, N. G., and Charles, H. P., 1979, Transfer of a gene for sucrose utilization into *Escherichia coli* K12, and consequent failure of expression of genes for D-serine utilization, *J. Gen. Microbiol.* **110**:47–59.

Alexander, J. K., 1980, Suppression of defects in cyclic adenosine 3′,5′-monophosphate metabolism in *Escherichia coli*, *J. Bacteriol.* **144**:205–209.

Alexander, J. K., and Tyler, B., 1975, Genetic analysis of succinate utilization of enzyme I mutants of the phosphoenolpyruvate: sugar phosphotransferase system in *Escherichia coli*, *J. Bacteriol.* **124**:252–261.

Alper, M. D., and Ames, B. N., 1978, Transport of antibiotics and metabolite analogs by systems under cyclic AMP control: Positive selection of *Salmonella typhimurium cya* and *crp* mutants, *J. Bacteriol.* **133**:149–157.

Amaral, D., and Kornberg, H. L., 1975, Regulation of fructose uptake by glucose in *Escherichia coli*, *J. Gen. Microbiol.* **90**:157–168.

Bachmann, B. J., 1983, Linkage map of *Escherichia coli* K-12, edition 7, *Microbiol. Rev.* **47**:180–230.

Bankaitis, V. A., and Bassford, P. J., Jr., 1982, Regulation and adenylate cyclase synthesis in *Escherichia coli*: Studies with *cya-lac* operon and protein fusion strains, *J. Bacteriol.* **151**:1346–1357.

Begley, G. S., Hansen, D. E., Jacobson, G. R., and Knowles, J. R., 1982, Stereochemical course of the reactions catalyzed by the bacterial phosphoenolpyruvate: glucose phosphotransferase system, *Biochemistry* **21**:5552–5556.

Beneski, D. A., Misko, T. P., and Roseman, S., 1982, Sugar transport by the bacterial phosphotransferase system. Preparation and characterization of membrane vesicles from mutant and wild type *Salmonella typhimurium*, *J. Biol. Chem.* **257**:14565–14575.

Berman, M., and Lin, E. C. C., 1971, Glycerol-specific revertants of a phosphoenolpyruvate-phosphotransferase mutant: Suppression by desensitization of glycerol kinase to feedback inhibition, *J. Bacteriol.* **105**:113–120.

Berman, M., Zwaig, N., and Lin, E. C. C., 1970, Suppression of a pleiotropic mutant affecting glycerol dissimilation, *Biochem. Biophys. Res. Commun.* **38**:272–278.

Beyreuther, K., Raufuss, H., Schrecker, O., and Hengstenberg, W., 1977, The phosphoenolpyruvate–dependent phosphotransferase system of *Staphylococcus aureus*. 1. Amino-acid sequence of the phosphocarrier protein HPr, *Eur. J. Biochem.* **75**:275–286.

Bitoun, R., de Reuse, H., Touati-Schwartz, D., and Danchin, A., 1983, The phosphoenolpyruvate dependent carbohydrate phosphotransferase system of *Escherichia coli*: Cloning of the *ptsHI-crr* region and studies with a *pts-lac* operon fusion, *FEMS Microbiol. Lett.* **16**:163–167.

Bolshakova, T. N., Gabrielyan, T. R., Bourd, G. I., and Gershanovitch, V. N., 1978, Involvement of the *Escherichia coli* phosphoenolpyruvate-dependent phosphotransferase system in regulation of transcription of catabolic enzymes, *Eur. J. Biochem.* **89**:483–490.

Botsford, J. L., 1981, Cyclic nucleotides in prokaryotes, *Microbiol. Rev.* **45**:620–642.

Botsford, J. L., and Drexler, M., 1978, The cyclic 3′,5′-adenosine monophosphate receptor protein and regulation of cyclic 3′,5′-adenosine monophosphate synthesis in *Escherichia coli*, *Mol. Gen. Genet.* **165**:47–56.

Bourd, G. I., Erlagaeva, R. S., Bolshakova, T. N., and Gershanovitch, V. N., 1975, Glucose catabolite repression in *Escherichia coli* K12 mutants defective in methyl-α-D-glucoside transport, *Eur. J. Biochem.* **53**:419–427.

Britton, P., Boronat, A., Hartley, D. A., Jones-Mortimer, M. C., Kornberg, H. L., and Parra, F., 1983, Phosphotransferase-mediated regulation of carbohydrate utilization in *Escherichia coli* K12: Location of the *gsr* (*tgs*) and *iex* (*crr*) genes by specialized transduction, *J. Gen. Microbiol.* **129**:349–358.

Britton, P., Lee, L. G., Murfitt, D., Boronat, A., Jones-Mortimer, M. C., and Kornberg, H. L., 1984, Location and direction of the *ptsH* and *ptsI* genes on the *Escherichia coli* K12 genome, *J. Gen. Microbiol.* **130:**861–868.

Brouwer, M., Elferink, M. G. L., and Robillard, G. T., 1982, Phosphoenolpyruvate-dependent fructose phosphotransferase system of *Rhodopseudomonas sphaeroides:* Purification and physicochemical and immunochemical characterization of a membrane-associated enzyme I, *Biochemistry* **21:**82–88.

Clark, B., and Holms, W. H., 1976, Control of the sequential utilization of glucose and fructose by *Escherichia coli, J. Gen. Microbiol.* **95:**191–201.

Cohn, M., and Horibata, K., 1959, Inhibition by glucose of the induced synthesis of the β-galactoside enzyme system of *Escherichia coli:* Analysis of maintenance, *J. Bacteriol.* **78:**601–612.

Cordaro, C., 1976, Genetics of the bacterial phosphoenolpyruvate: glycose phosphotransferase system, *Annu. Rev. Genet.* **10:**341–359.

Cordaro, J. C., and Roseman, S., 1972, Deletion mapping of the genes coding for HPr and enzyme I of the phosphoenolpyruvate: sugar phosphotransferase system in *Salmonella typhimurium, J. Bacteriol.* **112:**17–29.

Cordaro, J. C., Anderson, R. P., Grogran, E. W., Wenzel, D. J., Engler, M., and Roseman, S., 1974, Promoter-like mutation affecting HPr and enzyme I of the phosphoenolpyruvate: sugar phosphotransferase system in *Salmonella typhimurium, J. Bacteriol.* **120:**245–252.

Cordaro, J. C., Melton, T., Stratis, J. P., Atagün, M., Gladding, C., Hartman, P. E., and Roseman, S., 1976, Fosfomycin resistance: Selection method for internal and extended deletions of the phosphoenolpyruvate: sugar phosphotransferase genes of *Salmonella typhimurium, J. Bacteriol.* **128:**785–793.

Curtis, S. J., and Epstein, W., 1975, Phosphorylation of D-glucose in *Escherichia coli* mutants defective in glucose phosphotransferase, mannose phosphotransferase, and glucokinase, *J. Bacteriol.* **122:**1189–1199.

Deutscher, J., and Saier, M. H., Jr., 1983, ATP-dependent protein kinase-catalyzed phosphorylation of a seryl residue in HPr, a phosphate carrier protein of the phosphotransferase system in *Streptococcus pyogenes, Proc. Natl. Acad. Sci. USA* **80:**6790–6794.

Deutscher, J., Kessler, U., Alpert, C. A., and Hengstenberg, W., 1984, The bacterial phosphoenolpyruvate dependent phosphotransferase system: P-ser–HPr and its possible regulatory function, *Biochemistry* **23:**4455–4460.

Dills, S. S., Apperson, A., Schmidt, M. R., and Saier, M. H., Jr., 1980, Carbohydrate transport in bacteria, *Microbiol. Rev.* **44:**385–418.

Dobrogosz, W. J., Hall, G. W., Sherba, D. K., Silva, D. O., Harman, J. G., and Melton, T., 1983, Regulatory interactions among the *cya, crp* and *pts* gene products in *Salmonella typhimurium, Mol. Gen. Genet.* **192:**477–486.

Durham, D. R., and Phibbs, P. V., Jr., 1982, Fractionation and characterization of the phosphoenolpyruvate: fructose 1-phosphotransferase system from *Pseudomonas aeruginosa, J. Bacteriol.* **149:**534–541.

Edwards, V. H., 1969, Correlations of lags in the utilization of mixed sugars in continuous fermentation, *Biotechnol. Bioeng.* **11:**99–102.

Elvin, C. M., and Kornberg, H. L., 1982, A mutant β-D-glucoside transport system of *Escherichia coli* resistant to catabolite inhibition, *FEBS Lett.* **147:**137–142.

Entian, K.-D., 1980, Genetic and biochemical evidence for hexokinase PII as a key enzyme involved in carbon catabolite repression in yeast, *Mol. Gen. Genet.* **178:**633–637.

Entian, K.-D., 1981, A carbon catabolite repression mutant of *Saccharomyces cerevisiae* with elevated hexokinase activity: Evidence for regulatory control of hexokinase PII synthesis, *Mol. Gen. Genet.* **184:**278–282.

Entian, K.-D., and Fröhlich, K.-U., 1984, *Saccharomyces cerevisiae* mutants provide evidence of hexokinase PII as a bifunctional enzyme with catalytic and regulatory domains for triggering carbon catabolite repression, *J. Bacteriol.* **158**:29–35.

Entian, K.-D., and Mecke, D., 1982, Genetic evidence for a role of hexokinase isoenzyme PII in carbon catabolite repression in *Saccharomyces cerevisiae, J. Biol. Chem.* **257**:870–874.

Epstein, W., Jewett, S., and Fox, C. F., 1970, Isolation and mapping of phosphotransferase mutants in *Escherichia coli, J. Bacteriol.* **104**:793–797.

Erni, B., Trachsel, H., Postma, P. W., and Rosenbuch, J. P., 1982, Bacterial phosphotransferase system. Solubilization and purification of the glucose-specific enzyme II from membranes of *Salmonella typhimurium, J. Biol. Chem.* **257**:13726–13730.

Ferenci, T., and Kornberg, H. L., 1974, The role of phosphotransferase syntheses of fructose 1-phosphate and fructose 6-phosphate in the growth of *Escherichia coli* on fructose, *Proc. R. Soc. London Ser. B* **187**:105–119.

Feucht, B. U., and Saier, M. H., Jr., 1980, Fine control of adenylate cyclase by the phosphoenolpyruvate: sugar phosphotransferase system in *Escherichia coli* and *Salmonella typhimurium, J. Bacteriol.* **141**:603–610.

Fox, C. F., and Wilson, G., 1968, The role of a phosphoenolpyruvate dependent kinase system in β-glucoside catabolism in *Escherichia coli, Proc. Natl. Acad. Sci. USA* **59**:988–995.

Fraser, A. D. E., and Yamazaki, H., 1982, Significance of β-galactosidase repression in glucose inhibition of lactose utilization in *Escherichia coli, Curr. Microbiol.* **7**:241–244.

Gachelin, G., 1970, Studies on the α-methylglucoside permease of *Escherichia coli.* A two-step mechanism for the accumulation of α-methylglucoside 6-phosphate, *Eur. J. Biochem.* **16**:342–357.

Garnak, M., and Reeves, H. C., 1979, Purification and properties of phosphorylated isocitrate dehydrogenase from *Escherichia coli, J. Biol. Chem.* **254**:7915–7920.

Gay, P., Cordier, P., Marquet, M., and Delobbe, A,, 1973, Carbohydrate metabolism and transport in *Bacillus subtilis.* A study of *ctr* mutations, *Mol. Gen. Genet.* **121**:355–368.

Gershanovitch, V. N., Bourd, G. I., Jurovitzkaya, N. V., Skavronskaya, A. G., Klyutchova, V. V., and Shabolenko, V. P., 1967, β-Galactosidase induction in cells of *Escherichia coli* not utilizing glucose, *Biochim. Biophys. Acta* **134**:188–190.

Gershanovitch, V. N., Ilyina, T. S., Rusina, O. Y., Yourovitskaya, N. V., and Bolshakova, T. N., 1977, Repression of inducible enzyme synthesis in a mutant of *Escherichia coli* K12 deleted for the *ptsH* gene, *Mol. Gen. Genet.* **153**:185–190.

Gilman, A. G., 1984, G proteins and dual control of adenylate cyclase, *Cell* **36**:577–579.

Glesyna, M. L., Bolshakova, T. N., and Gershanovitch, V. N., 1983, Effect of *ptsI* and *ptsH* mutations on initiation of transcription of the *Escherichia coli* lactose operon, *Mol. Gen. Genet.* **190**:417–420.

Goldbeter, A., and Koshland, D. E., Jr., 1981, An amplified sensitivity arising from covalent modification in biological systems, *Proc. Natl. Acad. Sci. USA* **78**:6840–6844.

Guidi-Rontani, C., and Gicquel-Sanzey, B., 1981, Expression of the maltose regulon in strain lacking the cyclic AMP receptor protein, *FEMS Microbiol. Lett.* **10**:383–387.

Guidi-Rontani, C., Danchin, A., and Ullmann, A., 1980, Catabolite repression in *Escherichia coli* mutants lacking cyclic AMP receptor protein, *Proc. Natl. Acad. Sci. USA* **77**:5799–5801.

Guiso, N., and Blazy, B., 1980, Regulatory aspects of the cyclic AMP receptor protein in *Escherichia coli* K-12, *Biochem. Biophys. Res. Commun.* **94**:278–283.

Hagihara, H., Wilson, T. H., and Lin, E. C. C., 1963, Studies on the glucose-transport system in *Escherichia coli* with α-methylglucoside as substrate, *Biochim. Biophys. Acta* **78**:505–515.

Haguenauer, R., and Kepes, A., 1971, The cycle of renewal of intracellular α-methyl glucoside accumulation by the glucose permease of *E. coli, Biochimie* **53**:99–107.

Harman, J. G., and Botsford, J. L., 1979, Synthesis of 3′:5′-cyclic monophosphate in *Salmonella typhimurium* growing in continuous culture, *J. Gen. Microbiol.* **110**:243–246.

Harte, M. J., and Webb, F. C., 1967, Utilization of mixed sugars in continuous fermentation. II, *Biotechnol. Bioeng.* **9**:205–221.

Harwood, J. P., and Peterkofsky, A., 1975, Glucose-sensitive adenylate cyclase in toluene-treated cells of *Escherichia coli* B, *J. Biol. Chem.* **250**:4656–4662.

Harwood, J. P., Gazdar, C., Prasad, C., Peterkofsky, A., Curtis, S. J., and Epstein, W., 1976, Involvement of the glucose enzymes II of the sugar phosphotransferase system in the regulation of adenylate cyclase by glucose in *Escherichia coli*, *J. Biol. Chem.* **251**:2462–2468.

Heller, K. B., Lin, E. C. C., and Wilson, T. H., 1980, Substrate specificity and transport properties of the glycerol facilitator of *Escherichia coli*, *J. Bacteriol.* **144**, 274–278.

Hoffee, P., Englesberg, E., and Lamy, F., 1964, The glucose permease system in bacteria, *Biochim. Biophys. Acta* **79**:337–350.

Hommes, R. W. J., Postma, P. W., Neijssel, O. M., Tempest, D. W., Dokter, P., and Duine, J. A., 1984, Evidence of a quinoprotein glucose dehydrogenase apoenzyme in several strains of *Escherichia coli*, *FEMS Microbiol. Lett.* **24**:329–333.

Huber, R. E., Pisko-Dubienski, R., and Hurlburt, K. L., 1980, Immediate stoichiometric appearance of β-galactosidase products in the medium of *Escherichia coli* incubated with lactose, *Biochem. Biophys. Res. Commun.* **96**:656–661.

Hudig, H., and Hengstenberg, W., 1980, The bacterial phosphoenolpyruvate dependent phosphotransferase system (PTS). Solubilisation and kinetic parameters of the glucose-specific membrane-bound enzyme II component of *Streptococcus faecalis, FEBS Lett.* **114**:103–106.

Ingebritsen, T. S., and Cohen, P., 1983, Protein phosphatases: Properties and role in cellular regulation, *Science* **221**:331–338.

Jablonski, E. G., Brand, L., and Roseman, S., 1983, Sugar transport by the bacterial phosphotransferase system. Preparation of a fluorescein derivative of the glucose-specific phosphocarrier protein III[Glc] and its binding to the phosphocarrier protein HPr, *J. Biol. Chem.* **258**:9690–9699.

Jacobson, G. R., Lee, C. A., and Saier, M. H., Jr., 1979, Purification of the mannitol-specific enzyme II of the *Escherichia coli* phosphoenolpyruvate: sugar phosphotransferase system, *J. Biol. Chem.* **254**:249–252.

Jacobson, G. R., Lee, C. A., Leonard, J. E., and Saier, M. H., Jr., 1983a, Mannitol-specific enzyme II of the bacterial phosphotransferase system. I. Properties of the purified permease, *J. Biol. Chem.* **258**:10748–10756.

Jacobson, G. R., Kelly, D. M., and Finlay, D. R., 1983b, The intramembrane topography of the mannitol-specific enzyme II of the *Escherichia coli* phosphotransferase system, *J. Biol. Chem.* **258**:2955–2959.

Jin, R. Z., and Lin, E. C. C., 1984, An inducible phosphoenolpyruvate: dihydroxyacetone phosphotransferase system in *Escherichia coli*, *J. Gen. Microbiol.* **130**:83–88.

Jones-Mortimer, M. C., and Kornberg, H. L., 1974, Genetical analysis of fructose utilization by *Escherichia coli, Proc. R. Soc. London Ser. B* **187**:121–131.

Jones-Mortimer, M. C., and Kornberg, H. L., 1980, Amino-sugar transport systems of *Escherichia coli* K12, *J. Gen. Microbiol.* **117**:369–376.

Joseph, E., Bernsley, C., Guiso, N., and Ullmann, A., 1982, Multiple regulation of the activity of adenylate cyclase in *Escherichia coli, Mol. Gen. Genet.* **185**:262–268.

Kaback, H. R., 1968, The role of the phosphoenolpyruvate-phosphotransferase system in the transport of sugars by isolated membrane preparations of *Escherichia coli, J. Biol. Chem.* **243**:3711–3724.

Kalachev, I. Y., Gershanovitch, V. N., and Bourd, G. I., 1980, Transmembrane phosphorylation of α-methylglucoside and regulation of the activity of β-galactoside permease in the bacterium *E. coli* K12, *Biokhimiya* **45**:873–882.

Kalachev, I. Y., Umyaroz, A. M., and Bourd, G. I., 1981, Interaction of membrane transport proteins in *E. coli* K12, *Biokhimiya* **46**:732–743.

Kalbitzer, H. R., Hengstenberg, W., Rösch, P., Muss, P., Bernsmann, P., Engelmann, R., Dörschug, M., and Deutscher, J., 1982, HPr proteins of different microorganisms studied by hydrogen-1 high-resolution nuclear resonance: Similarities of structures and mechanism, Biochemistry 21:2879-2885.

Kepes, A., 1960, Etudes cinétiques sur la galactoside-permease d'*Escherichia coli, Biochim. Biophys. Acta* 40:70-84.

Koch, A. L., 1964, The role of permease in transport, *Biochim. Biophys. Acta* 79:177-200.

Kolb, A., Spassky, A., Chapon, C., Blazy, B., and Buc, H., 1983, On the different binding affinities of CRP at the *lac, gal* and *malT* promoter regions, *Nucleic Acids Res.* 11:7833-7852.

Konings, W. N., and Robillard, G. T., 1982, Physical mechanism for regulation of proton solute transport in *Escherichia coli, Proc. Natl. Acad. Sci. USA* 79:5480-5484.

Koop, A. H., Hartley, M., and Bourgeois, S., 1984, Analysis of the *cya* locus of *Escherichia coli, Gene* 28:133-146.

Kornberg, H. L., and Reeves, R. E., 1972, Inducible phosphoenolpyruvate-dependent hexose phosphotransferase activities in *Escherichia coli, Biochem. J.* 128:1339-1344.

Kornberg, H. L., and Riordan, C., 1976, Uptake of galactose into *Escherichia coli* by facilitated diffusion, *J. Gen. Microbiol.* 94:75-89.

Kornberg, H. L., and Watts, P. D., 1978, Roles of *crr*-gene products in regulating carbohydrate uptake by *Escherichia coli, FEBS Lett.* 89:329-332.

Kornberg, H. L., and Watts, P. D., 1979, *tgs* and *crr:* Genes involved in catabolite inhibition and inducer exclusion in *Escherichia coli, FEBS Lett.* 104:313-316.

Kornberg, H. L., Watts, P. D., and Brown, K., 1980, Mechanisms of "inducer exclusion" by glucose, *FEBS Lett.* 117(Suppl.):K28-K36.

Koshland, D. E., Jr., Goldbeter, A., and Stock, J. B., 1982, Amplification and adaptation in regulatory and sensory systems, *Science* 217:220-225.

Kubota, Y., Iuchi, S., Fujisawa, A., and Tanaka, S., 1979, Separation of four components of the phosphoenolpyruvate: glucose phosphotransferase system in *Vibrio parahaemolyticus, Microbiol. Immunol.* 23:131-146.

Kukuruzinska, M. A., Harrington, W. F., and Roseman, S., 1982, Sugar transport by the bacterial phosphotransferase system. Studies on the molecular weight and association of enzyme I, *J. Biol. Chem.* 257:14470-14476.

Kundig, W., and Roseman, S., 1971, Sugar transport. II. Characterization of constitutive membrane-bound enzymes II of the *Escherichia coli* phosphotransferase system, *J. Biol. Chem.* 246:1407-1418.

Kundig, W., Ghosh, S., and Roseman, S., 1964, Phosphate bound to histidine in a protein as an intermediate in a novel phosphotransferase system, *Proc. Natl. Acad. Sci. USA* 52:1067-1074.

Lee, C. A., and Saier, M. H., Jr., 1983, Mannitol-specific enzyme II of the bacterial phosphotransferase system. III. The nucleotide sequence of the permease gene, *J. Biol. Chem.* 258:10761-10767.

Lee, C. A., Jacobson, G. R., and Saier, M. H., Jr., 1981, Plasmid-directed synthesis of enzymes required for D-mannitol transport and utilization in *Escherichia coli, Proc. Natl. Acad. Sci. USA* 78:7336-7340.

Lee, L. G., Britton, P., Parra, F., Boronat, A., and Kornberg, H. L., 1982, Expression of the *ptsH*+ gene of *Escherichia coli* cloned on plasmid pBR322: A convenient means for obtaining the histidine-containing carrier protein HPr, *FEBS Lett.* 149:288-292.

Lengeler, J., 1975a, Mutations affecting transport of the hexitols D-mannitol, D-glucitol, and galactitol in *Escherichia coli* K-12: Isolation and mapping, *J. Bacteriol.* 124:26-38.

Lengeler, J., 1975b, Nature and properties of hexitol transport systems in *Escherichia coli, J. Bacteriol.* 124:39-47.

Lengeler, J., and Steinberger, H., 1978a, Analysis of the regulatory mechanisms controlling the activity of the hexitol transport systems in *Escherichia coli* K12, *Mol. Gen. Genet.* 167:75-82.

Lengeler, J., and Steinberger, H., 1978b, Analysis of the regulatory mechanisms controlling the synthesis of the hexitol transport systems in *Escherichia coli* K12, *Mol. Gen. Genet.* **164:**163–169.

Lengeler, J., Auburger, A.-M., Mayer, R., and Pecher, A., 1981, The phosphoenolpyruvate-dependent carbohydrate: phosphotransferase system enzymes II as chemoreceptors in chemotaxis of *Escherichia coli* K12, *Mol. Gen. Genet.* **183:**163–170.

Lengeler, J., Mayer, R. J., and Schmid, K., 1982, The phosphoenolpyruvate-dependent phosphotransferase system enzyme III and plasmid-encoded sucrose transport in *Escherichia coli*, *J. Bacteriol.* **151:**468–471.

Leonard, J. E., and Saier, M. H., Jr., 1983, Mannitol-specific enzyme II of the bacterial phosphotransferase system. II. Reconstitution of vectorial transphosphorylation in phospholipid vesicles. *J. Biol. Chem.* **258:**10757–10760.

Lin, E. C. C., 1970, The genetics of bacterial transport systems, *Annu. Rev. Genet.* **4:**225–262.

Link, C. D., and Reiner, A., 1982, Inverted repeats surround the ribitol–arabitol genes of *E. coli* C, *Nature* **298:**94–96.

Link, C. D., and Reiner, A. M., 1983, Genotypic exclusion: A novel relationship between the ribitol–arabitol and galactitol genes of *E. coli*, *Mol. Gen. Genet.* **189:**337–339.

Magasanik, B., 1970, Glucose effects: Inducer exclusion and repression, in: *The Lactose Operon* (J. R. Beckwith and D. Zipser, eds.), pp. 189–219, Cold Spring Harbor Laboratory, Cold Spring Harbor, N.Y.

Mattoo, R. L., and Waygood, E. B., 1983, Determination of the levels of HPr and enzyme I of the phosphoenolpyruvate–sugar phosphotransferase system in *Escherichia coli* and *Salmonella typhimurium, Can. J. Biochem. Cell. Biol.* **61:**29–37.

Mattoo, R. L., Khandelval, R. L., and Waygood, E. B., 1984, Isoelectrophoretic separation and the detection of soluble proteins containing acid-labile phosphate: Use of the phosphoenolpyruvate: sugar phosphotransferase system as a model system for N^1-P-histidine- and N^3-P-histidine-containing proteins, *Anal. Biochem.* **139:**1–16.

Meadow, N. D., and Roseman, S., 1982, Sugar transport by the bacterial phosphotransferase system. Isolation and characterization of a glucose-specific protein (III^{Glc}) from *Salmonella typhimurium, J. Biol. Chem.* **257:**14526–14537.

Meadow, N. D., Saffen, D. W., Dottin, R. P., and Roseman, S., 1982a, Molecular cloning of the *crr* gene and evidence that it is the structural gene for III^{Glc}, a phosphocarrier protein of the bacterial phosphotransferase system, *Proc. Natl. Acad. Sci. USA* **79:**2528–2532.

Meadow, N. D., Rosenberg, J. M., Pinkert, H. M., and Roseman, S., 1982b, Sugar transport by the bacterial phosphotransferase system. Evidence that *crr* is the structural gene for the *Salmonella typhimurium* glucose-specific phosphocarrier protein III^{Glc}, *J. Biol. Chem.* **257:**14538–14542.

Misset, O., and Robillard, G. T., 1982, *Escherichia coli* phosphoenolpyruvate-dependent phosphotransferase system: Mechanism of phosphoryl-group transfer from phosphoenolpyruvate to HPr, *Biochemistry* **21:**3136–3142.

Misset, O., Brouwer, M., and Robillard, G. T., 1980, *Escherichia coli* phosphoenolpyruvate-dependent phosphotransferase system. Evidence that the dimer is the active form of enzyme I, *Biochemistry* **19:**883–890.

Misset, O., Blaauw, M., Postma, P. W., and Robillard, G. T., 1983, Bacterial phosphoenolpyruvate-dependent phosphotransferase system. Mechanisms of the transmembrane sugar translocation and phosphorylation, *Biochemistry* **22:**6163–6170.

Mitchell, W. J., Misko, T. P., and Roseman, S., 1982, Sugar transport by the bacterial phosphotransferase system. Regulation of other transport systems (lactose and melibiose), *J. Biol. Chem.* **257:**14553–14564.

Nelson, S. O., and Postma, P. W., 1984, Interactions in vivo between III^{Glc} of the phosphoenolpyruvate: sugar phosphotransferase system and the glycerol and maltose uptake systems of *Salmonella typhimurium, Eur. J. Biochem.* **139:**29–34.

Nelson, S. O., Scholte, B. J., and Postma, P. W., 1982, Phosphoenolpyruvate: sugar phosphotransferase system-mediated regulation of carbohydrate metabolism in *Salmonella typhimurium, J. Bacteriol.* **150:**604–615.

Nelson, S. O., Wright, J. K., and Postma, P. W., 1983, The mechanism of inducer exclusion: Direct interaction between purified III^{Glc} of the phosphoenolpyruvate: sugar phosphotransferase system and the lactose carrier of *Escherichia coli, EMBO J.* **2:**715–720.

Nelson, S. O., Schuitema, A. R. J., Benne, R., van der Ploeg, L. H. T., Plijter, J. J., Aan, F., and Postma, P. W., 1984a, Molecular cloning, sequencing and expression of the *crr* gene: The structural gene for III^{Glc} of the bacterial PEP: glucose phosphotransferase system, *EMBO J.* **3:**1587–1593.

Nelson, S. O., Lengeler, J., and Postma, P. W., 1984b, The role of III^{Glc} of the PEP: glucose phosphotransferase system in inducer exclusion in *Escherichia coli, J. Bacteriol.* **160:**360–364.

Nestler, E. J., and Greengard, P., 1983, Protein phosphorylation in the brain, *Nature* **305:**583–588.

Neuhaus, J. M., and Wright, J. K., 1983, Chemical modification of the lactose carrier of *Escherichia coli* by plumbagin, phenyl arsinoxide or diethylpyrocarbonate affects the binding of galactoside, *Eur. J. Biochem.* **137:**615–621.

Newman, M. J., Foster, D. L., Wilson, T. H., and Kaback, H. R., 1981, Purification and reconstitution of functional lactose carrier from *Escherichia coli, J. Biol. Chem.* **256:**11804–11808.

Neyssel, O. M., Tempest, D. W., Postma, P. W., Duine, J. A., and Frank Jzn, J., 1983, Glucose metabolism by K$^+$-limited *Klebsiella aerogenes:* Evidence for the involvement of a quinoprotein glucose dehydrogenase, *FEMS Microbiol. Lett.* **20:**35–39.

Niaudet, B., Gay, P., and Dedonder, R., 1975, Identification of the structural gene of the PEP-phosphotransferase enzyme I in *Bacillus subtilis* Marburg, *Mol. Gen. Genet.* **136:**337–349.

Niwano, M., and Taylor, B. L., 1982, Novel sensory adaptation mechanism in bacterial chemotaxis to oxygen and phosphotransferase substrates, *Proc. Natl. Acad. Sci. USA* **79:**11–15.

O'Brien, R. W., Neyssel, O. M., and Tempest, D. W., 1980, Glucose phosphoenolpyruvate phosphotransferase activity and glucose uptake rate of *Klebsiella aerogenes* growing in chemostat cultures, *J. Gen. Microbiol.* **116:**305–314.

Okada, T., Ueyama, K., Niiya, S., Kanazawa, H., Futai, M., and Tsuchiya, T., 1981, Role of inducer exclusion in preferential utilization of glucose over melibiose in diauxic growth of *Escherichia coli, J. Bacteriol.* **146:**1030–1037.

Osumi, T., and Saier, M. H., Jr., 1982, Regulation of lactose permease activity by the phosphoenolpyruvate: sugar phosphotransferase system: Evidence for direct binding of the glucose-specific enzyme III to the lactose permease, *Proc. Natl. Acad. Sci. USA* **79:**1457–1461.

Paigen, K., and Williams, B., 1970, Catabolite repression and other control mechanisms in carbohydrate utilization, *Adv. Microbiol. Physiol.* **4:**251–324.

Parra, F., Jones-Mortimer, M. C., and Kornberg, H. L., 1983, Phosphotransferase-mediated regulation of carbohydrate utilization in *Escherichia coli* K12: The nature of the *iex* (*crr*) and *gsr* (*tgs*) mutations, *J. Gen. Microbiol.* **129:**337–348.

Pastan, I., and Perlman, R. L., 1969, Repression of β-galactosidase synthesis by glucose in phosphotransferase mutants of *Escherichia coli:* Repression in the absence of glucose phosphorylation, *J. Biol. Chem.* **244:**5836–5842.

Peri, K. G., Kornberg, H. L., and Waygood, E. B., 1984, Evidence for the phosphorylation of enzyme II^{Glucose} of the phosphoenolpyruvate sugar phosphotransferase system of *Escherichia coli* and *Salmonella typhimurium, FEBS Lett.* **178:**55–58.

Perret, J., and Gay, P., 1979, Kinetic study of a phosphoryl exchange reaction between fructose and fructose 1-phosphate catalyzed by the membrane-bound enzyme II of the phosphoenolpyruvate–fructose 1-phosphotransferase system of *Bacillus subtilis, Eur. J. Biochem.* **102:**237–246.

Peterkofsky, A., and Gazdar, C., 1973, Measurements of rates of adenosine 3':5'-cyclic monophosphate synthesis in intact *Escherichia coli* B, *Proc. Natl. Acad. Sci. USA* **70:**2149–2152.

Peterkofsky, A., and Gazdar, C., 1975, Interaction of enzyme I of the phosphoenolpyruvate: sugar phosphotransferase system with adenylate cyclase of *Escherichia coli, Proc. Natl. Acad. Sci. USA* **72**:2920–2924.

Postma, P. W., 1976, Involvement of the phosphotransferase system in galactose transport in *Salmonella typhimurium, FEBS Lett.* **61**:49–53.

Postma, P. W., 1977, Galactose transport in *Salmonella typhimurium, J. Bacteriol.* **129**:630–639.

Postma, P. W., 1981, Defective enzyme II-B^Glucose of the phosphoenolpyruvate: sugar phosphotransferase system leading to uncoupling of transport and phosphorylation in *Salmonella typhimurium, J. Bacteriol.* **147**:382–389.

Postma, P. W., 1982, Regulation of sugar transport in *Salmonella typhimurium, Ann. Microbiol.* **133A**:261–267.

Postma, P. W., and Lengeler, J. W., 1985, Phosphoenolpyruvate: carbohydrate phosphotransferase system of bacteria, *Microbiol Rev.* **49**:232–269.

Postma, P. W., and Roseman, S., 1976, The bacterial phosphoenolpyruvate:sugar phosphotransferase system, *Biochim. B. Biophys. Acta* 457:213–257.

Postma, P. W., and Scholte, B. J., 1979, Regulation of sugar transport in *Salmonella typhimurium,* in: *Function and Molecular Aspects of Biomembrane Transport* (E. Quagliariello, F. Palmieri, S. Papa, and M. Klingenberg, eds.), pp. 249–257, Elsevier, Amsterdam.

Postma, P. W., and Stock, J. B., 1980, Enzymes II of the phosphotransferase system do not catalyze sugar transport in the absence of phosphorylation, *J. Bacteriol.* **141**:476–484.

Postma, P. W., and van Thienen, G. M., 1978, Energization of sugar transport in *Salmonella typhimurium,* in: *The Proton and Calcium Pumps* (M. Avron, G. F. Azzone, J. C. Metcalfe, E. Quagliariello, and N. Siliprandi, eds.), pp. 149–159, Elsevier, Amsterdam.

Postma, P. W., Schuitema, A., and Kwa, C., 1981, Regulation of methyl β-galactoside permease activity in *pts* and *crr* mutants of *Salmonella typhimurium, Mol. Gen. Genet.* **181**:448–453.

Postma, P. W., Neyssel, O. M., and van Ree, R., 1982, Glucose transport in *Salmonella typhimurium* and *Escherichia coli, Eur. J. Biochem.* **123**:113–119.

Postma, P. W., Epstein, W., Schuitema, A. R. J., and Nelson, S. O., 1984, Interaction between III^Glc of the PEP: sugar phosphotransferase system and glycerol kinase of *Salmonella typhimurium, J. Bacteriol.* **158**:351–353.

Reizer, J., and Saier, M. H., Jr., 1983, Involvement of lactose enzyme II of the phosphotransferase system in rapid expulsion of free galactosides from *Streptococcus pyogenes, J. Bacteriol.* **156**:236–242.

Reizer, J., Novotny, M. J., Panos, C., and Saier, M. H., Jr., 1983, Mechanism of inducer expulsion in *Streptococcus pyogenes:* A two-step process activated by ATP, *J. Bacteriol.* **156**:354–361.

Reizer, J., Novotny, M. J., Stuiver, I., and Saier, M. H., Jr., 1984, Regulation of glycerol uptake by the phosphoenolpyruvate: sugar phosphotransferase system in *Bacillus subtilis, J. Bacteriol.* **159**:243–250.

Rephaeli, A. W., and Saier, M. H., Jr., 1978, Kinetic analyses of the sugar phosphate: sugar transphosphorylation reaction catalyzed by the glucose enzyme II complex of the bacterial phosphotransferase system, *J. Biol. Chem.* **253**:7595–7597.

Rephaeli, A. W., and Saier, M. H., Jr., 1980a, Substrate specificity and kinetic characterization of sugar uptake and phosphorylation, catalyzed by the phosphotransferase system in *Salmonella typhimurium, J. Biol. Chem.* **255**:8585–8591.

Rephaeli, A. W., and Saier, M. H., Jr., 1980b, Regulation of genes coding for enzyme constituents of the bacterial phosphotransferase system, *J. Bacteriol.* **141**:658–663.

Reynolds, A. E., Felton, J., and Wright, A., 1981, Insertion of DNA activates the cryptic *bgl* operon in *E. coli* K12, *Nature* **293**:625–629.

Robillard, G. T., 1982, The enzymology of the bacterial phosphoenolpyruvate-dependent sugar transport system, *Mol. Cell. Biochem.* **46**:3–24.

Robillard, G. T., and Konings, W. N., 1981, Physical mechanism for regulation of phosphoenolpyruvate-dependent glucose transport activity in *Escherichia coli*, *Biochemistry*, **20**:5025–5032.

Robillard, G. T., and Konings, W. N., 1982, A hypothesis for the role of dithiol–disulfide interchange in solute transport and energy-transducing processes, *Eur. J. Biochem.* **127**:597–604.

Robillard, G. T., Dooyewaard, G., and Lolkema, J., 1979, *Escherichia coli* phosphoenolpyruvate dependent phosphotransferase system. Complete purification of enzyme I by hydrophobic interaction chromatography, *Biochemistry* **18**:2984–2989.

Roehl, R. A., and Vinopal, T., 1980, Genetic locus, distant from *ptsM*, affecting enzyme IIA/IIB function in *Escherichia coli* K-12, *J. Bacteriol.* **142**:120–130.

Rose, S. P., and Fox, C. F., 1971, The β-glucoside system of *Escherichia coli*. II. Kinetic evidence for a phosphoryl-enzyme II intermediate. *Biochem. Biophys. Res. Commun.* **45**:376–380.

Rotman, B., Ganesan, A. K., and Guzman, R., 1968, Transport systems for galactose and galactosides in *Escherichia coli*. II. Substrate and inducer specificities, *J. Mol. Biol.* **36**:247–260.

Roy, A., Haziza, C., and Danchin, A., 1983a, Regulation of adenylate cyclase synthesis in *Escherichia coli*: Nucleotide sequence of the control region, *EMBO J.* **2**:791–797.

Roy, A., Danchin, A., Joseph, E., and Ullmann, A., 1983b, Two functional domains in adenylate cyclase of *Escherichia coli*, *J. Mol. Biol.* **165**:197–202.

Rusina, O. Y., and Gershanovitch, V. N., 1983, Mapping of mutations within genes coding for enzyme I and HPr protein of the phosphoenolpyruvate-dependent phosphotransferase system of *Escherichia coli* K-12. II. Mapping of *ptsH* mutations within the gene, *Genetika* **19**:397–405.

Rusina, O. Y., and Gershanovitch, V. N., 1983, Mapping of mutations within genes coding for enzyme I and HPr protein of the phosphoenolpyruvate-dependent phosphotransferase system of *Escherichia coli* K-12. II. Mapping of *ptsH* mutations within the gene, *Genetika* **19**:397–405.

Saier, M. H., Jr., 1977, Bacterial phosphoenolpyruvate: sugar phosphotransferase systems: Structural, functional and evolutionary interrelationships, *Bacteriol. Rev.* **41**:856–871.

Saier, M. H., Jr., and Feucht, B. U., 1975, Coordinate regulation of adenylate cyclase and carbohydrate permeases by the phosphoenolpyruvate: sugar phosphotransferase system in *Salmonella typhimurium*, *J. Biol. Chem.* **250**:7078–7080.

Saier, M. H., Jr., and Roseman, S., 1972, Inducer exclusion and repression of enzyme synthesis in mutants of *Salmonella typhimurium* defective in enzyme I of the phosphoenolpyruvate: sugar phosphotransferase system, *J. Biol. Chem.* **247**:972–975.

Saier, M. H., Jr., and Roseman, S., 1976a, Sugar transport. The *crr* mutation: Its effect on repression of enzyme synthesis, *J. Biol. Chem.* **251**:6598–6605.

Saier, M. H., Jr., and Roseman, S., 1976b, Sugar transport. Inducer exclusion and regulation of the melibiose, maltose, glycerol, and lactose transport systems by the phosphoenolpyruvate: sugar phosphotransferase system, *J. Biol. Chem.* **251**:6606–6615.

Saier, M. H., Jr., and Stiles, C. D., 1975, *Molecular Dynamics in Biological Membranes*, Springer-Verlag, Berlin.

Saier, M. H., Jr., Simoni, R. D., and Roseman, S., 1970, The physiological behaviour of enzyme I and heat-stable protein mutants of a bacterial phosphotransferase system, *J. Biol. Chem.* **245**:5870–5873.

Saier, M. H., Jr., Young, W. S., and Roseman, S., 1971a, Utilization and transport of hexoses by mutant strains of *Salmonella typhimurium* lacking enzyme I of the phosphoenolpyruvate-dependent phosphotransferase system, *J. Biol. Chem.* **246**:5838–5840.

Saier, M. H., Jr., Feucht, B. U., and Roseman, S., 1971b, Phosphoenolpyruvate-dependent fructose phosphorylation in photosynthetic bacteria, *J. Biol. Chem.* **246**:7819–7821.

Saier, M. H., Jr., Bromberg, F. G., and Roseman, S., 1973, Characterization of constitutive galactose permease mutants in *Salmonella typhimurium*, *J. Bacteriol*, **113**:512–514.

Saier, M. H., Jr., Feucht, B. U., and McCaman, M. T., 1975, Regulation of intracellular adenosine cyclic 3':5'-monophosphate levels in *Escherichia coli* and *Salmonella typhimurium:* Evidence for energy-dependent excretion of the cyclic nucleotide, *J. Biol. Chem.* **250**:7593–7601.

Saier, M. H., Jr., Simoni, R. D., and Roseman, S., 1976, Sugar transport. Properties of mutant bacteria defective in proteins of the phosphoenolpyruvate: sugar phosphotransferase system, *J. Biol. Chem.* **251**:6584–6597.

Saier, M. H., Jr., Feucht, B. U., and Mora, W. K., 1977a, Sugar phosphate: sugar transphosphorylation and exchange group translocation catalyzed by the enzyme II complexes of the bacterial phosphoenolpyruvate: sugar phosphotransferase system, *J. Biol. Chem.* **252**:8899–8907.

Saier, M. H., Jr., Cox, D. F., and Moczydlowski, E. G., 1977b, Sugar phosphate: sugar transphosphorylation coupled to exchange group translocation catalyzed by the enzyme II complexes of the phosphoenolpyruvate: sugar phosphotransferase system in membrane vesicles of *Escherichia coli*, *J. Biol. Chem.* **252**:8908–8916.

Saier, M. H., Jr., Straud, H., Massman, L. S., Judice, J. J., Newman, M. J., and Feucht, B. U., 1978, Permease-specific mutations in *Salmonella typhimurium* and *Escherichia coli* that release the glycerol, maltose, melibiose and lactose transport systems from regulation by the phosphoenolpyruvate: sugar phosphotransferase system, *J. Bacteriol.* **133**:1358–1367.

Saier, M. H., Jr., Keeler, D. K., and Feucht, B. U., 1982, Physiological desensitization of carbohydrate permeases and adenylate cyclase to regulation by the phosphoenolpyruvate: sugar phosphotransferase system in *Escherichia coli* and *Salmonella typhimurium*, *J. Biol. Chem.* **257**:2509–2517.

Saier, M. H., Jr., Novotny, M. J., Comeau-Fuhrman, D., Osumi, T., and Desai, J. D., 1983, Cooperative binding of the sugar substrates and allosteric regulatory protein (enzyme III[Glc] of the phosphotransferase system) to the lactose and melibiose permeases in *Escherichia coli* and *Salmonella typhimurium*, *J. Bacteriol.* **155**:1351–1357.

Sanderson, K. E., and Roth, J. R., 1983, Linkage map of *Salmonella typhimurium:* Edition VI, *Microbiol. Rev.* **47**:410–453.

Sarno, N. V., Tenn, L. G., Desai, A., Chin, A. M., Grenier, F. C., and Saier, M. H., Jr., 1984, Genetic evidence for glucitol-specific enzyme III, an essential phosphocarrier protein of the *Salmonella typhimurium* glucitol phosphotransferase system, *J. Bacteriol.* **157**:953–955.

Schaefler, S., 1967, Inducible system for the utilization of β-glucoside in *Escherichia coli*, *J. Bacteriol.* **93**:254–263.

Scholte, B. J., and Postma, P. W., 1980, Mutation in the *crp* gene of *Salmonella typhimurium* which interferes with inducer exclusion, *J. Bacteriol.* **141**:751–757.

Scholte, B. J., and Postma, P. W., 1981, Competition between two pathways for sugar uptake by the phosphoenolpyruvate-dependent sugar phosphotransferase system in *Salmonella typhimurium*, *Eur. J. Biochem.* **114**:51–58.

Scholte, B. J., Schuitema, A. R., and Postma, P. W., 1981, Isolation of III[Glc] of the phosphoenolpyruvate-dependent glucose phosphotransferase system of *Salmonella typhimurium*, *J. Bacteriol.* **148**:257–264.

Scholte, B. J., Schuitema, A. R., and Postma, P. W., 1982, Characterization of factor III[Glc] in catabolite repression resistant (*crr*) mutants of *Salmonella typhimurium*, *J. Bacteriol.* **149**:576–586.

Simoni, R. D., Nakazawa, T., Hays, J. B., and Roseman, S., 1973, Sugar transport. IV. Isolation and characterization of the lactose phosphotransferase system in *Staphylococcus aureus*, *J. Biol. Chem.* **248**:932–940.

Slater, A. C., Jones-Mortimer, M. C., and Kornberg, H. L., 1981, L-Sorbose phosphorylation in *Escherichia coli* K-12, *Biochim. Biophys. Acta* **646**:365–367.

Solomon, E., Miyai, K., and Lin, E. C. C., 1973, Membrane translocation of mannitol in *Escherichia coli* without phosphorylation, *J. Bacteriol.* **114**:723–728.

Stock, J. B., Waygood, E. B., Meadow, N. D., Postma, P. W., and Roseman, S., 1982, Sugar transport by the bacterial phosphotransferase system. The glucose receptors of the *Salmonella typhimurium* phosphotransferase system, *J. Biol. Chem.* **257**:14543–14552.

Tanaka, S., and Lin, E. C. C., 1967, Two classes of pleiotropic mutants of *A. aerogenes* lacking components of a PEP phosphotransferase system, *Proc. Natl. Acad. Sci. USA* **57**:913–919.

Tanaka, S., Lerner, S. A., and Lin, E. C. C., 1967, Replacement of a phosphoenolpyruvate-dependent phosphotransferase by a nicotinamide adenine dinucleotide-linked dehydrogenase for the utilization of mannitol. *J. Bacteriol.* **93**:642–648.

Tyler, B., and Magasanik, B., 1970, Physiological basis of transient repression of catabolic enzymes in *Escherichia coli, J. Bacteriol.* **102**:411–422.

Ullmann, A., and Danchin, A., 1983, Role of cyclic AMP in bacteria, *Adv. Cyclic Nucleotide Res.* **15**:32–53.

Umyarov, A. M., Voloshin, A. G., Bolshakova, T. N., and Gershanovitch, V. N., 1978, Effect of *ptsI* and *ptsH* gene dosages on manifestation of glucose catabolite repression of β-galactosidase synthesis in *Escherichia coli* K12, *FEBS Lett.* **96**:31–33.

Voloshin, A. G., Shulgina, M. V., and Bourd, G. I., 1981, Insensitivity of the *Escherichia coli* K12 adenylate cyclase to mannose under the conditions of catabolite repression, *FEMS Microbiol. Lett.* **10**:291–293.

Wagner, E. F., Fabricant, J. D., and Schweiger, M., 1979, A novel ATP-driven glucose transport system in *Escherichia coli, Eur. J. Biochem.* **102**:231–236.

Walter, R. W., Jr., and Anderson, R. L., 1973, Evidence that the inducible phosphoenolpyruvate: D-fructose 1-phosphate phosphotransferase system of *Aerobacter aerogenes* does not require "HPr," *Biochem. Biophys. Res. Commun.* **52**:93–97.

Wang, J. Y. J., and Koshland, D. E., Jr., 1978, Evidence for protein kinase activities in the prokaryote *Salmonella typhimurium, J. Biol. Chem.* **253**:7605–7608.

Wang, J. Y. J., and Koshland, D. E., Jr., 1981, The identification of distinct protein kinases and phosphatases in the prokaryote *Salmonella typhimurium, J. Biol. Chem.* **256**:4640–4648.

Wang, R. J., and Morse, M. L., 1968, Carbohydrate accumulation and metabolism in *Escherichia coli.* I. Description of pleiotropic mutants, *J. Mol. Biol.* **32**:59–66.

Wang, R. J., Morse, H. G., and Morse, M. L., 1970, Carbohydrate accumulation and metabolism in *Escherichia coli:* Characterization of the reversions of *ctr* mutations, *J. Bacteriol.* **104**:1318–1324.

Waygood, E. B., 1980, Resolution of the phosphoenolpyruvate: fructose phosphotransferase system of *Escherichia coli* into two components; enzyme II[Fructose] and fructose-induced HPr-like protein (FPr), *Can. J. Biochem.* **58**:1144–1146.

Waygood, E. B., and Steeves, T., 1980, Enzyme I of the phosphoenolpyruvate: sugar phosphotransferase system (PTS) of *Escherichia coli*—Purification to homogeneity and some properties. *Can. J. Biochem.* **58**:40–48.

Waygood, E. B., Meadow, N.D., and Roseman, S., 1979, Modified assay procedures for the phosphotransferase system in enteric bacteria, *Anal. Biochem.* **95**:293–304.

Weigel, N., Waygood, E. B., Kukuruzinska, M. A., Nakazawa, A., and Roseman, S., 1982a, Sugar transport by the bacterial phosphotransferase system. Isolation and characterization of enzyme I from *Salmonella typhimurium, J. Biol. Chem.* **257**:14461–14469.

Weigel, N., Kukuruzinska, M. A., Nakazawa, A., Waygood, E. B., and Roseman, S., 1982b, Sugar transport by the bacterial phosphotransferase system. Phosphoryl transfer reactions catalyzed by enzyme I of *Salmonella typhimurium, J. Biol. Chem.* **257**:14477–14491.

Weigel, N., Powers, D. A., and Roseman, S., 1982c, Sugar transport by the bacterial phosphotransferase system. Primary structure and active site of a general phosphocarrier protein (HPr) from *Salmonella typhimurium, J. Biol. Chem.* **257**:14499–14509.

White, R. J., 1970, The role of the phosphoenolpyruvate phosphotransferase system in the transport of N-acetyl-D-glucosamine by *Escherichia coli, Biochem. J.* **118**:89–92.

Woodward, M. J., and Charles, H. P., 1983, Polymorphism in *Escherichia coli: rtl, atl* and *gat* regions behave as chromosomal alternatives, *J. Gen. Microbiol.* **129**:75–84.

Wright, J. K., Teather, R. M., and Overath, P., 1983, Lactose permease of *Escherichia coli, Methods Enzymol.* **97**:158–175.

Yang, J. K., and Epstein, W., 1983, Purification and characterization of adenylate cyclase from *Escherichia coli* K12, *J. Biol. Chem.* **258**:3750–3758.

Yang, J. K., Bloom, R. W., and Epstein, W., 1979, Catabolite and transient repression in *Escherichia coli* do not require enzyme I of the phosphotransferase system, *J. Bacteriol.* **138**:275–279.

Zimmerman, F. K., and Scheel, I., 1979, Mutants of *Saccharomyces cerevisiae* resistant to carbon catabolite repression, *Mol. Gen. Genet.* **154**:75–82.

11

Active Transport of Sugars into Escherichia coli

PETER J. F. HENDERSON

1. INTRODUCTION

Bacteria often inhabit environments where nutrients are in short supply, and different species must compete with each other for the available carbohydrates. Accordingly, they expend metabolic energy in order to sequester the sugars and achieve intracellular concentrations sufficient for optimal growth rates. *Escherichia coli* can grow on at least 20 different carbohydrates or related compounds (Hays, 1978; Silhavy *et al.*, 1978), but the strategies for energizing their initial transport across the cytoplasmic membrane fall into four general classes. One of these, where the sugar is phosphorylated using phosphoenolpyruvate during the translocation, is described in detail in Chapter 10. The others, where transport is energized either by a gradient of H^+, by a gradient of Na^+, or by a phosphorylated compound, are discussed in this chapter. The properties of the individual sugar transport systems will be reviewed, but they will be grouped according to these three classes of energization mechanism. First, however, the underlying concepts and experimental strategies involved in investigations of bacterial transport will be outlined (Sections 1 and 2). This chapter is aimed at the newcomer to the field, but it is hoped that the arrangement of the sections will enable the specialist to turn directly to topical areas of interest.

My overall objective is to illustrate how the combination of biochemical and genetic approaches has facilitated progress in elucidating the nature of transport proteins. Although their location in biological membranes may blunt the investi-

PETER J. F. HENDERSON • Department of Biochemistry, University of Cambridge, Cambridge CB2 1QW, United Kingdom.

gator's enthusiasm, given the associated difficulties with assay, hydrophobicity, protein instability, and so on, much progress has been made, usually by perspiration and occasionally by inspiration, in response to these challenging difficulties. The substantial advantage of working with E. coli (or working against it, as often seems to be the case) is the extensive opportunity for genetic manipulation, so that the number and nature of the transport proteins can be established via their DNA. This contributes at all the following levels of a progressive investigation of any novel transport activity:

1. Characterize the energetics of the transport system *in vivo*, isolating its activity by means of specific substrates or in appropriate mutants
2. Isolate and sequence the DNA
3. Identify and purify the protein(s) involved; the next steps are now the most challenging:
4. Establish the three-dimensional structure of the protein
5. Deduce its detailed molecular mechanism, just as do enzymologists/ protein chemists/X-ray crystallographers who work with water-soluble proteins

Like climbing mountains, the route can be easy, severe, v. diff., or sometimes impassable, but the view from the top should be inspiring.

1.1. The "Active Transport" Systems for Carbohydrates in E. coli

The term *active transport* is used to describe the net transport of a chemically unmodified solute across a biological membrane from a low electrochemical potential to a higher potential (see, e.g., West, 1983). For this transport, input of metabolic energy is required. This energization is achieved in at least three different ways for active transport of sugars into E. coli. The first to be considered below is the linkage of lactose, galactose, arabinose, xylose, and fucose transport to the electrochemical gradient of protons established by respiration or ATP hydrolysis (Section 3). The second is the linkage of melibiose transport to the gradient of Na^+ ions (Section 4). The third is the direct energization of transport of maltose, galactose, arabinose, xylose, and ribose by a phosphorylated metabolite, possibly ATP; these systems all involve a periplasmic binding protein with high affinity for the substrate (Section 5). The characteristics of each type of system will be described in detail; some of their general properties are summarized in Table I. This also includes the accepted phenotype name for each, with which it is helpful to be familiar.

1.2. Lactose Transport—A Special Case

Studies on the lactose "permease" of E. coli by Monod and his colleagues (reviewed in Beckwith and Zipser, 1970) pioneered our understanding of trans-

Table I. The Active Transport Systems for Sugars in E. coli

Sugar substrate	Phenotype name	Gene location (min)	K_m for sugar (μM)	Mechanism of energization
Lactose	LacY	8	50–900	Sugar/H$^+$ symport
Galactose	GalP	64	50–450	Sugar/H$^+$ symport
Arabinose	AraE	61	140–320	Sugar/H$^+$ symport
Xylose	XylE	91	70–170	Sugar/H$^+$ symport
L-Fucose	Fuc	?	?	Sugar/H$^+$ symport
Melibiose	MelB	93	200	Sugar/Na$^+$ symport
Maltose	MalB	92	1.0–2.0	Phosphate bond?
Ribose	RbsB?	84	0.3–2.3	Phosphate bond?
Galactose	MglP	45	0.4–0.7	Phosphate bond?
Arabinose	AraF	45	4.0–6.0	Phosphate bond?
Xylose	XylF	80	0.2–2.0	Phosphate bond?

port processes in general, and elegantly illuminated the regulation of gene expression in bacteria. Progress with the lactose transport system has continued to dominate the field. Recent breakthroughs include reconstitution of the active protein into liposomes (Newman and Wilson, 1980), its purification (Newman *et al.*, 1981; Foster *et al.*, 1982; Viitanen *et al.*, 1984; Costello *et al.*, 1984), and the complete sequencing of the gene (Büchel *et al.*, 1980; Ehring *et al.*, 1980). I will refer to these advances extensively. However, a number of reviews of lactose transport are available (West, 1980; Lombardi, 1981; Hengge and Boos, 1983; Kaback, 1983), so the emphasis here will be on the other sugar transport systems of *E. coli*. Progress with these was often a direct spin-off of the investigations on lactose transport.

1.3. Primary and Secondary Transport, Symport, and Antiport

The definition of active transport above obviously embraces a variety of molecular mechanisms. It is conceptually helpful to divide them into ''primary'' and ''secondary'' (see review by Rosen and Kashket, 1978) and the ''secondary'' into ''symport'' and ''antiport,'' terms introduced by Mitchell (1963, 1973).

Primary transport involves the direct conversion of chemical or photosynthetic energy into an electrochemical potential of transport substrate across the membrane barrier. Thus, translocation of protons by the respiratory enzymes (Haddock and Jones, 1977; Ingledew and Poole, 1984), by the ATPase (Kagawa, 1978; Futai and Kanazawa, 1984), or by bacteriorhodopsin (Henderson and Unwin, 1975), all fall into this class, as does the phosphotransferase mechanism of carbohydrate transport (Chapter 10).

Secondary transport involves the conversion of a preexisting electrochemical gradient, e.g., of protons or Na^+ ions, into a new electrochemical gradient of the transported species. Thus, the energy stored across the membrane as the primary gradient is used to generate a second gradient of a different species. More than one secondary system can be coupled to the primary gradient, and the ultimate energy source for all of them is the primary chemical or photochemical conversion. For example, in *E. coli* primary proton ejection powers secondary Na^+/H^+ antiport (Na^+ ejection; West and Mitchell, 1974) and then secondary Na^+/melibiose symport (melibiose influx; Section 4). In mammalian mitochondria an extensive proliferation of such transport systems for phosphate, adenine nucleotides, amino acids, and citric acid cycle intermediates, is energized initially by redox reactions driving protons across the membrane (Chappell, 1968).

1.4. Amino Acid Transport

Carbohydrate transport has not been studied in isolation from other nutrient transport systems of *E. coli* (Rosen, 1978). In particular, uptake of amino acids displays similar general features, and studies of glutamine (Berger and Heppel, 1974; Hunt and Hong, 1981), proline (Lombardi and Kaback, 1972; Stalmach *et al.*, 1983; Mogi and Anraku, 1984; Cairney *et al.*, 1984), histidine (Higgins *et al.*, 1982), and leucine (Rahmanian *et al.*, 1973; Amanuma and Anraku, 1974) transport especially contributed to our understanding of the energetics, genetics, and involvement of binding proteins in both amino acid and carbohydrate transport systems. The references listed should serve as entry points to the literature on amino acid transport, together with reviews by Rosen (1978), Anraku (1978), and Landick and Oxender (1982).

1.5. The Cell Membranes

E. coli, a gram-negative organism, has two lipid bilayer membranes in its cell wall, the structure of which has been nicely illustrated by Di Renzio *et al.* (1978) (see Fig. 1). The outer membrane is freely permeable to mono- and disaccharides, by virtue of protein-formed unspecific pores, permitting diffusion of substrates of up to an M_r of about 900 (Inouye, 1979; West and Page, 1984). By contrast, the inner cytoplasmic membrane does not permit passage of carbohydrates unless a protein(s) specific for each individual sugar is present. It is here that the energization of transport is effected. Other components of the cell wall (e.g., lipopolysaccharide, peptidoglycans) are thought not to influence the transport process.

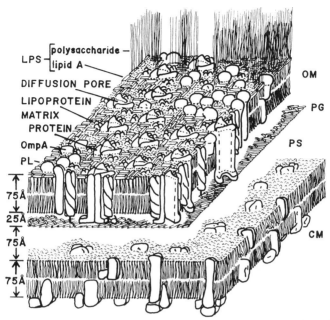

Figure 1. A schematic representation of the components in the cell envelope of *E. coli*. Abbreviations: PL, phospholipid; OM, outer membrane; PG, peptidoglycan; PS, periplasmic space; CM, cytoplasmic membrane. The porin proteins, e.g., OmpA, LamB, are located in the outer membrane (Sections 1.5, 5.3.1). The binding proteins involved in chemotaxis and transport (Section 5) are in the periplasmic space. Auxiliary proteins associated with binding protein-mediated transport are in the cytoplasmic membrane and on its cytoplasmic face (Section 5). The proton- or sodium-linked sugar transport proteins (Sections 3, 4) are in the cytoplasmic membrane. (From Di Renzio *et al.*, 1978.) Reproduced, with permission, from the *Annual Review of Biochemistry, Volume 47,* © 1978 by Annual Reviews Inc.

2. EXPERIMENTAL SYSTEMS FOR MEASURING TRANSPORT

The measurement of carbohydrate transport is an easy proposition with many species of free-living bacteria, because the cells are relatively tough and withstand washing and filtration under mild vacuum without losing accumulated nutrient. Obvious exceptions are obligate anaerobes, in which the maintenance of anaerobiosis presents special problems; halophiles, where high salt concentrations are required to prevent lysis; and organisms producing proteases, which may self-destruct. Any new procedure may be checked by demonstrating that intracellular components are not released; the easiest to measure are nucleotides with their absorbance at 260 nm or proteins by a specific assay (Schaffner and Weissman, 1973), or by measuring viability. If damage is found, the first possible causes are incorrect osmotic strength of the supporting medium, temperature shock during isolation, and too rigorous homogenization during resuspension. These may be overcome, but autolysis by indigenous proteases is a more difficult problem. The following procedures have proved satisfactory with *E. coli*.

2.1. Carbohydrate Transport into Intact Cells of E. coli

2.1.1. Growth and Harvesting of Bacteria

The bacteria are grown on defined minimal salts medium of pH 7.0–7.2 supplemented with the carbohydrate under investigation (10 mM monosaccharide or 5 mM disaccharide) and any auxotrophic requirements such as amino acids or vitamins. It is often useful to have glycerol as a supplementary carbon source (see Section 2.1.2). Complex media containing undefined components are best avoided because any covert carbohydrate would introduce errors of interpretation.

Importantly, after growth the residual carbohydrate must be removed from the suspension. Otherwise it will interfere with subsequent transport measurements. Removal is achieved by sedimenting the bacteria (6000 g for 7–10 min) and resuspending them in minimal salts medium to a dilution of 100-fold or more; the washing procedure is repeated three times. If this seems excessive, remember that the K_m of certain transport systems is less than 1 μM (see below). The washing is performed at room temperature or even at the growth temperature (30–37°C) for two reasons: continued metabolism of the sugar will help deplete its concentration; and exposure to temperatures of 0–4°C may release periplasmic proteins that are components of transport systems (Neu and Heppel, 1965). Finally, the bacteria are resuspended to a concentration of 1–1.4 mg dry mass/ml.

In our experience with *E. coli*, the minimal salts medium used for washing can be replaced by 150 mM KCl plus buffer, e.g., 5 mM 2[N-morpholino]ethanesulphonate (MES) or 5 mM N-2-hydroxyethyl piperazine-N-2-ethane sulfonate (Hepes). A pH of 6.5 rather than 7.0 has some advantages in the investigation of proton-linked transport systems. It was customary to include chloramphenicol during harvesting and washing of bacterial cultures in order to inhibit any further growth. However, growth is prevented anyway by the removal of the carbon source and auxotrophic requirements, and there is a danger that excessive concentrations of chloramphenicol may affect the permeability of the cell membrane.

2.1.2. The Importance of an Energy Source

Since transport consumes metabolic energy, it is vital that an energy source is available in order to achieve maximal rates. There are at least three possible energy sources: endogenous metabolites retained despite the washing procedure; the carbohydrate added as the substrate for transport; and the deliberate addition of a second metabolizable carbon compound, e.g., glycerol, lactate, or succinate. Since retention of endogenous metabolites is rather variable, depending for example on the *E. coli* strain and the stage of growth at harvesting, and since transport is often investigated with nonmetabolizable substrates (see below), we prefer the third option. Accordingly, our *E. coli* strains are generally grown on carbohydrate plus glycerol, and 3 min before each transport assay glycerol is added back to the cells to replete their energy stores; at this time and throughout the transport assay the suspension is bubbled with air. Measurements of respiration rate and end-product excretion confirmed that this procedure was necessary and successful. Furthermore, there was a separate facilitated diffusion transport system for glycerol (Heller *et al.*, 1980), which did not appear to compete with the pentose, hexose, or disaccharide transport systems.

2.1.3. Assay Procedure

A discontinuous procedure is mandatory with the great majority of bacterial transport assays. The suspension is exposed to radioisotope-labeled sugar at an appropriate concentration, samples are taken at measured time intervals, the bacteria are separated from the ambient medium, and the radioactivity accumulated in the cells is measured. Exceptions are the use of *o*-nitrophenyl-β-D-galactoside for assay of the lactose transporter (Rickenberg *et al.*, 1956; Kennedy, 1970), and *o*-nitrophenyl-β-D-glucoside for assay of the β-glucoside phosphotransferase system (Schaefler, 1967; Schaefler and Maas, 1967). A detailed protocol is as follows (Henderson *et al.*, 1977).

The final suspension (0.5 ml) of bacteria (A_{680} = 1.8–2.2, 1.2–1.5 mg dry mass/ml) is mixed with 5 μl of 1 M glycerol and bubbled with air for 3 min at 25°C, and then 12.5 μl of 2 mM radioactive sugar (final concentration 50 μM, 2.5–12.5 C_i/mole) added with mixing. Portions (0.1 ml) are withdrawn and discharged onto cellulose acetate or cellulose nitrate (0.45 μm pore size) filters after 15 sec, 1 min, 2 min, and 3 min. The timing and volumes of samples can be adjusted as desired. Each filtered sample is washed immediately and completely with 5–10 ml of medium (room-temperature). The damp-dry filter is transferred to a vial where the radioactivity retained in the filtered bacteria is measured by liquid scintillation counting.

Background measurements are made with identical scintillant-containing damp filters not exposed to radioisotopes; similarly, 8 nmoles of the radioisotope-labeled sugar with filter is counted to find its specific radioactivity, which is used to convert counts in each vial to nanomoles of sugar per milligram of bacteria.

2.1.4. Isolation of Transport from Metabolism Using Mutants or Structural Analogues of the Substrate

The immediate problem of using the growth sugar substrate for the assay described above is that it is quickly metabolized. The radioactivity accumulating inside the cells then represents many metabolites minus loss from efflux of radioactive end products such as CO_2, H_2O, lactate, acetate, or succinate (though immediate metabolism of the transported sugar has the advantage of prolonging the linear initial rate). The potential ambiguities can be avoided in two ways. The first is to use a mutant defective in the first enzyme of metabolism; the cells would then be grown on a "neutral" substrate, e.g., glycerol, plus a lower concentration of the transport substrate, which is usually needed as inducer. The second is to employ a structural analogue previously shown to be a substrate for transport but not for the first enzyme of metabolism.

These strategies have been invaluable in studies of carbohydrate transport in *E. coli*. Examples include the following: β-galactosidase-negative (*lacZ*) strains or methyl-β-D-thiogalactosidase (TMG) as analogue for lactose transport (Rickenberg *et al.*, 1956; Kennedy, 1970); galactokinase-negative (*galK*) strains or 6-deoxy-D-galactose/2-deoxy-D-galactose/D-talose as analogues for D-galactose transport (Horecker *et al.*, 1960; Wu, 1967; Rotman *et al.*, 1968; Wilson, 1974; Henderson and Giddens, 1977); L-arabinose isomerase-negative (*araA*) strains or 6-deoxy-D-galactose as analogue for L-arabinose transport (Novotny and Englesberg, 1966; Brown and Hogg, 1972); xylose isomerase-negative (*xylA*) strains or 6-deoxy-D-glucose as analogue for D-xylose transport (Davis, unpublished results); maltose phosphorylase (*malP*) mutants or 5-thio-maltose/methyl-α-maltoside as analogues for maltose transport (Ferenci, 1980); and methyl-β-D-galactoside as substrate analogue for the other galactose transport system, Mgl (Boos, 1969; Section 5.3.3).

2.1.5. Separation of Multiple Transport Systems for One Carbohydrate Using Mutants

In the case of galactose, arabinose, and xylose transport into *E. coli*, there are at least two biochemically and genetically distinguishable transport systems for each sugar (see Sections 3 and 5; Henderson *et al.*, 1984). It seems likely that for lactose there is only one, and for maltose only one (Hengge and Boos, 1983), but for other sugars that support growth the number of transport systems for each is not established. Therefore, measurements using the normal sugar and wild-type strains may reflect the combined activities of two (or more) transport systems. This problem may be recognized by nonhyperbolic steady-state kinetic patterns or by finding that more than one genetic locus is involved in total transport. In order to investigate unambiguously a single transport system, mutants must be made impaired in all but one. [Alternatively, only one of the activities may be retained in vesicles (see Section 2.3), where it can be investigated, but this still leaves the problem of investigating the other(s).] The isolation of such mutants may require ingenuity and luck, and a review of successful strategies is beyond the scope of this chapter. However, the advent of mutagenesis by Mud (Ap^{r}lac) and λplacMu phages (Casadaban and Cohen, 1979; Bremer *et al.*, 1984) has made the process much easier (see reviews by Silhavy *et al.*, 1983; Jones-Mortimer and Henderson, 1986). In our experience, the required mutants can be obtained in a few weeks, where 10 years ago, the available genetic technology meant up to a year's work.

Sometimes there exists a structural analogue that is a substrate for only one of the transport systems for a sugar. Proving that this is true requires the mutants described above, but, once characterized, such an analogue is invaluable for the investigation of a single transport system even in wild-type strains. One example is the use of methyl-β-D-galactoside to study the high-affinity galactose transport system Mgl (Table I; Section 5.3.3).

2.2. Variations on the Transport Theme—Equilibrium Exchange, Efflux, and Overshoot

Once the transport process was isolated from metabolism, the initial rate of carbohydrate influx was found to decelerate until a plateau level of accumulation was achieved. In the cases of lactose, galactose, and arabinose transport, this seemed to represent a steady state where uptake (+ energization) and efflux rates were balanced (e.g., see Rickenberg *et al.*, 1956; Horecker *et al.*, 1960; Henderson and Kornberg, 1975). The efflux process could be investigated by diluting the cells and/or terminating the energy supply, and filtering samples at timed intervals just as for influx (e.g., Winkler and Wilson, 1966; Vorisek and Kepes, 1972; Wilson, 1976). Alternatively, a high concentration of unlabeled sugar could be added outside the cells, and the rate at which it discharged the internal isotope-labeled material investigated (Horecker *et al.*, 1960; Henderson and Kornberg, 1975). [The corollary of this experiment was to use unlabeled sugar for the initial influx, and then add a negligible concentration of high-specific-activity sugar and monitor its appearance inside the cells as a measure of the steady-state influx rate (Vorisek and Kepes, 1972).]

An interesting alternative method is the so-called "overshoot" assay, originally devised by Widdas (1952) for any exchange-diffusion process, and exploited in bacterial transport investigations by Wilson and co-workers (e.g., Winkler and Wilson, 1966; Wong and Wilson, 1970). Its principal advantage is the independence of energization, and hence its suitability for reconstitution assays (Newman and Wilson, 1980). The theory was recently reviewed by Heinz and Weinstein (1984). The procedure is to "load" deenergized bacteria with a high concentration of internal unlabeled sugar by incubating a concentrated suspension with inhibitors of energization, e.g., uncoupling agents, and the sugar. For each assay a sample of this is diluted (at least 50-fold) into medium containing a negligible concentration of high-specific-activity labeled sugar. If the sugar carrier catalyzes exchange diffusion, the label then rapidly accumulates inside the cells, followed by a slower discharge, a process that can be followed by filtration/washing as described above.

Such varied experimental strategies serve to define steady-state kinetic parameters—apparent K_m and V_{max}—for inward, outward, energized, and nonenergized transport, as well as substrate exchange. Any molecular model of the process must account satisfactorily for the rates of all these measured activities and any binding or rate constants that can be established (e.g., see West, 1980, 1983; Lombardi, 1981; Lancaster, 1982; Kaback, 1983; Wright et al., 1983).

2.3. Carbohydrate Transport in Subcellular Vesicles

Kaback (1971) devised a procedure for making right-side-out subcellular vesicles capable of energized nutrient transport. This achievement was important because use of vesicles conferred the following advantages. The absence of cytoplasmic enzymes isolated transport from metabolism without the need to make mutants. Potentially competing transport processes depending on periplasmic binding proteins (see below) or cytoplasmic proteins (i.e., phosphotransferase systems) were absent. Removal of the cell wall rendered the energization process susceptible to ionophores, which could be used to investigate the relationship of transport to the transmembrane electrochemical gradient of protons in a controlled way. Endogenous energy stores were negligible so that energization was readily controlled by addition of appropriate respiratory substrates, inhibitors, or uncoupling agents (Kaback, 1972, 1974). The vesicles also made an ideal starting material for identifying, purifying, and reconstituting transport proteins. There may be some contamination by inverted vesicles (Futai, 1978), but the level was not significant for most experimental purposes.

2.3.1. Preparation of Subcellular Vesicles

Spheroplasts are made from intact cells by the method of Witholt et al. (1976; Witholt and Boekhout, 1978). They are subjected to controlled osmotic lysis and the vesicles isolated by differential centrifugation as described by Kaback (1971). [I will provide a detailed protocol for E. coli on request. The procedure probably needs adaptations when used for other species of bacteria (Konings and Boonstra, 1977; Futai, 1978).] The vesicles can be stored at $-80°C$, provided that the freezing is performed rapidly by immersing the vesicle container in liquid nitrogen.

2.3.2. Measurement of Transport into Right-Side-Out Subcellular Vesicles

The procedure is different from that in intact cells, and the use of the recommended resuspension and washing media is critical. The following protocol is based entirely on that evolved by Kaback (1971). Frozen membrane vesicles of known protein concentration in 100 mM potassium phosphate, pH 6.5, are thawed at 25–30°C using a water bath and virtually continuous vibration with a vortex mixer. The thawed stock suspension is kept on ice. A sample containing 0.5–1.0 mg protein for each transport assay is diluted with an equal volume of water, 50 mM potassium phosphate if required, and 5 μl of 1 M $MgSO_4$ to give a final volume of 0.5 ml 50 mM KP_i, 10 mM $MgSO_4$. The mixture is incubated at 25°C and bubbled with oxygen. After 2.5 min 10 μl of 10 mM phenazine methosulfate is added and after 2.8 min, 10 μl of 1 M potassium ascorbate, pH 6.6. These constitute the optimum substrate for respiration (Kaback, 1971, 1972; Table II). Transport is initiated by the addition of 10 μl (or more) of 2 mM radioactive sugar at 3 min. Timed samples of 0.1 or 0.2 ml are withdrawn, filtered (0.45-μm cellulose acetate or nitrate), washed (4 ml of 0.1 M LiCl), and the radioactivity in the retained vesicles determined as described above.

Table IIa. Respiratory Substrates That Energize Galactose Transport into Subcellular Vesicles of E. coli[a]

Substrate	Initial rate (nmoles/mg per min)	Extent (nmoles/mg per 2 min)
Ascorbate	1.4	1.4
Ascorbate/TMPD	2.6	2.7
Ascorbate/PMS	7.0	6.3
Formate	6.4	5.2
Formate + UQ-1	4.7	3.4
NADH	2.3	2.4
NADH + UQ-1	4.8	3.8
D-Lactate	2.7	2.8
D,L-α-Glycerophosphate	2.0	1.9
Succinate	1.6	1.5

[a]Vesicles were prepared from cells of *E. coli* S183-27T, grown on glycerol, nitrate, and selenate under relatively anaerobic conditions (Horne and Henderson, 1983). All substrates (20 mM, except NADH, which was 5 mM final concentration) were added 30 sec before the labeled sugar. Where added, ubiquinone-1 (UQ-1), phenazine methosulfate (PMS), and tetramethyl-*p*-phenylenediamine (TMPD) were present at concentrations of 0.08, 1.00, and 10 mM, respectively. And air was bubbled through the suspensions. All values shown are means of duplicate measurements (experiments of Horne, 1980; see also Stroobant and Kaback, 1975; Ramos *et al.*, 1976).

Table IIb. Electron Acceptors That Support Energization of Galactose Transport into Subcellular Vesicles of E. coli[a]

Substrate	Initial rate (nmoles/mg per min)	Extent (nmoles/mg per 2 min)
None + NO_3^- + air	1.8	0.9
Ascorbate/PMS + NO_3^- + air	13.9	7.4
Ascorbate/PMS + NO_3^- + argon	8.4	3.5
Ascorbate/PMS + argon	5.7	2.9
Formate + NO_3^- + air	17.0	10.5
Formate + NO_3^- + argon	18.2	8.2
Formate + air	13.5	6.8
Formate + argon	10.6	5.7

[a]Conditions were as for Table IIa, and air or argon was bubbled through vesicle suspensions as indicated. Either oxygen or NO_3^- was an efficient electron acceptor for the energization of transport (Horne, 1980; Boonstra et al., 1975; Konings and Boonstra, 1977).

2.3.3. Inverted Vesicles

When *E. coli* cells are fragmented by explosive decompression in a French press, inverted vesicles of the inner membrane are formed. These can be separated from cytoplasmic contents and cell debris by differential centrifugation (Herzberg and Hinkle, 1974; Futai, 1978). Such vesicles are convenient for measuring *uptake* of nutrients that are normally *excreted* by the intact cell (e.g., see Rosen and McLees, 1974; Silver, 1978), and so their value for measuring sugar transport is limited. However, they have two important uses.

The first is for investigating the symmetry of the transport process and/or proteins in cytoplasmic membranes (Lancaster and Hinkle, 1977; Futai, 1978). For example, by comparing the pattern of labeling or protease susceptibility of a particular membrane protein in right-side-out vesicles with the pattern in inverted vesicles, the topology of labeled residues can be deduced (Goldkorn et al., 1983; Carrasco et al., 1984a).

Second, the French pressure treatment is the most convenient method of preparing relatively large amounts of inner membrane proteins, because it uses very concentrated suspensions of bacteria and fewer centrifugation steps compared with the dilution and centrifugations required for preparation of right-side-out vesicles.

2.4. Flow Dialysis

Flow dialysis was devised to measure the binding of ligands to proteins under conditions where sampling of the ambient medium would not require filtration (Colowick and Womack, 1969). This permitted virtually continuous monitoring of binding and precise measurements of time-dependence. The procedure was ingeniously adapted by Kaback and co-workers to provide a near-continuous measurement of transport activity and its energization in right-side-out vesicles (Ramos et al., 1976; Ramos and Kaback, 1977a,b). The only disadvantage was that it required relatively large amounts of biological material and radioisotope-labeled compounds.

3. PROTON-LINKED SUGAR TRANSPORT SYSTEMS

3.1. The Chemiosmotic Theory

Among the earliest observations on lactose transport in E. coli, the following were reminiscent of the process of oxidative phosphorylation studied in mitochondria. Energization of transport was impaired by agents that uncoupled respiration from ATP synthesis, e.g., 2,4-dinitrophenol (Pavlasova and Harold, 1969), and by conditions that inhibited respiration (Rickenberg et al., 1956; Kennedy, 1970; Schairer and Haddock, 1972). In 1961, Mitchell proposed that the hitherto mysterious "high-energy intermediate" that coupled mitochondrial respiration to ATP synthesis was a transmembrane electrochemical gradient of protons (Mitchell, 1961). He realized (Mitchell, 1963, 1966) that such a "protonmotive force" could similarly link bacterial respiration (or ATP hydrolysis) to lactose transport, *provided that the lactose transport system catalyzed an obligatory lactose/H^+ symport* (or the equivalent lactose/OH^- antiport). West (1970) obtained direct experimental evidence that the lactose carrier indeed catalyzed obligatory translocation of protons with sugar, and this has been substantially confirmed (West and Mitchell, 1972, 1973; Foster et al., 1982). This discovery led to the observation that many sugar, amino acid, and other nutrient transport systems were energized by the protonmotive force, though, as we shall see in Section 5, it was not the only mechanism. Despite much initial opposition, the chemiosmotic theory eventually received wide-ranging experimental support (reviewed by Mitchell, 1976), which led to the award of a Nobel Prize to its progenitor in 1978.

The proton-coupled transport mechanism is illustrated in Fig. 2. Respiration (Haddock and Jones, 1977; Ingledew and Poole, 1984) or ATP hydrolysis (Kagawa, 1978, 1984) first generates the proton gradient by primary transport processes (Section 1.3). The tendency of the proton to escape back down the gradient is retarded by the impermeability of the lipid membrane (Mitchell, 1961, 1963), but specific proton-conducting pathways—e.g., the ATP synthase, transport systems, flagellar motor—allow proton reequilibration while utilizing some of the energy released for (e.g.) ATP synthesis, nutrient uptake against the concentration gradient, or locomotion. The addition of artificial protonophores,

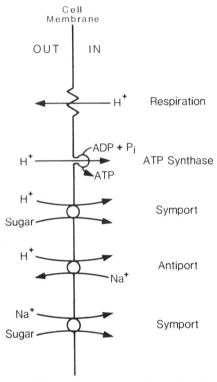

Figure 2. Scheme of proton-coupled transport systems in the cytoplasmic membrane of *E. coli*.

the classical uncoupling agents, would short-circuit these processes by rendering the lipid membrane permeable to protons, and their ability to so act has become a criterion for the operation of proton-linked mechanisms (reviewed by Harold, 1972, 1977; Hamilton, 1975).

3.2. Experimental Evidence for Sugar/H^+ Symport

3.2.1. Measurement of Sugar-Promoted pH Changes

Under anaerobic conditions, ATP-depleted cells of *E. coli* should be unable to generate a protonmotive force. Addition of a substrate of an induced sugar transport system outside the cells would lead only to its movement *down* the concentration gradient until internal and external concentrations were balanced. If the mechanism of the transport process obligatorily linked sugar influx to inward H^+ translocation, then an alkaline pH change would be observed outside the cells. In order to detect this, the medium has to be of low buffering capacity and relatively high concentrations of sugar and bacteria must be used to maximize the extent of sugar/H^+ movement. Also, respiration and ATP hydrolysis must be prevented. West achieved the required experimental conditions and showed that addition of substrates of the lactose transporter promoted an alkaline pH change in the medium (West, 1970; West and Mitchell, 1972, 1973). The appearance of the pH change was susceptible to uncoupling agents, supporting the idea that it represented a transmembrane proton gradient rather than the same shift of pH inside and outside the bacteria. Furthermore, it happened with substrate analogues or mutants in which transport occurred but metabolism of the sugar was prevented.

Subsequently, such sugar-promoted pH changes were observed using right-side-out subcellular vesicles containing amplified *lacY* activity (Patel *et al.*, 1982). Of particular significance was the recent observation that they also occurred in liposomes into which a *purified* lactose protein had been reconstituted (Foster *et al.*, 1982). Thus, the ability to catalyze sugar/H^+ symport resided in a single protein (see also Viitanen *et al.*, 1984).

Subsequently, galactose (Henderson, 1974), arabinose (Henderson, 1974), xylose (Lam *et al.*, 1980), and L-fucose (Bradley *et al.*, submitted) were shown to promote alkaline pH changes with intact cells under the same experimental conditions. Galactose and arabinose were also effective in subcellular vesicles (Daruwalla *et al.*, 1981; Horne and Henderson, 1983), but the same experiment was not successful with xylose and L-fucose, probably because the vesicles were from wild-type strains without amplified transport activity. In none of these cases has sufficient purification of the transport protein been achieved to attempt the sugar/H^+ measurements in reconstituted proteoliposomes.

3.2.2. Susceptibility of Transport to Agents That Modify the Protonmotive Force

3.2.2a. Inhibitors of Respiration. Unlike the mitochondrial respiratory chain, respiration in *E. coli* is not impaired by rotenone or antimycin; instead, cyanide or azide, which inhibit the terminal oxidases (Haddock and Jones, 1977; Ingledew and Poole, 1984), must be used. The tendency of their solutions to become alkaline must be countered by adequate buffer. Lactose transport, or that of its analogue TMG, into intact cells was partially inhibited by cyanide (about 50%; Schairer and Haddock, 1972). By contrast, energization of lactose transport into subcellular vesicles was completely dependent on the presence of respiratory substrate and oxygen, and could be abolished by cyanide (Kaback, 1972; Schuldiner and Kaback, 1975). Similarly, the energized transport of galactose, arabinose, xylose, or L-fucose into subcellular vesicles required a functional respiratory chain (Kaback, 1972; Daruwalla *et al.*, 1981; Horne and Henderson, 1983; Bradley *et al.*, 1986).

Kaback (1977) and co-workers extensively investigated the ability of different respiratory substrates and alternative electron acceptors such as nitrate or fumarate to support transport activity into vesicles. Similar results for galactose transport are shown in Table II, in which it was clear that a rank order of substrate efficiency existed. This was partly related to their efficiency in generating the protonmotive force, as measured by the transmembrane distribution of, e.g., radioactive triphenylmethylphosphonium ion, weak acids, or weak bases, and was probably also a consequence of energy used to translocate the respiratory substrate itself (Stroobant and Kaback, 1975; Ramos *et al.*, 1976; Konings and Boonstra, 1977; Kaback, 1977).

Why, then, was proton-linked secondary transport into intact cells relatively resistant to inhibitors of respiration (Schairer and Haddock, 1972)? The answer lay in considering the potential role of the H^+-ATPase.

3.2.2b. Inhibition of the H^+-ATPase.   The normally considered role of the H^+-ATPase in the framework of the chemiosmotic theory is to absorb H^+, previously translocated across the membrane by respiration, and synthesize ATP (Fig. 2; Mitchell, 1961; Harold, 1972; reviewed by Kagawa, 1978, 1984). However, it was well recognized to be reversible (Mitchell, 1961; Kagawa, 1978, 1984) so that ATP hydrolysis inside the cell could be used to generate a transmembrane protonmotive force. In the absence of respiration (prevented by anaerobiosis or cyanide or a mutation), the ATPase activity, using ATP generated by intracellular glycolysis, was shown to energize transport (Schairer and Haddock, 1972). In an ATPase-negative proton-leaky mutant, β-galactoside transport was impaired, but could be restored by dicyclohexylcarbodiimide, which had the effect of "repairing" the leaks (Rosen, 1973a,b). Since subcellular vesicles were depleted of, and impermeable to, ATP, respiratory activity was a *sine qua non* for their energization of transport. However, the ability of the H^+-ATPase to support transport into vesicles was elegantly demonstrated in the special case where PEP could enter to generate ATP inside via pyruvate kinase activity (Hugenholtz *et al.*, 1981).

In these and other ways (reviewed by Harold, 1977; Rosen and Kashket, 1978; West, 1980), evidence accumulated that energization of transport of β-galactosides, and of other sugars, was linked to the respiratory chain and H^+-ATPase via a common high-energy intermediate, the protonmotive force.

3.2.2c. Inhibition by Uncoupling Agents.   As already mentioned, the classical uncoupling agents inhibited transport of all sugars into wild-type *E. coli*, except those sugars taken up by a phosphotransferase mechanism (Chapter 10). However, as Berger showed in his investigations of amino acid transport, this observation in itself did not prove that the transport activity was energized by the protonmotive force (Berger, 1973; Berger and Heppel, 1974). For example, ATPase-negative mutants treated with cyanide could not generate a protonmotive force; yet if their intracellular ATP concentration was maintained by a suitable glycolytic substrate, then energization of transport of *some* amino acids (or sugars), but *not* others, was maintained, i.e., energization was still effective in the total absence of the protonmotive force. Provided the ATP concentration remained high, even uncoupling agents did not affect such transport. Those systems retaining activity all appeared to involve a periplasmic binding protein, i.e., they were shock-sensitive, rather than "shock-insensitive" as classified by Berger and Heppel (1974; reviewed by Silhavy et al., 1978; Rosen and Kashket, 1978). These uncoupler-insensitive systems were dependent on an intracellular phosphorylated compound for energization, as explained in Section 5. The apparent sensitivity to uncouplers in wild-type cells was due to loss of the phosphorylated compound as a result of accelerated ATPase activity and did not mean that the protonmotive force energized transport.

The use of uncoupling agents in subcellular vesicles involved no such ambiguity. If transport into vesicles was susceptible to uncouplers, then it was linked to the protonmotive force.

3.2.2d. Energization of Transport by Artificial Generation of a Transmembrane Protonmotive Force. One experiment that convincingly affirmed the chemiosmotic theory was the observation that ATP synthesis could be supported by an artificially generated pH gradient in chloroplasts (Jagendorf and Uribe, 1966); similarly, it was shown that artificially generated membrane potentials supported ATP synthesis in mitochondria (Cockrell et al., 1967; Glynn, 1967). In both organelles, electron transfer had been eliminated. Accordingly, it was predicted that lactose transport into *E. coli* should be driven by a pH gradient and/or membrane potential generated in the absence of respiration. Rigorously energy-depleted bacteria were required in which respiration and ATP synthesis were eliminated by use of appropriate inhibitors, mutants, and a substrate analogue. Also, concentrated bacterial suspensions were necessary to detect the expectedly low levels of transport.

Under these conditions a pH gradient, set up by acid added outside, drove inward sugar (β-galactoside) transport (Flagg and Wilson, 1977); an electrical gradient, set up by efflux of positive ions or influx of negative ions, similarly drove inward substrate transport (Flagg and Wilson, 1977). Several types of control eliminated the participation of ATP or other routes of ion transport, and confirmed the participation of the protonmotive force, e.g., sensitivity to uncoupling agents (Flagg and Wilson, 1977).

Kaback and co-workers have elegantly extended these observations by using right-side-out subcellular vesicles, taking advantage of their energy-depleted state and susceptibility to ionophores. Either or both components of the protonmotive force, ΔpH and $\Delta\psi$, contributed to the energization of transport, and their quantitative relationships were established (e.g., see Schuldiner and Kaback, 1975; Ramos and Kaback, 1977a,b; Kaczorowski and Kaback, 1979). The purified protein reconstituted into liposomes displayed the same interrelationship of transport with ΔpH and $\Delta\psi$ (Garcia et al., 1983).

A most interesting result was that a similar membrane potential, positive outside, drove substrate transport *into* everted membrane vesicles made with a French Press (Lancaster and Hinkle, 1977; Teather et al., 1977). This implied that the carrier was functionally symmetrical—the net direction of transport was dictated by the prevailing gradients of sugar, pH, and electrical potential, not by the topology of the protein(s) in the membrane.

Such experiments have not been done with galactose, arabinose, xylose, or L-fucose transport, partly because of the existence of a second transport system for at least the first three of these (see Section 3.3). However, some preliminary data indicated that galactose transport into intact cells could be driven by an artificial ΔpH, and that its transport into vesicles was driven by an artificial ΔpH or $\Delta\psi$ (Horne, 1980).

3.3. Examples of Sugar/H⁺ Symport in E. coli

3.3.1. Lactose

From all the above it will be apparent that lactose transport is the best characterized of the sugar/H$^+$ symport systems in *E. coli*. It is encoded by a single structural gene *lacY* located at 8 min on the *E. coli* linkage map (Bachmann, 1983), the DNA of which has been sequenced (Büchel et al., 1980; Ehring et al., 1980). The gene has been cloned, and amplified levels of expression achieved (Teather et al., 1978, 1980). This important step and the development of a reconstitution assay (Newman and Wilson, 1980) led to the purification of relatively large amounts of the protein to apparent homogeneity (Newman et al., 1981; Foster et al., 1982; Wright et al., 1983). It has an apparent M_r of about 30,000 on SDS–PAGE (Jones and Kennedy, 1969; Kaback, 1983) and a true M_r of 46,504 (Büchel et al., 1980; König and Sandermann, 1982). There is some evidence that it may function as a dimer (Goldkorn et al., 1984).

The range of synthetic substrates for the lactose transporter is quite broad because, although the β-D-galactoside moiety is critical for substrate activity, substitutions can be made for the glucose (examples are -OH, -SCH$_3$, S-galactose, S-phenyl, -O-phenyl, -O-nitrophenyl) without abolition of binding/translocation (Rickenberg et al., 1956; Kennedy, 1970). This enabled the discovery of a specific photoactivatable labeling reagent, p-nitrophenyl-α-D-galactoside, which was invaluable for purification (Kaczorowski and Kaback, 1979; Kaczorowski et al., 1980; Kaback, 1983).

Some of these compounds, e.g., isopropyl-β-D-thiogalactoside or methyl-β-D-thiogalactoside, are gratuitous inducers of the lac operon (Beckwith and Zipser, 1970); the physiological inducer is allolactose (Jobe and Bourgeois, 1972). Since galactose is a substrate for the lactose transporter, mutations in the galactose transport systems can be circumvented by induction, or mutation to constitutivity, of the lactose transport system.

Analogues linking the galactose moiety to its β-D-substituent by an oxygen atom are generally substrates for β-galactosidase, whereas those linked by a sulfur atom are not (Sinnott and Viratelle, 1973). "X-gal" (5-bromo-4-chloro-3-indolyl-β-D-galactoside), a colorless compound transported by LacY and hydrolyzed by β-galactosidase to a derivative that turns blue, is of great value in genetic manipulation and sequencing reactions (Sanger et al., 1980; Maniatis et al., 1982; Silhavy et al., 1984).

The apparent K_m for energized lactose transport was in the range 240–900 μM (Kennedy, 1970; Wright et al., 1981, 1983). The reaction involved two substrates, and so the apparent K_m for sugar was modified by changes in [H$^+$] (Kaczorowski and Kaback, 1979; Page and West, 1981, 1982, 1984). There may be an obligatory order of addition of sugar and proton to the enzyme during the transport process (reviewed by Kaback, 1983; West, 1983; Garcia et al., 1983); it would also be interesting to know which is the rate-limiting step in the postulated cyclic processes effecting translocation (Garcia et al., 1983). Determination of the number and nature of such steps is an important goal of steady-state and non-steady-state investigation of any such transport system (e.g., see West, 1980, 1983; Lombardi, 1981; Kaback, 1983; Page and West, 1984). In vesicles the optimum pH for energized transport was about 5.5 (Ramos and Kaback, 1977b), the pH at which the total protonmotive force was maximal (Ramos et al., 1976).

3.3.2. Galactose

Evidence for galactose/H$^+$ symport was first obtained by showing that galactose (or the substrate analogue D-fucose; see below) elicited an alkaline pH change when added to energy-depleted suspensions of appropriately induced E. coli cells (Henderson, 1974; Henderson et al., 1975). Mutations in the galP gene located at 64 min on the E. coli chromosome (Riordan and Kornberg, 1977; Bachmann, 1983) abolished this galactose/H$^+$ symport activity (Henderson et al., 1977). Furthermore, no sugar/H$^+$ symport occurred with methyl-β-D-galactoside, the substrate for the Mgl galactose transport system, when its activity was shown to be present by measuring transport of radioactive methyl-β-D-galactoside (Henderson et al., 1977). By making subcellular vesicles from a galP constitutive strain, the additional criteria for energization of GalP by the proton-motive force—respiration dependence, uncoupler sensitivity, ionophore effects, pH dependence—could be tested and confirmed (Horne and Henderson, 1983; see also Kaback, 1972). Furthermore, such vesicles exhibited alkaline pH changes evoked only by sugar substrates of the GalP system (Horne and Henderson, 1983). These and subsequent studies demonstrated that galactose/H^+ symport in E. coli was effected by the product of the gene(s) galP located at about 64 min on the linkage map (Riordan and Kornberg, 1977; Henderson et al., 1977; Bachmann, 1983). The available biochemical evidence indicated that one protein was involved, but the possibility of more was not eliminated (Macpherson et al., 1983). The gene is in the process of being cloned (D. C. M. Moore, unpublished work). Amplified expression (about fivefold) has been found in constitutive strains and a mutant (Henderson and Giddens, 1977; Macpherson et al., 1983).

Some chemical analogues of galactose were substrates, including glucose, 6-deoxygalactose, 6-deoxyglucose, 2-deoxygalactose, 2-deoxyglucose, talose, mannose, and halogen substituents of galactose or glucose in the 6 or 4 positions (Rotman et al., 1968; Wilson, 1974; Henderson et al., 1977). Talose and 2-deoxygalactose were relatively specific for the GalP transport system, and so their radioactive derivatives were useful for its assay in wild-type strains (Henderson and Giddens, 1977). Arabinose and xylose were not substrates or inhibitors, indicative of the importance of C-6 for binding of sugar to GalP, and substitution of -OCH_3 for -OH at the 1 position also abolished binding; this latter observation enabled the use of radioactive methyl-β-D-galactoside to assay MglP activity in wild-type strains (Rotman et al., 1968; Boos, 1969).

Intracellular unmodified galactose was the physiological inducer of the GalP system, expression of which was probably regulated by the product of the galR gene (Wu, 1967; Buttin, 1968; Wu et al., 1969; Boos, 1969; Von Wilcken-Bergmann and Müller-Hill, 1982). 6-Deoxy-D-galactose (D-fucose) and 6-fluoro-6-deoxy-D-galactose were gratuitous inducers (Rotman et al., 1968; Henderson et al., 1977; Bradley and Henderson, unpublished results).

The apparent K_m for energized galactose transport into intact cells via GalP (measured in mgl mutants) was in the range 240–450 μM (Wilson, 1974; Henderson and Giddens, 1977). In vesicles the value was 50–70 μM (Horne, 1980). Its dependence on [H^+] was not systematically investigated, though the pH optimum for galactose transport was about 5.5 (Horne and Henderson, 1983).

3.3.3. Arabinose

Evidence for arabinose/H^+ symport was first obtained by showing that arabinose (or its substrate analogue, D-fucose) elicited an alkaline pH change when added to energy-depleted suspensions of appropriately induced E. coli cells

(Henderson, 1974; Henderson and Skinner, 1974; Henderson et al., 1975). Mutations in the araE gene abolished the sugar/H^+ symport activity and no sugar/H^+ symport could be detected in AraF$^+$ AraE$^-$ strains (Daruwalla, 1979; Daruwalla et al., 1981). Energized arabinose transport into subcellular vesicles derived from a strain with high AraE activity was respiration dependent, uncoupler sensitive, and modified by ionophores (Daruwalla et al., 1981; see also Kaback, 1972). Furthermore, only substrates of AraE promoted alkaline pH changes with such vesicles (Daruwalla et al., 1981). These and subsequent studies (Maiden and Moore, unpublished data) demonstrated that arabinose/H^+ symport in E. coli was effected by the product of the araE gene, which is located at about 61 min on the linkage map (Heffernan et al., 1976; Kolodrubetz and Schleif, 1981a; Macpherson et al., 1981; Bachmann, 1983). The gene has been cloned and is being sequenced. Both genetic (Kolodrubetz and Schleif, 1981a) and biochemical (Macpherson et al., 1981) evidence indicated that only one protein was involved. Amplified expression (about fivefold) occurred in araA (isomerase-negative) mutants (Katz and Englesberg, 1971; Beverin et al., 1971; Daruwalla et al., 1981).

The only alternative substrate for the AraE transport system appeared to be D-fucose (Novotny and Englesberg, 1966; Brown and Hogg, 1972; Daruwalla et al., 1981), which turned out to be an anti-inducer in wild-type strains (Englesberg et al., 1965); resistant mutants expressed the ara genes constitutively, or were hyperinducible by D-fucose (Beverin et al., 1971). The physiological inducer was internal free arabinose, which interacted with the araC gene product (1 min) to effect postive regulation of the araE (61 min), araF,G (45 min), and araB,A,D (1 min) operons (Englesberg et al., 1965; Kolodrubetz and Schleif, 1981b; Kosiba and Schleif, 1982; Stoner and Schleif, 1983; Hendrickson and Schleif, 1984).

The apparent K_m for energized arabinose transport into intact cells via AraE (measured in araF mutants) was in the range 100–170 μM (reviewed by Daruwalla et al., 1981). In vesicles the value was 317 μM (Daruwalla et al., 1981). The dependence of K_m on [H^+] was not systematically studied, though the pH optimum was about 5.5.

3.3.4. Xylose

It should now suffice to say that xylose/H^+ symport activity in E. coli was established by the usual experiments with intact cells and right-side-out vesicles (Lam et al., 1980). The gene(s) for xylose transport, xylT, had been tentatively located (David and Wiesmeyer, 1970a), but, unlike those for galactose or arabinose transport, had not been resolved into two discrete systems. Recent results (Davis et al., 1984; Davis, unpublished observations) showed that xylose/H^+ symport was associated with the gene xylE located at 91.4 min whereas a separate transport system that did not involve protons was coded for by gene(s) *xylF* near 80 min; this is probably identical with the *xylT* locus originally identified in *E. coli* by David and Wiesmeyer (1970a) and later in the closely related *Salmonella typhimurium* by Shamanna and Sanderson (1979a,b).

A single gene coding for xylose/H^+ symport has been cloned and the sequence of its DNA and reading frame established (Davis, unpublished results). Amplification of its expression has not been achieved. Since induction occurred in *xylA* mutants, *xylE* was presumably induced by internal unmodified xylose. 6-Deoxyglucose was found to be an alternative substrate for transport but it was not an inducer.

The apparent K_m for energized xylose transport into intact cells via XylE was in the range 53–170 μM (Davis, unpublished results). In vesicles the value was about 70 μM (Davis, unpublished results). Its pH dependence has not been investigated.

3.3.5. L-Fucose

Once again, the usual experimental strategies with intact cells and vesicles showed that an L-fucose/H^+ symport system operated in *E. coli* (Bradley *et al.*, 1986). Its gene location was not established, but it may be near the *fuc* operon at 60 min (Bachmann, 1983). L-Galactose and D-arabinose were alternative substrates but not inducers, There is no evidence to support the existence of a second type of transport system for L-fucose in *E. coli*.

4. CATION-LINKED MELIBIOSE TRANSPORT

Melibiose is the disaccharide 6-*O*-α-D-galactopyranosyl-D-glucose. Investigations of its transport into *E. coli* provide an excellent illustration of how substrate analogues and mutants first enabled identification of a separate transport system, the energization mechanism of which could then be established, followed by cloning and sequencing of the genes involved. It transpired that Na^+, rather than H^+, was the symport counterion (Stock and Roseman, 1971; Tsuchiya *et al.*, 1977), and melibiose remains the only sugar for which this is apparently the case in *E. coli,* although some amino acids use Na^+ as the counterion (Frank and Hopkins, 1969; Tsuchiya *et al.*, 1977; Anraku, 1978; Stalmach *et al.*, 1983; Cairney *et al.*, 1984; Mogi and Anraku, 1984). The success of a reconstitution procedure, combined with genetic manipulation, seems to have identified the melibiose transport protein (Tsuchiya *et al.*, 1982; Hanatani *et al.*, 1984), though to my knowledge its large-scale purification has not been achieved. A historical approach nicely illustrates the ways in which available biochemical and genetic technology facilitated or limited progress.

4.1. Characterization of a Separate Melibiose/Na⁺ Symport System Using Substrate Analogues

Prestige and Pardee (1965) found that galactinol (a *meso*-inositol-α-galactoside) induced a transport system for the artificial substrate methyl-1-thio-β-D-galactoside (TMG) without concomitant induction of β-galactosidase (i.e., the *lac* operon) in *E. coli* B grown at 37°C. A galactinol- or melibiose-induced "TMG II" transport system was similarly found in *E. coli* K12, provided that growth was at 25°C. The inducer and substrate specificity clearly distinguished the TMG II transport system from TMG I—the lactose permease. Furthermore, the TMG II system was active in *lacY* mutants (Prestige and Pardee, 1965; Rotman *et al.*, 1968). The identity of this separate transport system for melibiose in *E. coli* was further supported by observations of Rotman (1959).

S. typhimurium is closely related to *E. coli* and grew on melibiose at the usual growth temperature of 37°C but lacked the LacY transport system. It was therefore a convenient organism for investigation of melibiose transport. Its melibiose-inducible $[^{14}C]$-TMG uptake was found to be stimulated by Na^+ (and Li^+) but not by K^+, Cs^+, NH_4^+, Mg^{2+}, Mn^{2+}, Ca^{2+}, or Cl^- (Stock and Roseman, 1971). The complementary sugar-dependent Na^+ uptake was more difficult to demonstrate because of the existence of a Na^+-efflux system, but reasonably convincing results were obtained (Stock and Roseman, 1971). The Na^+ dependence of TMG transport was also demonstrated in subcellular vesicles from *S. typhimurium* and *E. coli* (Tokuda and Kaback, 1977; Cohn and Kaback, 1980).

Subsequently, Tsuchiya, Wilson, and colleagues confirmed the Na^+ requirement for transport of melibiose, TMG, and other substrates into *E. coli* (Tsuchiya *et al.*, 1977). Li^+ was an alternative counterion, but it inhibited growth (Tsuchiya *et al.*, 1978; Lopilato *et al.*, 1978). Unexpectedly, the cation requirement varied with the sugar substrate used, the important physiological conclusion being that melibiose normally entered with Na^+, but that H^+ could substitute for Na^+ when the latter was omitted from the incubation medium (Tsuchiya and Wilson, 1978; Tsuchiya *et al.*, 1980; see also Tsuchiya *et al.*, 1978). The variation in cation selectivity for different sugar substrate derivatives was an important observation that may eventually illuminate subtle conformational changes of the sugar/cation symporter (Niiya *et al.*, 1982).

4.2. Genetics of Melibiose Transport in E. coli

The region coding for melibiose transport, *melB*, and α-galactosidase activity, *melA*, was located at 93 min on the *E. coli* linkage map (Schmitt, 1968; Bachmann, 1983). It has been cloned into plasmid vectors and shown to have the following gene order with probably one more gene *melC* downstream (Tsuchiya *et al.*, 1982; Hanatani *et al.*, 1984; Yazyu *et al.*, 1984):

promoter	*melA*	*melB*	*melC*
	α-galactosidase	transport	

This is similar to the organization of the *lac* operon:

promoter	*lacZ*	*lacY*	*lacA*
	β-galactosidase	transport	

It was therefore possible that the two operons evolved from a common ancestral gene (Yazyu *et al.*, 1984), but the nucleotide sequence of the *melB* gene was not homologous to that of *lacY* (Büchel *et al.*, 1980; Yazyu *et al.*, 1984). From the DNA sequence, the melibiose carrier protein should contain 469 amino acid residues with an M_r of 52,059, compared with the lactose protein of 417 residues and predicted M_r 46,504. Although homology of their primary sequences was very low, the protein sequences did exhibit broad similarities of alternating hydrophilic and hydrophobic regions (Fig. 3), the significance of which will be discussed. The intercistronic region between *melA* and *melB*, and between *melB* and the presumed *melC*, contained dyad symmetry sequences homologous to intercistronic sequences of *lamB–malA* of *E. coli*, and *hisJ–hisQ* or *hisG–hisD* of *S. typhimurium* (Clément and Hofnung, 1981; Higgins *et al.*, 1982; Stern *et al.*, 1984; Froshauer and Beckwith, 1984). These and similar consensus sequences in other intercistronic regions of *E. coli* may be involved in regulating the relative copy number of each cistron in an operon (see discussion by Stern *et al.*, 1984).

4.3. Reconstitution of the Melibiose/Na⁺ Symporter

Tsuchiya *et al.* (1982) identified three members of the Clarke-Carbon (1976) collection of colicin E plasmid-containing cells that harbored the *melA melB* or *melA* genes. The presence of one of the plasmids enhanced the expression of the melibiose carrier by seven- to eightfold.

French press vesicles (Section 2.3.3) were made from the induced, amplified strain, and the membrane proteins extracted using 1.25% octylglucoside in the presence of *E. coli* lipids; the proteins were then reconstituted into proteoliposomes, just as described for the lactose carrier (Newman and Wilson, 1980; Newman *et al.*, 1981). The proteoliposomes in 50 mM KP_i were preloaded with 20 mM unlabeled melibiose, and then diluted into 40 mM KP_i, 10 mM NaP_i containing 50 μM [³H]melibiose, the conditions for an "overshoot" assay of transport activity (Section 2.2). Gratifyingly, the label accumulated inside the vesicles and then slowly effluxed, indicating the presence of a functional

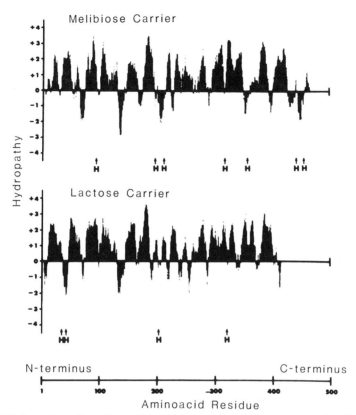

Figure 3. Hydropathic profiles of the melibiose/Na$^+$ and lactose/H$^+$ transport proteins. Hydropathy was calculated according to the method of Kyte and Doolittle (1982) with a segment span of seven residues. The portions above the line represent hydrophobic regions. The two proteins showed a broad similarity of alternating hydrophobic and hydrophilic regions, despite the absence of homology in their primary sequences (Büchel *et al.*, 1980; Yazyu *et al.*, 1984). The hydrophobic segments may be membrane-spanning α-helical structures (Henderson and Unwin, 1975; Kyte and Doolittle, 1982; Foster *et al.*, 1983; Yazyu *et al.*, 1984). (From Yazyu *et al.*, 1984.) Reproduced, with permission, from the *Journal of Biological Chemistry*, Volume 259, pages 4320–4326, copyright 1984.

melibiose transport system. This was confirmed by its sensitivity to the unlabeled alternative substrate *p*-nitrophenol-α-D-galactoside and by a reduction of about fourfold in the absence of Na$^+$ (Tsuchiya *et al.*, 1982); the residual activity was probably due to melibiose/H$^+$ symport rather than residual Na$^+$.

4.4. Identification of the melB Gene Product

Irradiated maxicells (Sancar *et al.*, 1979) harboring a plasmid encoding the *melAB* genes constitutively produced two polypeptides of M_r 50,000 and 31,000;

a second plasmid yielded only a faint appearance of a 50,000 polypeptide and two others of 31,000 and 27,000, which corresponded to the position of *amp* gene products in a control (Tsuchiya *et al.*, 1982). However, nonirradiated strains with the latter plasmid contained a melibiose-inducible protein of M_r 50,000 clearly visible in whole cell extracts but not in derived membranes (Hanatani *et al.*, 1984). Since α-galactosidase is a soluble protein of apparent M_r 20,000 (Burstein and Kepes, 1971), it was concluded that the *melA* gene coded for the subunit of a tetrameric α-galactosidase (Hanatani *et al.*, 1984). By elimination, the protein of M_r 31,000 in the constitutive strain should be the *melB* gene product. This conclusion was reinforced by SDS–PAGE of proteoliposomes made from a melibiose-induced or uninduced strain. The gel profiles were similar except for a polypeptide of apparent M_r 31,000 present only in the induced strain.

This result apparently contradicted the M_r of 51,000–52,000 estimated from a gene size of about 1.4 kb (Hanatani *et al.*, 1984) and 52,029 subsequently predicted from the detailed nucleotide sequence (Yazyu *et al.*, 1984). However, the *lacY* gene product was well established to migrate with an apparent M_r of 29,000–32,000 in SDS–PAGE despite its predicted M_r of 46,504 (Büchel *et al.*, 1980) and migration with M_r of about 45,000 in organic solvents (König and Sandermann, 1982). The anomalous migration in SDS–PAGE may be related to an unusually high content of nonpolar residues in both proteins—70% for the melibiose carrier and 71% for the lactose carrier (Büchel *et al.*, 1980, Yazyu *et al.*, 1984). The advent of methods for labeling the *melB* gene product specifically would help to confirm its identity.

4.5. Kinetic Constants for Melibiose Transport

The apparent K_m for melibiose or TMG measured in intact cells in the presence of 10 mM Na^+ was 0.2 mM (Tanaka *et al.*, 1980; Tsuchiya *et al.*, 1982). The V_{max} (at 25°C?) was 20 nmoles/min per mg cell protein. Unpublished work cited by Tsuchiya *et al.* (1982) reported that the K_m for Na^+ was 0.3 mM. A detailed steady-state analysis would be complicated by the presence of H^+ as an alternative substrate for Na^+, but would be interesting as a possible means of deducing the order of addition of substrates to the enzyme.

5. BINDING PROTEIN SUGAR TRANSPORT SYSTEMS

Neu and Heppel (1965) devised a "cold shock" procedure that removed periplasmic proteins from *E. coli* cells. Subsequently, transport of some amino acids or sugars was shown to be relatively insensitive to this procedure, whereas transport of others was substantially impaired (reviewed by Boos, 1974; Simoni and Postma, 1975; Anraku, 1978; Silhavy *et al.*, 1978). This effect was attributed to the presence in the periplasm of a protein that bound the sugar with a high affinity, a process that was eventually shown to be vital for efficient transport of the substrate into the cells by reconstitution experiments (Hunt and Hong, 1981; Brass *et al.*, 1981, 1983). A number of these "binding proteins" have now been purified (Section 5.3). In particular, an arabinose-binding protein has been crystallized, and its X-ray diffraction analysis refined to a level of resolution enabling a spectacular insight into the three-dimensional structure of the protein and its sugar-binding site (Quiocho and Vyas, 1984; Section 5.3.4). All sugar-binding proteins appear to have a single substrate-binding site, and the available evidence indicates a high degree of structural similarity (Quiocho and Pflugrath, 1980; Argos *et al.*, 1981; Vyas *et al.*, 1983; Duplay *et al.*, 1984).

Purification of the binding proteins was facilitated by their hydrophilic nature and location in the periplasmic space, but this raised the question of how such proteins could contribute to movement of the substrate through the hydrophobic cytoplasmic membrane. An involvement of cooperating membrane proteins was inferred, but evidence for their existence was only obtained recently, and relied substantially upon genetic technology (Section 5.3). Meanwhile, the binding proteins were found to be initial receptors in chemotaxis, the process whereby the microorganisms migrate toward a source of nutrients, including sugars (Adler *et al.*, 1973; Larsen *et al.*, 1974; Adler, 1975; Koshland, 1977). Their interaction with other proteins is therefore as much of interest as their interaction with substrate, but in this section I will primarily review their role in the transport process, and what is known of the mechanism of energization.

5.1. Detection of Binding Protein-Mediated Transport

An initial indication that sugar transport involved a binding protein would be its susceptibility (greater than 50% inhibition) to the "cold shock" procedure of Neu and Heppel (1965). However, even the lactose transport system, mediated by one very hydrophobic inner membrane protein (Section 3.3.1) was partly inhibited by this procedure, so that discrimination was best done by comparison with the effect on lactose transport into the same cells.

A better method is probably to measure binding protein activity directly. The "shock fluid" made by Heppel's procedure from a 0.5-liter culture of cells (or more), or the proteins released when spheroplasts are made (Witholt *et al.*, 1976) must first be concentrated by a factor of at least tenfold. This can be achieved by dialysis under pressure. The concentrated mixture of proteins is placed in a dialysis bag and exposed to 1–5 μM radioactive sugar substrate of high specific activity for a period of 18–24 hr (Lever, 1972; Clark and Hogg, 1981). In the absence of binding protein, the concentration of radioactive sugar inside the bag will be the same as that outside, but if binding protein is present

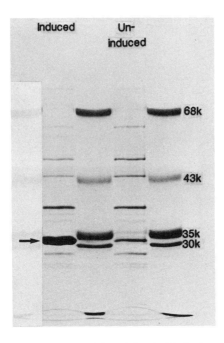

Figure 4. Detection of an arabinose-inducible periplasmic (binding) protein by SDS–PAGE. *E. coli* strain SB5314 was grown on glycerol in the presence (Induced) or absence (Uninduced) of arabinose. Spheroplasts were made from each (Witholt *et al.*, 1976; Macpherson *et al.*, 1981), and sedimented. A sample of each supernatant fluid, which contained proteins released from the periplasm, was solubilized in SDS-dissolving buffer at 100°C for 1 min and subjected to SDS–PAGE as described by Macpherson *et al.* (1981). A band of apparent M_r 33,000 was present in the induced but not in the uninduced culture (see arrow); this corresponded to the arabinose-binding protein (Section 5.3.4).

the apparent concentration of sugar inside will be higher than outside. It is important to measure the radioactivity with precise replicate samples, because detection depends on the difference between two fairly high numbers. The stability of the binding protein is obviously an important prerequisite for the method to be successful, and is the reason for conducting the isolation, concentration, and assay at 0–4°C. A successful substrate-binding assay can then be used to implement one of the established methods for purifying binding proteins (Anraku, 1968; Parsons and Hogg, 1974; Willis and Furlong, 1974; Ahlem *et al.*, 1982). An alternative procedure, in which periplasmic proteins are released by treatment of cells with chloroform, is suitable for small-scale preparations (Ames *et al.*, 1984). In some cases, particularly with hyperinducible mutants, the identity of the binding protein within the mixture of periplasmic proteins may be established by comparing SDS–polyacrylamide gels of samples from induced and uninduced strains (Fig. 4). Binding proteins can even be renatured after such separations (Copeland *et al.*, 1982).

Since the energization mechanism of binding protein-mediated transport is not proton-linked (Section 5.2), the failure of a transport system to display the characteristics described in Section 3.2 is also an indication that a binding protein may be involved, provided that the phosphotransferase mechanism (Chapter 10) has been eliminated. The most unequivocal demonstration of its transport role is to reconstitute purified binding protein into deficient cells, spheroplasts, or vesicles and show that transport is restored (reviewed by Hengge and Boos, 1983).

5.2. Energization of Binding Protein-Mediated Transport

Early work established that binding protein-mediated transport did not chemically modify the sugar (reviewed by Boos, 1974; Simoni and Postma, 1975; Silhavy *et al.*, 1978). Furthermore, the affinity for substrate was so high, dissociation constants of the order of 0.1–5 μM, that concentration gradients inside/outside in excess of 10,000 could be achieved (Vorisek and Kepes, 1972; Hengge and Boos, 1983). Initially, the energization appeared to be susceptible to uncoupling agents, but the work of Berger (1973; Berger and Heppel, 1974) changed this conclusion (reviewed by Hengge and Boos, 1983). The following observations led to the idea that a phosphate-containing product of glycolysis, probably ATP, directly energized the transport.

5.2.1. Energization of Transport Occurred in Cells Lacking a Protonmotive Force

Berger (1973) treated intact cells of *E. coli* with cyanide (or anaerobiosis) and DCCD to inhibit respiration and H^+-ATPase, respectively. Alternatively, respiration-and/or ATPase-negative mutants were used (Berger, 1973; Berger and Heppel, 1974). This prevented the cells from generating a protonmotive force. The transport of some amino acids, e.g., glutamine, into the cells was nevertheless still energized, provided that a substrate of glycolysis was provided—this may be a sugar substrate itself, of course. When a range of sugar transport systems were investigated, only those known to involve a binding protein could still be energized in the absence of the protonmotive force (reviewed by Boos, 1974; Simoni and Postma, 1975; Silhavy *et al.*, 1978). Provided that ATP hydrolysis was prevented, the transport was also relatively resistant to uncoupling agents.

5.2.2. Transport Was Sensitive to Arsenate

Treatment of intact *E. coli* cells with arsenate dramatically decreased the intracellular ATP concentration without much affecting the protonmotive force

(Klein and Boyer, 1972), provided the strain contained the phosphate transport system that facilitated arsenate entry (Rosenberg et al., 1977). Berger showed that arsenate inhibited transport of substrates of binding protein-containing systems, e.g., glutamine, whereas energization of transport, e.g.,of proline, by the protonmotive force was relatively unimpaired (Berger, 1973; Berger and Heppel, 1974; see also Klein and Boyer, 1972). Similarly, ribose transport was susceptible to arsenate, whereas proline transport was not (Curtis, 1974).

By using appropriate mutants or substrate analogues (Section 2.1.4), Daruwalla et al. (1981) confirmed that the binding protein transport systems for arabinose (AraF) and galactose (Mgl) were more sensitive to arsenate than the corresponding proton-linked transport systems (AraE and GalP, respectively, Sections 3.3.3 and 3.3.2). Similar experiments showed that the XylF transport system for xylose, which almost certainly involved a binding protein (Ahlem et al., 1982), was more sensitive to arsenate than the proton-linked xylose transport system XylE (Davis et al., 1984; Section 3.3.4).

5.2.3. Failure to Observe Sugar-Promoted pH Changes

An appropriately induced anaerobic, energy-depleted suspension of E. coli cells is set up as outlined in Section 3.2.1. Addition of a sugar substrate may not elicit an alkaline pH change, though acidification indicates that the sugar has entered the cells and is being metabolized via anaerobic fermentation pathways. Of course, the acidification may obscure any initial alkaline pH change (this is rarely a problem in fact) so the experiment is repeated with a metabolism-negative transport-positive mutant, or with a nonmetabolizable analogue of the normal transport substrate (Section 2.1.4). If addition of transport substrate still fails to elicit an alkaline pH change, then the transport may involve a binding protein system or a phosphotransferase mechanism (Chapter 10).

Failure to observe an alkaline pH change in wild-type cells, metabolism-negative mutants, or using substrate analogues, was a clear indication, in our experience, that a sugar transport system did not operate by a chemiosmotic mechanism. By these criteria the binding-protein transport system Mgl (galactose), AraF (arabinose), and XylF (xylose) were not energized by the proton-motive force (Henderson et al., 1977; Daruwalla et al., 1981; Davis et al., 1984).

5.2.4. Occurrence of Nucleotide-Binding Sites in Binding Protein Transport Systems

Walker et al. (1982) identifed a consensus amino acid sequence involved in the common binding of nucleotides to enzymes with different catalytic functions. Higgins et al. (1985) reported that a homologous sequence existed in the MalK component of the maltose transport system of E. coli, which also involved a binding protein (Section 5). Furthermore, the HisP and OppD components of binding protein transport systems for histidine and oligopeptide, respectively, contained similar sequences (Higgins *et al.*, 1985). Hobson *et al.* (1984) substantiated this observation by concluding that 8-azido-ATP, a photoaffinity labeling reagent, reacted with the HisP and HisM components of the histidine transport system. Higgins *et al.* (1985) obtained some evidence that OppD reacted with 5-*p*-fluorosulfonylbenzyladenosine and was retained on a Cibacron Blue column, which interacts with many nucleotide-binding proteins. Gilson *et al.* (1982) had already pointed out that HisP and MalK showed extensive homology with each other, and that MalK exhibited homologies with NADH dehydrogenase, which, of course, also binds nucleotides.

5.2.5. *Other Observations*

The evidence outlined in Sections 5.2.2 and 5.2.4 would seem to support Berger and Heppel's original conclusion (Sections 5.2.1 and 5.2.2) that ATP, and not the protonmotive force, was the source of energy for binding protein-mediated transport. However, an involvement of the protonmotive force has been suggested (Plate *et al.*, 1974; Ferenci *et al.*, 1977; Plate, 1979) despite the fact that the potential gradient of internal substrate that can be achieved far exceeds the energy available from proton-pumping (Hengge and Boos, 1983). Probably the transmembrane proton gradient has some other influence on non-proton-linked transport systems.

Hong and co-workers suggested that other compounds such as acetyl phosphate (Hong *et al.*, 1979) or NAD + pyruvate (Hunt and Hong, 1981) might be involved. Furthermore, 5-methoxyindole-1-carboxylate, an inhibitor of α-keto-acid dehydrogenase, specifically inhibited binding protein-dependent transport (Richarme, 1985). A final resolution of this question probably awaits successful reconstitution of a completely purified binding protein system in vesicles where the various energy sources can be investigated in a controlled way.

5.3. *Examples of Sugar Transport Systems of E. coli That Contain a Binding Protein*

The following examples contain only a summary of the biochemical and genetic characteristics of each sugar transport system. The restricted number of references listed should nevertheless enable the interested reader to pursue the properties of each system in more depth.

5.3.1. *Maltose*

This system transports maltose or higher α(1→4) glucose polymers (Wiesmeyer and Cohn, 1960). Its properties have been reviewed in detail by Hengge and Boos (1983), and the following constitutes a brief summary for comparison with the other transport systems discussed in this chapter.

The maltose transport system contains five proteins. One of these (LamB) is located in the outer membrane and contains a trimer of subunits (M_r 49,000), which also acts as the surface receptor for λ phage (Ferenci and Boos, 1980). It operates like a porin (Section 1.5; von Meyenburg and Nikaido, 1977; Ferenci and Boos, 1980). The second is a periplasmic binding protein (MalE) of M_r 40,661 (370 residues; Kellerman and Szmelcman, 1974; Ferenci and Klotz, 1978; Duplay et al., 1984). It binds 1 mole of sugar per mole protein, and must presumably also interact with the inner membrane protein(s), the chemotactic sensory system, and perhaps the outer membrane protein LamB; the evidence for these multiple interactions is discussed by Hengge and Boos (1983). There is at least one inner membrane protein (MalF) of apparent M_r 40,000 (true M_r 56,947; Froshauer and Beckwith, 1984) identified by combining gene fusion and immunological techniques (Shuman et al., 1980; Shuman and Silhavy, 1981) and probably a second (MalG) of apparent M_r 24,000 (Hengge and Boos, 1983; true $M_r = 32188$, 296 residues; Dassa and Hoffnung, 1985). Finally, a cytoplasmic protein (MalK) of apparent M_r 40,000 had the significant property of binding to the inner membrane only when a functional MalG protein was present (Bavoil et al., 1980; Shuman and Silhavy, 1981).

The genes for these proteins are located in a divergent operon (malB) located at 91.5 min on the chromosome with the order (Raibaud et al., 1980; Bachmann, 1983):

malG malF malE malK lamB

Its expression is positively regulated by the product of the *malT* gene (in the *malA* operon at 74 min), which binds between *malE* and *malK* (Raibaud and Schwartz, 1980). Since the products of *lamB* and *malE* are much more abundant than those of *malF, malG* and *malK*, there is presumably differential regulation of gene expression, perhaps mediated by intergenic sequences (Clément and Hofnung, 1981).

The primary sequences of the LamB, MalK, MalE, and MalF proteins have been determined (Clément and Hofnung, 1981; Gilson et al., 1982; Duplay et al., 1984; Froshauer and Beckwith, 1984) and in the case of LamB, at least, the exploitation of monoclonal antibodies with controlled mutagenesis is proving a viable way of elucidating the conformation of the protein and its topology in the membrane (Gabay et al., 1985). The purified maltose-binding protein can restore transport to mutants carrying a nonpolar deletion in the *malE* gene (Brass et al., 1983). It should be possible to immobilize one of the protein components in a liposome membrane, or by covalent attachment to a suitable support, so that its interaction with other components, including genetically engineered variants, may be investigated.

The natural substrate specificity of the maltose transport system extends up to, at least, the maltohexaose (Kellerman and Szmelcman, 1974; Schwartz *et al.*, 1976; Hengge and Boos, 1983) and to two structural analogues, 5-thiomaltose and methyl-β-D-maltoside (Ferenci, 1980). Intracellular unmodified maltose was the physiological inducer of both the *malA* and *malB* operons (Raibaud and Schwartz, 1980; Debarbouillé and Schwartz, 1980).

The apparent K_m for energized maltose transport was 1–2 μM, corresponding closely to the dissociation constant for interactions with the binding protein (Hengge and Boos, 1983). The energization appeared to be independent of the protonmotive force and dependent on glycolysis, but its insensitivity to arsenate raised a question as to whether ATP was involved (Ferenci *et al.*, 1977).

5.3.2. Ribose

The biochemistry and genetics of ribose catabolism in *E. coli* were first characterized by Anderson and Cooper (1970) and David and Wiesmeyer (1970b). Subsequently, a ribose-binding protein of about M_r 30,000 was purified from the periplasm (Willis and Furlong, 1974) and its amino acid sequence determined (271 residues; Groarke *et al.*, 1983). It bound 1 mole of ribose per mole protein with a dissociation constant of 0.13 μM, and was required for both high-affinity ribose transport and chemotaxis (Willis and Furlong, 1974; Galloway and Furlong, 1977).

The early work located an operon containing genes for ribose transport and ribokinase activity at 84 min on the chromosome (Anderson and Cooper, 1970; David and Wiesmeyer, 1970b). Recently, it has been analyzed in more detail, to give the probable gene order (Iida *et al.*, 1984; Lopilato *et al.*, 1984).

rbsp/o rbsA rbsC rbsB rbsK

rbsB codes for the binding protein and *rbsK* for ribokinase. Iida *et al.* (1984) inferred that *rbsA* and *rbsC* coded for the other components, with approximate M_r values 50,000 and 27,000, respectively, of the ribose transport systems, analogous to the membrane-located or cytoplasmic components of the maltose and other binding protein systems (Sections 5.3.1, 5.3.3–5.3.5). There may be another, as yet unidentified, gene in this region (Iida *et al.*, 1984; Lopilato *et al.*, 1984).

Expression of *rbs* genes appeared to be inducible in some *E. coli* strains and constitutive in others (David and Wiesmeyer, 1970b; Abou-Sabé and Richman, 1973a,b; see also Lopilato *et al.*, 1984). The only structural analogue of ribose that interacted with the binding protein was ribulose (Willis and Furlong, 1974).

The apparent K_m for ribose transport was about 0.3–2.3 μM (Willis and Furlong, 1974; Curtis, 1974), and its energization exhibited characteristics similar to those of the other binding protein transport systems (Curtis, 1974; Section

5.2). Interestingly, in some strains of *E. coli*, mutants lacking the binding transport system still grew on higher concentrations of ribose, implying the existence of a second ribose transport system (David and Wiesmeyer, 1970b; Lopilato *et al.*, 1984; Iida *et al.*, 1984). Whether this is of the proton-linked type remains to be elucidated.

5.3.3. Galactose

The existence of two separate transport systems for galactose in *E. coli* was first inferred by Rotman and co-workers (Rotman, 1959; Rotman *et al.*, 1968), who showed that only one of them transported methyl-β-D-galactoside. The phenotype was accordingly designated Mgl. This transport activity was associated with a periplasmic galactose (and methyl-β-D-galactoside)-binding protein (Boos, 1969, 1974), which was essential for both transport and chemotaxis (Boos, 1969, 1974; Hazelbauer and Adler, 1971; Adler, 1975). The genetic locus, *mgl*, was located at 45 min on the chromosome (Ganesan and Rotman, 1966; Boos *et al.*, 1982), and a complementation analysis of mutants impaired in galactose chemotaxis elegantly revealed at least four genes, *mglA, mglB, mglC, mglD* (Ordal and Adler, 1974), of which *mglD* appeared to be a regulatory region (Robbins, 1975; Robbins and Rotman, 1975; Robbins *et al.*, 1976). The mechanism for regulating expression of *mgl* is still unclear, though it seems to be independent of *galR*, the regulatory gene for *galP* and *gal (Lengeler et al.*, 1971; Wilson, 1974; von Wilcken-Bergmann and Müller-Hill, 1982). Recently, *mglB*, the structural gene for the galactose-binding protein, has been mapped more precisely (Boos *et al.*, 1982) and cloned (Müller-Hill *et al.*, 1982).

The galactose-binding protein of M_r about 35,000 was purified (Anraku, 1968), subjected to X-ray crystallography (Quiocho *et al.*, 1979; Quiocho and Pflugrath, 1980), and its amino acid sequence determined (309 residues; Mahoney *et al.*, 1981). Its overall shape of two globular domains with a cleft between them was similar to the structure of the L-arabinose-binding protein (Quiocho and Pflugrath, 1980; Section 5.3.4). All binding proteins may have such a tertiary structure, though the galactose-binding protein showed greater homology with the primary sequence of the ribose-binding protein than with any other (Groarke *et al.*, 1983; Iida *et al.*, 1984; Duplay *et al.*, 1984). Both the galactose and ribose-binding proteins interact with the same chemotactic transducer, the Trg protein (Hazelbauer and Harayama, 1979; Iida *et al.*, 1984).

The presence of both products of the *mglA* and *mglC* genes was shown to be necessary for transport function (Robbins, 1975; Robbins and Rotman, 1975; Robbins *et al.*, 1976). The *mglA* gene product was identified as a membrane protein of M_r about 50,000 (Harayama *et al.*, 1983; Rotman and Guzman, 1982), while the identity of the *mglC* product was less certain but it may have a M_r of 38,000 (Harayama *et al.*, 1983). It seems possible that *mglA mglC* serve analogous functions to *rbsA rbsC* of the ribose transport system (Iida *et al.*, 1984) and *malF malG malK* of the maltose transport system (see Iida *et al.*, 1984, and Sections 5.3.1 and 5.3.2). It will be interesting to determine the extent of their sequence homology and whether a nucleotide recognition site occurs in the two former systems.

Several chemical analogues of galactose were substrates for the galactose-binding protein transport system, including glucose, 6-deoxygalactose, 6-fluorogalactose, 6-chlorogalactose, methyl-β-D-galactoside, and glycerol-β-D-galactoside (Rotman *et al.*, 1968; Boos, 1969; Wilson, 1974; Henderson *et al.*, 1977). Alterations in the 2 position, i.e., 2-deoxygalactose, 2-deoxyglucose, talose, mannose, abolished transport by Mgl though these compounds were substrates for GalP (Henderson *et al.*, 1977; Henderson and Giddens, 1977; Section 3.3.2). Galactose, 6-deoxygalactose, 6-fluorogalactose, and 6-chlorogalactose were all inducers of the Mgl system (Lengeler *et al.*, 1971; Bradley, Taylor, and Henderson, unpublished results). Methyl-β-D-galactoside was a substrate specific for Mgl (Rotman *et al.*, 1968; Boos, 1969), and 6-chlorogalactose was a relatively specific inducer (Bradley, Taylor, and Henderson, unpublished results).

The use of specific substrates and appropriate mutants established that the Mgl system did not catalyze sugar/H^+ symport (Henderson *et al.*, 1977). Furthermore, its activity was relatively sensitive to arsenate (Daruwalla *et al.*, 1981) and relatively insensitive to uncoupling agents (Daruwalla, 1979). Together with what is known of its similarity to the structure of other binding protein systems, it seems likely that Mgl was similarly energized by a phosphorylated product of glycolysis, possibly ATP (Section 5.2).

The affinity of the binding protein for galactose was very high, $K_d < 1$ μM, similar to the value of K_m for transport (reviewed by Boos, 1974; Silhavy *et al.*, 1978). Thus, like other binding protein transport systems, Mgl can operate efficiently at low ambient substrate concentrations.

5.3.4. Arabinose

The existence of a transport system for arabinose in *E. coli*, which involved a binding protein, was first deduced by Schleif (1969) and Hogg and Englesberg (1969). The genetic locus, designated *araF*, was located at 45 min on the chromosome (Clark and Hogg, 1981; Kolodrubetz and Schleif, 1981a; Kosiba and Schleif, 1982) and shown by complementation analysis to involve at least two genes, *araF araG* (in that order from the promoter), the former of which coded for the binding protein (Kolodrubetz and Schleif, 1981a). Schleif and co-workers have cloned and sequenced the regulatory region to show how its expression is effected by inducer (free arabinose) and by binding of the *araC* gene product and *crp* protein (Kolodrubetz and Schleif, 1981b; Kosiba and Schleif, 1982; Hendrickson and Schleif, 1984). Their elegant approach is elucidating the

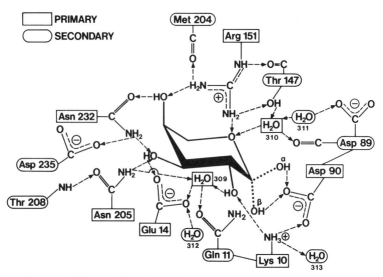

Figure 5. Structure of the sugar-binding site in the arabinose-binding protein. (From Quiocho and Vyas, 1984.) Reprinted by permission from *Nature*, Vol. 310, pp. 381–386, Copyright © 1984 Macmillan Journals Limited.

mechanism by which three separate regions of the *E. coli* genome—*ara* (1 min), *araE* (61 min; Section 3.3.3), and *araF, G* (45 min)—are coordinately regulated and yet may respond differently to influences such as catabolite repression (see especially Kolodrubetz and Schleif, 1981b; Stoner and Schleif, 1983; Hendrickson and Schleif, 1984).

The arabinose-binding protein of M_r 33,170 was purified by Parsons and Hogg (1974), its amino acid sequence determined (306 residues; Hogg and Hermodson, 1977), and its three-dimensional structure elucidated by X-ray crystallography (Quiocho *et al.*, 1974, 1977). Quiocho and co-workers refined the crystallographic analysis to locate the arabinose-binding site near the only cysteine residue in a cleft formed by the "hinged" packing of two globular domains of the protein, and even to resolve the precise interaction of the arabinose -OH groups with individual amino acid residues and water molecules (Quiocho and Vyas, 1984). The arrangement beautifully accommodated either the α- or the β-anomer (Fig. 5). Future work will perhaps illuminate which parts of the protein interact with the membrane-bound transport component (*araG*?) and also with the chemotactic transducer.

A preliminary identification of the *araG* gene product was reported, using two-dimensional electrophoresis and an appropriate mutant to detect a membrane protein of apparent M_r 38,000 and apparent pI 7.0 (Kolodrubetz and Schleif, 1981a). Only one genetic complementation group was detected, despite the clear existence of two functions (Kolodrubetz and Schleif, 1981a), so the exact number of components in the arabinose transport system is still open to question.

6-Deoxy-D-galactose (D-fucose) would appear to be the only sugar transported by AraF, G in addition to L-arabinose and, since it was not metabolized by arabinose isomerase, it provided a very useful means of isolating transport activity (Novotny and Englesberg, 1966). Interestingly, 6-deoxy-D-galactose was an anti-inducer of the L-arabinose operons, a phenomenon that could be exploited for mutant selection (Englesberg *et al.*, 1965; Beverin *et al.*, 1971).

Use of the two substrates and appropriate mutants established that the AraF, G system did not catalyze arabinose/H^+ symport (Daruwalla *et al.*, 1981). Furthermore, its activity was relatively sensitive to arsenate (Daruwalla *et al.*, 1981) and insensitive to uncoupling agents (Daruwalla, 1979). Again, it seemed likely that transport on AraF,G was energized by a phosphorylated product of glycolysis, possibly ATP (Section 5.2).

The affinity of the binding protein for arabinose was very high, $K_d \approx 1$ μM, similar to K_m values of 1–6 μM for transport by AraF,G (Hogg and Englesberg, 1969; Schleif, 1969; Brown and Hogg, 1972; Hogg, 1977; Daruwalla *et al.*, 1981). Thus, this transport system would operate efficiently at low ambient substrate concentrations.

5.3.5. Xylose

Genes coding for xylose transport and metabolism in *E. coli* were first mapped at 79 min (Bachmann, 1983) by David and Wiesmeyer (1970a). A similar system was identified in *S. typhimurium* by Shamanna and Sanderson (1979a,b), who also performed kinetic measurements that indicated the existence of two xylose transport systems in *E. coli* (Shamanna and Sanderson, 1979a). It now seems clear that one of these is proton-linked with a single gene located at 91.4 min (Lam *et al.*, 1980; Davis *et al.*, 1984; Section 3.3.4), whereas the other involves a binding protein (Ahlem *et al.*, 1982). Preliminary data of ours indicate that there are at least two genes for the xylose-binding protein transport system, provisionally designated *xylF,G,* which map at 79 min, near the genes *xylA,B* coding for the isomerase and kinase, respectively (see also Shamanna and Sanderson, 1979b). The order of all the genes in the operon is not established, though the DNA of *xylA* (and part of *xylB*) has been sequenced (Briggs *et al.*, 1984; Lawlis *et al.*, 1984; Rosenfeld *et al.*, 1984).

The xylose-binding protein was purified and comprehensively characterized by Ahlem *et al.* (1982). It had an apparent M_r of 37,000 and an apparent pI of 7.4. Its amino acid composition was determined but not the sequence; no cysteine residues were detected. One mole of xylose was bound per mole of binding protein, with a K_d value of 0.6 μM. An enhancement of intrinsic tryptophan fluorescence was effected by binding of xylose, presumably reflecting some conformational change in the protein.

Among other sugars, L-arabinose, D-ribose, D-galactose, and D-glucose failed to decrease xylose binding to the isolated protein, so it displayed a high degree of specificity (Ahlem *et al.*, 1982). Only D-xylose was an inducer.

The apparent K_m for xylose transport by the binding protein system was in the range 0.2–4.0 μM, in contrast to higher values of 53–170 μM for the proton-linked system (E. O. Davis, unpublished results). Since its activity was not linked to proton movement and was sensitive to arsenate (Davis *et al.*, 1984), its mechanism of energization appeared similar to that of other binding protein systems.

It seems likely that the biochemistry and genetics of this xylose transport system will soon be established to the level of detail accomplished for the other sugar-binding protein systems.

6. NEW DEVELOPMENTS

In all areas of biology an explosion of information has occurred since techniques were devised for the recombination and sequencing of DNA (e.g., Sanger *et al.*, 1977, 1980, 1982; Maniatis *et al.*, 1982). Thus, despite the low abundance and intransigent hydrophobic nature of transport proteins, it is now possible to determine their amino acid sequence via the DNA sequence of the gene and to amplify gene expression sufficiently for easy purification of the protein. As usual, the first *E. coli* sugar transport system for which this was achieved was the H^+/lactose symporter (Teather *et al.*, 1978, 1980; Büchel *et al.*, 1980; Newman *et al.*, 1981; Wright *et al.*, 1983). Other symporters have now been sequenced—those for melibiose, arabinose, and xylose—and soon there will be more, so I will briefly review the information obtained and the future strategies that should illuminate their structure–function relationships.

The primary sequences of the LacY, MelB, AraE, and XylE proteins displayed two striking similarities: a high proportion of hydrophobic residues and the alternation of hydrophobic and hydrophilic segments (Fig. 3; Forster *et al.*, 1983; Yazyu *et al.*, 1984; E. O. Davis, unpublished results; M. C. M. Maiden and D. C. M. Moore, unpublished results). It is a compelling idea that the hydrophobic regions are intramembrane sections of loops that zigzag across the membrane like the topology of bacteriorhodopsin (Henderson and Unwin, 1975; Forster *et al.*, 1983). Circular dichroism measurements indicated that the LacY protein was approximately 85% helical, so the hydrophobic segments would presumably have an α-helix conformation (Forster *et al.*, 1983). Thus, the amino acid sequences of cation/sugar symporters are so far consistent with their having a structure similar to that of bacteriorhodopsin.

Despite this similarity, the nucleotide sequence of the *lacY* gene DNA, and the amino acid sequence of the protein, did not display significant regions of homology with the melbiose (*melB*), arabinose (*araE*), or xylose (*xylE*) cation/sugar transport genes and proteins (Büchel *et al.*, 1980; Yazyu *et al.*, 1984; E. O. Davis, unpublished results; M. C. M. Maiden and D. C. M. Moore, unpublished results). However, there were significant homologies between the *xylE* and *araE* sequences. Perhaps one of these evolved from the other, whereas the *lacY* and *melB* systems evolved independently, despite the similarity of their function. Perhaps all the cation/sugar symporters do have identical residues involved in the common function of H^+-translocation, but the disposition of these residues in the primary sequence is too scattered to be recognized by present comparative techniques. One might expect an arrangement of residues for sugar binding similar to that in the arabinose-binding protein (Fig. 5; Quiocho and Vyas 1984), but, if present, the arrangement is too subtle to have been recognized.

　　Since comparative inspection of primary sequences has not identified residues involved in sugar- or cation-binding, what techniques will facilitate their identification? One approach is that of site-directed mutagenesis (Winter *et al.*, 1982) to alter individual amino acid residues that are thought to be critical. For example, Cys-148 in the LacY sequence was shown to be the only cysteine protected by substrate against reaction with *N*-ethylmaleimide, indicating that it was at or near the sugar-binding site (Beyreuther *et al.*, 1981; Mitaku *et al.*, 1984). However, conversion of Cys-148 to a Gly residue by site-directed mutagenesis did not abolish active transport of lactose, though it did reduce the initial rate and removed the substrate protection (Trumble *et al.*, 1984). Thus, Cys-148 was not essential for transport function. A second, less speculative, approach is to isolate point mutants impaired in transport function and resequence the DNA to establish which amino acid in the structure has been altered. In this way, a *lacY* mutant with a sugar specificity altered from lactose to maltose was found to have Thr-266 changed to Ile, implying that this residue was located in or near the sugar-binding site (Markgraf *et al.*, 1985). Five independent *melB* mutants with a cation specificity altered from H^+ to Li^+ were all found to have Pro-122 replaced with Ser, implying that this residue was involved in cation binding (Yazyu *et al.*, 1985; see also Niiya *et al.*, 1982). These approaches may appear laborious, but they offer the most promising way of elucidating the structure–function relationship in the absence of either crystallized pure protein suitable for X-ray or electron diffraction analysis or accurate methods of predicting the tertiary folding of a protein from a knowledge of its primary sequence.

　　Given some amplification and purification, the amounts of LacY protein in the membrane became sufficient to determine which residues were accessible from either the outside (right-side-out vesicles) or inside (inside-out vesicles) of the cytoplasmic membrane. The techniques involved the reaction of Cys-148 with radioisotope- or fluorescence-labeled maleimides (Beyreuther *et al.*, 1981; Mitaku *et al.*, 1984), binding of antibodies to the protein (Herzlinger *et al.*,

1984, 1985; Carrasco *et al.*, 1984a,b), and controlled proteolysis (Goldkorn *et al.*, 1983). These studies indicated, for example, that the -CO_2H terminus of the protein was on the inside of the cytoplasmic membrane and that the sugar-binding site was embedded within the protein.

Similar approaches should illuminate the topologies of the membrane-located proteins in the binding protein transport systems, now that their sequences are being determined, e.g., for MalB (Sections 5.2.4, 5.3.1). Here there is the added complexity of protein–protein interactions, and the careful scrutiny of ATPase subunit sequences described by Walker and co-workers (reviewed by Walker *et al.*, 1984) may be a useful guide for analyzing multisubunit transport systems.

The combination of genetic manipulation with biochemical analysis in *E. coli* should continue to illuminate the molecular mechanisms of transport processes, and, indeed, enzyme mechanisms in general. *E. coli* has been a versatile tool in the long-suffering experimenter's quest for insights and fun. Long may their collaboration continue.

ACKNOWLEDGMENTS. Research in the author's laboratory is supported by the SERC, the Wellcome Trust, the SmithKline Foundation, and the Royal Society, whose help is gratefully acknowledged.

REFERENCES

Abou-Sabé, M., and Richman, J., 1973a, On the regulation of D-ribose metabolism in *Escherichia coli* B/r I, *Mol. Gen. Genet.* **122**:291–301.

Abou-Sabé, M., and Richman, J., 1973b, On the regulation of D-ribose metabolism in *Escherichia coli* B/r II, *Mol. Gen. Genet.* **122**:303–312.

Adler, J., 1975, Chemotaxis in bacteria, *Annu. Rev. Biochem.* **44**:341–356.

Adler, J., Hazelbauer, G. L., and Dahl, M. M., 1973, Chemotaxis toward sugars in *Escherichia coli*, *J. Bacteriol.* **115**:824–847.

Ahlem, C., Huisman, W., Neslund, G., and Dahms, A. S., 1982, Purification and properties of a periplasmic D-xylose-binding protein from *Escherichia coli* K-12, *J. Biol. Chem.* **257**:2926–2931.

Amanuma, H., and Anraku, Y., 1974, Transport of sugars and amino acids in bacteria. XII. Substrate specificities of the branched chain amino acid-binding proteins of *Escherichia coli*, *J. Biochem.* **76**:1165–1173.

Ames, G. F.-L., Prody, C., and Kustu, S., 1984, Simple, rapid and quantitative release of periplasmic proteins by chloroform, *J. Bacteriol.* **160**:1181–1183.

Anderson, A., and Cooper, R. A., 1970, Biochemical and genetical studies on ribose catabolism in *Escherichia coli* K-12, *J. Gen. Microbiol.* **62**:335–339.

Anraku, Y., 1968, Purification and specificity of the galactose- and leucine-binding proteins, *J. Biol. Chem.* **243**:3116–3122.

Anraku, Y., 1978, Active transport of amino acids, in: *Bacterial Transport* (B. P. Rosen, ed.), pp. 171–219, Dekker, New York.

Argos, P., Mahoney, W. C., Hermodson, M. A., and Hanei, M., 1981, Structural prediction of sugar-binding proteins functional in chemotaxis and transport, *J. Biol. Chem.* **256**:4357–4361.

Bachmann, B. J., 1983, Linkage map of *Escherichia coli* K-12, edition 7, *Microbiol. Rev.* **47**:180–230.

Bavoil, S., Hofnung, M., and Nikaido, H., 1980, Identification of a cytoplasmic membrane-associated component of the maltose transport system of *Escherichia coli*, *J. Biol. Chem.* **255**:8366–8369.

Beckwith, J. R., and Zipser, D. (eds.), 1970, *The Lactose Operon*, Cold Spring Harbor Laboratory, Cold Spring Harbor, N.Y.

Berger, E. A., 1973, Different mechanisms of energy coupling for the active transport of proline and glutamine in *Escherichia coli*, *Proc. Natl. Acad. Sci. USA* **70**:1514–1518.

Berger, E. A., and Heppel, L. A., 1974, Different mechanisms of energy coupling for the shock-sensitive and shock-resistant amino acid permeases of *Escherichia coli*, *J. Biol. Chem.* **249**:7747–7755.

Beverin, S., Sheppard, D. E., and Park, S. S., 1971, D-Fucose as a gratuitous inducer of the L-arabinose operon in strains of *Escherichia coli* B/r mutant in gene *araC*, *J. Bacteriol.* **107**:79–86.

Beyreuther, K., Bieseler, B., Ehring, R., and Müller-Hill, B., 1981, Identification of internal residues of lactose permease of *Escherichia coli* by radiolabel sequencing of peptide mixtures, in: *Methods in Protein Sequence Analysis* (M. Elzina, ed.), pp. 139–148, Humana Press, Clifton, N.J.

Boonstra, J., Huttunen, M. T., Konings, W. N., and Kaback, H. R., 1975, Anaerobic transport in *Escherichia coli* membrane vesicles, *J. Biol. Chem.* **250**:6792–6798.

Boos, W., 1969, The galactose binding protein and its relationship to the β-methylgalactoside permease from *Escherichia coli*, *Eur. J. Biochem.* **10**:66–73.

Boos, W., 1974, Bacterial transport, *Annu. Rev. Biochem.* **43**:123–146.

Boos, W., Steinacher, I., and Engelhardt-Altendorf, D., 1982, Mapping of *mglB*, the structural gene of the galactose-binding protein of *E. coli*, *Mol. Gen. Genet.* **184**:508–518.

Bradley, S. A., Tinsley, C. R., and Henderson, P. J. F., 1986, Proton-linked L-fucose transport in *Escherichia coli*, submitted for publication.

Brass, J. M., Boos, W., and Hengge, R., 1981, Reconstitution of maltose transport in malB mutants of *Escherichia coli* through calcium-induced disruptions of the outer membrane, *J. Bacteriol.* **146**:10–17.

Brass, J. M., Ehmann, U., and Bukau, B., 1983, Reconstitution of maltose transport in *Escherichia coli*, *J. Bacteriol.* **155**:97–106.

Bremer, E., Silhavy, T. J., Weisemann, J. M., and Weinstock, G. M., 1984, λ*plac*Mu: A transposable derivative of bacteriophage lambda for creating *lacZ* protein fusions in a single step, *J. Bacteriol.* **158**:1084–1093.

Briggs, K. A., Lancashire, W. E., and Hartley, B. S., 1984, Molecular cloning: DNA structure and expression of the *Escherichia coli* D-xylose isomerase, *EMBO J.* **3**:611–616.

Brown, C. E., and Hogg, R. W., 1972, A second transport system for L-arabinose in *Escherichia coli* B/r controlled by the *araC* gene, *J. Bacteriol.* **111**:606–613.

Büchel, D. E., Gronenborn, B., and Müller-Hill, B., 1980, Sequence of the lactose permease gene, *Nature* **283**:541–545.

Burstein, C., and Kepes, A., 1971, The α-galactosidase from *Escherichia coli* K12, *Biochim. Biophys. Acta* **230**:52–63.

Buttin, G., 1968, Les systémes enzymatiques inducibles du metabolisme des oses chez *Escherichia coli*, *Adv. Enzymol.* **30**:81–137.

Cairney, J., Higgins, C. F., and Booth, I. R., 1984, Proline uptake through the major transport system (PP-I: *putP*) of *Salmonella typhimurium* is coupled to sodium ions, *J. Bacteriol.* **160**:22–27.

Carrasco, N., Herzlinger, D., Mitchell, R., Dechiara, S., Danho, W., Gabriel, T. F., and Kaback, H. R., 1984a, Intramolecular dislocation of the -CO_2H terminus of the *lac* carrier protein in reconstituted proteoliposomes, *Proc. Natl. Acad. Sci. USA* **81**:4672–4676.

Carrasco, N., Viitanen, P., Herzlinger, D., and Kaback, H. R., 1984b, Monoclonal antibodies against the lac carrier protein for *Escherichia coli*. 1. Functional studies, *Biochemistry* **23**:3681–3687.

Casadaban, M. J., and Cohen, S. N., 1979, Lactose genes fused to exogenous promoters in one step using a Mu-*lac* bacteriophage: *In vivo* probe for transcriptional control sequences, *Proc. Natl. Acad. Sci. USA* **76**:4530–4533.

Chappell, J. B., 1968, Systems used for the transport of substrates into mitochondria, *Br. Med. Bull.* **24**:150–157.

Clark, A. F., and Hogg, R. W., 1981, High-affinity arabinose transport mutants of *Escherichia coli*: Isolation and gene location, *J. Bacteriol.* **147**:920–924.

Clarke, L., and Carbon, J., 1976, A colony bank containing synthetic Col E1 hybrid plasmids representative of the entire *E. coli* genome, *Cell* **9**:91–99.

Clément, J. M., and Hofnung, M., 1981, Gene sequences of the λ receptor, an outer membrane protein of *Escherichia coli* K12, *Cell* **27**:507–514.

Cockrell, R. S., Harris, E. J., and Pressman, B. C., 1967, Synthesis of ATP driven by a potassium gradient in mitochondria, *Nature* **215**:1487–1488.

Cohn, D. E., and Kaback, H. R., 1980, Mechanism of the melibiose porter in membrane vesicles of *Escherichia coli*, *Biochemistry* **19**:4237–4243.

Colowick, S. P., and Womack, F. C., 1969, Binding of diffusible molecules by macromolecules: Rapid movement by rate of dialysis, *J. Biol. Chem.* **244**:774–777.

Copeland, B. R., Richter, R. J., and Furlong, C. E., 1982, Renaturation and identification of periplasmic proteins in two-dimensional gels of *Escherichia coli*, *J. Biol. Chem.* **257**:15065–15071.

Costello, M. J., Viitanen, P., Carrasco, N., Foster, D. L., and Kaback, H. R., 1984, Morphology of proteoliposomes reconstituted with purified *lac* carrier protein from *Escherichia coli*, *J. Biol. Chem.* **259**:15579–15586.

Curtis, S. J., 1974, Mechanism of energy coupling for transport of D-ribose in *Escherichia coli*, *J. Bacteriol.* **120**:295–303.

Daruwalla, K., 1979, The energisation of sugar transport systems in bacteria, Ph.D. thesis, University of Cambridge.

Daruwalla, K. R., Paxton, A. T., and Henderson, P. J. F., 1981, Energization of the transport systems for arabinose and comparison with galactose transport in *Escherichia coli*, *Biochem. J.* **200**:611–627.

Dassa, E., and Hofnung, M., 1985, Sequence of gene *malG* in *E. coli* K12, *EMBO J.* **4**:2287–2293.

David, J. D., and Wiesmeyer, H., 1970a, Control of xylose metabolism in *Escherichia coli*, *Biochim. Biophys. Acta* **201**:497–499.

David, J. D., and Wiesmeyer, H., 1970b, Regulation of ribose catabolism in *Escherichia coli*: The ribose catabolic pathway, *Biochim. Biophys. Acta* **208**:45–55.

Davis, E. O., Jones-Mortimer, M. C., and Henderson, P. J. F., 1984, Location of a structural gene for xylose-H^+ symport at 91 min on the linkage map of *Escherichia coli* K12, *J. Biol. Chem.* **259**:1520–1525.

Debarbouillé, M., and Schwartz, M., 1980, Mutants which make more *malT* product, the activator of the maltose regulon in *Escherichia coli*, *Mol. Gen. Genet.* **178**:589–595.

Di Renzio, J. M., Nakamura, K., and Inouye, M., 1978, The outer membrane proteins of gram-negative bacteria: Biosynthesis, assembly, and functions, *Annu. Rev. Biochem.* **47**:481–532.

Duplay, P., Bedouelle, H., Fowler, A., Zabin, I., Saurin, W., and Hofnung, M., 1984, Sequences of the *malE* gene and of its product, the maltose-binding protein of *Escherichia coli* K12, *J. Biol. Chem.* **259**:10606–10613.

Ehring, R., Beyreuther, K., Wright, J. K., and Overath, P., 1980, *In vitro* and *in vivo* products of *E. coli* lactose permease gene are identical, *Nature* **283**:537–540.

Englesberg, E., Irr, J., Power, J., and Lee, N., 1965, Positive control of enzyme synthesis by gene C in the L-arabinose system, *J. Bacteriol.* **90**:946–957.

Ferenci, T., 1980, Methyl-α-maltoside and 5-thiomaltose; analogues transported by the *Escherichia coli* maltose transport system, *J. Bacteriol.* **144**:7–11.

Ferenci, T., and Boos, W., 1980, The role of the *Escherichia coli* λ receptor in the transport of maltose and maltodextrins, *J. Supramol. Struct.* **13**:101–116.

Ferenci, T., and Klotz, U., 1978, Affinity chromatography isolation of the periplasmic maltose binding protein of *Escherichia coli*, *FEBS Lett.* **94**:213–217.

Ferenci, T., Boos, W., Schwartz, M., and Szmelcman, S., 1977, Energy coupling of the transport system of *Escherichia coli* dependent on maltose-binding protein, *Eur. J. Biochem.* **75**:187–195.

Flagg, J. L., and Wilson, T. H., 1977, A protonmotive force as the source of energy for galactoside transport in energy depleted *Escherichia coli*, *J. Membr. Biol.* **31**:233–255.

Foster, D., Boublik, M., and Kaback, H. R., 1983, Structure of the *lac* carrier protein of *Escherichia coli*, *J. Biol. Chem.* **258**:31–34.

Foster, D. L., Garcia, M. L., Newman, M. J., Patel, L., and Kaback, H. R., 1982, Lactose–proton symport by purified *lac* carrier protein, *Biochemistry* **21**:5634–5638.

Frank, L., and Hopkins, I., 1969, Sodium-stimulated transport of glutamate in *Escherichia coli*, *J. Bacteriol.* **100**:329–336.

Froshauer, S., and Beckwith, J., 1984, The nucleotide sequence of the gene for *malF* protein, an inner membrane component of the maltose transport system of *Escherichia coli*, *J. Biol. Chem.* **259**:10896–10903.

Futai, M., 1978, Experimental systems for the study of active transport in bacteria, in: *Bacterial Transport* (B. P. Rosen, ed.), pp. 7–41, Dekker, New York.

Futai, M., and Kanazawa, H., 1983, Structure and function of proton-translocating adenosine triphosphatase (F_0F_1): Biochemical and molecular biological approaches, *Microbiol. Rev.* **47**:285–312.

Gabay, J., Schenkman, S., Desaymard, C., and Schwartz, M., 1985, Monoclonal antibodies and the structure of bacterial membrane proteins, in: *Monoclonal Antibodies Against Bacteria* (A. S. L. Macario and C. de Macario, eds.), pp. 249–282, Academic Press, New York.

Galloway, D. R., and Furlong, C. E., 1977, The role of ribose-binding protein in transport and chemotaxis in *Escherichia coli* K-12, *Arch. Biochem. Biophys.* **184**:496–504.

Ganesan, A. K., and Rotman, B., 1966, Genetic determination and regulation of the methylgalactoside permease, *J. Mol. Biol.* **16**:42–50.

Garcia, M. L., Viitanen, P., Foster, D. L., and Kaback, H. R., 1983, Mechanism of lactose translocation in proteoliposomes reconstituted with *lac* carrier protein purified from *Escherichia coli*, *Biochemistry* **22**:2524–2531.

Gilson, E., Higgins, C. F., Hofnung, M., Ames, G. F.-L., and Nikaido, J., 1982, Extensive homology between membrane-associated components of histidine and maltose transport systems of *Salmonella typhimurium* and *Escherichia coli*, *J. Biol. Chem.* **257**:9915–9918.

Glynn, I. M., 1967, Involvement of a membrane potential in the synthesis of ATP by mitochondria, *Nature* **216**:1318–1319.

Goldkorn, T., Rimon, G., and Kaback, H. R., 1983, Topology of the Lac carrier protein in the membrane of *Escherichia coli*, *Proc. Natl. Acad. Sci. USA* **80**:3322–3326.

Goldkorn, T., Rimon, G., Kempner, E. S., and Kaback, H. R., 1984, Functional molecular weight of the *lac* carrier protein from *Escherichia coli* as studied by radiation inactivation analysis, *Proc. Natl. Acad. Sci. USA* **81**:1021–1025.

Groarke, J. M., Mahoney, W. C., Hope, J. N., Furlong, C. E., Robb, F. T., Zalkin, H., and

Hermodson, M. A., 1983, The amino acid sequence of D-ribose-binding protein from *Escherichia coli* K12, J. Biol. Chem. **258:**12952–12956.

Haddock, B. A., and Jones, C. W., 1977, Bacterial respiration, *Bacteriol. Rev.* **41:**47–99.

Hamilton, W. A., 1975, Energy coupling in microbial transport, *Adv. Microb. Physiol.* **12:**1–53.

Hanatani, M., Yazyu, H., Shiota-Niiya, S., Moriyama, Y., Kanazawa, H., Futai, M., and Tsuchiya, T., 1984, Physical and genetic characterization of the melibiose operon and identification of the gene products in *Escherichia* coli, *J. Biol. Chem.* **259:**1807–1812.

Harayama, S., Bollinger, J., Iino, T., and Hazelbauer, G. L., 1983, Characterization of the *mgl* operon of *Escherichia coli* by transposon mutagenesis and molecular cloning, *J. Bacteriol.* **153:**408–415.

Harold, F. M., 1972, Conservation and transformation of energy by bacterial membranes, *Bacteriol. Rev.* **36:**172–230.

Harold, F. M., 1977, Membranes and energy transduction in bacteria, *Curr. Top. Bioenerg.* **6:**83–149.

Hays, J. B., 1978, Group translocation transport systems, in: *Bacterial Transport* (B. P. Rosen, ed.), pp. 43–102, Dekker, New York.

Hazelbauer, G. L., and Adler, J., 1971, Role of galactose-binding protein in chemotaxis of *Escherichia coli* toward galactose, *Nature New Biol.* **230:**101–104.

Hazelbauer, G. L., and Harayama, S., 1979, Mutants in transmission of chemotactic signals from two independent receptors of *E. coli*, *Cell* **16:**617–625.

Heffernan, L., Bass, R., and Englesberg, E., 1976, Mutations affecting catabolite repression of the L-arabinose regulon in *Escherichia coli* B/r, *J. Bacteriol.* **126:**1119–1131.

Heinz, E., and Weinstein, A. M., 1984, The overshoot phenomenon in cotransport, *Biochim. Biophys. Acta* **776:**83–91.

Heller, K. B., Lin, E. C. C., and Wilson, T. H., 1980, Substrate specificity and transport properties of the glycerol facilitator of *Escherichia coli*, *J. Bacteriol.* **144:**274–278.

Henderson, P. J. F., 1974, Application of the chemiosmotic theory to the transport of lactose, D-galactose, and L-arabinose by *Escherichia coli*, in: *Comparative Biochemistry and Physiology of Transport* (L. Bolis, K. Bloch, S. E. Luria, and F. Lynen, eds.), pp. 409–424, North-Holland, Amsterdam.

Henderson, P. J. F., and Giddens, R. A., 1977, 2-Deoxy-D-galactose, a substrate for the galactose-transport system of *Escherichia coli*, *Biochem. J.* **168:**15–22.

Henderson, P. J. F., and Kornberg, H. L., 1975, The active transport of carbohydrates of *Escherichia coli*, *Ciba Found. Symp.* **31:**243–269.

Henderson, P. J. F., and Skinner, A., 1974, Association of proton movements with the galactose and arabinose transport systems of *Escherichia coli*, *Biochem. Soc. Trans.* **2:**543–545.

Henderson, P. J. F., Dilks, S. N., and Giddens, R. A., 1975, pH changes associated with the transport of sugars by *Escherichia coli*, *Proc. 10th FEBS Meet.*, pp. 43–53.

Henderson, P. J. F., Giddens, R. A., and Jones-Mortimer, M. C., 1977, Transport of galactose, glucose and their molecular analogues by *Escherichia coli* K12, *Biochem. J.* **162:**309–320.

Henderson, P. J. F., Bradley, S., MacPherson, A. J. S., Horne, P., Davis, E. O., Daruwalla, K. T., and Jones-Mortimer, M. C., 1984, Sugar–proton transport systems of *Escherichia coli*, *Biochem. Soc. Trans.* **12:**146–148.

Henderson, R., and Unwin, P. N. T., 1975, Three-dimensional model of purple membrane obtained by electron microscopy, *Nature New Biol.* **257:**28–32.

Hendrickson, W., and Schleif, R. F., 1984, Regulation of the *Escherichia coli* L-arabinose operon studied by gel electrophoresis DNA binding assay, *J. Mol. Biol.* **174:**611–628.

Hengge, R., and Boos, W., 1983, Maltose and lactose transport in *Escherichia coli*, examples of two different types of concentrative transport systems, *Biochim. Biophys. Acta* **737:**443–478.

Herzberg, E., and Hinkle, P., 1974, Oxidative phosphorylation and proton translocation in membrane vesicles prepared from *Escherichia coli*, *Biochem. Biophys. Res. Commun.* **58:**178–184.

Herzlinger, D., Viitanan, P., Carrasco, N., and Kaback, H. R., 1984, Monoclonal antibodies against the *lac* carrier protein from *Escherichia coli*. 2. Binding studies with membrane vesicles and proteoliposomes reconstituted with purified *lac* carrier protein, *Biochemistry* **23**:3688–3693.

Herzlinger, D., Carrasco, N., and Kaback, H. R., 1985, Functional and immunochemical characterization of a mutant of *Escherichia coli* energy uncoupled for lactose transport, *Biochemistry* **24**:221–229.

Higgins, C. F., Haag, P. D., Nikaido, K., Ardeshir, F., Garcia, G., and Ames, G., F.-L., 1982, Complete nucleotide sequence and identification of membrane components of the histidine transport operon of *S. typhimurium*, *Nature* **298**:723–727.

Higgins, C. F., Hiles, I. D., Whalley, K., and Jamieson, D. J., 1985, Nucleotide binding by membrane components of bacterial periplasmic binding protein-dependent transport systems, *EMBO J.* **4**:1033–1040.

Hobson, A. C., Weatherwax, R., and Ames, G. F.-L., 1984, ATP-binding sites in the membrane components of histidine permease, a periplasmic transport system, *Proc. Natl. Acad. Sci. USA* **81**:7333–7337.

Hogg, R. W., 1977, L-Arabinose transport and the L-arabinose binding protein of *Escherichia coli*, *J. Supramol. Struct.* **6**:411–417.

Hogg, R. W., and Englesberg, E., 1969, L-Arabinose binding protein from *Escherichia coli* B/r, *J. Bacteriol.* **100**:423–431.

Hogg, R. W., and Hermodson, M. A., 1977, Amino acid sequence of the L-arabinose-binding protein from *Escherichia coli* B/r, *J. Biol. Chem.* **252**:5135–5141.

Hong, J.-S., Hunt, A. G., Masters, P. S., and Lieberman, M. A., 1979, Requirement of acetyl phosphate for the binding protein-dependent transport systems in *Escherichia coli*, *Proc. Natl. Acad. Sci. USA* **76**:1213–1217.

Horecker, B. L., Thomas, J., and Monod, J., 1960, Galactose transport in *Escherichia coli*, *J. Biol. Chem.* **235**:1580–1590.

Horne, P., 1980, Galactose transport into membrane vesicles, Ph.D. thesis, University of Cambridge.

Horne, P., and Henderson, P. J. F., 1983, The association of proton movement with galactose transport into subcellular membrane vesicles of *Escherichia coli*, *Biochem. J.* **210**:699–705.

Hugenholtz, J., Hong, J.-S., and Kaback, M. R., 1981, ATP-driven transport in right-side-out bacterial membrane vesicles, *Proc. Nat. Acad. Sci. USA* **78**:3446–3449.

Hunt, A. G., and Hong, J.-S., 1981, The reconstitution of binding protein-dependent active transport of glutamine in isolated membrane vesicles from *Escherichia coli*, *J. Biol. Chem.* **256**:11988–11991.

Iida, A., Harayama, S., Iino, T., and Hazelbauer, G. L., 1984, Molecular cloning and characterization of genes required for ribose transport and utilization in *Escherichia coli* K-12, *J. Bacteriol.* **158**:674–682.

Ingledew, W. S., and Poole, R. K., 1984, The respiratory chains of *Escherichia coli*, *Microbiol. Rev.* **48**:222–271.

Inouye, M. (ed.), 1979, *Bacterial Outer Membranes*, Wiley, New York.

Jagendorf, A. T., and Uribe, E., 1966, ATP formation caused by acid–base transition of spinach chloroplasts, *Proc. Natl. Acad. Sci. USA* **55**:170–177.

Jobe, A., and Bourgeois, S., 1972, Lac repressor operator interaction. VI. The natural inducer of the *lac* operon, *J. Mol. Biol.* **69**:397–408.

Jones, T. H. D., and Kennedy, E. P., 1969, Characterization of the membrane protein of the lactose transport system of *Escherichia coli*, *J. Biol. Chem.* **244**:5981–5987.

Jones-Mortimer, M. C., and Henderson, P. J. F., 1986, Use of transposons to isolate and characterise mutants lacking membrane proteins, illustrated by the sugar transport systems of *Escherichia coli*, *Methods Enzymol.* **125**:157–180.

Kaback, H. R., 1971, Bacterial membranes, *Methods Enzymol.* **22**:99–120.

Kaback, H. R., 1972, Transport across isolated bacterial cytoplasmic membranes, *Biochim. Biophys. Acta* **265**:367–416.

Kaback, H. R., 1974, Transport studies in bacterial membrane vesicles, *Science* **186**:882–892.

Kaback, H. R., 1977, Molecular biology and energetics of membrane transport, in: *Biochemistry of Membrane Transport* (G. Semenza and E. Carafoli, eds.), pp. 598–625, Springer-Verlag, Berlin.

Kaback, H. R., 1983, The *Lac* carrier protein in *Escherichia coli*, *J. Membr. Biol.* **76**:95–112.

Kaczorowski, G. J., and Kaback, H. R., 1979, Effect of pH on efflux, exchange and counterflow of lactose translocation in membrane vesicles from *Escherichia coli*, *Biochemistry* **18**:3691–3697.

Kaczorowski, G. J., LeBlanc, G., and Kaback, H. R., 1980, Specific labelling of the *lac* carrier protein in membrane vesicles of *Escherichia coli* by a photoaffinity reagent, *Proc. Natl. Acad. Sci. USA* **77**:6319–6323.

Kagawa, Y., 1978, Reconstitution of the energy transformer, gate and channel, subunit reassembly, crystalline ATPase and ATP synthesis, *Biochim. Biophys. Acta* **505**:45–93.

Kagawa, Y., 1984, Proton motive ATP synthesis, in: *Bioenergetics* (L. Ernster, ed.), pp. 149–186, Elsevier, Amsterdam.

Katz, L., and Englesberg, E., 1971, Hyperinducibility as a result of mutation in structural genes and self-catabolite repression in the *ara* operon, *J. Bacteriol.* **107**:34–52.

Kellerman, O., and Szmelcman, S., 1974, Involvement of a "periplasmic" binding protein in active transport of maltose by *Escherichia coli* K12, *Eur. J. Biochem.* **47**:139–149.

Kennedy, E. P., 1970, The lactose permease system of *Escherichia coli*, in: *The Lactose Operon* (J. R. Beckwith and D. Zipser, eds.), pp. 49–92, Cold Spring Harbor Laboratory, Cold Spring Harbor, N.Y.

Klein, W. L., and Boyer, P. D., 1972, Energisation of active transport by *Escherichia coli*, *J. Biol. Chem.* **247**:7257–7265.

Kolodrubetz, D., and Schleif, R., 1981a, L-Arabinose transport systems in *Escherichia coli* K-12, *J. Bacteriol.* **148**:472–479.

Kolodrubetz, D., and Schleif, R., 1981b, Regulation of the L-arabinose transport operons in *Escherichia coli*, *J. Mol. Biol.* **151**:215–227.

König, B., and Sandermann, H., 1982, β-D-Galactoside transport in *Escherichia coli*, M_r determination of the transport protein in organic solvent, *FEBS Lett.* **147**:31–34.

Konings, W. N., and Boonstra, J., 1977, Anaerobic electron transfer and active transport in bacteria, *Curr. Top. Membr. Transp.* **9**:177–231.

Koshland, D. E., 1977, Bacterial chemotaxis and some enzymes in energy metabolism, *Symp. Soc. Gen. Microbiol.* **27**:317–331.

Kosiba, B. E., and Schleif, R., 1982, Arabinose-inducible promoter from *Escherichia coli*, *J. Mol. Biol.* **156**:53–56.

Kyte, J., and Doolittle, R. F., 1982, A simple method for displaying the hydropathic character of a protein, *J. Mol. Biol.* **157**:105–132.

Lam, V. M. S., Daruwalla, K. R., Henderson, P. J. F., and Jones-Mortimer, M. C., 1980, Proton-linked D-xylose transport in *Escherichia coli*, *J. Bacteriol.* **143**:396–402.

Lancaster, J. R., 1982, Mechanism of lactose–proton cotransport in *Escherichia coli*, *FEBS Lett.* **150**:9–18.

Lancaster, J. R., and Hinkle, P. C., 1977, Studies of the β-galactoside transporter in inverted membrane vesicles of *Escherichia coli*, *J. Biol. Chem.* **252**:7657–7661.

Landick, R., and Oxender, D. L., 1982, Bacterial periplasmic binding proteins, in: *Membranes and Transport* (A. N. Martonosi, ed.), Vol. 2, pp. 81–88, Plenum Press, New York.

Larsen, S. H., Adler, J., Gargus, J. J., and Hogg, R. W., 1974, Chemomechanical coupling without ATP: The source of energy for motility and chemotaxis in bacteria, *Proc. Natl. Acad. Sci. USA* **71**:1239–1243.

Lawlis, V. B., Dennis, M. S., Chen, E. Y., Smith, D. H., and Henner, D. J., 1984, DNA sequence of the xylA and xylB genes of Escherichia coli, Appl. Environ. Microbiol. 47:15–21.

Lengeler, J., Hermann, K. O., Unsöld, H. J., and Boos, W., 1971, The regulation of the β-methylgalactoside transport system and of the galactose binding protein of Escherichia coli K-12, Eur. J. Biochem. 19:457–470.

Lever, J. E., 1972, Quantitative assay of the binding of small molecules to protein: Comparison of dialysis and membrane filter assays, Anal. Biochem. 50:73–83.

Lombardi, F. J., 1981, Lactose–H^+ (OH^-) transport system of Escherichia coli, Biochim. Biophys. Acta 649:661–679.

Lombardi, F. J., and Kaback, H. R., 1972, The transport of amino acids by membranes prepared from Escherichia coli, J. Biol. Chem. 247:7844–7857.

Lopilato, J. E., Tsuchiya, T., and Wilson, T. H., 1978, Role of Na^+ and Li^+ in thiomethylgalactoside transport by the melibiose transport system of Escherichia coli, J. Bacteriol. 134:147–156.

Lopilato, J. E., Garwin, J. L., Emr, S. D., Silhavy, T. J., and Beckwith, J. R., 1984, D-Ribose metabolism in Escherichia coli K-12; Genetics, regulation, and transport, J. Bacteriol. 158:665–673.

Macpherson, A. J. S., Jones-Mortimer, M. C., and Henderson, P. J. F., 1981, Identification of the AraE transport protein of Escherichia coli, Biochem. J. 196:269–283.

Macpherson, A. J. S., Jones-Mortimer, M. C., Horne, P., and Henderson, P. J. F., 1983, Identification of the GalP galactose transport protein of Escherichia coli, J. Biol. Chem. 258:4390–4396.

Mahoney, W. C., Hogg, R. W., and Hermodson, M. A., 1981, The amino acid sequence of the D-galactose-binding protein from E. coli B/r, J. Biol. Chem. 256:4350–4356.

Maniatis, T., Fritsch, E. F., and Sambrook, J., 1982, Molecular Cloning: A Laboratory Manual, Cold Spring Harbor Laboratory, Cold Spring Harbor, N.Y.

Markgraf, M., Bocklage, H., and Müller-Hill, B., 1985, A change of threonine 266 to isoleucine in the lac permease of Escherichia coli diminishes the transport of lactose and increases the transport of maltose, Mol. Gen. Genet. 198:473–475.

Mitaku, S., Wright, J. K., Best, L., and Jähnig, F., 1984, Localization of the galactoside binding site in the lactose carrier of Escherichia coli, Biochim. Biophys. Acta 776:247–258.

Mitchell, P., 1961, Coupling of phosphorylation to electron and hydrogen transfer by a chemiosmotic type of mechanism, Nature 191:144–148.

Mitchell, P., 1963, Molecule, group and electron translocation through natural membranes, Biochem. Soc. Symp. 22:142–169.

Mitchell, P., 1966, Chemiosmotic coupling in oxidative and photosynthetic phosphorylation, Biol. Rev. 41:445–502.

Mitchell, P., 1973, Performance and conservation of osmotic work by proton-coupled solute porter systems, Bioenergetics 4:63–91.

Mitchell, P., 1976, Vectorial chemistry and the molecular mechanics of chemiosmotic coupling: Power transmission by proticity, Biochem. Soc. Trans. 4:399–430.

Mogi, T., and Anraku, Y., 1984, Mechanism of proline transport in Escherichia coli K12, J. Biol. Chem. 259:7791–7796.

Müller, N., Heine, H.-G., and Boos, W., 1982, Cloning of mglB, the structural gene for the galactose-binding protein of Salmonella typhimurium and Escherichia coli, Mol. Gen. Genet. 185:473–480.

Neu, H. C., and Heppel, L. A., 1965, The release of enzymes from Escherichia coli by osmotic shock and during the formation of spheroplasts, J. Biol. Chem. 240:3685–3692.

Newman, M. J., and Wilson, T. H., 1980, Solubilization and reconstitution of the lactose transport system from Escherichia coli, J. Biol. Chem. 255:10583–10586.

Newman, M. J., Foster, D. L., Wilson, T. H., and Kaback, H. R., 1981, Purification and Reconstitution of functional lactose carrier from Escherichia coli, J. Biol. Chem. 256:11804–11808.

Niiya, S., Yamasaki, K., Wilson, T. H., and Tsuchiya, T., 1982, Altered cation coupling to melibiose transport in mutants of *Escherichia coli*, *J. Biol. Chem.* **257**:8902-8906.

Novotny, C. P., and Englesberg, E., 1966, The L-arabinose permease system in *Escherichia coli* B/r, *Biochim. Biophys. Acta* **117**:217-230.

Ordal, G. W., and Adler, J., 1974, Isolation and complementation of mutants in galactose taxis and transport, *J. Bacteriol.* **117**:509-516.

Page, M. G. P., and West, I. C., 1981, The kinetics of the β-galactoside–proton symport of *Escherichia coli*, *Biochem. J.* **196**:721-731.

Page, M. G. P., and West, I. C., 1982, Alternative-substrate inhibition and the kinetic mechanism of the β-galactoside proton symport of *Escherichia coli*, *Biochem. J.* **204**:681-688.

Page, M. G. P., and West, I. C., 1984, The transient kinetics of uptake of galactosides into *Escherichia coli*, *Biochem. J.* **223**:723-731.

Parsons, R. G., and Hogg, R. W., 1974, Crystallisation and characterisation of the L-arabinose-binding protein of *Escherichia coli* B/r, *J. Biol. Chem.* **249**:3602-3607.

Patel, L., Garcia, M. L., and Kaback, H. R., 1982, Direct measurement of lactose/proton symport in *Escherichia coli*, *Biochemistry* **21**:5805-5810.

Pavlasova, E., and Harold, F. M., 1969, Energy coupling in the transport of β-galactosides by *Escherichia coli*: Effect of proton conductors, *J. Bacteriol.* **98**:198-204.

Plate, C. A., 1979, Requirement for membrane potential in active transport of glutamine by *Escherichia coli*, *J. Bacteriol.* **137**:221-225.

Plate, C. A., Suit, J. L., Jetten, A. M., and Luria, S. E., 1974, Effects of colicin K on a mutant of *Escherichia coli* deficient in Ca^{++}, Mg^{++}-ATPase, *J. Biol. Chem.* **249**:6138-6143.

Prestidge, L. S., and Pardee, A. B., 1965, A second permease for methylthio-β-D-galactoside in *Escherichia coli*, *Biochim. Biophys. Acta* **100**:591-593.

Quiocho, F. A., and Pflugrath, J. W., 1980, The structure of D-galactoside-binding protein at 4.1Å resolution looks like L-arabinose-binding protein, *J. Biol. Chem.* **255**:6559-6561.

Quiocho, F. A., and Vyas, N. K., 1984, Novel stereospecificity of the L-arabinose-binding protein, *Nature* **310**:381-386.

Quiocho, F. A., Phillips, G. N., Parsons, R. G., and Hogg, R. W., 1974, Crystallographic data of an L-arabinose binding protein from *Escherichia coli*, *J. Mol. Biol.* **86**:491-493.

Quiocho, F. A., Gilliland, G. L., and Phillips, G. N., 1977, The 2.8Å resolution structure of the L-arabinose binding protein from *Escherichia coli*, *J. Biol. Chem.* **252**:5142-5149.

Quiocho, F. A., Meador, W. E., and Pflugrath, J. W., 1979, Preliminary crystallographic data of receptors for transport and chemotaxis in *Escherichia coli*: D-Galactose and maltose-binding proteins, *J. Mol. Biol.* **133**:181-184.

Rahmanian, M., Claus, D. R., and Oxender, D. L., 1973, Multiplicity of leucine transport systems in *Escherichia coli* K12, *J. Bacteriol.* **116**:1258-1266.

Raibaud, O., and Schwartz, M., 1980, Restriction map of the *Escherichia coli* malA region and identification of the malT product, *J. Bacteriol.* **143**:761-771.

Raibaud, O., Roa, M., Braun-Breton, C., and Schwartz, M., 1980, Genetic map of the malK-lamB operon, *Mol. Gen. Genet.* **174**:241-248.

Ramos, S., and Kaback, H. R., 1977a, The electrochemical proton gradient in *Escherichia coli* membrane vesicles, *Biochemistry* **16**:848-854.

Ramos, S., and Kaback, H. R., 1977b, The relationship between the electrochemical proton gradient and active transport in *Escherichia coli* membrane vesicles, *Biochemistry* **16**:854-859.

Ramos, S., Schuldiner, S., and Kaback, H. R., 1976, The electrochemical gradient of protons and its relationship to active transport in *Escherichia coli* membrane vesicles, *Proc. Natl. Acad. Sci. USA* **73**:1892-1896.

Richarme, G., 1985, 5-Methoxyindole-2-carboxylic acid, a potent inhibitor of binding protein dependent transport in *Escherichia coli*, *Biochim. Biophys. Acta* **815**:37-43.

Rickenberg, H. W., Cohen, G. N., Buttin, G., and Monod, J., 1956, La galactoside-permease d'*Escherichia coli, Ann. Int. Pasteur Paris* **91**:829.

Riordan, C., and Kornberg, H. L., 1977, Location of *galP,* a gene which specifies galactose permease activity, on the *Escherichia coli* linkage map, *Proc. R. Soc. London Ser. B* **198**:401–410.

Robbins, A. R., 1975, Regulation of the *Escherichia coli* methylgalactoside transport system by gene *mglD, J. Bacteriol.* **123**:69–74.

Robbins, A., and Rotman, B., 1975, Evidence for binding protein-independent substrate translocation by the methylgalactoside transport system of *Escherichia coli* K12, *Proc. Natl. Acad. Sci. USA* **72**:423–427.

Robbins, A. R., Guzman, R., and Rotman, B., 1976, Roles of individual *mgl* gene products in the β-methyl-galactoside transport system of *Escherichia coli* K12, *J. Biol. Chem.* **251**:3112–3116.

Rosen, B. P., 1973a, β-galactoside transport and proton movements in an ATPase-deficient mutant of *Escherichia coli, Biochem. Biophys. Res. Commun.* **53**:1289–1296.

Rosen, B. P., 1973b, Restoration of active transport in an Mg^{++}-ATPase-deficient mutant of *Escherichia coli, J. Bacteriol.* **116**:1124–1129.

Rosen, B. P. (ed.), 1978, *Bacterial Transport,* Marcel Dekker, New York.

Rosen, B. P., and Kashket, E. R., 1978, Energetics of active transport, in: *Bacterial Transport* (B. P. Rosen, ed.), pp. 559–620, Marcel Dekker, New York.

Rosen, B. P., and McLees, J. J., 1974, Active transport of calcium in inverted membrane vesicles of *Escherichia coli, Proc. Natl. Acad. Sci. USA* **71**:5042–5046.

Rosenberg, H., Gerdes, R. G., and Chegwidden, K., 1977, Two systems for the uptake of phosphate in *Escherichia coli, J. Bacteriol.* **131**:505–511.

Rosenfeld, S. A., Stevis, P. E., and Ho, N. W. Y., 1984, Cloning and characterisation of the *xyl* genes from *Escherichia coli, Mol. Gen. Genet.* **194**:410–415.

Rotman, B., 1959, Separate permeases for the accumulation of methyl-β-D-galactoside and methyl-β-D-thiogalactoside in *Escherichia coli, Biochim. Biophys. Acta* **32**:599–601.

Rotman, B., and Guzman, R., 1982, Identification of the *mglA* gene product in the β-methylgalactoside transport system of *Escherichia coli* using plasmid DNA deletions generated *in vitro, J. Biol. Chem.* **257**:9030–9034.

Rotman, B., Ganesan, A. K., and Guzman, R., 1968, Transport systems for galactose and galactosides in *Escherichia coli:* Substrate and inducer specificities, *J. Mol. Biol.* **36**:247–260.

Sancar, A., Hack, A. M., and Rupp, W. D., 1979, Simple method for identification of plasmid-coded proteins, *J. Bacteriol.* **137**:692–993.

Sanger, F., Nicklen, S., and Coulson, A. R., 1977, DNA sequencing with chain-terminating inhibitors, *Proc. Natl. Acad. Sci. USA* **74**:5463–5467.

Sanger, F., Coulson, A. R., Barrell, B. G., Smith, A. J. H., and Roe, B. A., 1980, Cloning in single-stranded bacteriophage as an aid to rapid DNA sequencing, *J. Mol. Biol.* **143**:161–178.

Sanger, F., Coulson, A. R., Hong, G. F., Hill, D. F., and Petersen, G. B., 1982, Nucleotide sequence of bacteriophage λ DNA, *J. Mol. Biol.* **162**:729–773.

Schaefler, S., 1967, Inducible system for the utilization of β-glucosides in *Escherichia coli.* I. Active transport and utilization of β-glucosides, *J. Bacteriol.* **93**:254–263.

Schaefler, S., and Maas, W. K., 1967, Inducible system for the utilization of β-glucosides in *Escherichia coli.* II. Description of mutant types and genetic analysis, *J. Bacteriol.* **93**:264–272.

Schaffner, W., and Weissman, C., 1973, A rapid, sensitive and specific method for the determination of protein in dilute solution, *Anal. Biochem.* **56**:502–514.

Schairer, H. U., and Haddock, B. A., 1972, β-Galactoside accumulation in a Mg^{2+}/Ca^{2+} activated ATPase deficient mutant of *Escherichia coli, Biochem. Biophys. Res. Commun.* **48**:544–551.

Schleif, R., 1969, An L-arabinose binding protein and arabinose permeation in *Escherichia coli*, *J. Mol. Biol.* **46**:185–196.

Schmitt, R., 1968, Analysis of melibiose mutants deficient in α-galactosidase and thiomethylgalactoside permease II in *Escherichia coli* K12, *J. Bacteriol.* **96**:462–471.

Schuldiner, S., and Kaback, H. R., 1975, Membrane potential and active transport in membrane vesicles from *Escherichia coli*, *Biochemistry* **14**:5451–5461.

Schwartz, M., Kellerman, O., Szmelcman, S., and Hazelbauer, C. L., 1976, Further studies on the binding of maltose to the maltose-binding protein of *Escherichia coli*, *Eur. J. Biochem.* **71**:167–170.

Shamanna, D. K., and Sanderson, K. E., 1979a, Uptake and catabolism of D-xylose in *Salmonella typhimurium* LT2, *J. Bacteriol.* **139**:64–70.

Shamanna, D. K., and Sanderson, K. E., 1979b, Genetics and regulation of D-xylose utilization of *Salmonella typhimurium* LT2, *J. Bacteriol.* **139**:71–79.

Shuman, H. A., and Silhavy, T. J., 1981, Identification of the *malK* gene product: A peripheral membrane component of the *Escherichia coli* maltose transport system, *J. Biol. Chem.* **256**:560–562.

Shuman, H. A., Silhavy, T. J., and Beckwith, J. R., 1980, Labeling of proteins with β-galactosidase by gene fusion, *J. Biol. Chem.* **255**:168–174.

Silhavy, T. J., Ferenci, T., and Boos, W., 1978, Sugar transport systems in *Escherichia coli*, in: *Bacterial Transport* (B. P. Rosen, ed.), pp. 127–169, Dekker, New York.

Silhavy, T. J., Benson, S. A., and Emr, S. D., 1983, Mechanisms of protein localization, *Microbiol. Rev.* **47**:313.

Silhavy, T. J., Berman, M. L., and Enquist, L. W., 1984, *Experiments with Gene Fusions*, Cold Spring Harbor Laboratory, Cold Spring Harbor, N.Y.

Silver, S., 1978, Transport of cations and anions, in: *Bacterial Transport* (B. P. Rosen, ed.), pp. 221–324, Dekker, New York.

Simoni, R. D., and Postma, P. W., 1975, The energetics of bacterial active transport, *Annu. Rev. Biochem.* **44**:523–554.

Sinnott, M. L., and Viratelle, O. M., 1973, The pH dependence of galactosylation and degalactosylation with β-galactosidase, *Biochem. J.* **133**:81–87.

Stalmach, M. E., Grothe, S., and Wood, J. S., 1983, Two proline porters in *Escherichia coli* K12, *J. Bacteriol.* **156**:481–486.

Stern, M. J., Ames, G. F.-L., Smith, N. H., Robinson, E. C., and Higgins, C. F., 1984, Repetitive extragenic palindromic sequences: A major component of the bacterial genome, *Cell* **37**:1015–1026.

Stock, J., and Roseman, S., 1971, A sodium-dependent sugar cotransport system in bacteria, *Biochem. Biophys. Res. Commun.* **44**:132–138.

Stoner, C., and Schleif, R., 1983, The *araE* low affinity L-arabinose transport promoter, *J. Mol. Biol.* **171**:369–381.

Stroobant, P., and Kaback, H. R., 1975, Ubiquinone-mediated coupling of NADH dehydrogenase to active transport in membrane vesicles from *Escherichia coli*, *Proc. Natl. Acad. Sci. USA* **72**:3970–3974.

Tanaka, K., Niiya, S., and Tsuchiya, T., 1980, Melibiose transport in *Escherichia coli*, *J. Bacteriol.* **141**:1031–1036.

Teather, R. M., Hamelin, O., Schwarz, H., and Overath, P., 1977, Functional Symmetry of the β-galactoside carrier in *Escherichia coli*, *Biochim. Biophys. Acta* **467**:386–395.

Teather, R. M., Müller-Hill, B., Abrutsch, U., Aichele, G., and Overath, P., 1978, Amplification of the lactose carrier protein in *Escherichia coli* using a plasmid vector, *Mol. Gen. Genet.* **159**:239–248.

Teather, R. M., Bramhall, J., Riede, I., Wright, J. K., Fürst, M., Aichele, G., Wilhelm, U., and Overath, P., 1980, Lactose carrier protein of *Escherichia coli*, *Eur. J. Biochem.* **108**:223–231.

Tokuda, H., and Kaback, H. R., 1977, Sodium-dependent methyl-1-thio-β-D-galactopyranoside transport in membrane vesicles isolated from *Salmonella typhimurium*, *Biochemistry* **16**:2130–2136.

Trumble, W. R., Viitanen, P. V., Sarkar, H. K., Poonian, M. S., and Kaback, H. R., 1984, Site-directed mutagenesis of Cys$_{148}$ in the *lac* carrier protein of *Escherichia coli*, *Biochem. Biophys. Res. Commun.* **119**:860–867.

Tsuchiya, T., and Wilson, T. H., 1978, Cation–sugar cotransport in the melibiose transport system of *Escherichia coli*, *Membr. Biochem.* **2**:63–79.

Tsuchiya, T., Raven, J., and Wilson, T. H., 1977, Co-transport of Na$^+$ and methyl-β-D-thiogalactopyranoside mediated by the melibiose transport system of *Escherichia coli*, *Biochem. Biophys. Res. Commun.* **76**:26–31.

Tsuchiya, T., Lopilato, J., and Wilson, T. H., 1978, Effect of lithium ion on melibiose transport in *Escherichia coli*, *J. Membr. Biol.* **42**:45–59.

Tsuchiya, T., Takeda, K., and Wilson, T. H., 1980, H$^+$–substrate cotransport by the melibiose membrane carrier in *Escherichia coli*, *Membr. Biochem.* **3**:131–146.

Tsuchiya, T., Ottina, K., Moriyama, Y., Newman, M. J., and Wilson, T. H., 1982, Solubilization and reconstitution of the melibiose carrier from a plasmid-carrying strain of *Escherichia coli*, *J. Biol. Chem.* **257**:5125–5128.

Viitanen, P., Garcia, M. L., and Kaback, H. R., 1984, Purified reconstituted *lac* carrier protein from *Escherichia coli* is fully functional, *Proc. Natl. Acad. Sci. USA* **81**:1629–1633.

von Meyenburg, K., and Nikaido, H., 1977, Specificity of the transport process catalysed by the λ receptor protein, *Biochem. Biophys. Res. Commun.* **78**:1100–1107.

von Wilcken-Bergmann, B., and Müller-Hill, B., 1982, Sequence of *galR* gene indicates a common evolutionary origin of *lac* and *gal* repressor in *Escherichia coli*, *Proc. Natl. Acad. Sci. USA* **79**:2427–2431.

Vorisek, J., and Kepes, A., 1972, Galactose transport in *Escherichia coli* and the galactose-binding protein, *Eur. J. Biochem.* **28**:364–372.

Vyas, N. K., Vyas, M. N., and Quiocho, F. A., 1983, The 3Å resolution structure of a D-galactose-binding protein for transport and chemotaxis in *Escherichia coli*, *Proc. Natl. Acad. Sci. USA* **80**:1792–1796.

Walker, J. E., Saraste, M., Runswick, M. J., and Gay, N. J., 1982, Distantly related sequences in the α- and β-subunits of ATP synthase, myosin, kinases and other ATP-requiring enzymes and a common nucleotide binding fold, *EMBO J.* **1**:945–951.

Walker, J. E., Saraste, M., and Gay, N. J., 1984, Nucleotide sequence, regulation and structure of ATP-synthase, *Biochim. Biophys. Acta* **768**:164–200.

West, I. C., 1970, Lactose transport coupled to proton movements in *Escherichia coli*, *Biochem. Biophys. Res. Commun.* **41**:655–661.

West, I. C., 1980, Energy coupling in secondary active transport, *Biochim. Biophys. Acta* **604**:91–126.

West, I. C., 1983, *The Biochemistry of Membrane Transport*, Chapman & Hall, London.

West, I. C., and Mitchell, P., 1972, Proton-coupled β-galactoside translocation in non-metabolizing *Escherichia coli*, *Bioenergetics* **3**:445–462.

West, I. C., and Mitchell, P., 1973, Stoichiometry of lactose–proton symport across the plasma membrane of *Escherichia coli*, *Biochem. J.* **132**:587–592.

West, I. C., and Mitchell, P., 1974, Proton/sodium ion antiport in *Escherichia coli*, *Biochem. J.* **144**:87–90.

West, I. C., and Page, M. G. P., 1984, When is the outer membrane of *Escherichia coli* rate-limiting for uptake of galactosides?, *J. Theor. Biol.* **110**:11–19.

Widdas, W. F., 1952, Inability of diffusion to account for placental glucose transfer in the sheep and consideration of the kinetics of a possible carrier transfer, *J. Physiol. (London)* **11**:23–39.

Wiesmeyer, H., and Cohn, H., 1960, The characterisation of the pathway of maltose utilisation by *Escherichia coli*, *Biochim. Biophys. Acta* **39**:440–447.

Willis, R. C., and Furlong, C. E., 1974, Purification and properties of a ribose binding protein from *Escherichia coli*, *J. Biol. Chem.* **249**:6926–6929.

Wilson, D. B., 1974, The regulation and properties of the galactose transport system in *Escherichia coli* K12, *J. Biol. Chem.* **249**:553–558.

Wilson, D. B., 1976, Properties of the entry and exit reactions of the beta-methyl galactoside transport system in *Escherichia coli*, *J. Bacteriol.* **126**:1156–1165.

Winkler, H. H., and Wilson, T. H., 1966, The role of energy coupling in the transport of β-galactosides by *Escherichia coli*, *J. Biol. Chem.* **241**:2200–2211.

Winter, G., Fersht, A. R., Wilkinson, A. J., Zoller, M., and Smith, M., 1982, Redesigning enzyme structure by site-directed mutagenesis: Tyrosyl tRNA synthetase and ATP binding, *Nature* **299**:756–758.

Witholt, B., and Boekhout, M., 1978, The effect of osmotic shock on the accessibility of the murein layer of exponentially growing *Escherichia coli* to lysozyme, *Biochim. Biophys. Acta* **508**:296–305.

Witholt, B., Boekhout, M., Brock, M., Kingma, T., van Heerikhuizen, H., and de Leij, L., 1976, An efficient and reproducible procedure for the formation of spheroplasts from variously grown *Escherichia coli*, *Anal. Biochem.* **74**:160–170.

Wong, P. T. S., and Wilson, T. H., 1970, Counterflow of galactosides in *Escherichia coli*, *Biochim. Biophys. Acta* **196**:336–350.

Wright, J. K., Riede, I., and Overath, P., 1981, Lactose carrier protein of *Escherichia coli*: Interaction with galactosides and protons, *Biochemistry* **20**:6404–6415.

Wright, J. K., Teather, R. M., and Overath, P., 1983, Lactose permease of *Escherichia coli*, *Methods Enzymol.* **97**:158–175.

Wu, H. C. P., 1967, Role of the galactose transport system in the establishment of endogenous induction of the galactose operon in *Escherichia coli*, *J. Mol. Biol.* **24**:213–223.

Wu, H. C. P., Boos, W., and Kalckar, H. M., 1969, Role of the galactose transport system in the retention of intracellular galactose in *Escherichia coli*, *J. Mol. Biol.* **41**:109–120.

Yazyu, H., Shiota-Niiya, S., Shimamoto, T., Kanazawa, H., Futai, M., and Tsuchiya, T., 1984, Nucleotide sequence of the *melB* gene and characteristics of deduced amino acid sequence of the melibiose carrier in *Escherichia coli*, *J. Biol. Chem.* **259**:4320–4326.

Yazyu, H., Shiota, S., Futai, M., and Tsuchiya, T., 1985, Alteration in cation specificity of the melibiose transport carrier of *Escherichia coli* due to replacement of proline 122 with serine, *J. Bacteriol.* **162**:933–937.

12

Convergent Pathways of Sugar Catabolism in Bacteria

RONALD A. COOPER

1. INTRODUCTION

At first sight, bacteria appear to have a multiplicity of pathways for sugar degradation since a wide variety of five- and six-carbon sugars, sugar alcohols, sugar acids, and amino sugars can be utilized. For example, the commonly studied organism *Escherichia coli* can use more than 30 different monosaccharides for growth. However, the situation is more straightforward than it seems since the pathways for the catabolism of these compounds are interlinked and, with very few exceptions, they all produce glyceraldehyde-3-phosphate, which is converted by a trunk pathway (Fig. 1) to a common product of all sugar degradation, pyruvate.

Since D-glucose is the most abundant compound in the biosphere, it is not surprising that the ability to catabolize this sugar is widespread in the bacterial world, although certain methylotrophs and archaebacteria are unable to utilize it.

Despite this general usage, D-glucose is not necessarily catabolized by the route characteristic of eukaryotic cells, the Embden–Meyerhof glycolytic pathway. Whereas that pathway is utilized by a variety of anaerobes and facultative anaerobes, many aerobic bacteria employ a more oxidative sequence, the Entner–Doudoroff pathway (Entner and Doudoroff, 1952). Certain species utilize the hexose monophosphate pathway (Gromet *et al.*, 1957) whereas others use some hexose monophosphate pathway reactions coupled to the phosphorolytic cleavage of pentose phosphates or hexose phosphates, as found in the pentose

RONALD A. COOPER • Department of Biochemistry, University of Leicester, Leicester LE1 7RH, United Kingdom.

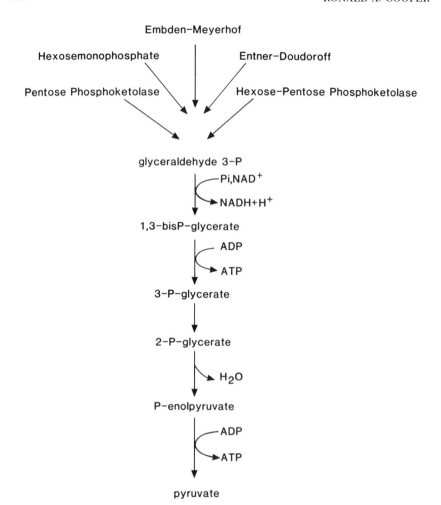

Figure 1. Trunk pathway of sugar catabolism. Various routes for glucose breakdown converge at glyceraldehyde-3-phosphate, which is then converted to pyruvate by the trunk pathway reactions.

phosphate phosphoketolase pathway (Heath *et al.*, 1956) and the hexose phosphate–pentose phosphate phosphoketolase pathway (Scardovi and Trovatelli, 1965; de Vries *et al.*, 1967). Nevertheless, these different pathways for D-glucose breakdown again all form glyceraldehyde-3-phosphate (Figs. 2, 4, and 5) as an intermediate and the almost ubiquitous occurrence of the trunk sequence from glyceraldehyde-3-phosphate to pyruvate is testimony to its importance in bacterial sugar catabolism.

While it is clear that these reactions of glyceraldehyde-3-phosphate con-

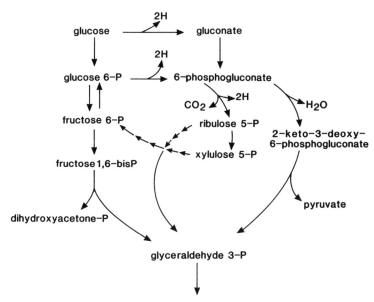

Figure 2. Convergent pathways for glucose catabolism. A schematic representation of the Embden–Meyerhof pathway, the Entner–Doudoroff pathway, and the hexose monophosphate pathway for the conversion of glucose to glyceraldehyde-3-phosphate.

stitute the true central pathway for sugar breakdown in the bacterial world, the sequence is not the only known route from triose phosphate to pyruvate since a potential alternative series of reactions, the methylglyoxal pathway (Fig. 3), has been observed in some bacteria (Cooper and Anderson, 1970).

The entry of exogenous sugar into the cell is usually the first step for all these catabolic sequences although an exception to this rule will be described later for pseudomonads. Sugar transport is an energy-requiring process in virtually all cases but not for the entry of the sugar alcohol glycerol (Lin, 1976). The catabolism of glycerol is unusual in another and possibly related way. In some bacteria the phosphorylation of internal glycerol by an ATP-dependent kinase to give α-glycerophosphate is subject to feedback inhibition by fructose-1,6-bisphosphate and this appears to regulate the overall rate of glycerol utilization (Lin, 1976). No similar feedback regulation of the first step in the internal catabolism of any other sugar has been reported and it may be that in all other cases, where transport is an energy-requiring process, regulation occurs at the point of entry into the cell.

Clearly, sugar transport is an integral part of sugar catabolism but since this topic is dealt with specifically in other chapters it will not be considered further here. Rather, this chapter will concern itself with the variety of reactions by which simple sugars are converted to pyruvate.

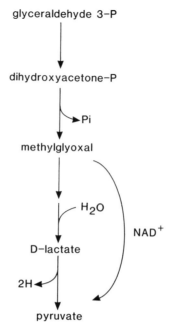

Figure 3. Methylglyoxal bypass of the trunk pathway. The formation of methylglyoxal from di-hydroxyacetone phosphate and its subsequent conversion to pyruvate bypasses the substrate-level phosphorylation steps of the trunk pathway and so uncouples pyruvate formation and ATP synthesis.

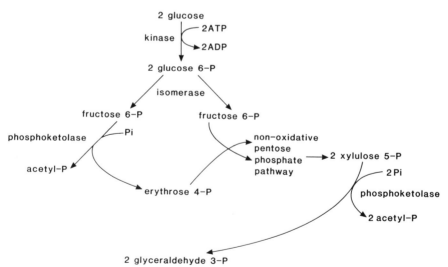

Figure 4. Hexose phosphate–pentose phosphate phosphoketolase pathway. Production of glyceraldehyde-3-phosphate requires certain Embden–Meyerhof and hexose monophosphate pathway reactions in addition to the key phosphorolytic cleavage of fructose-6-phosphate and xylulose-5-phosphate.

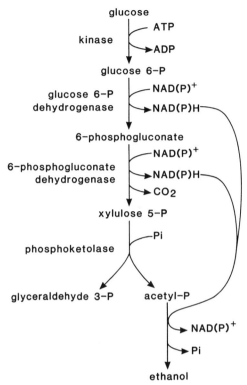

Figure 5. Pentose phosphate phosphoketolase pathway. The key reaction in this sequence is the phosphorolytic cleavage of xylulose-5-phosphate.

2. PATHWAYS FOR THE DEGRADATION OF GLUCOSE

Of the five sequences for glucose breakdown mentioned earlier, the Embden–Meyerhof and Entner–Doudoroff pathways are the most widely found and therefore the best documented.

2.1. *Embden–Meyerhof Pathway*

The sequence of reactions for the conversion of D-glucose to pyruvate via phosphorylated intermediates was established for yeast and muscle by the mid-1930s. Shortly thereafter, evidence in support of a similar sequence of reactions in bacteria began to emerge. Four criteria were generally applied in these studies: (1) production and utilization of the postulated intermediates; (2) conversion of D-[¹⁴C]glucose to products with the labeling patterns required for the operation of the Embden–Meyerhof pathway: (3) presence of the enzymes involved in the pathway of yeast and muscle; and (4) sensitivity of these enzymes and reactions to inhibitors known to be effective with yeast and muscle systems. However, in these early studies a number of different bacterial species were investigated in a rather fragmentary fashion and complete evidence for the Embden–Meyerhof pathway was not obtained for a particular organism. Eventually, evidence in support of the sequence for a number of bacteria was obtained (see Gunsalus *et al.*, 1955).

During the period 1968–1978, analysis of the pathway for D-glucose catabolism in *E. coli* by the isolation and characterization of mutants defective in D-glucose breakdown showed convincingly that the Embden–Meyerhof pathway is of major importance in this organism. Mutants defective in nine of the ten reactions of the Embden–Meyerhof pathway have been isolated (Wang and Morse, 1968; Vinopal *et al.*, 1975; Morrissey and Fraenkel, 1968; Bock and Neidhardt, 1966; Anderson and Cooper, 1969; Irani and Maitra, 1974; Garrido-Pertierra and Cooper, 1977); only phosphoglycerate mutase mutants have not been obtained. In all cases the absence of the particular enzyme has led to either a total or significant inhibition of D-glucose catabolism. A feature highlighted by this work with mutants was that accumulation of phosphorylated sugars as a consequence of a metabolic block was usually lethal to the cells (Bock and Neidhardt, 1966; Cozzarelli *et al.*, 1965; Fraenkel, 1968a).

Although for *E. coli* the overall sequence from D-glucose to pyruvate is the same as that described by Embden and Meyerhof, the first step, the phosphorylation of D-glucose, can occur by two mechanisms. When D-glucose is produced internally as a product of disaccharide hydrolysis, an ATP-dependent phosphorylation catalyzed by glucokinase is involved (Fraenkel *et al.*, 1964; Curtis and Epstein, 1975) and the sequence is exactly that described by Embden and Meyerhof. However, when D-glucose is supplied externally, its phosphorylation involves the phosphoenolpyruvate-dependent phosphotransferase system (PTS) (see Chapters 10 and 11) and is thus mechanistically distinct from the corresponding Embden–Meyerhof reaction.

A large number of anaerobes and facultative anaerobes appear to utilize the Embden–Meyerhof pathway for D-glucose breakdown but the evidence is not as complete as that for *E. coli*.

2.2. Entner–Doudoroff Pathway

A key enzyme of the Embden–Meyerhof pathway, 6-phosphofructokinase, is absent from many strictly aerobic bacteria such as the pseudomonads (Tiwari and Campbell, 1969) so the normal Embden–Meyerhof sequence cannot operate in such organisms. In 1952, Entner and Doudoroff identified a novel pathway for D-glucose breakdown in *Pseudomonas saccharophila* in which glucose-6-phosphate was first oxidized to give 6-phosphogluconate before conversion to glyceraldehyde-3-phosphate and pyruvate (Entner and Doudoroff, 1952). They proposed that 2-keto-3-deoxy-6-phosphogluconate (KDPG) was an intermediate in the reaction and this was soon confirmed (MacGee and Doudoroff, 1954). By use of D-[^{14}C]glucose, enzyme assays, and inhibitors, the pathway shown in Fig. 2 was established. Subsequent studies have indicated that the Entner–Doudoroff pathway is present in a variety of aerobic bacteria (DeLey, 1960; Baumann and Baumann, 1973) and, curiously, it is also found in the anaerobic zymomonads (McGill and Dawes, 1971). In the latter organisms, the NADH produced in the oxidative reactions is reoxidized at the expense of acetaldehyde formed from pyruvate and thus gives end products identical to those of the classic yeast fermentation involving the Embden–Meyerhof sequence. However, only 1 mole of ATP is produced from each mole of D-glucose fermented by the Entner–Doudoroff pathway.

Recent studies involving mutants defective in Entner–Doudoroff pathway and related enzymes, as well as considerations of patterns of enzyme induction during growth on various carbohydrates, have suggested a further role of the Entner–Doudoroff reactions. It has been proposed that in some pseudomonads, glyceraldehyde-3-phosphate formed in the cleavage of KDPG is converted by the reactions of gluconeogenesis to glucose-6-phosphate, thereby producing an ox-idative cycle that converts D-glucose to pyruvate and reduced NAD(P)H (Fig. 2) without any of the substrate-level phosphorylation reactions that occur in the usual trunk sequence of glyceraldehyde-3-phosphate catabolism (Fig. 1) (Lessie and Phibbs, 1984). However, the extent to which glyceraldehyde-3-phosphate is recycled into the Entner–Doudoroff pathway or converted to pyruvate via the trunk pathway reactions is at present unclear and may vary from organism to organism. Obviously, patterns of enzyme induction have to be treated with some caution because of the problems of gratuitous induction. While it is clear that the Entner–Doudoroff pathway enzymes are induced during growth on glycerol, so is the enzyme for the extracellular oxidation of D-glucose to D-gluconate (Hylemon and Phibbs, 1972). Similarly, the interpretation of mutant phenotypes is not always unambiguous. For example, it was considered that phosphoglyce-rate kinase-negative mutants of *P. putida,* which grew normally on D-glucose and glycerol but which failed to grow on a variety of gluconeogenic carbon sources, might recycle glyceraldehyde-3-phosphate through the Entner–Dou-doroff pathway. However, since mutants lacking the key Entner–Doudoroff pathway enzyme 6-phosphogluconate dehydrase still grew on glycerol, though not now on D-glucose, the proposal was dismissed (Aparicio *et al.,* 1971). It is, however, possible that normally glyceraldehyde-3-phosphate is catabolized by both routes and when one is blocked the other is able to compensate for the loss. It may also be that each route operates specifically under particular physiological circumstances, as appears to be the case in some pseudomonads for the reactions by which D-glucose is initially catabolized (see below). The situation with *E. coli* may also be pertinent here. Mutants defective in glyceraldehyde-3-phosphate dehydrogenase should still to be able to grow on D-gluconate if the glyceraldehyde-3-phosphate formed from KDPG were recycled to give D-glucose-6-phosphate. However, such mutants are unable to grow on D-gluconate (Irani and Maitra, 1974; Hillman and Fraenkel, 1975).

Two distinct reaction sequences for the oxidation of D-glucose, eventually to produce 6-phosphogluconate, have been identified (see Fig. 2) (Midgley and Dawes, 1973; Lessie and Phibbs, 1984). The first acts directly on extracellular D-glucose with glucose dehydrogenase and in some cases gluconate dehydrogenase, membrane-associated, pyridine nucleotide-independent enzymes acting in the periplasmic space to form gluconate and 2-ketogluconate, respectively. These aldonic acids are then transported through the cytoplasmic membrane and phosphorylated by ATP-dependent kinases eventually to produce 6-phosphogluconate. The second sequence involves the direct transport of D-glucose through the cytoplasmic membrane and its phosphorylation by an ATP-dependent kinase to give glucose-6-phosphate. This compound is oxidized internally by a pyridine nucleotide-dependent enzyme to give 6-phosphogluconate.

Mutants of *P. aeruginosa* blocked in the external oxidative route because of a deficiency in glucose dehydrogenase still grow normally on D-glucose using the internal oxidative route (Midgley and Dawes, 1973). However, similar mutants of *P. putida* fail to grow on D-glucose because they do not have the alternative internal pathway (Vicente and Canovas, 1973). Curiously, *P. aeruginosa* grown anaerobically with nitrate as the electron acceptor has no detectable activity of glucose dehydrogenase (Hunt and Phibbs, 1983) and growth now involves only the internal oxidation of glucose-6-phosphate. However, such cells still make normal amounts of glucose dehydrogenase apoenzyme but this is nonfunctional because they fail to make the necessary cofactor, pyrroloquinoline quinone (PQQ), under these conditions (van Schie *et al.,* 1984). If PQQ is provided in the environment, the apoenzyme can be activated readily.

Although *E. coli* has the genetic capability to make the Entner–Doudoroff pathway enzymes and the genes specifying glucose-6-phosphate dehydrogenase, 6-phosphogluconate dehydrase, and KDPG aldolase are close together on the chromosome (Bachmann, 1983), it does not normally catabolize D-glucose by the Entner–Doudoroff pathway. Mutational blocks in the Embden–Meyerhof pathway at phosphoglucose isomerase and in the hexose monophosphate pathway at 6-phosphogluconate dehydrogenase are required before D-glucose is catabolized by the Entner–Doudoroff sequence in *E. coli* (Kornberg and Soutar, 1973). However, growth on D-gluconate induces the Entner–Doudoroff pathway enzymes from 6-phosphogluconate and the pathway does operate for D-gluconate catabolism in this organism.

In an interesting recent development, it has been reported that during growth on D-glucose *E. coli* makes a high level of apo-glucose dehydrogenase but since the organism is apparently unable to make PQQ the enzyme is inactive. This was confirmed physiologically when a mutant defective in enzyme I of the PTS would not grow on D-glucose even though it still made the glucose dehydrogenase apoenzyme and could also grow normally on D-gluconate. However, when PQQ was added to the growth medium the mutant now grew on D-glucose, presumably via gluconate, at a rate only 10% slower than that of the parent organism (Hommes *et al.*, 1984).

The potential operation of both Embden–Meyerhof and Entner–Doudoroff pathways for D-glucose catabolism in certain organisms is a further feature that has been identified recently. *Klebsiella aerogenes* growing in a chemostat under D-glucose-limited conditions appears to utilize the PTS for D-glucose transport and presumably the Embden–Meyerhof pathway for glucose-6-phosphate catabolism. When some other component such as K^+ was growth-limiting in a D-glucose-sufficient culture, D-glucose utilization was much greater than the measured rate of the PTS and D-gluconate and /or D-2-ketogluconate were detected in the growth medium. This suggested that D-glucose was catabolized by reactions in addition to the PTS and a PQQ-dependent glucose dehydrogenase was detected at high activity in such cells (Neijssel *et al.*, 1983). Since a PTS-deficient mutant could still catabolize D-glucose at a fast rate, though slightly slower than the parent, it seemed that a pathway of D-glucose catabolism involving its oxidation to D-gluconate and presumably metabolism of this compound by the Entner–Doudoroff pathway could operate, alongside the Embden–Meyerhof pathway, in *K. aerogenes* under particular physiological conditions.

2.3. Hexose Monophosphate Pathway

The oxidation of glucose-6-phosphate to 6-phosphogluconate by yeast and muscle extracts and the further conversion of 6-phosphogluconate to pentose phosphate and CO_2 was first discovered in the 1930s by Warburg and Dickens (see Racker, 1954). By the early 1950s, similar reactions had been reported for a number of bacteria, including *E. coli,* in which a reaction sequence for the conversion of glucose-6-phosphate to pentose phosphate and the further transformation of the pentose phosphate to triose phosphate was proposed (Racker, 1948). These reactions by which 6-phosphogluconate is oxidized further and converted to pentose phosphate are clearly different from the dehydration of 6-phosphogluconate to give KDPG described in the preceding section. Investigation of the reactions by which the pentose was used led to the discovery of a sequence of nonoxidative reactions that produced fructose-6-phosphate and triose phosphate (Fig. 2). It was then proposed that the oxidative reactions starting with glucose-6-phosphate could be coupled to the nonoxidative reactions of the pentose phosphates to produce a cycle that, theoretically, would permit the total oxidation of glucose to CO_2 and H_2O and could thus replace the combination of the Embden–Meyerhof pathway and the tricarboxylic acid cycle for the total combustion of D-glucose (Racker, 1957).

Although the necessary enzymes are present in a large number of bacteria, it now seems clear that these reactions do not operate generally as a cycle for D-glucose oxidation. Rather, their main purpose is to produce the NADPH, ribose-5-phosphate, and erythrose-4-phosphate needed for biosynthesis. Moreover, since there is no simple quantitative equivalence in the demand for these particular intermediates, organisms must have reactions by which any "surplus" compounds can be returned to the central metabolic pathways and the nonoxidative reactions serve this role.

Nevertheless, it has been suggested that the hexose monophosphate (HMP) pathway is a major route for D-glucose degradation in the acetic acid bacteria (Gromet et al., 1957). Acetobacter xylinum is apparently devoid of phosphofructokinase activity (Benziman, 1969) so that the fructose-6-phosphate formed by the pathway may be isomerized to glucose-6-phosphate and thus reenter the pathway to complete a cyclic sequence (Fig. 2). The glyceraldehyde-3-phosphate can be converted to pyruvate by the trunk pathway reactions already described for glyceraldehyde-3-phosphate catabolism. However, A. xylinum contains an enzyme, fructose-6-phosphate phosphoketolase, that cleaves fructose-6-phosphate in the presence of inorganic orthophosphate to give acetylphosphate and erythrose-4-phosphate (Schramm and Racker, 1957). The physiological role of this enzyme is unclear but it may contribute to fructose-6-phosphate catabolism. However, radiorespirometric studies with A. suboxydans suggested that for this organism the HMP pathway was the only significant sequence that led to CO_2 formation from D-glucose (Kitos et al., 1958).

The demonstration of HMP pathway enzymes in organisms possessing either the Embden–Meyerhof or the Entner–Doudoroff pathway enzymes raises the problem of assessing the contribution each pathway makes to D-glucose catabolism. This question has been investigated most extensively for organisms with the Embden–Meyerhof and HMP pathway enzymes using methods that take advantage of the difference in the rate of release of $^{14}CO_2$ from D-[1-^{14}C]glucose and D-[6-^{14}C]glucose that is a consequence of the different reactions involved in the two pathways. The experiments suggested that for a variety of bacteria the HMP pathway might account for 25–30% of the D-glucose catabolized (Wang et al., 1958). More recent estimates for E. coli suggest that 13% of D-glucose is catabolized by the HMP pathway (Orthner and Pizer, 1974). However, such calculations of the partitioning of D-glucose are subject to many sources of error and need to be interpreted with great caution (Katz and Wood, 1960). Similarly, while early respirometric experiments suggested that the HMP pathways played a significant role in the D-glucose catabolism of certain pseudomonads, the possible recycling of glyceraldehyde-3-phosphate into the Entner–Doudoroff sequence was not considered or allowed for in the calculations. Since the enzyme 6-phosphogluconate dehydrogenase appears to be absent from several pseudomonads (Lessie and Phibbs, 1984), it seems likely that the HMP pathway does not play a significant role in the D-glucose catabolism of all pseudomonads.

Some information on the operation of the HMP pathway in *E. coli* has been obtained by studies using mutants. No observable alteration in the rate of growth on D-glucose occurred when 6-phosphogluconate dehydrogenase was missing and ribose-5-phosphate was formed exclusively by reversal of the nonoxidative reactions of the pathway (Fraenkel, 1968b). The absence of phosphoglucoseisomerase prevented the operation of the Embden–Meyerhof pathway and this caused a 66% reduction in the growth rate on D-glucose (Fraenkel and Levisohn, 1967). In these circumstances, glucose-6-phosphate was degraded exclusively by the HMP pathway. Such findings suggest that the operation of the full HMP pathway is not essential for *E. coli* but it can contribute significantly to D-glucose catabolism.

A *P. cepacia* mutant defective in 6-phosphogluconate dehydrase failed to utilize D-glucose or D-gluconate by the HMP pathway despite the presence of significant 6-phosphogluconate dehydrogenase activity (Allenza and Lessie, 1982). Secondary mutants with increased activity of the NAD-dependent 6-phosphogluconate dehydrogenase regained the ability to utilize D-gluconate albeit more slowly than the wild-type organism. It was thought likely that this growth deficiency was due to the toxicity of 6-phosphogluconate accumulating as a result of the block in the Entner–Doudoroff pathway rather than a general inability to utilize the HMP pathway.

2.4. Pentose Phosphate Phosphoketolase Pathway

Some of the anaerobic lactic acid bacteria break down D-glucose to produce pyruvate and then lactate by the same kind of reactions that occur in animal muscle, whereas others produce quite different end products. In this latter group, D-glucose catabolism involves an initial oxidation of glucose-6-phosphate and the end products produced by *Leuconostoc mesenteroides* are equimolar amounts of lactate, ethanol, and CO_2. Catabolism of specifically radiolabeled D-glucose indicated that the CO_2 came from C-1 of glucose, the lactate from C-4, C-5, and C-6, and the methyl-carbon of ethanol from C-2 of D-glucose. Such results are incompatible with the operation of the Embden–Meyerhof pathway and the release of C-1 as CO_2 indicates an involvement of the oxidative HMP pathway reactions. Further, fructose bisphosphate aldolase, triose phosphate isomerase, and 6-phosphofructokinase activity could not be detected in this organism (De-Moss *et al.*, 1951). The pathway involved became apparent on discovery of a novel enzyme, xylulose-5-phosphate phosphoketolase, that catalyzes the phosphorolytic cleavage of xylulose-5-phosphate to produce acetylphosphate and glyceraldehyde-3-phosphate (Fig. 5) (Heath *et al.*, 1956). The glyceraldehyde-3-phosphate is converted to pyruvate by the trunk pathway reactions for glyceraldehyde-3-phosphate catabolism (Fig. 1). Since D-glucose is catabolized anaerobically and the initial reactions are oxidative, both pyruvate and acetaldehyde (from the acetylphosphate) must serve as electron acceptors for the reoxidation of the NADH formed and this gives the characteristic end products of the pathway. As a consequence, less ATP is produced (1 mole/mole D-glucose catabolized) than by the Embden–Meyerhof pathway. Although the xylulose-5-phosphate phosphoketolase pathway has been reported for only a few species of bacteria, it appears to be their sole route for D-glucose catabolism.

2.5. Hexose Phosphate–Pentose Phosphate Phosphoketolase Pathway

A Gram-positive strictly anaerobic organism found in the human intestine, *Bifidobacterium bifidum*, catabolizes 1 mole of D-glucose to 1 mole of lactate and 1.5 moles of acetate. The organism has no detectable glucose-6-phosphate dehydrogenase or fructose-1,6-bisphosphate aldolase and only low 6-phosphofructokinase activity, suggesting that neither the Embden–Meyerhof pathway nor the HMP pathway operates in this organism. Xylulose-5-phosphate phosphoketolase and the closely related enzyme fructose-6-phosphate phosphoketolase were present, however. The operation of these two enzymes in conjunction with the HMP pathway enzymes transaldolase and transketolase permits the conversion of fructose-6-phosphate to acetylphosphate and glyceraldehyde-3-phosphate (Fig. 4) (Scardovi and Trovatelli, 1965; de Vries *et al.*, 1967). Once again, the trunk pathway reactions (Fig. 1) serve to convert the glyceraldehyde-3-phosphate to pyruvate. In the strain of *B. bifidum* first studied, the lactate dehydrogenase that reduced pyruvate to lactate specifically required fructose-1,6-bisphosphate for activity, thereby explaining the need for low 6-phosphofructokinase activity. In other strains, pyruvate can be cleaved in a phosphoclastic reaction giving acetylphosphate and formate (de Vries and Stouthamer, 1968). Depending on which reaction the pyruvate undergoes, 2.5 or 3.0 moles of ATP are produced per mole of D-glucose catabolized.

2.6. Methylglyoxal Pathway

The five pathways for D-glucose catabolism described in the preceding sections produce glyceraldehyde-3-phosphate by different routes but the organisms concerned then have a common pathway to convert the triose phosphate to pyruvate. It may thus seem that this sequence is invariant in D-glucose catabolism but the discovery of the methylglyoxal pathway (Cooper and Anderson, 1970) has shown that a possible alternative route does exist (Fig. 3). In this sequence, dihydroxyacetone phosphate, an isomer of glyceraldehyde-3-phosphate, is converted to methylglyoxal, which then reacts via the glyoxalase system (Dakin and Dudley, 1913) and a flavin-linked dehydrogenase to produce pyruvate. The methylglyoxal pathway was first observed in *E. coli* but subsequently the necessary enzymes were detected in many Enterobacteriaceae and related organisms, in some clostridia, and in certain pseudomonads (Cooper, 1984). The glyoxalase enzyme system for the conversion of methylglyoxal to D-lactate was first described in 1913 but no physiological role could then be allocated to it since no enzymatic formation of methylglyoxal could be found. An alternative route for the methylglyoxal is its direct oxidation to pyruvate by methylglyoxal dehydrogenase (Taylor *et al.*, 1980) although this enzyme has not been detected in organisms producing methylglyoxal from dihydroxyacetone phosphate.

All the organisms that are known to possess the enzymes of the methylglyoxal pathway also have the trunk pathway enzymes for the catabolism of glyceraldehyde-3-phosphate to pyruvate. The enzymes of the methylglyoxal pathway are formed under both aerobic and anaerobic conditions (Teixeira de Mattos and Tempest, 1983; Cooper, 1984) but in contrast to the trunk pathway reactions, their operation does not lead to substrate-level ATP synthesis. In terms of energy provision, the methylglyoxal pathway is thus not an alternative to the trunk pathway reactions and in fact its function may be to uncouple ATP formation from D-glucose breakdown (Teixeira de Mattos and Tempest, 1983; Cooper, 1984).

Clearly, the problem of the extent to which each pathway contributes to D-glucose catabolism arises and as yet there is no wholly satisfactory way of investigating this. Studies with a triose phosphate isomerase-negative mutant that produces methylglyoxal in the growth medium during D-glucose catabolism have shown that D-[1-^{14}C]glucose gives rise to methyl-labeled methylglyoxal and hence to methyl-labeled lactate (Cooper, 1975). This finding has been confirmed by studies on the reaction mechanism of purified methylglyoxal synthase (Iyengar and Rose, 1983). Since D-[1-^{14}C]glucose also produces methyl-labeled lactate by the classical Embden–Meyerhof reactions, the analysis of lactate produced from radiolabeled D-glucose will not indicate its route of formation.

An alternative approach to assess the operation of the methylglyoxal pathway utilizes mutants defective in enzymes of the trunk pathway. Mutants defective in glyceraldehyde-3-phosphate dehydrogenase, phosphoglycerate kinase, or enolase (Irani and Maitra, 1974; Hillman and Fraenkel, 1975) are unable to grow on D-glucose as sole source of carbon and energy, suggesting that the methylglyoxal pathway cannot replace the trunk sequence under such conditions. However, this interpretation is complicated by the fact that the substrate of a missing enzyme is likely to accumulate and growth of *E. coli* is known to be inhibited by high concentrations of sugar phosphates (Bock and Neidhardt, 1966; Cozzarelli *et al.*, 1965; Fraenkel, 1968a). A similar problem was described earlier with experiments to assess the relative operation of the Entner–Doudoroff and HMP pathways in *P. cepacia*. Growth on exogenous D-glucose is almost normal for *E. coli* mutants lacking pyruvate kinase activity since the operation of the PTS allows pyruvate to be formed from phosphoenolpyruvate (Garrido-Pertierra and Cooper, 1977). When lactose or maltose is used for growth, pyruvate kinase mutants grow very slowly since D-glucose is formed internally from these disaccharides and is thus not a substrate for the PTS. Again with this mutant the methylglyoxal pathway does not seem to be a simple alternative to the trunk sequence but since low concentrations of methylglyoxal have been found in the medium during growth on lactose and maltose the methylglyoxal pathway may be responsible for the slow growth seen.

Methylglyoxal is present at 0.1–0.2 mM concentration in the growth medium when wild-type cells grow on D-glucose-6-phosphate (but not when they grow on D-glucose) and in a mutant lacking D-lactate dehydrogenase 1 mM D-lactate was found in the growth medium at the end of growth on 5 mM D-glucose-6-phosphate. Since lactate was not found in the medium when the mutant grew on D-glucose, the D-lactate was unlikely to have been formed by reduction of pyruvate and it seems that 10% of the D-glucose-6-phosphate was catabolized by the methylglyoxal pathway (Cooper, 1975). In the triose phosphate isomerase mutant mentioned earlier where the dihydroxyacetone phosphate formed by D-glucose catabolism cannot be isomerized to glyceraldehyde-3-phosphate, 50% of D-glucose catabolism presumably occurs by the methylglyoxal pathway. This mutant, however, cannot grow for many generations on D-glucose, possibly because the high concentration of methylglyoxal produced is cytotoxic.

The regulatory properties of methylglyoxal synthase and glyceraldehyde-3-phosphate dehydrogenase are compatible with the partitioning of triose phosphate into the methylglyoxal pathway or the trunk sequence depending on physiological circumstances (Hopper and Cooper, 1972). Nevertheless, as yet there is no clear role for the methylglyoxal pathway in D-glucose catabolism. It may, in fact, serve a variety of purposes including the uncoupling of D-glucose catabolism from anabolism and providing intermediates needed by the cell. Methylglyoxal itself can serve as a regulator of cell growth (Egyud and Szent-Gyorgyi, 1966) and, like other low-molecular-weight substances such as α-ketobutyrate and guanosine tetraphosphate that affect cell growth, could be considered to function as an "alarmone." Additionally, the D-lactate produced by the methylglyoxal pathway may be an important metabolite providing the energy for the uptake of a wide range of substances into cells (Kaback, 1974).

3. INDIVIDUAL CATABOLIC PATHWAYS LEADING TO CENTRAL INTERMEDIATES

Although glucose is the most abundant sugar found in nature, many bacteria can utilize a wide variety of other sugars including some that do not occur

naturally. As mentioned earlier, these sugars are catabolized to glyceralde-
hyde-3-phosphate by reaction sequences that either produce an intermediate of
one of the pathways for glucose degradation already described or glyceralde-
hyde-3-phosphate itself. In this section, routes for the conversion of selected
monosaccharides will be considered.

3.1. D-Galactose

Three pathways for galactose utilization have been identified. Two of them,
the D-galactonate pathway [recently called the DeLey–Doudoroff pathway
(Lessie and Phibbs, 1984)] and the D-tagatose-6-phosphate pathway (Bissett and
Anderson, 1973), produce glyceraldehyde-3-phosphate, whereas the Leloir pathway (Kalckar, 1958) produces D-glucose-6-phosphate.

In the DeLey–Doudoroff pathway of *P. saccharophila*, D-galactose is catabolized by a series of reactions that almost parallel the Entner–Doudoroff pathway reactions for D-glucose catabolism (Fig. 2). D-Galactose is first oxidized to
D-galactonate by an NAD-dependent dehydrogenase and the D-galactonate is
dehydrated to 2-keto-3-deoxygalactonate before any phosphorylation occurs
(Fig. 6). The resulting 2-keto-3-deoxy-6-phosphogalactonate undergoes aldol
cleavage to give glyceraldehyde-3-phosphate and pyruvate. The enzymes in-

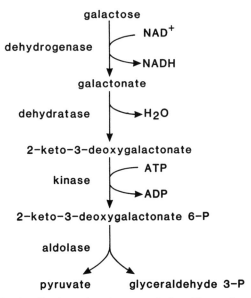

Figure 6. DeLey–Doudoroff pathway for galactose catabolism. The reactions have many features in
common with the Entner–Doudoroff pathway for glucose catabolism but different enzymes are
involved.

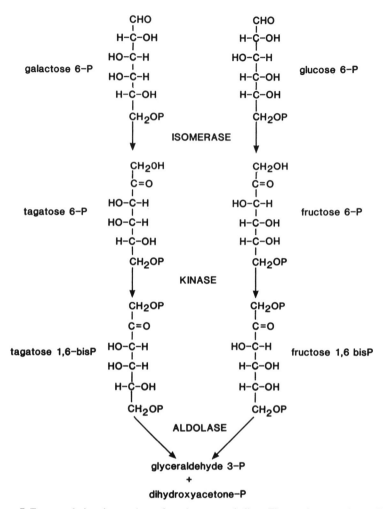

Figure 7. Tagatose-6-phosphate pathway for galactose catabolism. The reactions exactly parallel the Embden–Meyerhof pathway reactions for triose phosphate formation but quite separate enzymes are involved.

volved are inducible and quite distinct from the Entner–Doudoroff pathway enzymes (DeLey and Doudoroff, 1957).

The tagatose-6-phosphate pathway was first observed in *Staphylococcus aureus,* an organism that as a consequence of using the PTS for D-galactose uptake produces internal D-galactose-6-phosphate. The D-galactose-6-phosphate is isomerized to the ketose D-tagatose-6-phosphate and this in turn is phosphory-

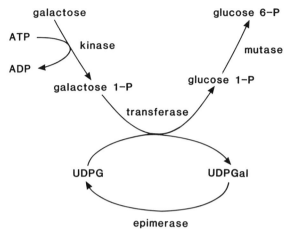

Figure 8. Leloir pathway for galactose catabolism. UDP-glucose plays a catalytic role in this sequence, which effects the formation of glucose-6-phosphate from galactose.

lated to D-tagatose-1,6-bisphosphate. In the final step, D-tagatose-1,6-bisphosphate undergoes aldol cleavage to yield glyceraldehyde-3-phosphate and dihydroxyacetone phosphate. Clearly, these reactions (Fig. 7) resemble very closely the Embden–Meyerhof reactions for the conversion of D-glucose-6-phosphate to triose phosphates. However, mutants lacking particular enzymes of the D-tagatose-6-phosphate pathway and thus unable to grow on D-galactose still grow on D-glucose and have the corresponding Embden–Meyerhof pathway enzymes, showing that the two sequences are quite distinct (Bissett and Anderson, 1974).

The Leloir pathway for D-galactose catabolism involves an ATP-dependent phosphorylation to give D-galactose-1-phosphate, which is then converted to D-glucose-1-phosphate via UDP-galactose and UDP-glucose. The D-glucose-1-phosphate is then converted to D-glucose-6-phosphate by phosphoglucomutase (Fig. 8). This pathway appears to be the sole route for D-galactose catabolism in *E. coli* since mutants defective in any of these steps are unable to utilize D-galactose.

Curiously, *E. coli* does have the enzymes of the DeLey–Doudoroff pathway for the utilization of D-galactonate (Fig. 6) (Cooper, 1978) but does not produce a D-galactose dehydrogenase to permit D-galactose catabolism by this route. However, experiments to construct such a pathway in *E. coli* by genetic manipulation have been reported (Buckel and Zehelein, 1981). DNA from *P. fluorescens,* an organism with D-galactose dehydrogenase, was cloned into a cosmid vector and used to transform a galactokinase-negative mutant of *E. coli.* Even after several days of incubation on plates with D-galactose as sole carbon and energy source, no colonies were seen. A clone carrying the D-galactose dehydrogenase gene was identified immunologically but subcloning the D-galactose dehydrogenase gene into a multicopy plasmid still did not permit growth of the galactokinase-negative host on D-galactose. However, by mutagenesis of the cloned DNA a plasmid was obtained in which the D-galactose dehydrogenase was now readily expressed and this plasmid permitted good growth of the galactokinase-negative host strain on D-galactose. Somewhat surprisingly, further manipulation of the cloned DNA to give even greater expression of D-galactose dehydrogenase resulted in weaker growth of the host *E. coli* on D-galactose, indicating that direct selection for substrate utilization does not necessarily yield the highest enzyme-producing strain.

Some streptococci possess both the Leloir pathway and the tagatose-6-phosphate pathway enzymes and growth on D-galactose induces both pathways (Bissett and Anderson, 1974; Hamilton and Lebtag, 1979). In shotgun cloning experiments aimed at demonstrating that genes of the cariogenic bacterium *Streptococcus mutans* PS14 (serotype c) were functionally expressed in *E. coli* K12, a recipient strain deleted for the Leloir pathway enzymes galactokinase, UDP-galactose epimerase, and glucose-1-phosphate uridyl transferase, and so unable to grow on D-galactose, was used. A number of D-galactose-positive clones were obtained and it was expected that these would consist of clustered *S. mutans* genes involved in the Leloir pathway. However, no galactokinase, UDP-galactose epimerase, or glucose-1-phosphate uridyl transferase activity could be detected. Unexpectedly, preliminary experiments indicated that the enzymes of the D-tagatose-6-phosphate pathway were present. So, remarkably, it seems that the tagatose-6-phosphate pathway genes had been introduced into *E. coli* and were being expressed sufficiently to allow growth on D-galactose (Smorawinska *et al.*, 1983).

3.2. D-*Fructose*

The bacterial catabolism of this widely distributed sugar shows unexpected complexities. An ATP-dependent fructokinase has long been known to convert D-fructose to D-fructose-6-phosphate, an intermediate in the Embden–Meyerhof pathway of D-glucose catabolism. However, in a variety of anaerobic and facultatively anaerobic bacteria, D-fructose is converted to D-fructose-1-phosphate, rather than to D-fructose-6-phosphate, by the phosphoenolpyruvate-dependent PTS (see Chapter 10). The further metabolism of D-fructose-1-phosphate requires an ATP-dependent 1-phosphofructokinase (Hanson and Anderson, 1966) to permit entry into the Embden–Meyerhof pathway as fructose-1,6-bisphosphate. This sequence has been studied extensively in *E. coli* and *Aerobacter aerogenes* and the physiological importance of these enzymes has been confirmed by investigation of mutants lacking them. One consequence of this sequence is that the gluconeogenic enzyme fructose-1,6-bisphosphatase should be required for D-fructose-6-phosphate production during growth on D-fructose and this was reported to be the case for *A. aerogenes* (Sapico *et al.,* 1968). Surprisingly, fructose bisphosphatase- negative mutants of *E. coli* grew readily on D-fructose and this paradox was resolved only when it was found that at high (10–50 mM) D-fructose concentrations, both D-fructose-1-phosphate and D-fructose-6-phosphate were formed by the PTS. At low (1 mM) fructose concentrations, only D-fructose-1-phosphate was formed and fructose bisphosphatase-negative mutants were unable to grow (Ferenci and Kornberg, 1971). Genetic analysis of *E. coli* mutants defective in both D-fructose-1-phosphate- and D-fructose-6-phosphate-forming phosphotransferase systems has shown that a D-fructose-specific enzyme II forms D-fructose-1-phosphate and that D-fructose-6-phosphate is formed by a quite separate enzyme II that is active in the phosphorylation of D-glucosamine, D-mannose, and D-glucose to produce the corresponding 6-phosphates (Kornberg and Jones-Mortimer, 1975). No growth on D-fructose was observed when both phosphotransferase systems were inactivated by mutation, suggesting that an ATP-dependent mannofructokinase (Sebastian and Asensio, 1972) does not normally play a significant role in D-fructose catabolism. However, in *A. aerogenes,* when D-fructose is formed internally from the hydrolysis of sucrose, D-fructose-6-phosphate is formed by an inducible ATP-dependent fructokinase that has no activity toward D-mannose and D-glucose (Kelker *et al.,* 1970).

D-Fructose utilization by *P. cepacia* (Allenza *et al.,* 1982) and *P. saccharophila* (Palleroni *et al.,* 1956) involves the operation of an inducible fructokinase to produce D-fructose-6-phosphate. This compound is then isomerized to D-glucose-6-phosphate and catabolized by the Entner–Doudoroff pathway. The Entner–Doudoroff pathway appears to be the only route available for *P. cepacia* to catabolize D-fructose since mutants defective in 6-phosphogluconate dehydratase or KDPG aldolase were unable to grow on D-fructose (Allenza and Lessie, 1982). Conversely, many other pseudomonads take up D-fructose, uniquely among the sugars they catabolize, by a PTS that produces D-fructose-1-phosphate (Baumann and Baumann, 1975; Roehl and Phibbs, 1982; Sawyer *et al.,* 1977; Van Dijken and Quayle, 1977). These organisms have a 1-phosphofructokinase that then forms D-fructose-1,6-bisphosphate. This key intermediate can be catabolized via fructose bisphosphate aldolase to give triose phosphates or converted to D-fructose-6-phosphate by fructose bisphosphatase. The D-fructose-6-phosphate so formed is isomerized to D-glucose-6-phosphate and so enters the Entner–Doudoroff pathway. In this way pseudomonads with the PTS can catabolize D-fructose by two separate sequences. For *P. aeruginosa,* studies with mutants deficient in Entner–Doudoroff pathway and related enzymes indicated that this sequence rather than the Embden–Meyerhof pathway was the major route for D-fructose utilization (Roehl *et al.,* 1983).

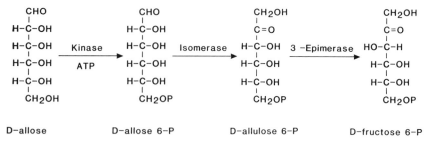

D-allose   D-allose 6-P   D-allulose 6-P   D-fructose 6-P

Figure 9. Pathway for D-allose utilization. The enzymes involved are not found in D-glucose- or D-ribose-grown cells.

3.3. D-*Allose*

Work by F. J. Simpson at the Prairie Regional Laboratory, Saskatoon, Canada, in the late 1950s showed that *A. aerogenes* strain PRL-R3 had a remarkable ability to grow on a very wide range of "unusual" D- and L-series sugars and D-allose is representative of these. This sugar is the 3-epimer of D-glucose and is catabolized by the sequential action of an ATP-dependent kinase that forms D-allose-6-phosphate, an isomerase that forms the ketose D-allulose-6-phosphate, and a 3-epimerase to convert the latter compound to D-fructose-6-phosphate (Fig. 9), which can then be metabolized by the Embden–Meyerhof pathway (Gibbins and Simpson, 1964). These enzymes for D-allose catabolism were not readily detectable after growth on D-glucose or D-ribose but it is not known whether they are specific to the catabolism of D-allose or a manifestation of the lack of specificity of some other sugar-catabolizing enzymes.

3.4. *Hexonic and Hexuronic Sugar Acids*

E. coli is capable of growing on at least 12 different sugar acids and 10 compounds: D-glucuronate, D-galacturonate, L-galactonate, L-gulonate, D-fructuronate, D-tagaturonate, D-mannonate, D-altronate, D-gluconate, and D-2-keto-3-deoxygluconate converge to produce 2-keto-3-deoxy-6-phosphogluconate (KDPG) with D-galactonate and D-2-keto-3-deoxygalactonate producing the isomeric 2-keto-3-deoxy-6-phosphogalactonate. These catabolic pathways (Fig. 10) are, apart from those for aromatic acid catabolism (Cooper and Skinner, 1980; Burlinghame and Chapman, 1983), the longest so far observed for *E. coli*. Because of the need to accommodate the degradation of a relatively large number of different sugars, the genetic control of the sequence is complex and thus is of particular interest.

An initially parallel and then converging reaction sequence for the cata-

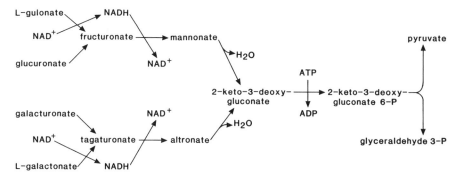

Figure 10. Sequence for the catabolism of various sugar acids by *E. coli*. Gluconate catabolism also produces 2-keto-3-deoxygluconate-6-phosphate and so joins this scheme at the last step.

bolism of D-glucuronate and D-galacturonate was first described almost 30 years ago (Ashwell *et al.*, 1958). In this sequence the alduronic acids are first isomerized to their corresponding keturonic acids D-fructuronate and D-tagaturonate and more recently it has been shown that oxidation of L-gulonate (Cooper, 1980) and L-galactonate (Cooper, 1979) also produces, respectively, D-fructuronate and D-tagaturonate, thus allowing these unusual L-series sugar acids to feed into the same metabolic pathway. Reduction of D-fructuronate gives D-mannonate and reduction of D-tagaturonate gives D-altronate. These two aldonic sugar acids are dehydrated to give D-2-keto-3-deoxygluconate (KDG) and this compound (Lagarde *et al.*, 1973) as well as D-fructuronate, D-tagaturonate, D-mannonate, and D-altronate (Robert-Baudouy *et al.*, 1974) can support the growth of *E. coli* when provided exogenously. Finally, KDG is phosphorylated by an ATP-dependent kinase to give KDPG and so converge with the pathway of D-gluconate catabolism (Fig. 2). L-Gulonate utilization by this route is quite distinct from its catabolism in animals where it is converted to L-xylulose either directly or indirectly via L-ascorbic acid.

The isomerization of D-glucuronate and D-galacturonate is brought about by a single isomerase but the dehydrogenases for L-gulonate and L-galactonate, the reductases for D-fructuronate and D-tagaturonate, and the dehydratases for D-mannonate and D-altronate are distinct and specific enzymes, despite the close chemical similarity of the pairs of compounds. These conclusions from biochemical investigations have been confirmed and extended by studies on the biochemical genetics of the pathways in *E. coli*. Mutants lacking KDPG aldolase were unable to grow on any of these hexonic or hexuronic acids, while loss of the uronic acid isomerase prevented growth on D-glucuronate and D-galacturonate but had no effect on the growth on any of the other compounds. Loss of fructuronate reductase or mannonate dehydratase prevented growth on D-glucuronate or

L-gulonate and, likewise, loss of tagaturonate reductase or altronate dehydratase prevented growth on D-galacturonate or L-galactonate. All of the pathway enzymes are inducible and two distinct patterns of induction occur. D-Galacturonate causes induction of the common isomerase and the two enzymes leading from D-tagaturonate to D-KDG, while D-glucuronate induces all five enzymes for D-galacturonate and D-glucuronate conversion to D-KDG. The actual inducers of the system appear to be D-tagaturonate and D-fructuronate (Robert-Baudouy et al., 1974).

The genetic regulation of the pathway is complex. L-Galactonate dehydrogenase and L-gulonate dehydrogenase have not been characterized genetically but there are three operons for the five enzymes leading from the uronic acids to D-KDG and separate operons for D-KDG transport, KDG kinase, and KDPG aldolase. The last three operons are under the control of a common regulatory gene but strong decoordination occurs. Growth on D-gluconate induces only the aldolase, suggesting that there are differences in the operator structures of the three operons (Robert-Baudouy et al., 1974).

3.5. L-Fucose, D-Arabinose, L-Rhamnose, and L-Mannose

In E. coli K12, L-fucose (6-deoxy-L-galactose) is catabolized to dihydroxyacetone phosphate and L-lactaldehyde by an inducible pathway that involves the sequential action of L-fucose permease, L-fucose isomerase, L-fuculokinase, and L-fuculose-1-phosphate aldolase (Fig. 11) (Heath and Ghalambor, 1962). Under aerobic conditions the aldehyde is oxidized by an NAD-linked dehydrogenase to L-lactate, which in turn is further oxidized by a flavin-linked dehydrogenase to give pyruvate. During anaerobic growth the aldehyde serves as an electron acceptor to allow the energetically most effective catabolism of dihydroxyacetone phosphate and thus is reduced by an NADH-linked oxidoreductase to L-1,2-propanediol. This latter compound accumulates in the growth medium and is not used by the cells even if oxygen becomes available (Hacking and Lin, 1976). Interestingly, genetic analysis indicates that genes encoding the permease, isomerase, kinase, aldolase, propanediol oxidoreductase, and lactaldehyde dehydrogenase enzymes all map at 60.2 min. on the E. coli K12 genetic map (Chen et al., 1984). Somewhat surprisingly, these genes appear to be divided into at least two operons with the permease, isomerase, and kinase belonging to a single operon and the aldolase, oxidoreductase, and dehydrogenase belonging to a separate operon (Chakrabarti et al., 1984).

Mutations in genes regulating the L-fucose catabolic pathway allow the operation of two additional catabolic sequences in E. coli K12. Although, as already described, E. coli does not normally utilize L-1,2-propanediol, it can develop the ability to do so by mutations that result in the constitutive formation of L-1,2-propanediol oxidoreductase under aerobic conditions. This enzyme then

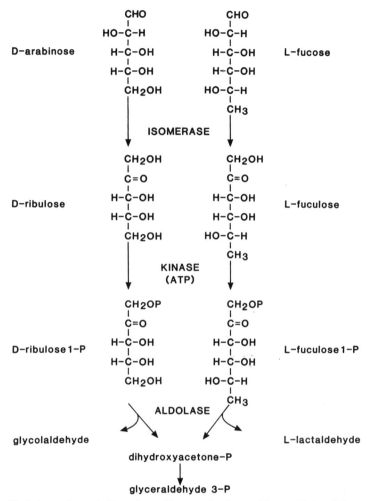

Figure 11. Pathway for catabolism of D-arabinose and L-fucose. The specificity of the enzymes allows both sugars to be utilized.

acts as a propanediol dehydrogenase to allow L-1,2-propanediol to be used as a growth substrate (Hacking and Lin, 1977). Invariably, with the full constitutivity of the oxidoreductase activity, the aldolase also becomes constitutive, while the permease, the kinase, and the isomerase become noninducible. Consequently, the ability to grow on L-fucose is lost. However, further mutations permit the constitutive synthesis of the previously noninducible enzymes without affecting the formation of the oxidoreductase or aldolase and growth on L-fucose is recovered (Chen *et al.*, 1984).

The second catabolic sequence involves the uncommon pentose D-arabinose. Wild-type strains are unable to utilize D-arabinose as sole carbon and energy source but mutants with this ability can be isolated (LeBlanc and Mortlock, 1971). The mutation involved is one that allows a regulatory protein of the L-fucose pathway also to recognize D-arabinose (or a derivative) as an inducer and thus leads to the inducible formation of the L-fucose catabolic enzymes in cells exposed to D-arabinose. The specificity of these enzymes is such that they also act on D-arabinose, so L-fucose isomerase converts D-arabinose to D-ribulose, which is then phosphorylated by L-fuculokinase to give D-ribulose-1-phosphate. Finally, L-fuculose-1-phosphate aldolase cleaves D-ribulose-1-phosphate to glycolaldehyde and dihydroxyacetone phosphate (Fig. 11).

K. *aerogenes* is also unable to utilize D-arabinose but can develop the ability to do so through mutation of a regulatory gene as described for *E. coli*. In addition, a second type of regulatory gene mutation that leads to the constitutive formation of L-fucose isomerase and L-fuculokinase (but not of L-fuculose-1-phosphate aldolase, which is apparently regulated separately) also allows *K. aerogenes* to grow on D-arabinose. The L-fucose isomerase is used to convert D-arabinose to D-ribulose, an intermediate and inducer of the ribitol catabolic pathway in this organism (Charnetsky and Mortlock, 1973). In turn, this leads to the induction of D-ribulokinase, which converts D-ribulose to D-ribulose-5-phosphate, thereby entering the nonoxidative reactions of the HMP pathway. This is the major route for D-arabinose catabolism in *K. aerogenes*. In the absence of D-ribulokinase, D-arabinose is utilized, but more slowly, by the enzymes of the L-fucose catabolic pathway as in *E. coli* K12. Conversely, many strains of *E. coli* including K12 lack the ribitol pathway and so cannot utilize D-arabinose by this D-ribulose-5-phosphate route (Scangos and Reiner, 1978).

L-Rhamnose (6-deoxy-L-mannose) is catabolized in *E. coli* by reactions that involve the sequential action of L-rhamnose isomerase, L-rhamnulokinase, and L-rhamnulose-1-phosphate aldolase to produce dihydroxyacetone phosphate and L-lactaldehyde (Fig. 12) (Takagi and Sawada, 1964). These reactions are very similar to those for L-fucose catabolism but involve distinct enzymes. Anaerobic growth on L-rhamnose also leads to formation of L-1,2-propanediol by reduction of L-lactaldehyde and it seems that the same L-1,2-propanediol oxidoreductase active during anaerobic growth on L-fucose is involved (Boronat and Aguilar, 1979). More surprisingly, it appears that while anaerobic growth on either L-fucose or L-rhamnose leads to the formation of active L-1,2-propanediol oxidoreductase, aerobic growth on L-fucose, but not on L-rhamnose, induces a form of the enzyme that is immunologically similar to the anaerobic enzyme but that has no catalytic activity (Boronat and Aguilar, 1981). The way in which catalytic activity of this enzyme is regulated is not known. Although the oxidoreductase is encoded by a gene at the *fuc* locus, the enzyme is inducible by both L-rhamnose and L-fucose. Induction by L-rhamnose is mediated through a

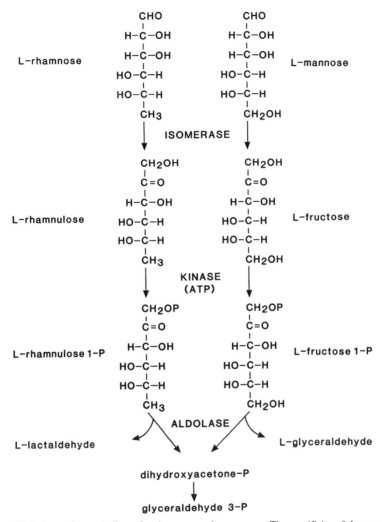

Figure 12. Pathway for catabolism of L-rhamnose and L-mannose. The specificity of the enzymes allows both sugars to be utilized.

gene at the *rha* locus but the way in which this gene interacts with the *fucO* gene is unknown. Nor is it known whether a single L-lactaldehyde dehydrogenase is shared by the L-rhamnose and L-fucose catabolic systems.

The L-rhamnose catabolic enzymes are also formally analogous to those for L-fucose catabolism in that they also can act on other sugars. In *K. aerogenes,* both L-rhamnose and the unnatural sugar L-mannose can serve as inducers for the

L-rhamnose catabolic enzymes (Mayo and Anderson, 1969). It was proposed that L-mannose catabolism involved its isomerization to L-fructose by the L-rhamnose isomerase, phosphorylation by L-rhamnulokinase to give L-fructose-1-phosphate, and cleavage of this compound by L-rhamnulose-1-phosphate aldolase to give L-glyceraldehyde and dihydroxyacetone phosphate. Despite the ability of L-mannose to induce the L-rhamnose catabolic enzymes, wild-type *K. aerogenes* could not grow on L-mannose and in fact this substance inhibited growth. However, it was possible to select mutants of *K. aerogenes* that could grow readily at the expense of L-mannose but the nature of the mutation permitting growth was not identified. Some *E. coli* K12 strains can grow on L-mannose without the need for a mutation and it seems as though the L-rhamnose catabolic enzymes are responsible for this growth (R. A. Cooper, unpublished observation).

3.6. Apologia

The sequences described in this section are representative rather than comprehensive and have been chosen to illustrate a number of general features that can be seen in the bacterial reactions for the conversion of a variety of sugars to central metabolites. First, different pathways can operate in different species (and sometimes in the same species) for the catabolism of a particular sugar. Second, different sugars can be catabolized by chemically parallel reactions that are catalyzed in some cases by very specific enzymes and in others by less specific enzymes. Finally, complex genetic control of enzyme formation makes it possible for a range of sugars to be catabolized by a minimum number of individual reactions.

4. EPILOGUE

Bacteria are traditionally thought of as organisms with very great metabolic versatility. As described in this chapter, the ways in which they achieve the catabolism of a wide range of sugars involve, overall, fewer individual reactions than might be supposed and the secret of success is the effective utilization of as few unique reactions as possible. In this sense there is, perhaps, less to their versatility than meets the eye!

REFERENCES

Allenza, P., and Lessie, T. G., 1982, *Pseudomonas cepacia* mutants blocked in the Entner–Doudoroff pathway, *J. Bacteriol.* **150:**1340–1347.

Allenza, P., Lee, Y. N., and Lessie, T. G., 1982, Enzymes related to fructose utilization in *Pseudomonas cepacia, J. Bacteriol.* **150:**1348–1356.

Anderson, A., and Cooper, R. A., 1969, Gluconeogenesis in *Escherichia coli:* The role of triosephosphate isomerase, *FEBS Lett.* **4:**19–20.

Aparicio, M. L., Ruiz-Amil, M., Vicente, M., and Canovas, J. L. 1971, The role of phosphoglycerate kinase in the metabolism of *Pseudomonas putida, FEBS Lett.* **14:**326–328.

Ashwell, G., Wahba, A. J., and Hickman, J., 1958, A new pathway of uronic acid metabolism, *Biochim. Biophys. Acta* **30:**186–187.

Bachmann, B. J., 1983, Linkage map of *Escherichia coli* K12, edition 7, *Microbiol. Rev.* **47:**180–230.

Baumann, L., and Baumann, P., 1973, Enzymes of glucose catabolism in cell-free extracts of nonfermentative marine eubacteria, *Can. J. Microbiol.* **19:**302–304.

Baumann, P., and Baumann, L., 1975, Catabolism of D-fructose and D-ribose by *Pseudomonas doudoroffii, Arch. Microbiol.* **105:**225–240.

Benziman, M., 1969, Factors affecting the activity of pyruvate kinase of *Acetobacter xylinum, Biochem. J.* **112:**631–636.

Bissett, D. L., and Anderson, R. L., 1973, Lactose and D-galactose metabolism in *Staphylococcus aureus:* Pathway of D-galactose 6-phosphate degradation, *Biochem. Biophys. Res. Commun.* **52:**641–647.

Bissett, D. L., and Anderson, R. L., 1974, Lactose and D-galactose metabolism in group N streptococci: Presence of enzymes for both the D-galactose 1-phosphate and D-tagatose 6-phosphate pathways, *J. Bacteriol.* **117:**318–320.

Bock, A., and Neidhardt, F. C., 1966, Isolation of a mutant of *Escherichia coli* with a temperature-sensitive fructose 1,6-diphosphate aldolase activity, *J. Bacteriol.* **92:**464–469.

Boronat, A., and Aguilar, J., 1979, Rhamnose induced propanediol oxidoreductase in *Escherichia coli:* Purification, properties, and comparison with the fucose-induced enzyme, *J. Bacteriol.* **140:**320–326.

Boronat, A., and Aguilar, J., 1981, Metabolism of L-fucose and L-rhamnose in *Escherichia coli:* Differences in the induction of propanediol oxidoreductase, *J. Bacteriol.* **147:**181–185.

Buckel, P., and Zehelein, E., 1981, Expression of *Pseudomonas fluorescens* D-galactose dehydrogenase in *E. coli, Gene* **16:**149–159.

Burlinghame, R., and Chapman, P. J., 1983, Catabolism of phenylpropionic acid and its 3-hydroxy derivative by *Escherichia coli, J. Bacteriol.* **155:**113–121.

Chakrabarti, T., Chen, Y.-M., and Lin, E. C. C., 1984, Clustering of genes for L-fucose dissimilation by *Escherichia coli, J. Bacteriol.* **157:**984–986.

Charnetzky, W. T., and Mortlock, R. P., 1973, Ribitol catabolic pathway in *Klebsiella aerogenes, J. Bacteriol.* **119:**162–169.

Chen, Y.-M., Chakrabarti, T., and Lin, E. C. C., 1984, Constitutive activation of L-fucose genes by an unlinked mutation in *Escherichia coli, J. Bacteriol.* **159:**725–729.

Cooper, R. A., 1975, The methylglyoxal by-pass of the Embden–Meyerhof pathway, *Biochem. Soc. Trans.* **3:**837–840.

Cooper, R. A., 1978, The utilization of D-galactonate and D-2-oxo-3-deoxygalactonate by *Escherichia coli* K12, *Arch. Microbiol.* **118:**199–206.

Cooper, R. A., 1979, The pathway for L-galactonate catabolism in *Escherichia coli* K12, *FEBS Lett.* **103:**216–220.

Cooper, R. A., 1980, The pathway for L-gulonate catabolism in *Escherichia coli* K12 and *Salmonella typhimurium* LT-2, *FEBS Lett.* **115:**63–67.

Cooper, R. A., 1984, Metabolism of methylglyoxal in micro-organisms, *Annu. Rev. Microbiol.* **38:**49–68.

Cooper, R. A., and Anderson, A., 1970, The formation and catabolism of methylglyoxal during glycolysis in *Escherichia coli, FEBS Lett.* **11:**273–276.

Cooper, R. A., and Skinner, M. A., 1980, Catabolism of 3- and 4-hydroxyphenylacetate by the 3,4-dihydroxyphenylacetate pathway in *Escherichia coli, J. Bacteriol.* **143:**302–306.

Cozzarelli, N. R., Koch, J. P., Hayashi, S., and Lin, E. C. C., 1965, Growth stasis by accumulated L-α-glycerophosphate in *Escherichia coli, J. Bacteriol.* **90:**1325–1329.

Curtis, S. J., and Epstein, W., 1975, Phosphorylation of D-glucose in *Escherichia coli* mutants defective in glucosephosphotransferase, mannosephosphotransferase and glucokinase, *J. Bacteriol.* **122**:1189–1199.

Dakin, H. D., and Dudley, H. W., 1913, On glyoxalase, *J. Biol. Chem.* **14**:423–431.

DeLey, J., 1960, Comparative carbohydrate metabolism and localization of enzymes in Pseudomonas and related microorganisms, *J. Appl. Bacteriol.* **23**:400–441.

DeLey, J., and Doudoroff, M., 1957, The metabolism of D-galactose in *Pseudomonas saccharophila, J. Biol. Chem.* **227**:745–757.

DeMoss, R. D., Bard, R. C., and Gunsalus, I. C., 1951, The mechanism of the heterolactic fermentation: a new route for ethanol formation, *J. Bacteriol.* **62**:499–511.

deVries, W., and Stouthamer, A. H., 1968, Fermentation of glucose, lactose, galactose, mannitol and xylose by bifidobacteria, *J. Bacteriol.* **96**:472–478.

deVries, W., Gerbrandy, S. J., and Stouthamer, A. H., 1967, Carbohydrate metabolism in *Bifidobacterium bifidum, Biochim. Biophys. Acta* **136**:415–425.

Egyud, L. G., and Szent-Gyorgyi, A., 1966, On the regulation of cell division, *Proc. Natl. Acad. Sci. USA* **56**:203–207.

Entner, N., and Doudoroff, M., 1952, Glucose and gluconic acid oxidation of *Pseudomonas saccharophila, J. Biol. Chem.* **196**:853–862.

Ferenci, T., and Kornberg, H. L., 1971, Role of fructose 1,6-diphosphate in fructose utilization by *Escherichia coli, FEBS Lett.* **14**:360–363.

Fraenkel, D. G., 1968a, The accumulation of glucose 6-phosphate from glucose and its effect in an *Escherichia coli* mutant lacking phosphoglucose isomerase and glucose 6-phosphate dehydrogenase, *J. Biol. Chem.* **243**:6451–6457.

Fraenkel, D. G., 1968b, Selection of *Escherichia coli* mutants lacking glucose 6-phosphate dehydrogenase or gluconate 6-phosphate dehydrogenase, *J. Bacteriol.* **95**:1267–1271.

Fraenkel, D. G., and Levisohn, S. R., 1967, Glucose and gluconate metabolism in an *Escherichia coli* mutant lacking phosphoglucose isomerase, *J. Bacteriol.* **93**:1571–1578.

Fraenkel, D. G., Falcoz-Kelly, F., and Horecker, B. L., 1964, The utilization of glucose 6-phosphate by glucokinaseless and wild-type strains of *Escherichia coli, Proc. Natl. Acad. Sci. USA* **52**:1207–1213.

Garrido-Pertierra, A., and Cooper, R. A., 1977, Pyruvate formation during the catabolism of simple hexose sugars by *Escherichia coli*: Studies with pyruvate kinase-negative mutants, *J. Bacteriol.* **129**:1208–1214.

Gibbins, L. N., and Simpson, F. J., 1964, The incorporation of D-allose into the glycolytic pathway of *Aerobacter aerogenes, Can. J. Microbiol.* **10**:829–836.

Gromet, Z., Schramm, M., and Hestrin, S., 1957, Synthesis of cellulose by *Acetobacter xylinum.* 4. Enzyme systems present in a crude extract of glucose-grown cells, *Biochem. J.* **67**:679–689.

Gunsalus, I. C., Horecker, B. L., and Wood, W. A., 1955, Pathways of carbohydrate metabolism in micro-organisms, *Bacteriol. Rev.* **19**:79–128.

Hacking, A. J., and Lin, E. C. C., 1976, Disruption of the fucose pathway as a consequence of genetic adaptation to propanediol as a carbon source in *Escherichia coli, J. Bacteriol.* **126**:1166–1172.

Hacking, A. J., and Lin, E. C. C., 1977, Regulatory changes in the fucose system associated with the evolution of a catabolic pathway for propanediol in *Escherichia coli, J. Bacteriol.* **130**:832–838.

Hamilton, I. R., and Lebtag, H., 1979, Lactose metabolism by *Streptococcus mutans:* Evidence for induction of the tagatose 6-phosphate pathway, *J. Bacteriol.* **140**:1102–1104.

Hanson, T. E., and Anderson, R. L., 1966, D-Fructose 1-phosphate kinase, a new enzyme instrumental in the metabolism of D-fructose, *J. Biol. Chem.* **241**:1644–1645.

Heath, E. C., and Ghalambor, M. A., 1962, The metabolism of L-fucose. 1. The purification and properties of L-fuculose kinase, *J. Biol. Chem.* **237**:2423–2426.

Heath, E. C., Hurwitz, J., and Horecker, B. L., 1956, Acetyl phosphate formation in the phos-phorolytic cleavage of pentose phosphate, *J. Am. Chem. Soc.* **78**:5449.

Hillman, J. D., and Fraenkel, D. G., 1975, Glyceraldehyde 3-phosphate dehydrogenase mutants of *Escherichia coli, J. Bacteriol.* **122**:1175–1179.

Hommes, R. W. J., Postma, P. W., Neijssel, O. M., Tempest, D. W., Dokter, P., and Duine, J. A., 1984, Evidence of a quinoprotein glucose dehydrogenase apoenzyme in several strains of *Escherichia coli, FEMS Microbiol. Lett.* **24**:329–333.

Hopper, D. J., and Cooper, R. A., 1972, The purification and properties of *Escherichia coli* methylglyoxal synthase, *Biochem. J.* **128**:321–329.

Hunt, J. C., and Phibbs, P. V., Jr., 1983, Regulation of alternate peripheral pathways of glucose catabolism during aerobic and anaerobic growth of *Pseudomonas aeruginosa, J. Bacteriol.* **154**:793–802.

Hylemon, P. B., and Phibbs, P. V., Jr., 1972, Independent regulation of hexose catabolizing enzymes and glucose transport activity in *Pseudomonas aeruginosa, Biochem. Biophys. Res. Commun.* **48**:1041–1048.

Irani, M., and Maitra, P. K., 1974, Isolation and characterisation of *Escherichia coli* mutants defective in enzymes of glycolysis, *Biochem. Biophys. Res. Commun.* **56**:127–133.

Iyengar, R., and Rose, I. A., 1983, Methylglyoxal synthase uses the trans-isomer or triose-1,2-enediol 3-phosphate, *J. Am. Chem. Soc.* **105**:3301–3303.

Kaback, H. R., 1974, Transport studies in bacterial membrane vesicles: Cytoplasmic membrane vesicles devoid of soluble constituents catalyze the transport of many metabolites, *Science* **186**:882–892.

Kalckar, H. M., 1958, Uridinediphosphogalactose: Metabolism, enzymology and biology, *Adv. Enzymol.* **20**:111–133.

Katz, J., and Wood, H. G., 1960, The use of glucose-C^{14} for the evaluation of the pathways of glucose metabolism, *J. Biol. Chem.* **235**:2165–2177.

Kelker, N. E., Hanson, T. E., and Anderson, R. L., 1970, Alternate pathways of D-fructose metabolism in *Aerobacter aerogenes:* A specific D-fructokinase and its preferential role in the metabolism of sucrose, *J. Biol. Chem.* **245**:2060–2065.

Kitos, P. A., Wang, C. H., Mohler, B. A., King, T. E., and Cheldelin, V. H., 1958, Glucose and gluconate dissimilation in *Acetobacter suboxydans, J. Biol. Chem.* **233**:1295–1298.

Kornberg, H. L., and Jones-Mortimer, M. C., 1975, *PtsX:* A gene involved in the uptake of glucose and fructose by *Escherichia coli, FEBS Lett.* **51**:1–4.

Kornberg, H. L., and Soutar, A. K., 1973, Utilization of gluconate by *Escherichia coli:* Induction of gluconate kinase and 6-phosphogluconate dehydratase activities, *Biochem. J.* **134**:489–498.

Lagarde, A. E., Pouysségur, J. M., and Stoeber, F. R., 1973, A transport system for 2-keto-3-deoxy-D-gluconate uptake in *Escherichia coli* K12: Biochemical and physiological studies in whole cells, *Eur. J. Biochem.* **36**:328–341.

LeBlanc, D. J., and Mortlock, R. P., 1971, Metabolism of D-arabinose: A new pathway in *Escherichia coli, J. Bacteriol.* **106**:90–96.

Lessie, T. G., and Phibbs, P. V., Jr., 1984, Alternative pathways of carbohydrate utilization in pseudomonads, *Annu. Rev. Microbiol.* **38**:359–387.

Lin, E. C. C., 1976, Glycerol dissimilation and its regulation in bacteria, *Annu. Rev. Microbiol.* **30**:535–578.

MacGee, J., and Doudoroff, M., 1954, A new phosphorylated intermediate in glucose oxidation, *J. Biol. Chem.* **210**:617–626.

McGill, D. J., and Dawes, E. A., 1971, Glucose and fructose metabolism in *Zymomonas anaerobia, Biochem. J.* **125**:1059–1068.

Mayo, J. W., and Anderson, R. L., 1969, Basis for the mutational acquisition of the ability of *Aerobacter aerogenes* to grow on L-mannose, *J. Bacteriol.* **100**:948–955.

Midgley, M., and Dawes, E. A., 1973, The regulation of transport of glucose and methyl α-glucoside in *Pseudomonas aeruginosa, Biochem. J.* **132**:141–154.

Morrissey, A. T. E., and Fraenkel, D. G., 1968, Selection of fructose 6-phosphate kinase mutants in *Escherichia coli, Biochem. Biophys. Res. Commun.* **32**:467–473.

Neijssel, O. M., Tempest, D. W., Postma, P. W., Duine, J. A., and Frank Jzn, J., 1983, Glucose metabolism by K⁺-limited *Klebsiella aerogenes:* Evidence for the involvement of a quinoprotein glucose dehydrogenase, *FEMS Microbiol. Lett.* **20**:35–39.

Orthner, C. L., and Pizer, L. I., 1974, An evaluation of regulation of the hexosemonophosphate shunt in *Escherichia coli, J. Biol. Chem.* **249**:3750–3755.

Palleroni, N. J., Contopoulou, R., and Doudoroff, M., 1956, Metabolism of carbohydrates by *Pseudomonas saccharophila.* II. Nature of the kinase reaction involving fructose, *J. Bacteriol.* **71**:202–207.

Racker, E., 1948, Enzymatic formation and breakdown of pentose phosphate, *Fed. Proc.* **7**:180.

Racker, E., 1954, Alternate pathways of glucose and fructose metabolism, *Adv. Enzymol.* **15**:141–182.

Racker, E., 1957, Micro- and macro-cycles in carbohydrate metabolism, *Harvey Lect.* **51**:143–174.

Robert-Baudouy, J. M., Portalier, R. C., and Stoeber, F. R., 1974, Regulation du metabolisme des hexuronates chez *Escherichia coli* K12: Modalites de l'induction des enzymes du systems hexuronate, *Eur. J. Biochem.* **43**:1–15.

Roehl, R. A., and Phibbs, P. V., Jr., 1982, Characterization and genetic mapping of fructose phosphotransferase mutations in *Pseudomonas aeruginosa, J. Bacteriol.* **149**:897–905.

Roehl, R. A., Feary, T. W., and Phibbs, P. V., Jr., 1983, Clustering of mutations affecting central pathway enzymes of carbohydrate catabolism in *Pseudomonas aeruginosa, J. Bacteriol.* **156**:1123–1129.

Sapico, V., Hanson, T. E., Walter, R. W., and Anderson, R. L., 1968, Metabolism of D-fructose in *Aerobacter aerogenes:* Analysis of mutants lacking D-fructose 6-phosphate kinase and D-fructose 1,6-diphosphatase, *J. Bacteriol.* **96**:51–54.

Sawyer, M. H., Baumann, P., Baumann, L., Berman, S. M., Canovas, J. L., and Berman, R. H., 1977, Pathways of D-fructose catabolism in species of Pseudomonas, *Arch. Microbiol.* **112**:49–55.

Scangos, G. A., and Reiner, A. M., 1978, Ribitol and D-arabitol catabolism in *Escherichia coli, J. Bacteriol.* **134**:492–500.

Scardovi, V., and Trovatelli, L. D., 1965, The fructose 6-phosphate shunt as peculiar pattern of hexose degradation in the genus *Bifidobacterium, Ann. Microbiol. Enzymol.* **15**:19–29.

Schramm, M., and Racker, E., 1957, Formation of erythrose 4-phosphate and acetyl phosphate by a phosphorolytic cleavage of fructose 6-phosphate, *Nature* **179**:1349–1350.

Sebastian, J., and Asensio, C., 1972, Purification and properties of the mannokinase from *Escherichia coli, Arch. Biochem. Biophys.* **151**:227–233.

Smorawinska, M., Hsu, J. C., Hansen, J. B., Jagusztyn-Krynicka, E. K., Abiko, Y., and Curtiss, R., III, 1983, Clustered genes for galactose metabolism from *Streptococcus mutans* cloned in *Escherichia coli, J. Bacteriol.* **153**:1095–1097.

Takagi, Y., and Sawada, H., 1964, The metabolism of L-rhamnose in *Escherichia coli.* I. L-Rhamnose isomerase, *Biochim. Biophys. Acta* **92**:10–17.

Taylor, D. G., Trudgill, P. W., Cripps, R. E., and Harris, P. R., 1980, The microbial metabolism of acetone, *J. Gen. Microbiol.* **118**:159–170.

Teixeira de Mattos, M. J., and Tempest, D. W., 1983, Metabolic and energetic aspects for the growth of *Klebsiella aerogenes* NCTC 418 on glucose in anaerobic chemostat culture, *Arch. Microbiol.* **134**:80–85.

Tiwari, N. P., and Campbell, J. J. R., 1969, Enzymatic control of the metabolic activity of *Pseudomonas aeruginosa* grown in glucose or succinate media, *Biochim. Biophys. Acta* **192**:395–401.

Van Dijken, J. P., and Quayle, J. R., 1977, Fructose metabolism in four *Pseudomonas* species, *Arch. Microbiol.* **114**:281–286.

van Schie, B. J., Van Dijken, J. P., and Kuenen, J. G., 1984, Noncoordinated synthesis of glucose dehydrogenase and its prosthetic group PQQ in *Acinetobacter* and *Pseudomonas* species, *FEMS Microbiol. Lett.* **24**:133–138.

Vicente, M., and Canovas, J. L., 1973, Glucolysis in *Pseudomonas putida:* Physiological role of alternative routes from the analysis of defective mutants, *J. Bacteriol.* **116**:908–914.

Vinopal, R. T., Hillman, J. D., Schulman, H., Reznikoff, W. S., and Fraenkel, D. G., 1975, New phosphoglucose isomerase mutants of *Escherichia coli, J. Bacteriol.* **122**:1172–1174.

Wang, C. H., Stern, I., Gilmour, C. M., Klungsoyr, S., Reed, D. J., Bialy, J. J., Christensen, B. E., and Cheldelin, V. H., 1958, Comparative study of glucose catabolism by the radio-respirometric method, *J. Bacteriol.* **76**:207–216.

Wang, R. J., and Morse, M. L., 1968, Carbohydrate accumulation and metabolism in *Escherichia coli.* I. Description of pleiotropic mutants, *J. Mol. Biol.* **32**:59–66.

Index

Maltose (*cont.*)
  PTS, 376
  transport, 263, 410, 416, 439–441, 443
  uptake, 232, 386
  variants, 63
Maltose phosphorylase, mutant, 416
Mammalian cells, 1–22, 29–68, 77–106,
      111–139, 188, 308
Mannitol, 262, 308, 310, 336, 375
  cycle, 312, 313, 328
  degradation, 313
  phosphorylation, 364
  synthesis, 311, 313
  transphosphorylation, 364, 369
  transport, 362
  utilization, 311, 313
Mannitol dehydrogenase, 311, 312
Mannitol-1-phosphatase, 311
Mannitol-1-phosphate, 311, 364, 369
Mannitol-1-phosphate dehydrogenase, 311, 312
D-Mannonate, 480
D-Mannonate dehydratase, 481
Manno(fructo)kinase, 375, 479
D-Mannose, 7, 8, 13, 18, 57, 226, 235, 236,
      238, 297, 428, 443
  adenylate cyclase, 380
  carbon source, 163, 174, 262
  fermentation, 269
  and glycoprotein, 295
  metabolism, 8, 36, 370
  phosphorylation, 479
  PTS, 363
  respiration, 191, 195
  toxicity, 35, 44, 57
  transport, 57, 262, 362
Mannose-6-phosphate, 7, 18, 57, 262
L-Mannose, 482
  catabolism, 485
  as inducer, 486
  isomerization, 486
  α-Mannosidase deficiency, 36
  α-Mannoside, 36
Medicago sativa, 159, 161, 171
Melarsen oxide, 214
Mel B protein, sequence, 446
Melibiose, 233, 372, 387
  carbon source, 159
  carrier, 383
    nucleotide sequence, 434
  carrier protein, 432
    sequence, 447

Melibiose (*cont.*)
  cation-linked transport, 430–434
  diauxie, 392
  genetics, 431–432
  hydrolysis, 263
  inducible protein, 434
  pigment production, 163
  PTS, 376, 377
  Na+ symport, 412, 431, 446
    reconstitution, 432–433
  transport, 410, 430–434
Membrane
  cell, 2, 268, 412
  conformation, 16
  cytoplasmic, 409, 468
  damage, 215
  glycosomal, 200, 217
  lipid bilayer, 412
  mitochondrial, 131, 132, 266
  permeabilize, 246
  plasma, 13, 14, 16, 19, 232, 255, 261
    depolarization, 261
  potential, 239, 240
  proteins, 1, 15–16
  PTS proteins, 363–365
  subcellular, 232
  translocation, 424
  transport proteins, 409
  vesicles, 14, 15, 19, 20, 240, 366, 367,
      382, 418, 419, 420, 423, 425, 426, 428
Memmoniella (Stachybotrys) echinata, 296
Menadiol, 196
Mercaptopicolinate, 128
mRNA, 239
  liver-specific, 52
Metabolic
  energy, 226
  flux, 247
  pathway, 6–8
Metabolism
  aerobic, 191
  anaerobic, 193, 195
  carbon, 290
  end-product, 191
  hexose, 304
  nitrogen, 290
Metabolite
  determination of level, 247–249
  regulatory, 236
  secondary, 151, 291
Methanol, 287